# www.wadsworth.com

*wadsworth.com* is the World Wide Web site for Wadsworth and is your direct source to dozens of online resources.

At *wadsworth.com* you can find out about supplements, demonstration software, and student resources. You can also send email to many of our authors and preview new publications and exciting new technologies.

**wadsworth.com**
Changing the way the world learns®

# Sociology

# Sociology

CONCEPTS AND CHARACTERISTICS

*Eleventh Edition*

JUDSON R. LANDIS
*California State University, Sacramento*

Australia • Canada • Mexico • Singapore • Spain • United Kingdom • United States

# WADSWORTH

## ✳ ™
## THOMSON LEARNING

**Publisher:** Eve Howard
**Assistant Editor:** Dee Dee Zobian
**Editorial Assistant:** Stephanie Monzon
**Marketing Manager:** Matthew Wright
**Project Editor:** Jerilyn Emori
**Print Buyer:** Tandra Jorgensen
**Permissions Editor:** Robert Kauser
**Production Service:** Sara Dovre Wudali/
  Gustafson Graphics

**Photo Researcher:** Terri Wright/
  Terri Wright Design
**Copy Editor:** Linda Ireland
**Cover Designer:** Bill Stanton/Stanton Design
**Cover Printer:** Phoenix Color Corp.
**Compositor:** Gustafson Graphics
**Printer:** The Maple-Vail Book
  Manufacturing Group

For more information about our products, contact us:
**Thomson Learning Academic Resource Center**
**1-800-423-0563**
http://www.wadsworth.com

**International Headquarters**
Thomson Learning
International Division
290 Harbor Drive, 2nd Floor
Stamford, CT 06902-7477
USA

**UK/Europe/Middle East/South Africa**
Thomson Learning
Berkshire House
168-173 High Holborn
London WC1V 7AA
United Kingdom

**Asia**
Thomson Learning
60 Albert Street, #15-01
Albert Complex
Singapore 189969

**Canada**
Nelson Thomson Learning
1120 Birchmount Road
Toronto, Ontario M1K 5G4
Canada

**Library of Congress Cataloging-in-Publication Data**
Landis, Judson R.
   Sociology: concepts and characteristics / Judson R. Landis.—11th ed.
      p. cm.
   Includes bibliographical references and index.
   ISBN 0-534-57861-6
      1. Sociology.   I. Title.
   HM586 .L35 2000
   301—dc21                                          00-043642

# Contents

**3    NORMS, ROLES, CULTURE, SOCIETY   69**

## *Part 2    Social Organization    104*

**4    GROUPS   106**

## *Part 3* *Social Change and Social Deviance* *367*

# *Preface*

THE PURPOSES of a first course in a discipline usually are several—to introduce an area of study, to communicate its unique perspective or way of looking at the world, and to offer the promise of secrets yet to be discovered—all in the hope that interested students will come back for more (maybe even thirty units more). That there are many different views of the best way to achieve these ambitions is apparent in sociology from some of the introductory textbooks available.

Many books provide an encyclopedic analysis of what a discipline is and does—all its tools, techniques, and substantive areas. I have decided not to try to do that in this book. It seems to me that a feeling for an area of study and its unique perspective may be gained by sampling the concepts of that field. Concepts are the building blocks—the language—of most disciplines. Once one has this language, the rest comes more easily and makes more sense. This book approaches the task of helping students develop a general understanding of the sociological perspective by focusing on the basic concepts of sociology. Through examining several substantive areas and the research connected with them, students can begin to see how the concepts contribute to sociological analysis.

At the end of most of the chapters, I have included readings to further illustrate the chapter's concepts. These readings are drawn from a variety of sources. Included are descriptions of research and excerpts from novels, autobiographies, and nonfiction works. Some of the selections were written by professional social scientists; some were not. For example, writings of both Helen Keller and Malcolm X are used to illustrate socialization and development of self. In addition, there are selections dealing with the Heaven's Gate suicides, kids selling cocaine, the "glass ceiling," the experiences of an illegal immigrant, the evils of automation, the history of the American Indian, and a debate on the "Jeweler's Dilemma." I have found the readings effective and stimulating, and I think they illustrate the concepts and ideas in an unusual way. It's always a challenge to find good readings, readings that are both illustrative and interesting. Four of the readings (those written by Helen Keller, Malcolm X, Ralph Linton, and Harper Lee) now have survived all eleven editions of this book.

In this edition, most chapters have been reworked, and all statistical information including graphs and tables has been updated to reflect the most recent data available. I have added a number of new boxes and readings, such as "It's Not What You Say, It's How You Say It," by Sarah Lyall and "My Exotic Tribe: Children in a School Hallway," by Donald Ratcliff, and I have deleted others. I have also added brief biographies or sketches of some people whose ideas have been important in the development of sociology.

Many people have helped me on this and earlier editions with ideas and criticism. These include especially my colleagues at CSU, Sacramento, from whom I have borrowed books and stolen ideas.

The following reviewers were also helpful:

> Jan Abu Shahkrah, Portland Community College
> Ida Cook, University of Central Florida
> Donna Lee King, University of North Carolina at Wilmington
> John Lie, University of Illinois
> Robert B. Townsend Jr., College of Lake County
> Constance Verdi, Prince George's Community College
> Tracy L. Wallace, Halifax Community College

I'd also like to thank Verica Dering, Steve Rutter, the people who worked on the book—Sara Dovre Wudali and Linda Ireland—and the people at Wadsworth—Dee Dee Zobian, Jerilyn Emori, Bob Kauser, and Eve Howard.

This is dedicated to Sheron, who helped whenever I asked, and to Jeffrey, Brian, and Kevin, who have continued to be patient with me.

# Sociology

# Introduction: Knowledge, Science, and Sociology

**Knowledge**

**Science**
*Types of Science* • *Social Science*

**Sociology**
*Sociological Theories*

**Research Methods in Sociology**
*Three Research Techniques* • *Problems in Social Science Investigation* • *Ethics*

THIS BOOK IS designed to acquaint you with a field of study called *sociology*. Generally speaking, sociology is the study of human society; it is the study of social behavior and the interaction of people in groups. Sociologists study various aspects of social life. For example, we study large organizations (what do Wal-Mart, Petsmart, and Microsoft have in common?); we study small groups (how are group leaders picked?); we study deviant behavior (do certain parts of the United States have high crime rates? why?); we study political institutions (who makes the crucial decisions in society, all of us, or a few people at the top?); we study the family (what are the most important changes in the American family in the last five years?); and we study religious institutions (what do Jews, Catholics, and Sikhs have in common?). We think that the best way to understand human behavior is by looking at the groups people belong to and the culture in which they live.

In this chapter, we look first at the idea of *knowledge*, then at a particular way of gaining knowledge called *science*, and finally at one of the social sciences called *sociology*. Before I get into these specific topics, however, I'd like to give you an idea of the way social scientists study the world by looking at several **questions** they might ask.

**1.** A white sports fan is fascinated by the fact that although the majority of our society is white, the majority of the athletes playing professional basketball are black. He concludes that because of his racial, genetic, and hereditary background factors, he had better take up tennis. How good is his logic?

**2.** A famous person (Marilyn Monroe, Ernest Hemingway, Freddie Prinz, and so on) commits suicide. Or a murder-suicide takes place in which a person kills at least two other people and then kills himself or herself. What, if anything, will happen among those in the rest of the society who know about these events?

**3.** A statistics fanatic finds that in a particular year, Mexico and Miami had a lot of murder but little suicide and that Hungary, Denmark, and western states like New Mexico, Nevada, and Wyoming had many suicides. Our numbers freak wonders if this is a one-time event or typical, and if typical, why?

**4.** A poll tells us that most people (80–85 percent) favor using the death penalty as punishment for serious crimes like murder. That same poll mentions that a good portion of those who favor the death penalty do so because they believe that it deters others from committing crime. Is that really so, and is the poll accurate?

**5.** We know what crime looks like these days—burglary, robbery, auto theft, and so on. But what about 300 years ago: Were there crime waves then like now?

**6.** A sane individual goes to the admitting officer of a mental hospital and gives that person an accurate case history (which in general states that he is as sane as you and I) and also says one thing that is not true: that he occasionally hears voices, unclear voices that seem to be saying words like "empty," "hollow," and "thud." What will happen to our sane person?

**7.** A teacher is led to believe (falsely) that several of his or her students are very sharp and are bound to do well this semester. How does the teacher behave toward the "sharp" students, and how do the students perform?

These are the types of questions that social scientists study and the types of questions that we try to answer. The questions listed actually have been studied, and some fairly interesting answers have emerged. Later in this chapter and in following chapters, I'll return to each of these and describe how we know what we know.

## KNOWLEDGE

Why do we want to study sociology? For ages people have been fascinated with their own existence, and they have persistently tried to find out more about themselves and the world in which they live. Their attempts have occasionally been bumbling and crude, sometimes amazingly sophisticated. For example, around 1900 B.C., a group of immigrants from what is now Holland and the Rhineland arrived in southern England. (They eventually became known as Beaker people because pottery drinking vessels were found in their graves.) Upon their arrival, they began building a structure of posts, mounds, holes, and stones on the Salisbury plain southwest of London that later became known as Stonehenge. Some of the stones they used were as long as 30 feet and weighed 50 tons. The stones were placed on end in holes in an intricate pattern of circles and alleys. It took the Beaker people and the Wessex people, who continued the work, some 300 years to complete what amounted to a giant calendar, a sort of Stone Age computer. The stones were set up so that sighting through cracks between stones would show sunrise, sunset, moonrise, and moonset on certain important days (for example, midsummer day, midwinter day, first day of summer, first day of winter). Other sightings allowed the Wessex people to predict eclipses of the moon.

Why was this elaborate structure built? As a calendar it was useful for judging the proper time to plant and harvest crops. It probably also proved helpful to priests, who could impress people with their apparent power to make the sun and moon rise and set wherever they wanted. And possibly the structure was built just because the Beaker and Wessex people were curious to see if it could be done.

Aside from the enormous physical work involved, the builders of Stonehenge had to collect and correlate a great deal of information: planting and harvesting times, times of sunrises and sunsets, and phases of the moon. They had to pass this information on from generation to generation. Archaeologists who have studied Stonehenge tell us that what looks like a single monument is actually several. The first structure didn't work as well as the builders wanted, and a second model was built over the first. It likewise needed improvement, and a third was constructed.

Unlike most primitive structures, Stonehenge has lasted to modern times. Some stones have weathered and fallen from their positions; others have been taken by farmers for stone fences. But much of the structure remains intact. For years, its presence and purpose had confounded people. And although some who studied Stonehenge felt that its existence had something to do with the positions of the sun and moon, the true complexity of Stonehenge was first revealed only in 1961, when astronomy professor Gerald Hawkins fed data on the positions and relationships of the stones, posts, and holes into a computer for analysis. Then modern people learned the significance of what the Beaker and Wessex people had learned 3,500 years earlier.[1]

In the late 1600s, some chemists who were attempting to explain what happens when substances are heated or burned came up with the theory of *phlogiston*. The theory stated that phlogiston is an invisible substance that exists in all combustible bodies and is released during combustion, the act of burning. Substances rich in phlogiston burned easily, and perhaps fire was a manifestation of phlogiston. The new theory was immediately accepted: It helped explain various phenomena, and it guided the famous scientists of the day. The problem, of course, is that it was totally wrong. Even when confronted with evidence that discredited the phlogiston theory, scientists in the 1780s were very slow to give up on it. Joseph Priestley, a famous scientist of the time, died in 1804 still believing in phlogiston.[2]

These are just two of countless examples of people's attempts to describe and understand their world, to obtain knowledge about their existence. Regardless of how it is collected and sometimes even regardless of its accuracy, knowledge tends to accumulate. Explanations seem to last if they are convincing; that is, they are believed as long as they seem to explain a part of human experience. If the explanations leave out important elements, they tend to be modified or even abandoned to be replaced by a better theory.

One way or another, knowledge accumulates, and it usually does so in one of three general ways: mysticism, rationalism, or empiricism. **Mysticism** refers to knowledge gained by intuition, revelation, inspiration, magic, visions, or spells. Societies often hold special places for those individuals who seem to possess unusual powers: the magician, the spiritualist, the priest, the witch doctor. The ceremonial use of drugs to produce visions is also highly valued among some peoples. In the 17th century, the Puritans of the Massachusetts Bay Colony were convinced of the presence of witchcraft in their midst. Their evidence came from several young girls, who, between dramatic seizures, pointed out fellow villagers who were aiding the devil in his work. The girls became so energetic in their attempts to cast out evil that 350 people were taken into custody and accused of witchcraft.[3]

Knowledge arrived at through mysticism tends to be private. Only the person experiencing the revelation or vision has it, and others must take the mystic's word for it. Even though this may enhance the appeal of the mystic, it tends to compromise the quality of the product. If no one else sees or feels it, can it really be true?

**Rationalism** refers to knowledge gained through common sense, logic, and reason. The writings of Aristotle and the dialogues of Plato in ancient times, and the ideas of Descartes, Spinoza, Voltaire, and Thomas Paine in the 17th and 18th centuries, provide examples of knowledge emerging through the careful use of logic and reason. In the 20th century, many continued to find the use of common sense, logic, and reason the most appropriate way to arrive at knowledge. As with mysticism, however, there are limitations. For example, at one time knowledge emerging from a common-sense analysis told us that the earth was flat ("Couldn't be round; people on the other side would fall off. . . ."), and that if two metal objects—a heavy one and a light one—were dropped from a height, the heavy object would fall much faster. We see that the intellectual approach alone can be as inaccurate as the mystical one.

**Empiricism,** on the other hand, refers to knowledge that is gained by sense observation—by observing or experiencing phenomena with the senses of touch, sight, hearing, smell, or taste. This is the basis of science. Instead of simply thinking about the heavy and light objects, we actually drop them from a tall building and observe whether the heavy one falls faster. Public knowledge then grows as we invite a crowd of observers to help us watch the fall of the objects, or we suggest that the experiment be tried by others in other places.

## SCIENCE

We have said that empiricism is the basis of science. A one-sentence definition of what scientists are up to could go something like this: **Science** is *understanding* and explanation through *description* by means of *measurement,* making possible *prediction* and thus *adjustment* to or *control* of the environment. For example, violent windstorms occasionally spring out of the Caribbean and the western North Atlantic. Huge waves caused by the storms are hazardous to shipping, and tremendous damage often results if a storm crosses land. Scientists try to understand these storms by collecting information about them and listing their common traits until a fairly consistent description emerges: Certain conditions of wind and atmospheric pressure lead to a situation in which a low-pressure cell is surrounded by winds circulating counterclockwise at speeds greater than 75 miles per hour. Then predictions about these storms, called hurricanes, become possible. The conditions that produce them, the time of year when they are most likely to occur (June to November, most frequent in September), their direction (west, then north, finally east), their speed (10–30 miles per hour), and their destructive power can all be predicted, allowing societies to control their impact. Advance knowledge of the hurricane's arrival means that clouds can be seeded with chemicals to reduce the force of the hurricane, and measures can be taken to move people and objects out of its way.

In a similar way, medical researchers observe a disease. Careful description of its characteristics allows predictions about its future occurrence and suggestions for its control through antibiotics, isolation, rest, and so on. Science (or empiricism) as a way of obtaining knowledge involves specific characteristics and assumptions. One such assumption is that the best way to know about the world is through the senses, aided when possible by mechanical means such as microscopes, scales, telescopes, and other such extensions of the senses. Scientists assume that knowledge obtained in this way is better than that of the rationalists and mystics before them. Knowledge gained through science seems to explain and predict phenomena more adequately than does that obtained by other methods.

Another basic assumption of most scientific endeavors is that phenomena are related causally. Science maintains that a **cause-effect relationship** exists in that each event has a prior cause or causes that can be discovered. Gravity (cause) makes an object fall (effect) from a building. If a heavy object falls faster or if objects fall at the same speed, these effects are also caused by phenomena that we can try to discover.

However, an event is usually caused by more than one factor. This is referred to as **multiple causation.** Someone runs into my car and crunches it. Why—what is the cause? It happened while I was trying to turn left at Watt and Arden, which happens to be the busiest and most poorly designed intersection in the county (top of a hill, a drop-off in three directions, poorly lit, many dogs), and traffic was backed up in all directions. It was near dusk and wet. The person who hit me coming the other way had just been in a fight with his boss and was a little bit tightened up about life. He had to pull around a drunk who had gone to sleep at the wheel; as he turned to talk to a guy about it, he hit me. It's possible, although unlikely, that I was slightly distracted by the exciting Lakers game that I was listening to. Now, what caused the accident?

Because isolating a single cause or even several causes becomes enormously difficult when working on a complex problem, scientists search for relationships and correlations. Look at **Question 1** at the beginning of the chapter. Blacks, although a small percentage of the population, represent a high percentage of those in professional sports, especially in professional basketball. Our sports fan who is about to turn to tennis thinks that blacks have inherited a predisposition toward basketball and sports. How's his logic—what's the answer? Sociologists look for a structural explanation and see a number of influences or relationships. Basketball is an inner-city playground game in the United States. As such, it is more likely to be played by inner-city residents who are more likely to be members of minority racial and ethnic groups. Playing playground basketball is inexpensive and doesn't require any special equipment or club memberships. Making it in professional sports is seen as a good path to upward mobility, especially if other ways (like having a middle-class or wealthy family background, inheriting a family business, or attending and graduating from a high-level university) are blocked. Also look at role models—the inner-city kid's hero, the person he models his behavior after, is probably a sports

## A SECOND OPINION

Stanford chemist/physicist Richard Zare, during an interview in September 1996 related to his study of the possibility of life on Mars, had the following to say about science: "Well, I must tell you about science and my feelings about it. I don't trust any scientific account all by itself. If it's really important but the work has only been done by one group—like ours—then I'm sorry, I wouldn't recommend it. What you want—just like a second opinion from a doctor—you want other experts to look too, because what I've learned from the years I've spent in science is that science is great at disproving things, but it's very poor at proving anything.

"It seems that if science can truly prove anything, it only does so by exhaustion—by having other people, other experts, repeat the research or the experiment, and then if they can't think of any other explanation that makes sense out of what the first group of experts has claimed, there's a good chance that the results are correct.

"So although I'm bold enough to tell you that it is our best understanding and belief that the evidence today says there was primitive life on Mars, I'm prepared to change my mind if somebody brings me new evidence that brings me to a better interpretation of what we have seen."

—From *"Sunday Interview: The Martian Chronicler,"* San Francisco Chronicle, *September 1, 1996.*

---

star, someone who has made big bucks in professional sports. Being a basketball star probably seems a lot more possible than does being a doctor, politician, or professor. It seems clear that *social* factors have a great deal to do with athletic success.

The concept of **control,** essential to science, has several meanings. As used earlier, it describes a product of knowledge. For example, if we know specific facts, such as when a hurricane will hit, then we can regulate or control events to a certain extent, such as moving people to safety. Or, as we learn about disease we can often control its effects with antibiotics. The second meaning of control refers to its use as an aid in the process of obtaining knowledge. Scientists attempt to control those factors in the environment that might have an effect on what they are studying. Our friends who are dropping objects from tall buildings may decide to perform their experiment only on calm, clear days. They are not studying wind and rain, but these factors might affect how the objects drop, and so they would want to make allowances for them. (Ideally, they should perform the experiment in a vacuum, if that were possible.)

Scientists are assumed to be *objective.* They are expected to record what they actually see, not what they hope to see or wish they had seen. They don't select or choose only those data that will fit or prove the hypothesis on which they are working while ignoring contrary evidence. One doesn't test the idea that sociology professors are absentminded by looking for a sociology professor who is thus afflicted and then saying, "I told you so." If the light object falls faster, or if both objects *rise* when dropped, then that is what the scientist reports.

Of course, absolute objectivity is impossible, and scientists who suggest that they are totally objective are misleading themselves and everyone else. Scientists are humans with beliefs and values that are bound to affect their work, in obvious and valuable ways. Their choice of topic stems from their values, as does their vigor and enthusiasm in pursuing it. Scientists have a passion for their research and are unlikely to study something they are not interested in. To ask them to set aside all nonscientific human emotions and biases is unreasonable. Objectivity does ask that scientists try to be aware of their biases and make them public so we can know of them.

Finally, **replication** is an essential characteristic of science. Studies are repeated by others in similar and in unique circumstances to see if the results are consistent, which, if they are, gives us confidence in their accuracy.

## Types of Science

The range of matters in which scientists are involved is probably far wider than most people think. There are numerous types or categories of science, and they involve widely differing methods. *Pure* and *applied* science represent two complementary approaches. **Pure science** attempts to discover facts and principles about the universe without regard to the possible practical uses the knowledge may have. **Applied science** concerns itself with making knowledge useful to people. The applied scientist devises practical and utilitarian uses of knowledge obtained through pure scientific endeavors. Splitting the atom was for the most part a product of pure science. Harnessing the atom to make bombs, to power submarines, and to fuel power plants is a product of applied science. The two—pure and applied—are closely linked, and the distinction between them is sometimes more artificial than real.

Science is also categorized by subject matter. Hence, you have the *physical* sciences like chemistry that study the physical universe; the *biological* sciences like zoology that study living organisms; and the *social* sciences like sociology that study various aspects of people and societies.

## Social Science

Social science in general and sociology in particular have developed as a result of studying certain characteristics observed in human beings. For example, humans tend to group together and to cooperate. People discovered long ago that they could accomplish much more working together than they could separately. Cooperation has obvious rewards. But we know that not all human behavior is cooperative: Conflict is also an important part of the human condition. One of the first cooperative group efforts probably involved a battle or conflict with another group. Conflict arises when people compete for the same resources, such as wealth or power. Very often when one party or group benefits, another party or group is deprived; this conflict is a major focal point of social science.

Another observable characteristic of human beings is that they tend in similar situations to behave in the same ways, time after time. We all know this. When a car comes toward us, we have a pretty good idea of which side of the road it will be on. Students can correctly predict the behavior of their instructor on the first day of class, just as the instructor can usually anticipate the questions and concerns the students will have about the class. The behavior of most people on almost any elevator is also predictable: Generally they will get in, push the button for their floor, stand as far as possible from others, face the front, stare at the numbers flashing, and be quiet and subdued. Amazing? Not really, for human behavior in its simpler aspects is not difficult to predict.

The complex aspects of behavior, however, are more difficult to predict. As people get together in groups, organizations, and societies, factors multiply, and behavior becomes more complicated. But behavior can be understood. Social scientists try to find the regularities and consistent patterns in human activities. They work through description, understanding, and explanation to prediction and perhaps control. Let's say, for instance, that we're worried about airplane hijacking. We first try to describe the condition. What sort of people are involved—first offenders or people with prior criminal histories? What motivates them—psychological problems, financial needs, a political viewpoint? Adequate description of the event may allow us to predict its occurrence. Careful screening of passengers to point out the hijacker type might, if we're very good at it, allow us to control the occurrence of the event. The point is that knowledge can be gained about human behavior using the general principles of science. This is the work of the social scientist.

The various social sciences approach the task of studying and recording human behavior in diverse ways. *Anthropologists* traditionally have gone about the task by examining artifacts and remains of long-extinct communities, or by living with and studying nonliterate tribes and societies. The study of culture is of central importance in anthropological analysis. *Historians* record an accurate chronology of past events. As events are placed in perspective, analysis of emerging trends can lead historians to make predictions about the future. *Political scientists* study the characteristics and patterns of political systems and the principles and conduct of government. Their topics include political parties, elections, systems of government, foreign policy, and the comparative structure of governments. *Economists* are interested in patterns of production, distribution, and consumption of goods and services. They study such topics as price and market theories; consumer behavior; merchandising and selling practices; money, banking, and credit; and economic growth and development. *Psychologists* deal with individuals; their adjustment and personalities; and their patterns of learning, motivation, and perception. *Sociologists* focus on groups, patterns of interaction, and descriptions of the institutions and social organization of society.

## SOCIOLOGY

*A sociologist is a person who goes to a football game and watches the crowd.*

The dictionary tells us that **sociology** is the "science or study of the origin, development, organization, and functioning of human society." Sociologists study human society and social behavior, and they focus on groups, institutions, and social organizations. The discipline of sociology includes several areas of study. For example, sociologists specialize in such areas as small groups, large-scale organizations, race relations, religion, marriage and the family, social problems, collective behavior, criminology and delinquency, social class, urban and rural sociology, population and demography, age and gender roles, political sociology, and the sociology of medicine and law.

In sociology, as in other disciplines, the process of discovery centers on the interplay of two elements: theory and research methods. A *theory*, which provides a framework for understanding phenomena, suggests specific hypotheses or predictions. Then, by using appropriate research methods, one can test the hypotheses empirically through controlled sense observation. The *research findings* often suggest modifications of the theory that in turn will make it more complete. This is the interplay of theory and research. For example, Einstein's theory of relativity provides a framework for making predictions about the nature of space and time under various conditions, assuming the speed of light is a constant. Physicists and astronomers test these predictions using sophisticated research methods that involve laser beams, high-speed aircraft, and complicated mathematics.

Data on cause of death do indeed tell us that year after year Miami has many homicides and that certain western states have high suicide rates as described in **Question 3** (see page 2). Explanations for Miami's murders easily come to mind—high percentage of gun ownership, high crime rate generally, and especially the high volume of narcotics traffic and the violent crime associated with it. But why so many suicides in New Mexico, Nevada, Wyoming, Arizona, and Colorado? Years ago, when he was studying suicide in western Europe, Émile Durkheim theorized that one type of suicide (egoistic) would be related to lack of social integration. Lack of social integration can be seen in such factors as high divorce rates, rapid population growth, rapid population turnover, and low church membership. Durkheim's theory presents us with a possible explanation for high suicide rates in the western states. Now we can go back to those states and see if they have the other characteristics that the theory suggests they should. Ideally, there is an interplay like this one between theory and method.

In the next few pages we will examine more closely the theory and research methods sociologists use in their efforts to understand the human condition. First, let's look at sociological theories.

## Sociological Theories

A *theory* is a coherent set of propositions used to explain a class of phenomena. Scientists develop theories to explain that aspect of the world they are studying. Next they attempt to test the theories with the information or data they subsequently collect. Theories are crucially important, for scientific research is organized around them. If the theory is reasonably accurate, it will explain many of the phenomena with which it is concerned. If it is not, then it will not be very helpful, and eventually a better theory—one that explains more about the phenomena or subject—will replace it.

In the second century A.D., the Greek scholar Ptolemy theorized that the earth was the center of the universe and the other planets rotated in complicated ways around it. This was the accepted theory, and it had strong support in the scientific and religious communities. But it did not explain the movements of the planets as well as it should have. Some 1,300 years later, the Polish astronomer Copernicus began publishing papers that outlined a new theory—that the sun was the center of the universe and the earth and other planets rotated around the sun. It was a revolutionary idea. In the years after Copernicus died in 1543, Kepler, Galileo, and others made continued observations enabling them to refine the theory that Copernicus had suggested. Substantial opposition arose to the new theory because it challenged some beliefs that were very important to the Catholic Church. In fact, Galileo was imprisoned for a time because of his support of the theory. It slowly became accepted, however, because it was a better explanation for astronomical phenomena than the Ptolemaic theory had been.

All disciplines have competing theories that attempt to explain the phenomena that are important to that discipline, and sociology is no exception. In the following paragraphs, we will briefly examine some of the major theories in sociology: functional analysis, conflict theory, and symbolic interaction.

**Functional Analysis**    The key terms in **functional analysis** are *structure* and *function*. A good way to understand functional analysis is to start with a biological analogy. If we look at the human body, we see that numerous functions must be performed—breathing and eating, for example—for the organism to survive. To perform these functions, structures like the nose, lungs, digestive tract, and so on have developed. Functional analysis holds that society can be analyzed in the same manner. Societies (and groups) need to perform certain functions to maintain their existence. They must populate themselves, they must care for the sick, the young must be socialized, goods and services must be distributed. Structures develop to perform these functions: a family system to control reproduction, an educational system to train the young, and economic and medical institutions to carry out other functions. This theory holds that structure and function are closely linked and that they are the crucial factors in understanding and explaining society. Functional analysis explains a given pattern of activity by defining its contribution, or function, to the group or society of which it is a part.

*SF: maintain*
*stability +*
*social*
*equil.*

*conservative,*
*emphasis on*
*social order*

Functional analysis tends to emphasize social equilibrium, stability, and the integration of the elements of society. One criticism of this theory is that it has a conservative bias; with its emphasis on social order, functional analysis tends to ignore conflict and social change.

Several terms commonly connected with functional analysis should be clarified. **Function** refers to an act that contributes to the existence of a unit, such as breathing and eating for the life of the organism. **Dysfunction** is the opposite—an act that leads to the change or destruction of a unit, like cancer in an organism. **Manifest function** is intended and recognized; **latent function** is unintended and unrecognized. According to Robert Merton, the Hopi Indians continue their ceremonial rain dancing even though the dances do not produce much rain. The latent function—reinforcing group identity by providing a periodic reason for getting together and engaging in a common activity—has become more important than the manifest function of producing rain.[4]

Try a functional analysis of a sociology class. Why *is* there such a class? It functions to pass on knowledge, viewpoints, and information. A structure has developed consisting of an instructor, some number of students, and a set of rules to facilitate this transmission of knowledge. It also functions to provide the teacher with a job. How well is the knowledge reaching the students? To answer this question, a structure—tests and a grading system—has developed. Attending class is functional if the student wants to get a good grade; cutting class often is dysfunctional. The first big test given in a class can be dysfunctional if it changes the mood of the class. The rapport and mutual understanding that develop between teacher and students over six weeks is suddenly compromised, and an adversary relationship develops. A manifest function of the typical classroom procedure—the teacher assigns work and students listen, read, obey, memorize, and give back on tests what the instructor has said—is to facilitate the learning process. A latent function of this procedure might be to train students to take orders and become good bureaucrats in the large organizations they will join in the future.

Now that you have some idea of how it works, try a functional analysis of some events or types of behavior—for example, prostitution, professional football, a funeral ceremony, or punishment for a crime. The discussions of family and religious institutions in Chapters Seven and Eight are clear examples of functional analysis.[5]

*CT: conflict + disord.*

**Conflict Theory**   In contrast with the social order and stability stressed in functional analysis, another theory focuses on conflict and discord. **Conflict theory** suggests that competition and conflict are common in social interaction and that the study of these processes is the most appropriate way to understand society. In the mid-19th century, Karl Marx wrote that the capitalists' or industrialists' exploitation of the workers would inevitably lead to a conflict that could be resolved only by a workers' revolution, resulting in a classless society. Marx felt that economic factors were the basic cause of conflict. Today, conflict theorists tell us that the potential for conflict exists in all social situations. In every grouping of people an imbalance of

*economic factors → conflict*

power and authority exists. Some lead, others follow; some make decisions, others take orders. It's true in the factory, the classroom, the family, the football team, the small group, and the large organization. People continually compete for scarce resources such as wealth, status, power, or authority, and this leads to conflict. Conflict also arises over differences in values and interests, for example, between superiors and subordinates, management and labor, and in-groups and out-groups. In examining social interaction, conflict theorists ask their basic questions: Who benefits? Who is deprived? Conflict theorists focus on how power is used by special-interest groups for their own benefit.

Some conflicts are minor and of little consequence. Others are deeply felt and can result in the division of whole societies into hostile classes as, for example, Marx suggested. Conflict can range from outright violence to hostility, tension, competition, and rivalry; even humor is often a form of conflict. Conflict tends to be perceived as negative, but its consequences can also be positive. Conflict with an outside force will probably bind the members of a group more closely together. Conflict over issues can lead to beneficial social change. Lengthy discussions between consumer advocates and manufacturers have led to improved auto safety. Likewise, improved wages and working conditions for labor and changes in women's rights and in the civil rights of minorities have emerged from conflict between groups with differing interests and conflicting viewpoints. Thus, conflict theory attempts to examine and explain how social change takes place, an issue with which functional analysis has some difficulty.

Getting back to our sociology class, what would a conflict theorist see in this situation? The testing and grading system necessary to facilitate the transmission of information creates a system of domination (instructor) and subordination (students) and sets up conflicting interests. Consequently, the conflict theorist would not be at all surprised with the sudden change in mood as the first test is given. Competition for scarce resources (As) leads to predictable types of behavior: cheating to get good grades, memorization of masses of soon-to-be-forgotten material, and rivalries with other students. The conflict theorist might also call attention to the tendency of the educational system to reinforce the current stratification system. Students from lower-class backgrounds, who often have less adequate educational preparation, may do less well in college classes, receive lower grades, or perhaps flunk out; because of this, they may obtain lower-level jobs and as a consequence stay near the bottom of the social-class ladder.

Try using conflict theory on some events or types of behavior like prostitution, professional football, or punishment for a crime. Ask who benefits and who is deprived. Conflict theory will come up several times in this book, particularly in Chapters Five, Six, and Nine in the discussions of stratification, power, prejudice, and discrimination.[6]

**Symbolic Interaction**   Symbolic interaction focuses on interactions among people and on the processes by which individuals develop viewpoints about themselves and to relate to their associates. In comparison with functional analysis and conflict theories, which examine social structure in groups, institutions, and societies, **symbolic interaction** theory narrows the focus to person-to-person interactions in everyday life. People interpret and define the symbols, gestures, and words of people around them, and they modify their own behavior accordingly. This activity is the fundamental concern of symbolic interactionists, who see social life as a *process*. They are interested in describing patterns of interaction; how parties to the interaction interpret what is going on; the use of signs, symbols, and other forms of communication; the meanings actions have for others; and the processes of socialization and development of the self.

Let's return once more to the sociology class, this time as symbolic interactionists. We would be interested in the type of interaction between instructor and students and would look at the signals—the verbal and nonverbal bits of communication—that pass back and forth. Suppose a difficult point is being made. The instructor attempts to evaluate how well the class is getting the point and adjusts the lecture according to his or her interpretations of the students' responses. Back in the corner is a student totally disgusted with this instructor, and yet class participation is required. Notice the way the student balances competing values—hostility, deference, and the desire for an A—in his gestures, tone of voice, body language, asides to other students, and so on.

The instructor has a particular self-concept; how is it expressed in his or her behavior? One instructor might be informal and innovative; another might be vastly self-confident, secure in the knowledge that he or she is the best instructor in the Western world. Imagine that a student in the classroom who has always seen herself as an A student gets a C on a test. What responses does she make that help her maintain her original self-concept? If a change in mood appears after the first test, the symbolic interactionist would be interested in how the students express it and how the instructor reacts to it. The symbolic interactionist, knowing that interaction is a two-way process, would be interested not only in changes in the students' behavior but also in how the instructor's behavior changes as a consequence of his or her interaction with the class.

The following examples of the symbolic interactionist approach come to mind: Chapter Two on socialization and self, parts of the discussion of groups in Chapter Four, discussions on self-fulfilling prophecy throughout, and much of Chapter Fourteen on deviation.[7]

It is important to recognize that these three theories represent contrasting approaches that we use to explain and organize the material of sociology, and they can be applied to any topic—racial discrimination, poverty, sports, crime, and so on. Theories provide a framework for looking at phenomena, a starting point for organizing ideas.

## RESEARCH METHODS IN SOCIOLOGY

Like other social sciences, sociology attempts to understand human behavior from the scientific point of view. This means that the sociologist is guided by the principles and assumptions of science that were outlined earlier in this chapter. To get some idea of what this means, let's consider some of the ways in which sociological research has been done. First, we must distinguish between two categories of research, *reactive* and *nonreactive*.[8] **Reactive research** refers to situations in which the observer or researcher is a part of the research situation. Much of social science research is based on surveys, specifically questionnaires and interviews. The researcher creates a questionnaire or interview, administers it to a group of people—the subjects—and measures the attitudes or behaviors of the subjects responding to the questionnaire or interview. These reactive measures are widely used because they are relatively inexpensive, are easy to administer, can be done on a group basis, and are less complicated than other methods. Also, in some cases they require less time to develop and to administer than do other methods.

**Question 4** describes the use of a reactive technique. Trained interviewers knock on doors or call people and ask them to answer certain questions in what are called public opinion polls. These days when we ask people what they think about the death penalty, we find that most favor it. The problem is that this type of approach tends to influence attitudes and behavior. Here's a simple illustration: If you *know* you're being observed and your behavior is being recorded, what happens? Well, you behave differently; you are uneasy, on guard, and tentative. The situation is not normal, and you don't behave the way you usually would. The researcher may be liked or disliked, the subjects may cooperate or not, and they may be ignorant or knowledgeable of the subject matter of the questionnaire; any of a number of things may happen. Whatever the result, the issue with reactive research remains the same: A foreign element is introduced into a situation, and attitudes are created as well as measured.

**Nonreactive research** has no intruding observer and doesn't use questionnaires or interviews. Instead, nonreactive research focuses on physical traces and signs left behind by people: on records and archives, such as hospital records, census data, and government records, and through simple observation in which the researcher observes but does not intrude. For example, you could find out which is the most popular radio station in your city by calling people on the phone and asking them (reactive), or by going to a busy gas station and as the cars come in, check the spot on the dial that their radio is tuned to (nonreactive). Or, you could find out who is the most liberal (or conservative) legislator in your district by doing an in-depth interview of legislators (reactive), or by looking up their voting records on a number of liberal versus conservative issues (nonreactive).

**Question 2** is an example of a nonreactive study. David Phillips's studies of "famous people" suicides used records. He looked at newspapers to get his sample of famous suicides. He obtained suicide rates in the nation from vital statistics records collected by the government. He marked the dates when the famous suicides occurred and then looked at the national data to see what happened then. He found that suicide rates in the country *increased sharply* in the period immediately after each famous suicide! In other studies, Phillips found that famous suicides and murder-suicides seemed to be related to other behaviors; that is, after famous people committed suicide

## KAI ERIKSON (1931–   )

Kai Erikson earned his Ph.D. in Sociology at the University of Chicago. In his studies of deviance and social change, Erikson emphasized the importance of history when conducting sociological inquiry. He argued that not only could social science make an important contribution to the understanding of history, but that history can also be helpful to sociologists puzzling over contemporary problems. This combination of history and sociology are evident in Erikson's book *Wayward Puritans,* in which Erikson draws on the records of the Massachusetts Bay Colony to illustrate the way in which deviant behavior fits into the texture of social life in general. Particularly interesting is his analysis of the major form of deviant behavior facing the colonists—witchcraft—and how the colonists dealt with it.

In his book *Encounters,* Erikson explains the impact an encounter with a famous person can have on an individual. "Most encounters have a special significance to those doing the remembering. They are retained in memory and later retold not just because they provide a moment of insight into the persons depicted, but because they had a shaping influence on the lives of the tellers." Erikson was a professor of sociology and American studies at Yale University, and is now professor emeritus.

Erikson appears in several places in this book. His nonreactive historical study of the Massachusetts Bay Colony is discussed in this chapter. Another of his books, *Everything in Its Path,* describes what happened when a flood in West Virginia destroyed a number of communities, and an excerpt from this book appears as a reading in Chapter Twelve. Erikson's views of the positive aspects of deviant behavior are covered in Chapter Fourteen.

and after highly publicized murder-suicides happened, certain other odd events took place. Can you imagine what they might be? (We'll return to this question later in this chapter.)

Another type of nonreactive study relates to **Question 5.** Kai Erikson decided to study crime in the past to see if it bore any resemblance to crime in the present. He used historical records and analyzed deviant behavior in a 17th-century Puritan settlement in Massachusetts. In his book *Wayward Puritans,* he tells us that the Puritans did indeed have crime waves—their "crime waves" were of a religious nature involving conflict with Quakers and the appearance of hysteria over supposed witchcraft. It is Erikson's view that although the crime waves sound quite different, the Puritans' way of viewing deviant behavior is not all that different from ours of today.

The value of nonreactive research is that it is unobtrusive. Either the traces or records of behavior are examined after the people have left them, or the act of observation is hidden so that the people don't know they are being observed. If people don't know they are being studied, it seems likely that we get a more accurate measurement of whatever it is we are studying. On the other hand, nonreactive studies tend to lack depth and flexibility. We may know about the *what* but be curious about the *why*, and nonreactive studies won't provide us with motives. The reactive study—the interview—is more flexible and can probe some of the answers inaccessible to the nonreactive study. Probably the most effective study is one that combines elements of each approach. Keep these two categories in mind as we look further at sociological research.

### Three Research Techniques

In the previous paragraphs we have mentioned several sources of research information: questionnaires, interviews, census data, hospital records, government records, and so on. To get a better idea of how sociologists obtain knowledge, let's look at three typical research techniques in more detail: participant observation, survey research, and the experiment.

**Participant observation,** now often called *field research*, is a study in which the researcher is or appears to be a participant in the activity or group that is being studied. *Street Corner Society* by William F. Whyte is a classic participant-observation study. While a student at Harvard some years ago, Whyte decided he wanted to study a slum district of a large city. He walked around Boston until he came upon an interesting area he called "Cornerville." How should he study the area? He first tried a door-to-door questionnaire, but he was totally dissatisfied with this approach. After several other false starts, Whyte sought help from settlement-house social workers in Cornerville, and they introduced him to "Doc." Doc was a native of Cornerville and knew everybody in the area. He took Whyte around as "my friend," and Whyte became involved in the community in ways he never could have achieved as an outsider. Shortly after, he rented a room in Cornerville, further establishing himself in the community. He found that as he got to know key members of groups, he could get information he wanted from other members of the groups. He learned how to join in street-corner conversations, and it wasn't long before Doc told him that he was as much a fixture on the street corner as the lamppost.

Everything did not go smoothly, however. On one occasion Whyte was talking about police payoffs to a gambling operator. When he became too inquisitive, the gambler became suspicious. On another occasion while talking with some Cornerville people, he decided to swear and use obscenities like they did. The conversation immediately stopped, and he was gently reminded that although they talked that way, they knew that he did not.

Whyte's relationship with Doc changed as his research continued. At first Doc was a friend and entrée into situations and groups. Later Doc became more of a collaborator as he helped Whyte interpret what he was seeing. In Whyte's case, many of the citizens in Cornerville knew he was something more than just another fellow on the street. It was reasonably well known that he was "working on a book." However, he was well enough integrated into the community that the knowledge that he was an observer as well as a participant didn't seem to change the way people behaved around him. By using participant observation, Whyte learned facts about Cornerville that he probably never could have learned otherwise. Through participant observation, he was able to analyze "the structure and functioning of the community through intensive examination of some of its parts—*in action.*"[9]

As described in **Question 6,** David Rosenhan and some friends individually went to mental hospitals, gave accurate case histories, and added (inaccurately) that they had been hearing some odd voices. What happened? Rosenhan and his friends were admitted to the hospitals, usually diagnosed as schizophrenic, and held from one to more than seven weeks. Although I'm not sure that's what he had in mind, it allowed Rosenhan and his friends to study the hospitals and mental-health care as participant observers. In the paper he wrote, Rosenhan comments on faulty diagnosis, inadequate care, misuse of medication, and the tendency of mental-health labels to go unquestioned once they've been applied.

We touched earlier on **survey research,** a very common social science research technique, in discussing reactive studies. The survey involves the systematic collection of information from or about people through the use of self-administered questionnaires or interviews. The researcher is interested in the general characteristics of a sample of people or in some experience or event in which they have been involved. A survey can be mainly descriptive or it can look for relationships among variables in the hope of getting to explanation and cause. Recently, I administered a short questionnaire to my criminology class that asked four background questions (gender, age, marital status, and major in school) and three opinion or attitude questions dealing with subjects we had discussed during the semester (death penalty, gun control, and major social problems). The items I asked about could be called variables. (A **variable** is a condition that changes or has varying values, such as age, I.Q., marital status, opinion about the death penalty, and so on.) The results are as follows:

| Gender | | Age | | Marital Status | | Major | |
|---|---|---|---|---|---|---|---|
| Male | 53 | Range | 20 to 50 | Single | 76% | Criminal justice | 74% |
| Female | 50 | Average | 24.8 | Married | 19% | Sociology/social science | 21% |
| | | | | Divorced/ separated | 5% | Other | 5% |

| | Yes | No |
|---|---|---|
| I am in favor of the death penalty (that is, I think death is an appropriate punishment for those who commit certain major crimes). | 80% | 20% |

The type of handgun control I favor is:*

| | |
|---|---|
| Registration | 68% |
| More severe punishment for use | 41% |
| Confiscation | 18% |
| None—I don't favor handgun control | 3% |

*Adds to more than 100% because more than one answer could be selected.

They ranked eight social problems from most serious to least serious as follows:

1. Violent crime
2. Poverty
3. AIDS
4. Pollution of the environment
5. Racism
6. Economic instability
7. Overpopulation
8. Political/religious terrorism

So far we have a simple descriptive survey that tells us that this class is about evenly divided between males and females, is a little bit older than the average college class, is mainly single, and contained mostly criminal justice majors (this class is required in their major). They support the death penalty, seem to favor several types of gun control, and, in their evaluation of eight social problems, find violent crime to be the most serious. We can leave it at this, but why not go a step further and look for other relationships that might exist? A simple thing would be to see if males differ from females, or if young students are different from older students. So, I compared the males with the females and the 25 oldest students with the 25 youngest ones:

*Males and females were similar on most items, but although they both favored the death penalty, females favored it less (74 percent) than did males (85 percent). And on ranking social problems, females considered poverty and economic instability to be more serious problems than the males did. Males viewed violent crime as the most serious.*

*The youngest and oldest students were similar on most variables except how they ranked social problems. The younger students saw violent crime and AIDS as serious problems, whereas the older students ranked poverty at the top and ranked AIDS fairly low in seriousness.*

Depending on the size of the sample and the number of variables in the survey, we can get more and more detailed, and perhaps doing so will help us move from description to cause.

Several times in this chapter we have mentioned *cause,* and this brings to mind the story (perhaps true) of the student who was comparing auto traffic over a bridge with water level of the bay under that bridge. After many careful measurements, the student discovered that there was a relationship, or correlation, between the two—specifically, the more cars on the bridge, the higher the water under the bridge. The student found this to be amazing and assumed the following cause-effect relationship: The heavy traffic on the bridge must be compacting the land somehow and forcing the water higher. A wise instructor intervened at this point and suggested that, contrary to the apparent cause-effect relationship, perhaps some other variables were involved. Further study indicated that traffic was heaviest during morning and evening rush hours and that high tides often occurred at about those same times. The lesson is that correlation doesn't necessarily mean cause, and relationship can come from sources other than those first assumed.

The **experiment** represents a third kind of research technique. In an experiment, the researcher manipulates one variable (the *causal* or **independent variable**) and watches for changes in the other variable (the *effect* or **dependent variable**). For example, a researcher might find two groups of people in good health, then have one group run several miles a day and have the other group do no physical exercise. Finally, the researcher would watch the two groups for incidence of heart disease. The independent variable, exercise, is manipulated for two groups, and then possible changes in the dependent variable, heart disease, are observed.

As you can see, the experiment searches for explanatory or causal relationships. It systematically alters or manipulates aspects of the world to see what changes follow. Suppose you are interested in the relationship between studying and academic performance; that is, you want to improve your grades. You could keep a record of how much you study for each test and the grades you receive. This would be a type of survey. Or you could systematically vary your studying—10 hours for one test, 20 hours for another, 40 hours for a third—and check the effect on your grades. This would be an experiment.

**Question 7** describes an experiment done by Robert Rosenthal and Lenore Jacobson (R&J). R&J were doing research in elementary-school classrooms in South San Francisco. At the beginning of the semester they casually told each teacher that five of her (his) students were very sharp—were "academic spurters" who could be expected to do well during the year. R&J named the students so that the teachers would have no doubt. In fact, R&J had misled the teachers—the so-called academic spurters had been *selected at random* from each class. R&J then observed the students and the teachers through the rest of the school year. In the first and second grades, the selected students showed great gains in I.Q., much greater than gains by nonselected students. Further, the teachers described these students as much more capable, more appealing, more likely to be successful, and so on. R&J report that the teachers treated the selected kids differently, not in obvious ways such as spending more time with them but in very subtle ways—tone of voice, facial expression, touch, and posture. This experiment illustrates a number of things, not the least of which is ethics. Do you think it was ethical for R&J to mislead the teachers the way they did? We'll discuss another aspect of this research in Chapter Two.

In the last few pages we have described three research techniques: participant observation, survey research, and the experiment. You might find it interesting and challenging to try out one or more of these techniques. For example, take one of the groups you belong to—such as an athletic team, a dorm group, a social club, or a political organization—and study it. A short questionnaire could give you a descriptive survey of attitudes on the death penalty, on political conservatism, on types of deviant behavior, or on presidential candidates. Or, more informally, as participant observer, watch the group in action. Who interacts with whom? Who are the leaders of the group and why? Is disruption caused by the same people, or is it spread throughout the group? Are some people left out by the rest of the group?

See whether, by using controlled observation, you can discover something new about your group.

## Problems in Social Science Investigation

We have been describing sociological research efforts—how sociologists come to know what they know. We have found that by working within the general guidelines of science, sociologists develop special techniques to study human behavior. All sciences do the same thing. Subject matter varies, of course, as do the problems sciences face. In the case of the social sciences, the subject is *people*. Variations in humans and their behavior are vast. Multiple causation is the rule, not the exception, and digging out all the related factors is always difficult and often impossible. One problem shared by social and natural scientists alike results from the *interactive* nature of research: The very act of observation can change the nature or performance of the subject matter. This is especially true in the social science situation: When people are studied, they usually know it, and this may affect their behavior in ways that are difficult to predict. (We commented on this earlier in discussing reactive and nonreactive studies.) For example, look at a study done some years ago at the Hawthorne Works of the Western Electric Company. Some researchers were interested in increasing production among women who wired electrical relays. They experimented with two groups. For one group, lighting was improved so they could see their work better. For the other group nothing was done. Output improved equally in *both* groups. Next, lighting was reduced to the point at which the workers had difficulty seeing what they were doing. But there was no decrease in efficiency or speed of production! What was going on? It became apparent that there was much more to working efficiency than just physical working conditions. In fact, the workers were responding to *being studied* by the researchers, and this had more effect on their output than changing supposedly important physical factors had. This finding, sometimes called the **Hawthorne effect,** was the impetus for a series of studies dealing with psychological and social aspects of the work situation.

The problem begins with multiple causation of human behavior, and now we find that one of these causes is one we introduce by trying to study it. It is difficult enough as it is, and *we* make it worse—that is, more complicated—by looking at it.

Social scientists are involved in the unique situation of *humans studying humans*, and this leads to problems. It is difficult to be objective under the best of circumstances, and in this field, scientists study people just like themselves. If it were oxygen or laser beams, we would be less likely to identify with the subject. But in social science the subject is people, and scientists might identify with their feelings, their reactions, and their beliefs. I suppose we are asking: Can people possibly be objective and unbiased when, in effect, they are studying *themselves?* This question has bothered social scientists for decades, and their responses to it vary. Some believe that one can put aside biases and beliefs and become a dispassionate observer.

## PREDICTING THE FUTURE

On August 11, 1999, we had the last solar eclipse of the millennium. Astronomers handled it beautifully. They told us when it would happen (11:03:04 UT in south-central Romania), how long it would last (2 minutes and 22 seconds in Bucharest), and what we could expect to see depending on where we were watching from. They compounded that achievement by telling us when the *next* solar eclipses would happen. They tell us that the next one in Europe won't happen until 2081, but there will be a decent one in the United States on August 21, 2017! Then a social scientist was asked to predict social conditions in the United States in 2017—family size, migration patterns, crime patterns, economic conditions, major social problems, and so on. The social scientist told them he would check out his data and get back to them in 2018.

Others say no—humans can't study other humans and be objective. If this last group is correct, there isn't much future for social science, or at least we will always have to bear in mind its limitations. Who's right? Well, certainly there are problems with objectivity when studying humans. Most social scientists probably see this as another variable in their studies that needs to be considered but one that by itself doesn't compromise the scientific effort.

### Ethics

Ethics is another tricky issue that social scientists have to consider, especially because they are working with human subjects.[10] A few paragraphs back (page 20), we learned that R&J were studying teachers and we wondered if it was ethical for them to mislead (lie to) the teachers the way that they did. Let's consider some of the relevant ethical issues that researchers have to consider: harm to subjects, informed consent, deception, and privacy. Clearly as a researcher you wouldn't want to expose subjects to harm. Sometimes, however, it is difficult to anticipate all the possible directions a project can take. In one classic study, students were asked to pretend to be guards and prisoners in a mock prison situation. Both groups got carried away—guards physically and psychologically abused prisoners, and prisoners broke down and became apathetic. In another study, a crime (a robbery) was staged to see if bystanders might intercede. One witness called the police who showed up with guns drawn to arrest the researchers. And Stanley Milgram has done several creative studies on obedience to authority in which he asked subjects to shock other subjects. Look at a brief summary of his work (see page 73) and see if you think there are any ethical problems.

*Informed consent* means that subjects should know enough about what they are getting into so they can decide whether they want to be involved in the project or not, and that their participation is voluntary. Of course, the researcher probably doesn't want to disclose everything for fear it will spoil the project. Think about Milgram's project—what could he tell his subjects without giving the venture away? In a controversial study of sexual encounters in public restrooms, participants' identities were traced through their auto license plates, and they were later interviewed. Their identities were kept confidential, but they had no informed consent and the potential for personal damage was spectacular. *Deception* is common in social research. One study found that 58 percent of the research studies reported in three major social psychological journals used some form of deception. This usually involves misleading the subjects as to what the study is about. In the R&J study mentioned on page 20, the teachers were misled about the focus of the study and about the abilities of their students. Why is there so much deception in research? The researcher hopes that the subject will behave normally and without preconceived ideas. Many researchers believe that numerous topics cannot be effectively studied *without* deceiving the subjects. One hope for offsetting the effects of deception is debriefing. In debriefing, the researcher tells the subjects the true nature of the study and why deception was necessary after the study is over. If this is done carefully, it might help the situation, but some social scientists hold the belief that deception is inappropriate in any research setting.

*Privacy* refers to being able to keep your behavior, attitudes, and feelings private if you want to. Ethical investigators guarantee anonymity or confidentiality to their subjects. Concerns about invasion of privacy are heightened by stories of unauthorized listening to or viewing of activities assumed to be private. In a research setting, this could refer to the use of questionnaires with a hidden identification number, one-way mirrors, and hidden microphones or cameras. In a case that was widely criticized, an effort to better understand how juries work led judges to give permission to secretly record actual jury deliberations. Another case involved the study of sexual behavior in public restrooms mentioned earlier. Though confidentiality might be guaranteed, it is felt that there are certain sensitive areas (jury deliberations, sexual behavior, one's home, a doctor's office, and so on) where absolute privacy is essential. Guarantees of confidentiality can lead to other problems for researchers. In 1991, animal rights activists were involved in break-ins at university labs in Washington state. Graduate student Rik Scarce had interviewed some activists and was thought by authorities to know about some of their illegal activities. He had guaranteed privacy to his sources, and therefore he refused to testify or give information and was jailed for six months.

### Is It Only Common Sense?

Students of introductory sociology, and many others as well, frequently describe sociology as mere common sense. Sociologists deal with contemporary human behavior and patterns of interaction between people. Atoms and rockets and relativity are one thing, but when it comes to understanding people, we're all experts.

In some aspects of human behavior, our experiences and our intuition might serve us well. None of us, however, has experienced all situations. Because we live in a particular society, area, social class, and community at a particular time in history, our experiences are necessarily limited. We actually know much less about human behavior than we think we do. For example, remember **Question 2?** Suicide rates in the nation went up after a famous-person suicide. What else happened after a famous suicide or after a publicized murder-suicide? Can you imagine? Well, auto accidents (especially single-car accidents) and auto fatalities increased! The more publicity for suicide, the more auto accidents there are, and the age of the people in the auto accidents is significantly correlated with the age of the suicide victim. Apparently, people imitate suicidal behavior, and a substantial proportion of certain types of auto accidents are suicide attempts. Would your intuition or common sense have told you that? If so, try this one: After highly publicized (front page of the paper, on network nightly news) murder-suicide cases, there is an increase in private and commercial *plane crashes!* Again, violence is imitated and "accidents" aren't necessarily accidents.[11]

A peculiarity of common sense is that it can provide a rationale for almost *any* response. It takes a more sophisticated analysis to predict human behavior accurately and to answer the complex questions of why people do what they do. To put what we know by means of common sense further in perspective, try this adaptation of Gerald Maxwell's social awareness test.[12] How well does your common sense serve you? (The answers appear at the end of this chapter in note 12.)

**Social Awareness**    Test Directions: Answer true or false to the following questions:

1. There are five times as many arrests of men as there are of women.

2. Because of discrimination and depressed living conditions, more blacks commit suicide proportionately than do whites.

3. With the exception of movie stars, people who divorce are slow to remarry.

4. Land is generally less expensive in the suburbs than in other parts of the city.

5. Women are more likely to commit suicide than are men.

6. There are more Hindus in the world than Protestants.

7. Cities are where the jobs are; consequently, there are generally more men than women in cities.

8. Children from divorced or unhappy homes are usually more careful in selecting a mate, and they make better marriages.

9. Panic is a common response for people confronted by disaster.

10. People with some college education are more likely to attend church than are people with only a high-school education.

## SUMMARY

This book focuses on the *concepts* of sociology. Concepts provide a way of generalizing about phenomena. Sociologists use concepts such as role, motivation, anticipatory socialization, culture, in-group, cult, institution, and mob to help them organize and understand the events they are studying. We could say that the major task of sociology (and probably of other disciplines as well) is to build concepts—that is, to isolate certain unifying, abstract qualities that underlie and thus "explain" behavior. This book attempts to gather some of the major concepts of sociology and to define and illustrate them. The discussion of concepts is divided into three major sections. These sections discuss the individual in society (Socialization and Culture), the organization of society (Social Organization), and society in flux (Social Change and Social Deviance).

This chapter has looked at the sources of knowledge—how do we know what we know? The reading that follows explores this problem. It is a description of the steps involved in the research process, using as an example a recent study of drug use and race/ethnicity.

## TERMS FOR STUDY

applied science (8)

cause-effect relationship (6)

conflict theory (12)

control (7)

dependent variable (20)

dysfunction (12)

empiricism (5)

experiment (20)

function (12)

functional analysis (11)

Hawthorne effect (21)

hypothesis (27)

independent variable (20)

latent function (12) *more meaningful purpose*

manifest function (12)

multiple causation (6)

mysticism (4)

nonreactive research (15)

operational definition (27)

participant observation (17)

pure science (8)

random sample (28)

rationalism (5)

reactive research (15)

replication (8)

science (5)

sociology (10)

survey research (18)

symbolic interaction (14)

variable (18)

## INFOTRAC COLLEGE EDITION
**Search Word Summary**

To learn more about the topics from this chapter, you can use the following words to conduct an electronic search on InfoTrac College Edition, an online library of journals. Here you will find a multitude of articles from various sources and perspectives:

**www.infotrac-college.com/wadsworth/access.html**

| | |
|---|---|
| empiricism | Aristotle |
| functional analysis | Political Science |
| mysticism | Copernicus |
| capital punishment | Puritans |
| Stonehenge | |

---

## Reading 1.1

---

# THE RESEARCH PROCESS— RACE, DRUGS, AND CRIMINAL SENTENCING

In this chapter we have focused on examining the processes through which we gain knowledge—how we know what we know. This involves an interplay between theory and research and the use of a reasonably precise research methodology. Researchers go through a series of steps when they undertake a project—these steps could be called the research process. These steps can vary slightly from one project to the next but the overall pattern will be similar. In the following paragraphs, we will summarize the steps that two colleagues of mine went through in a study they just completed.

*Background:* Drugs were a major concern in the 1990s. Citizens believed that illegal drug use was related to other crime, and police found enforcement of existing drug laws to be difficult. Some criminologists have focused on a curious characteristic of drug laws: Specifically, striking differences exist in the severity of laws dealing with different drugs. For example, the punishment for use of methamphetamine and marijuana is far less severe than the punishment for use of crack cocaine, powder cocaine, and heroin. This situation as well as the popular perception that there is a relationship between race/ethnicity and drug of choice intrigued Carole Barnes and Rodney Kingsnorth (B&K) and raised several questions that they wanted to answer. Many, maybe most, research efforts seem to start this way.

1. **Problem** The first step in the research process is the statement of a problem. This is usually a general statement that is, at the same time, specific enough to allow the variables in the statement to be studied. To say that "sociologists have more fun than psychologists" might be a fascinating statement but it's hardly a research problem—at least one concept ("more fun") is fuzzy and difficult to measure. The research problem for B&K was this: Is the response of the criminal justice system to those charged with drug offenses based on the drug used or on the race/ethnicity of the user?

2. **Identify Concepts** The major concepts suggested in the problem must be identified. B&K's major concepts were race/ethnicity, drug of choice, offense level (possession, possession for sale, and sale), prosecutorial decision, and severity and length of sentence.

3. **Operational Definitions** The elements in the study have to be precisely defined. **Operational definitions** (sometimes called *working definitions*) allow others to know exactly how something was measured. Imagine if someone told you that the distance between *A* and *B* was 38. You'd say, "38 what? Inches, yards, meters, miles, cubits???" You need to know exactly how the variables in the study were defined. Most studies have a number of operational definitions. For B&K, "drug use" was defined as being charged, processed, and sentenced with a single drug offense involving crack or powder cocaine, heroin, methamphetamine, or marijuana.

4. **Hypothesis** Hypotheses are specific testable statements or propositions that are used to guide the study. An **hypothesis** can predict a relationship between variables—"a person with a lower level of education will have a higher level of racial prejudice." Another type of statement, called a *null hypothesis,* predicts no relationship—"there is no relationship between level of education and degree of racial prejudice." B&K's hypothesis was as follows: Are observed race/ethnicity differences in drug sentences caused by discrimination or by differences in the laws governing different drugs?

5. **Research Design** Several tools are available to the researcher. The design or method of study typically follows from or is a consequence of the way the problem and hypothesis are stated. Possible designs include a case study (study one group at one point in time), a comparative study (compare two or more groups at a single point in time), a longitudinal study (study one group at two or more points in time), or an experiment (study a group before and after a stimulus is introduced). Data sources can include survey research, participant observation, or making use of already available data. The study can be reactive or nonreactive. B&K used a case study design that was nonreactive and called for use of already available data. In a technique called *historical analysis,* they examined case files from a county district attorney's office.

6. **Sample Design** Sampling design refers to how you select cases to be studied. In many studies, the total population of whatever it is you want to study is very large—all blond-haired people, all basketball players, all vegetarians. So the researcher selects a sample from the larger population. How this sample is selected is crucial if you want it "to speak for" (to be able to generalize to) the population as a whole. A probability sample is one in which each subject in the population has an equal chance of being selected. Suppose we want to study blond-haired vegetarians (BVs) of which there are many. We need a sample, so we give each of our BVs a

number, scramble the numbers or put them in a hat, and then select a sample of 100. If we do it carefully, we would have a type of probability sample called a **random sample.** If, on the other hand, we took the first 100 BVs we could find (because they live on our block, or happen to be in our sociology class), that wouldn't be so good. We wouldn't be able to generalize beyond our block or our sociology class. We would have come up with a nonprobability sample, in this case, an accidental sample.

The population in B&K's study was the 1,369 drug offenders charged in 1987 in Sacramento County, California. B&K felt this to be a manageable number, so they studied the whole group rather than take a sample. B&K realized in selecting their population that they could only "speak for" Sacramento County; their sample was certainly not representative of drug offenders in the United States, the West, or even in California.

7. **Data Collection** This step involves collecting information to answer questions posed earlier in the problem and in the hypotheses. B&K studied the case files of people arrested and charged with drug offenses, and they looked specifically for information on (1) the race/ethnicity of the person charged, (2) the type of drug the person was charged with using, (3) how the prosecutor pursued the case (what charges were brought), and (4) what sentence the offender received.

8. **Data Analysis** In this step we analyze the data we have collected and, if things turn out right, begin to answer the questions posed in the problem and hypotheses. B&K discovered that (1) the drug involved in the offense varied by race/ethnicity: whites were arrested for marijuana and methamphetamine use, blacks were arrested for crack cocaine use, and Latinos were arrested for heroin and methamphetamine use; (2) blacks were more likely than whites or Latinos to have their cases dismissed (not prosecuted); (3) whites were most likely to have their cases reduced to misdemeanors (minor offenses), and blacks were most likely to receive a prison term; and (4) when sentenced to prison, blacks and Latinos received substantially longer terms than whites.

9. **Conclusions** In this step, the researchers make conclusions from the data they have collected and analyzed and relate these conclusions back to the original problem and hypotheses. In B&K's study, clear race/ethnicity differences existed in the drug cases they studied. B&K concluded that the best explanation for what they found was the character of inner-city drug markets and the drug suppression strategies that emerge to deal with these markets. There are apparently different drugs of choice—marijuana and methamphetamine for whites, crack cocaine for blacks, methamphetamine and heroin for Latinos (see data analysis 1). The crack cocaine drug market (which involves black offenders) is more public and visible because it takes place on street corners, in parks, and around known crack houses. Law enforcement can and does focus much of their energy getting large numbers of relatively easy arrests. Quantity becomes more important than quality, and prosecutors dismiss many of the cases because of insufficient evidence (see data analysis 2).

The law allows methamphetamine use to be treated more leniently than use of cocaine or heroin. Many experts find this to be biased and indefensible—they wonder if the legal penalties are a response not to the dangers of the drug itself, but to the perceived social danger of the user population. Be that as it may, the law permits methamphetamine cases to be reduced to misdemeanors, whereas crack cocaine and heroin cases are prosecuted as serious crimes or felonies (see data analysis 3). Finally, blacks receive longer prison terms because being charged with possession, possession for sale, or sale of crack cocaine (their drug of choice) is viewed by the law as a more serious crime and carries more severe punishment than does being charged with possession, possession for sale, or sale of methamphetamine or marijuana, the drugs of choice for whites (see data analysis 4).

B&K's study was guided by a hypothesis that asked the question: Are observed race/ethnicity differences in drug sentences caused by discrimination or by differences in the laws governing different drugs? B&K's conclusion: Given differences in drug of choice, differential sentences prescribed by law are clearly more important than the race/ethnicity of the offender in explaining longer prison terms received by blacks and Latinos. Indeed, B&K observed that when different race/ethnicity groups (whites, blacks, Latinos) used the same drug, they received the same treatment.

10. **Report Writing** The final step in the research process is to let others know what you have done. Other social scientists might be skeptical of the findings or the conclusions. They will want to verify the results, to try the study in other areas and on other samples. So, the study must be "published" in a public forum of some sort. This can mean a paper delivered at a scholarly meeting or an article in a professional journal. B&K did both—they presented a paper at a meeting of the Pacific Sociological Association, and they published a report of their research in an article that appeared in the *Journal of Criminal Justice* in early 1996.[13]

---

## QUESTIONS 1.1

**1.** Think of two or three questions that you would like to study. Then, "do the research." That is, take your question through the first seven steps of the research process (up to data analysis). Show how you would attack the problem using the steps in the research process.

**2.** This is an example of the research process with a subject from social

science. Would the process be different for the natural or physical sciences? Explain why or why not.

**3.** What is the difference between a probability and a nonprobability sample?

**4.** Why is "publishing" the results necessary?

---

## NOTES

1. For more detail on the fascinating story of Stonehenge, see Gerald Hawkins, *Stonehenge Decoded* (New York: Dell, 1965).

2. There is a short discussion of phlogiston in *On Understanding Science*, by James Conant (New York: New American Library-Mentor, 1951), chapter 3.

3. For a more detailed discussion of the Puritans and their problems, see Kai

Erikson, *Wayward Puritans* (New York: Wiley, 1966).

4. Robert Merton, *Social Theory and Social Structure*, 2d ed. (New York: Free Press, 1957), pp. 64–65.

5. From among the numerous sources available on functional analysis, the following have been especially helpful: Neil Smelser, ed., *Sociology* (New York: Wiley, 1967), pp. 706–708; and Robert Merton,

*Social Theory and Social Structure,* 2d ed. (New York: Free Press, 1957). See the section on functional analysis in the *International Encyclopedia of the Social Sciences* for a summary and an excellent bibliography.

6. For a background on conflict theory, see the works of Karl Marx; Lewis Coser, *The Functions of Social Conflict* (Glencoe, Ill.: Free Press, 1956); Ralf Dahrendorf, *Class and Class Conflict in Industrial Society* (Stanford, Calif.: Stanford University Press, 1959); see also the section on conflict in the *International Encyclopedia of the Social Sciences* for an overview and bibliography.

7. For more on symbolic interaction, see the works of Charles Cooley or George H. Mead (see index) or a social psychology text such as *Social Psychology,* 4th ed., by Alfred Lindesmith, Anselm Strauss, and Norman Denzin (Hinsdale, Ill.: Dryden Press, 1975).

8. This distinction is best made in *Unobtrusive Measures: Nonreactive Research in the Social Sciences,* by Eugene Webb, Donald Campbell, Richard Schwartz, and Lee Sechrest (Chicago: Rand McNally, 1966).

9. William Foote Whyte, *Street Corner Society,* 2d ed. (Chicago: University of Chicago Press, 1955). Also see a good summary of Whyte's work in *The Origins of Scientific Sociology,* by John Madge (New York: Free Press, 1962), chapter 7.

10. The discussion of ethics in the following paragraphs is taken from *Approaches to Social Research,* 2d ed., by Royce Singleton, Jr., Bruce Straits, and Margaret Miller Straits (New York: Oxford University Press, 1993), chapter 16.

11. A number of papers by David Phillips deal with various aspects of imitation of

violent behavior. For example, see *American Sociological Review* 39 (1974), pp. 340–354; *Science* 196 (1977), pp. 464–465; *American Journal of Sociology* 84 (1979), pp. 150–173; *Social Forces* 58 (1980), pp. 1001–1024; *American Sociological Review* 48 (1983), pp. 560–568.

12. Gerald Maxwell describes a social awareness test that he has devised (*The American Sociologist,* November 1966, pp. 253–254). I have used his title as well as one or two of his questions.

Answers to the social awareness test:

1. True. (See Uniform Crime Reports, 1998.)

2. False. (See p. 189.)

3. False. Divorced persons are generally more likely to marry than are single persons of the same age.

4. True. (See p. 343.)

5. False. (See p. 433.)

6. True. (See p. 265.)

7. False. Most cities have more women than men.

8. False. Couples whose parents had unhappy marriages are more prone to divorce.

9. False. (See pp. 405–406.)

10. True. (See pp. 270–271.)

13. Carole Barnes and Rodney Kingsnorth, "Race, Drug, and Criminal Sentencing: Hidden Effects of the Criminal Law," *Journal of Criminal Justice* 24 (1996), pp. 39–55.

# PART ONE

# *Socialization and Culture*

In this section, we discuss how the infant develops into a social being. The socialization process is of major importance. Through this process, values are transmitted from one generation to the next, and infants grow and learn to adapt to their environments. The self, the person's conception of what and who he or she is, slowly emerges from interaction with others. Through this interaction, individuals are introduced to norms as well as to roles—those behaviors that are expected of them in specific situations and positions. This personal and social development takes place within the context of a specific culture. The culture provides a set of behaviors, traditions, customs, habits, and skills that also become a part of the individual. These processes—socialization, development of the self, internalization of norms, roles, and other aspects of the culture—are particularly evident in young children, although they continue to be important throughout life.

# Socialization and Self

## TWO RESEARCH QUESTIONS

**1.** *Is a child's mental age (I.Q.) affected by the number of brothers and sisters he or she has? In families with two or more children, are there predictable mental-age differences among the children? Whose I.Q. is higher—firstborn, last born, in between? Or is there any consistent pattern?*

**2.** *Students are assigned the task of teaching rats to learn a maze. Some students are given rats that are said to be stupid, poor at learning mazes. Other students are given rats said to be very sharp, fast learners. In fact, all the rats are alike and have equal capabilities. How will the two groups of rats perform in learning mazes. Why?*

FEW PEOPLE SPEND much time thinking about the transformation they went through in the process of becoming mature human beings. Not only are we not concerned with the steps required to carry us from an early blob of protoplasm to a complicated, interacting individual, but it is even difficult for us to believe that we were ever at the stage of egg and sperm. Beyond a certain idle curiosity as we watch a baby grow up, we take these changes in the developing human for granted. They are like the phases of the moon or the coming of summer—they seem just to happen.

However, if we decide to analyze the development of the individual in some detail, several approaches are available. A biological approach would emphasize physiological maturation. As a man's hair changes from brown to gray to gone, so the person grows and matures in other ways. This approach might suggest that maturity is merely a matter of cell changes and that, given enough time, these changes produce a mature social being.

But there is much more to development of the organism than just biological and physiological changes. Findings from research on animals suggest that being around and having contact with beings like yourself is a crucial part of normal development. In classic studies, Harry and Margaret Harlow gave baby monkeys artificial "mothers"—some were constructed of wire mesh but were capable of providing food through a feeding tube; others had no feeding tube but were covered with soft, comfortable, terry cloth. When hungry, the baby monkeys went for the food, but when frightened, they went for the terry cloth, demonstrating to the Harlows the importance of "intimate physical contact." Monkeys raised in isolation, especially for periods of longer than three months, were likely to be permanently damaged emotionally and had difficulty adjusting to other monkeys.

Likewise, interaction with other humans is crucial for normal human development. Isolation from (or inadequate) human interaction can affect our growth in a number of ways. Examples of extreme isolation are unusual, but occasionally such cases are described. Some years ago, Kingsley Davis described the cases of Anna and Isabelle, who, because they were illegitimate children, were kept in nearly total isolation the first years of their lives.[1] Anna's mother was mentally retarded and left Anna alone in an attic room most of the time. Isabelle's mother was a deaf mute who stayed in a dark room with Isabelle, shut off from the rest of the family. When Anna was discovered, she was about six years old and could not walk, talk, or do anything that showed intelligence. She was in bad physical condition as well. After about two years in a county home, she had learned to walk, understand simple commands, feed herself, and achieve some neatness, but she still did not speak. She spent the next three years of her life (she died at the age of ten and a half) in a private home for retarded children, and there she made some further progress. Her hearing and vision were normal, and she walked and ran fairly well. She could bounce and catch a ball, string beads, identify a few colors, and build with blocks. And she finally began to develop speech. She attempted conversations with others, although she spoke in phrases rather than in sentences. In summary, when Anna was found at the age of six, she had the mental capacity of a newborn infant. Four and a half years later she had progressed to a mental level of two and a half or three years.

Isabelle was six and a half years old when she was found. Like Anna, Isabelle was in bad shape both physically and mentally. For speech she made a strange croaking sound. She reacted to strangers, especially to men, with much fear and hostility. In many ways she behaved like a deaf child, and her mental capacity was no more than that of a six-month-old baby. Specialists working with her at first believed her to be feebleminded and uneducable. An intensive training program was started. At first it seemed hopeless, but gradually Isabelle began to respond. Then suddenly she began to learn rapidly. Two months after starting to speak, she was putting sentences together. Sixteen months later she had a vocabulary of 1,500 to 2,000 words and was asking complicated questions. She covered in two years the stages of learning that ordinarily require six. Her I.Q. tripled in a year and a half. Davis concluded his description of Isabelle by noting that she was over fourteen years old and had passed the sixth grade. She was bright, cheerful, and energetic, and her teachers reported that she participated in school activities as normally as other children.

What do these cases tell us? It seems apparent that inadequate social interaction, near isolation in these cases, inhibits the development of the individual. Although they were six years old, neither Anna nor Isabelle had developed beyond the infant level, and both were believed to be feebleminded. Yet when effective interaction with others began, both girls progressed. In Isabelle's case the development was remarkable. This is probably because she had concentrated, expert training from the beginning and because she learned language quickly. Perhaps this is the key. Or perhaps Anna was more isolated—Isabelle's mother stayed with her, whereas Anna was often by herself. Or perhaps Isabelle's mental capacity was superior to Anna's to begin with. This we don't know. But in each case, the effect of inadequate social interaction on human development is clearly demonstrated. Sociologists are convinced that the development of the individual is, at least in part, a *social process*.

## SOCIALIZATION: A SOCIAL PROCESS

The process by which an organism becomes a social being is called socialization. **Socialization** refers to the learning of expectations, habits, skills, values, beliefs, and other requirements necessary for effective participation in social groups. Biological maturity is necessary, hereditary factors might set limits, and the process takes place in a physical environment, but the crucial aspect of socialization is that it is a social process. The individual develops through interaction in the social environment; the social environment determines the result. This is not to imply that if baby Michael Jordan and baby Michael Jackson had been switched in the nursery shortly after birth, Jackson would have been a spectacular professional basketball player slam-dunking with his tongue sticking out and Jordan would have been a famous singer with a high-pitched voice and one glove. We have to look at other factors as well, but biology and heredity certainly play an important part.

The exact extent of the role of biological factors is difficult to determine. In fact, there is a heated and interesting debate going on now about this very issue—biology versus social environment. Edward Wilson, an entomologist from Harvard, proposed in a book published in 1975 that a new science called sociobiology be established. **Sociobiology** is defined as "the systematic study of the biological basis of all forms of social behavior, including sexual and parental behavior, in all kinds of organisms, including man." Sociobiologists are convinced that the behavior of lower organisms is biologically determined, and they believe that the same may be true of much of human behavior. In their writings they cite apparent genetic constraints on human social behavior in such areas as choice of sexual partners (the incest taboo and homosexuality), gender roles, aggression, kinship rules, and infant development.[2] The ideas of the sociobiologists are controversial, some say inflammatory; they have delighted some social scientists and threatened and angered others. Probably most social scientists recognize that biology places certain limits or conditions on human development. Beyond this, however, social scientists and especially sociologists believe that the crucial elements for understanding human behavior are the social environment and social interaction. By and large, according to sociologists, these are the primary determinants of who we are.

By **social interaction** we mean the process of being aware of others when we act, of modifying our behavior according to others' responses. Social interaction occurs in a variety of ways, and its patterns are complex. When a student asks a teacher a question, she speaks differently than she does when she talks to the student next to her in class. The teacher evaluates the question as fairly sophisticated, and this affects the teacher's response to the student. If the interaction continues, each response is modified and determined by the response of the other immediately preceding it. The pattern of interaction can be affected by manner, body language, deference, relative status, degree of acquaintance, eye contact, and numerous other factors in addition to the spoken words. Social interaction occurs between two people—husband and wife, strangers on the street—and in groups and even in large organizations. Try watching some examples of social interaction, and see whether you can identify the more subtle aspects of communication. How do the parties show agreement and disagreement, interest and boredom, and happiness and unhappiness through the use of body language, spatial distance, and facial expressions?

Socialization is continuous: It takes place throughout life, starting as soon as the infant leaves the womb and continuing until death. It occurs through interaction with other humans. We could call these other humans *agents of socialization.* Who these agents will be depends on one's place in the life span. For the baby, the socialization agents are the parents, who begin very early to communicate accepted and expected modes of behavior to the infant. Brothers and sisters also become socialization agents. An aspect of the socialization process can be observed in children as they pretend to be mothers, fathers, airplane pilots, or truck drivers. They are already anticipating and playing future roles. Recently, I watched a 2-year-old boy "shooting" baskets while his 12-year-old brother watched. Again and again, the little boy held the "ball" in front of his face with two hands, jumped, and pushed as hard as he could, trying to get the ball to the basket nearly ten feet over his head. Except that there was no ball—it was imaginary! A real ball would have been much too big for him to hold. But he kept at it, just like he'd learned from watching his brother do it with a real basketball.

Later, peers become socialization agents. Regardless of whether they are school friends, members of a street gang, or teammates on a football team, their influence becomes very important. Still later, one's colleagues on the job, one's spouse, and even one's own children become agents of socialization. Today, with the focus on instant communication and computerized processing of information, new socialization sources have appeared. Advertising seems to determine tastes in what to wear. Popular charismatic "stars" in film, music, and sport have great influence. Television and print media get interested in issues that immediately become a part of all of our lives—eating healthier, dangers of smoking, good and bad cholesterol. The exploding presence of the Internet has led to concern about what new messages (new agents of socialization) are suddenly readily available, and how they should be controlled.

As you can see, socialization is extremely important, and it comes from a variety of sources. Through this lifelong process we gradually take specific points of view. Our views of religion, politics, sports, style of family life, how to commit particular crimes, and how to raise children are all learned through the socialization process. I throw a volleyball to my son. He catches it in his hands and throws it back. Very easy; he doesn't even think about it. But suppose I throw the ball to a French or English boy—what happens? If the throw is above his shoulders, he will probably "head" it back to me. If it's lower, he will control the ball with his legs and feet and kick it back to me. Each child has been socialized—has learned—to respond in a specific way. Children respond differently from each other because different factors are emphasized where each child lives. But again, the socialization process "tracks" them into a certain response.

The first research question posed at the beginning of this chapter asked whether mental age or I.Q. might be related to family size and birth order. It turns out that sociologists have been looking at this for some time and have come to some interesting conclusions. They have found that mental age tends to decline as family size increases. They have also found that mental age declines as birth order increases. In other words, later-born children tend to have lower I.Q.s than their older brothers and sisters. The researchers suggest that perhaps the confluence of two factors—family intellectual environment and a teaching function—may in part explain the differences. They suggest that the family's intellectual environment—its average mental age—declines with each new baby, especially if babies are closely spaced. The more young children around, the less intellectually rich the environment, at least during the time that they are young. The "teaching function" effect refers to the fact that last-born children lack the opportunity to teach children coming after them. Teaching a younger brother or sister stimulates the intellectual development of the older child, and because last-born children have no one to teach, they suffer from a "last-born handicap."[3]

## GENDER SOCIALIZATION

When we think about socialization, we tend to think on an intimate and personal level about parents, school friends, teammates—those closest to us who seem to have the greatest impact on us. For a moment, let's move to another, more general level. Earlier we noticed that how a child returns a thrown ball (heads it, or catches and throws it) is determined through the socialization patterns of the child's culture. Or imagine the differences in life experiences for the child raised in a wealthy family in an affluent suburb, compared with a child raised in a poor or homeless family in the inner city. The point is that families, peers, and schools operate within a larger, more abstract setting. That setting is one of nationality, culture, social class, and gender. These more general concepts produce a sorting, sifting, and tracking of their own. The consequence of this sorting is great, though sometimes hardly noticed.

As with other types of socialization, gender socialization emerges through interaction with the family, the school, the peer group, and mass media. Studies have shown the following: Fathers are more likely to "rough it up" with boys than with girls. Parents of daughters are more likely than parents of sons to describe their babies as "dainty," "pretty," "beautiful," and "cute" even though actual differences might not exist. Parents of boys tend to choose masculine toys more than feminine toys for their sons, whereas parents of girls are more likely to choose neutral (rather than masculine or feminine) toys for their daughters.

Schools tend to recognize and encourage males for assertiveness, curiosity, independence, and initiative, but are less supportive of these traits in females. Male peer groups continue to socialize boys into the same traits: competitiveness, aggression, self-reliance, violence, and masculinity. It should be no surprise then if men and women seem to "speak with different voices." Women's sense of self and worth is based generally on their ability to make and maintain relationships, whereas for men, independence, autonomy, and achievement seem to be the main focus. Deborah Tannen has observed men and women in the workplace and reports that there is much miscommunication—men and women are often at cross purposes. A man and a woman return from a conference in which they both gave papers. She compliments his; he gives a detailed critique of hers. Although she had problems with his, she felt uncomfortable critiquing it. Women are more likely than men to speak in a way that builds participation, consensus, and support. Tannen thinks this comes from socialization in childhood play groups. Male groups teach boys to dominate, go for the spotlight, and to use ridicule and put-downs to control others. Female play groups are based more on pairs of best friends, sharing, and building intimacy. As adults, males are used to standing in front of large groups, commanding attention, taking charge, and being overbearing and intimidating if necessary. Women have been socialized to build relationships, use a softer touch, and be conciliatory. What is viewed as important and basic to women can be seen as a sign of weakness to a man.

In a final example, note the effect that gender socialization has on physical appearance. Males' attractiveness is related to physical abilities with bodies valued for being active, whereas females' bodies are judged on the basis of current standards of beauty. Body shape becomes an important consideration, and socialization into a "cult of thinness" can follow. Beauty queens, fashion models, and female television and movie characters are predominantly slender. One consequence is that young women are much more likely to diet and to use diet pills and amphetamines to control weight than are young men.[4]

Awareness of and interest in gender socialization has increased dramatically during the last two decades, influenced greatly by the women's movement. Much social science research is focusing on the topic, and we are trying to better understand the processes at work. Important questions remain. For example, male and female math scores are very similar through the early school years. Then they begin to diverge with girls performing less well. Interestingly, this does not seem to happen for girls in all-girl classes—their scores stay more equal with males (based on national averages). What is happening? Something about the socialization process: Perhaps young women's expectations (and their performances) change as they get older, especially when they are competing directly with males. Or perhaps the gender of their teachers (more female teachers in elementary school than in secondary) has an effect on how certain subjects are learned.

## PRIMARY AND SECONDARY SOCIALIZATION

Socialization may be better understood as occurring at two levels: primary and secondary. According to Berger and Luckmann, **primary socialization** refers to the first socialization an individual experiences in childhood, through which he or she becomes a member of society. **Secondary socialization** refers to any subsequent process that inducts an already socialized individual into new sectors of his or her society.[5] Primary socialization usually takes place in the family. Here, the individual has no choice about the important or significant socializing agents; the child almost automatically and inevitably accepts and internalizes the family's view of the world. As primary socialization proceeds, the child's referents move from specific to general. A progressive abstraction occurs whereby at one stage the child understands that her parent specifically wants her to do certain things (for example, not to throw her food on the floor), and at a later stage she understands that people in general expect her to behave in certain ways (to use correct table manners). Through primary socialization, the individual's first world is constructed. Consequently, according to Berger and Luckmann, this first level of socialization is the most important, for the basic structure of all secondary socialization must resemble that of primary socialization.

Secondary socialization, which takes over where primary socialization leaves off, involves the individual's moving into and internalizing knowledge of new areas or sectors of life. Secondary socialization takes place when the individual decides to learn to read or to write, to take up skydiving, to become a police officer, or to raise a family. This level of socialization does not need an emotionally charged atmosphere to succeed, nor does it presuppose a high degree of identification, and it does not possess the quality of inevitability. In secondary socialization, the individual can be more objective than in primary socialization.

A person entering a new profession illustrates adult secondary socialization; an example might be the steps taken in the socialization of new police officers, outlined by Arthur Niederhoffer. Rookie officers proudly accept the symbols of their new job—uniform, revolver, handcuffs, and badge. At the police academy they learn about law, police procedures, rules and regulations, first aid, and how to fire weapons. They graduate, become police officers, perhaps face a hostile and unappreciative public, and reality shock occurs. More experienced officers now tell the rookies that if they want to be successful on the street, they will need to forget everything they learned at the academy. They learn to react to people's language, attitudes, and dress as well as to whether someone committed a crime or not. Police officers learn when to enforce the law and when to look the other way. They learn how to make an effective arrest, and they learn about bribes and graft. According to Niederhoffer, police officers' attitudes change the longer they are on the force; experiences on the street make them cynical and suspicious of other people. Officers who have a college education but are slow to get promoted become more and more cynical. And generally as officers get close to retirement, they become less cynical.[6]

## RESOCIALIZATION AND ANTICIPATORY SOCIALIZATION

Socialization at the adult level (secondary socialization) is usually a gradual process; that is, the changes in the individual are more minor than are those that occurred during primary socialization. Occasionally, however, major modifications or reconstructions of people are attempted. This could be called **resocialization.** It usually represents a concentrated effort by a group or organization. Consider prisons as an example. Individuals commit crimes; they are caught, convicted, and sent to reformatories or prisons. The idea is that the discomfort of their confinement will help them see the error of their ways, counseling and training programs there will help them change their behavior, and criminals will be *resocialized* into good citizens.

What actually happens? Resocialization certainly does take place, but its effects are far more diversified than one might imagine. Prisons are very effective crime schools for some prisoners. Regardless of the inmates' expertise on entering, they can improve their crafts while inside. Through association with other inmates, they can learn new skills and techniques of crime. They can set up networks of important criminal friends and contacts for when they are released. Inmates learn a new language in prison. For example, "gleaning," "low riders," "bonaroos," "shots," "politicians," "merchants," "hoods," and "toughs" are prison slang for types of convicts.[7] The inmates learn new sexual practices, sometimes against their will, and they must learn how to get what they want in a society that is not at all like the one they are used to. Some inmates even *like* prison; they have found a new home. These inmates find it easier to get along inside than outside, and they take steps to make sure they can return quickly once they are released. These people, too, have been resocialized.

When soldiers go through basic training, they are being resocialized. Recruits are taught to get up at times when they would ordinarily go to bed, to cut their hair in peculiar ways, and to spend enormous amounts of time on the maintenance and appearance of their shoes, clothing, and equipment. They are told that a patch of dirt, rock, and sand outside the barracks is in fact a plot of grass that must be raked and cleaned every morning. They are taught to obey authority unquestioningly, to believe that the individual is unimportant, and to act as a unit with their fellow recruits at all times.

There are other examples of resocialization. Parents whose children have joined religious cults are usually astounded at the abrupt changes in people they thought they knew well. Brainwashing of war prisoners through isolation and sensory deprivation is another type of resocialization. One technique used by the Chinese in Korea to lower morale was to allow only collection notices, divorce subpoenas, "Dear John" letters, and other demoralizing mail to reach the prisoners. They held back positive letters and let the negative letters through.[8] Organizations such as Alcoholics Anonymous, dedicated to helping people recover from dependence on alcohol, are also involved in attempts at resocialization.

When a person adopts the values, behavior, or viewpoints of a group he or she would like to belong to, this is called **anticipatory socialization.** This is a common occurrence and explains the behavior of the lower-class person who develops middle-class values, or the behavior of the "gung-ho" military recruit who wants to become an officer, or the new behavior and viewpoints of law or medical students who anticipate becoming lawyers or doctors. Anticipatory socialization has the interesting function of easing the

## SIGMUND FREUD (1856–1939)

Sigmund Freud earned his MD in Neurology at the University of Vienna. Freud was trained in natural science, but his major work focused on the psychological rather than the physiological explanations for mental disorders. In 1896, he named his area of interest "psychoanalysis." Freud's patients were upset, anxious, disturbed, perhaps neurotic, and they sought him out for help. One aspect of psychoanalysis was the investigation of the patient's spontaneous flow of thoughts, called *free association,* to reveal the unconscious mental processes at the root of the neurotic disturbance. Freud created an entirely new approach to the understanding of human personality by his demonstration of the existence and role of the unconscious. Freud also worked on the psychology of human sexuality and dream interpretation. Although his research covered a wide variety of issues, it was his perspective on the conflict between the individual's instincts and society's requirements that was particularly influential in sociology. We will discuss Freud several times in this book: in this chapter on personality and in Chapter Fourteen on deviant behavior.

transition from one stage in life to another. When individuals practice for a new role ahead of time, they often assume that the new role is much less difficult than it actually is. On the other hand, if the new stage in life is never reached, the trauma can be even greater than a sudden, unexpected change would have been.

## SELF AND PERSONALITY

Psychology and sociology are most closely linked when they attempt to define the individual and to explain developmental processes. A separate discipline called *social psychology,* which deals with theories of socialization and individual development, has evolved. *Self* and *personality* are the major concepts in this analysis. The words are similar in meaning, but **personality** is a somewhat more general term and may be defined as the sum total of the physical, mental, emotional, social, and behavioral characteristics of an individual. **Self** refers to one's awareness of and ideas and attitudes about one's own personal and social identity. Psychologists are perhaps more inclined to use the term *personality;* sociologists usually prefer to use *self.*

Generally, psychologists have placed more emphasis on hereditary and biological factors in personality development, whereas sociologists focus on social interaction and the social environment. Psychologists place more emphasis on early childhood than do sociologists. Although numerous psychological theories have been developed, the most famous concepts are those of Sigmund Freud. Freud was a Viennese physician and founder of psychoanalysis. According to Freud, the personality is made up of three major components: the *id*, the *ego*, and the *superego.* The **id** is the most primitive aspect of the personality and represents the basic instinctual drives with which a person is born. Sexual and aggressive desires in the id constantly seek expression. You might think of the id as the "do it" part of your personality. You wake up on a school day and what does id tell you? Roll over, go back to sleep, maybe get up tomorrow. The **superego** is an internalized set of rules and regulations that represents the values and ideas of society as initially interpreted for individuals by their parents. It is sometimes referred to as the *conscience,* and is the aspect of personality that controls behavior. You might think of the superego as the "don't do it" or "do the right thing" part of your personality. Id says roll over and go back to sleep, but superego says no! no! no! Get up, go to class, study hard, parents are paying good money for this. The **ego** is the acting self, the mediator or referee between the id, superego, and the outside world. Id works to maximize pleasure for the individual, whereas ego is more tuned in to reality and the conditions in the immediate environment. So what do you do? Your ego determines that it would probably be best for you to go to class (and sleep in this weekend).

Freud also believed that the individual passes through a series of psychosexual stages and that personality is fixed very early in life, around the age of six. Sociologists believe that although the early years are extremely important, socialization and development of self are lifelong processes. Because Freud was a psychoanalyst and was working with people who were mentally ill, he emphasized the restricting nature of parents and society and the conflicting aspects of group life. Sociologists have focused more on the positive roles played by parents and groups in socialization and development of self.

With this brief view of Freud as an introduction, let's turn to sociological theories of the self. Sociologists believe that self-development is rooted in social behavior and not in biological, hereditary, or instinctual factors. The self is developed during the socialization process. Through interaction and association with others, you as an individual develop an image of *what you are.* This involves a perception of your role requirements and of your position and behavioral expectations in the various social groupings with which you identify. As sociologist Charles H. Cooley put it, the word *self* means simply that which is designated in common speech by the pronouns of the first person singular: *I, me, mine,* and *myself.*[9] Aspects of one's self might include the following: I am male, I am a college student, I am of medium height, I am athletic, I am usually happy, I am intelligent. The self might include the following as well: I can never do anything right, I am not worth much, nobody listens to me, I am inadequate, I am unloved, I am not as capable as my older brother.

Individuals develop these images or viewpoints about themselves from the way others respond to them. More specifically, people develop these viewpoints from the way *they think* others respond to them. Development of the self is a two-way interactive process that takes place throughout the period of socialization, the entire life span. Development of the self is an extremely subjective process involving interpretations of others' evaluations.

Current evaluations are based on past evaluations, so that earlier mistakes become additive. A father may continually tell his daughter that she is extremely intelligent. This becomes part of her self, one of the ways she views herself, and she behaves and has certain expectations because of it. If she flunks bonehead English as a college freshman, contradictory information comes in. What happens next? She makes excuses and rationalizes her failure to support the view of self that she held earlier. She will probably seek reinforcement from other sources (father, girlfriend, I.Q.-test scores) to support her self-view. If evidence is lacking or if reinforcement becomes more difficult to find (other bad grades, low I.Q.-test scores), this part of her self-conception will probably begin to change.

Aspects of self constantly undergo change. Often these changes occur slowly. Occasionally changes occur more rapidly, suddenly, and dramatically, as we mentioned earlier, in cases of resocialization of inmates of prisons and concentration camps and of recruits in military basic training. All these changes are a result of our evaluations of others' evaluations of us.

Cooley described the development of the self in his concept, the **looking-glass self.** The looking-glass self contains three elements: the imagination of our appearance to the other person, the imagination of his or her judgment of that appearance, and some sort of self-feeling, such as pride or mortification. Suppose, for example, a person eating in a crowded restaurant accidentally knocks his plate off the table. It makes a huge crash, and the food spills all over him. According to Cooley, he first "steps outside himself" and observes himself from the viewpoint of others in the room (". . . a well-dressed fellow with spaghetti all over him . . ."). Next, still examining himself as object, he imagines that others are evaluating his behavior (". . . must be a rather clumsy and awkward person . . ."). Finally, the individual as subject develops feelings and reactions to these imaginary evaluations of himself as object. He gets embarrassed, his face reddens, and he tries unsuccessfully to pretend it didn't happen. The process is made more interesting when we note that probably the same things would have happened if the person had been the only one in the restaurant. We are carrying our judges around with us in our heads, and they are constantly evaluating our behavior.

George Herbert Mead[10] described the development of the self this way: Individuals will conceive of themselves as they believe significant others conceive of them. They will then tend to act in accordance with expectations they impute to these significant others concerning the way "people like them" should act. As children develop the ability to examine and control their behavior in accordance with others' views and attitudes about their behavior, they are learning to "take the role of the other." At this first stage of development, called *primary socialization* by Berger and Luckmann, children do not cooperate with others; they relate only to specific individuals such as parents and perhaps an older brother or sister or a teacher—people whom sociologists call **significant others.** Later, as individuals develop, they are less likely to continue to react to *individual others*, and they learn instead to react more to a less personalized *grouping of others*. This grouping

# ★GEORGE HERBERT MEAD (1863–1931)

George H. Mead earned his BA in Philosophy at Oberlin College in Ohio. Considering that Mead was a philosopher and not a sociologist, it is interesting that he had a profound impact on the development of symbolic interaction in American sociology. Mead taught philosophy and social psychology at the University of Chicago, and sociology graduate students often took the courses. He published little during his lifetime, but his students gathered his lecture notes after he died and published them in several major volumes, for example, *Mind, Self and Society* in 1943. For Mead, social psychology "studies the activity or behavior of the individual as it lies within the social process. The behavior of an individual can be understood only in terms of the behavior of the whole social group of which he is a member, since his individual acts are involved in larger, social acts which go beyond himself and which implicate the other members of the group." Mead's social psychology stood in stark contrast to the primarily societal theories offered by most of the major European theorists. It is interesting that Mead never received any graduate degrees. He taught philosophy at the University of Chicago from 1894 until his death in 1931. We will see Mead's ideas several places in this book: in Chapter One (and nearly every other chapter) on symbolic interaction, and in this chapter on self and on the importance of language.

---

of others, which Mead called the **generalized other,** represents the sum of the viewpoints of the social group or community of people to which one belongs. The second stage of development, *secondary socialization,* begins when children are mature enough to cooperate with others in joint activities and to be able to react to the idea of people in general, rather than to specific others.

Although our internalized attitudes of others are crucial, they do not determine all behavior. Mead made a distinction between two aspects of the self, the "I" and the "me." The "I" represents the subjective, acting self, which may initiate spontaneous and original behavior. The "**me**" sees self as object and represents a conception of others' attitudes and viewpoints toward self. The "I" is freer, more innovative; the "me" is more conventional. "I" behaves; "me" judges and evaluates. "I" is in part governed by "me," but not completely. The abstract scrawling on the wall or in the dirt by the young child might represent "I" behavior; the same child's in-school drawing of the houses or trees that seem to please the teacher so much would be closer to "me" behavior. The driver involved in an auto accident who leaps out of the car and punches the other driver, swears at the passenger, and kicks a passing dog is probably showing us the "I" aspect of self; another passenger who goes to the nearest phone to call the highway patrol illustrates the "me." Or imagine the football player about to catch a punt as a group of monstrous players from the other team race down the field to smash him. The player's "I" says, "I don't want to catch that thing. It's coming down hard and is going to hurt my hands; look at all those guys going to tackle me—I think I'll let the ball bounce off my helmet." The "me," however, says, "Oh no you don't—there are a lot of people watching. You're supposed to *catch* that ball and *run* with it. The team and the coach are depending on you. *Do* it!"

## THE IMPORTANCE OF LANGUAGE

Mead also first emphasized that socialization and development of the self could not occur without *language.* Indeed, many argue that for the most part language distinguishes humans from other animals. The essential factor in language and symbolic communication is that a symbol arouses in one's self the same meaning it arouses in another. Primitive communication begins with a conversation of gestures, such as the mating dance of birds, the snarl of dogs, the roar of lions, or the cry of an infant. At this level, meanings of gestures are not shared. To be sure, the gestures may bring forth a response from some observer, such as from another dog or lion or from the infant's parent. But the important fact, according to Mead, is that the gesture does not arouse in the actor the same response it arouses in the observer. For example, a very young baby feels discomfort and cries. The parent interprets this as hunger and feeds the baby. If the parent is correct, this is communication of a sort, but for the baby, the act of crying had no communicative intent; it was merely a biological response. This represents primitive communication because the meanings were not shared by both parties.

As children mature they learn to use symbols and words. The words are poorly formed at first, and probably only a parent could understand them, but they are the beginnings of language—something beyond primitive communication. As their use of language improves, children learn that what they say and do elicits responses from others. Language allows them to replace behavior with ideas; they can now say they are hungry rather than acting it out by crying or pointing at food. Language allows them to think, makes possible the internalization of attitudes of others, and allows individuals to control their responses to others. We can imagine a child's internal conversation: "I would really like to have ice cream and cake, but my parents seem to like me better if I eat that other stuff they put out. . . ." Mead suggests that the use of symbolic communication continues to develop as a child grows, and through language, individuals are able to think, to develop shared social meanings, to take themselves as objects, and to evaluate their own behavior as they think others do. In summary, then, language makes possible the development of mind and self.

Language is important in another way. The language and symbols we use influence us to "see" the world and to "think" in particular ways. Imagine trying to balance your checkbook or to do your algebra or geometry homework using Roman numerals instead of the number system we are used to; much modern math probably could not have developed using the Roman system. The Inuit has many ways of expressing the phenomenon we

## SHOOTING AN ELEPHANT

George Orwell tells the story of the English police officer on duty in a Burmese town who is told one day that an elephant has run amok, is ravaging the bazaar, and has killed a native—"caught him with its trunk, put its foot on his back and ground him into the earth." The officer gets a rifle and seeks out the elephant. When he finds the elephant, he realizes immediately that he ought not to shoot him—the elephant is eating peacefully and is not dangerous. "But at that moment I glanced round at the crowd that had followed me. It was an immense crowd, two thousand at the least and growing every minute. I looked at the sea of yellow faces all happy and excited over this bit of fun, all certain that the elephant was going to be shot. They did not like me, but with the magical rifle in my hands I was momentarily worth watching. And suddenly I realized that I should have to shoot the elephant after all. The people expected it of me and I had got to do it; I could feel their two thousand wills pressing me forward, irresistibly. Here was I, the white man with his gun, standing in front of the unarmed native crowd—seemingly the leading actor of the piece; but in reality I was only an absurd puppet pushed to and fro by the will of those yellow faces behind."

The officer did not want to shoot the elephant, but he did, many times, and the elephant finally died. Later there was much discussion about whether he had done the right thing. "The older men said I was right, the younger men said it was a damn shame to shoot an elephant for killing a coolie. And afterwards I was very glad that the coolie had been killed; it put me legally in the right and it gave me a sufficient pretext for shooting the elephant. I often wondered whether any of the others grasped that I had done it solely to avoid looking a fool."

*Excerpts from "Shooting an Elephant" in* Shooting an Elephant and Other Essays *by George Orwell are reprinted by permission of Harcourt Brace Jovanovich, Inc.*

describe by the single term *snow.* Navaho language is more literal, concrete, and specific than is English. Spanish has two verbs for *to be* whose different uses reflect a type of thinking unique to people who speak the language. French makes distinctions between a familiar and a formal *you.* Even color is "seen" differently in different societies depending on the words or symbols available; some societies recognize only two or three colors. Numerous examples make this same interesting point: The language and symbols we use act as a "filter" through which we experience the world. It is a filter *we* construct, which nevertheless leads us to see what we see in particular ways, to focus on certain elements, and to miss others entirely.[11]

The relationship between biology and social environment in the development of the human is extremely complex. We mentioned this earlier in the section on sociobiology (page 35), and we will return to it again in the section on debates and issues (pages 49–50). Recent research on the developmental processes in children illustrate the interactive nature of this relationship. The circuits or series of brain cells that process sounds that define words that lead to language are operative in most children by the age of six months to one year. The more words a young child hears, the larger that child's vocabulary will be. Further, children become functionally deaf to sounds that are not part of their native language. Japanese-speaking, English-speaking, and Swedish-speaking children, for example, will all have different "brain-wiring" based on the sounds they have heard. They don't hear or process sounds with which they are not familiar. This circuitry develops early, and that's why learning a second language becomes increasingly difficult after the age of ten.

In related studies of math and logic and how these skills are developed, it was found that babies taught simple concepts like "one" and "many" will do better in math later on. Children involved in music (singing, playing the piano) seem to do better in spatial reasoning (called the "Mozart effect"). This is perhaps because the "wiring" for music and for math (spatial reasoning) are near each other in the brain. The lesson is that the early years of socialization are important. A baby whose eyes are clouded by cataracts from birth will, even after cataracts are removed at the age of two, be forever blind. This reminds us of Anna at the beginning of this chapter—the absence of social interaction and language in her early years had a lasting effect.[12]

## DEFINITION OF THE SITUATION

An important principle of socialization and self-concept is that our understanding of reality is subjective and socially structured. We respond to what we *think* is so, rather than what really is so. When we peer into Cooley's looking glass the view might be quite distorted, but we believe what we see. Two ideas, *definition of the situation* and *self-fulfilling prophecy*, help illustrate this point.

Sociologist W. I. Thomas introduced **definition of the situation** by saying, "If men define situations as real, they are real in their consequences." The point he was making is that reality is socially structured and that people respond as much or more to the meaning a situation has for them than to the objective features of a situation. If a student defines passing a certain exam as the most important thing in the world, then for her it is, and she goes to great extremes to ensure that she does pass it. People define certain races as inferior and certain cultural beliefs or practices (cannibalism, polygamy) as peculiar, regardless of the objective reality of the situation. Or recall the old club initiation stunt: The victim is blindfolded and fed cold cooked spaghetti. Halfway through the meal he is told he is eating worms. What happens? He gags and is sick to his stomach. Why? He responds to what he thinks is true. That response has been socially defined and includes what his feeders tell him, how important initiation is to him, and what his background has taught him regarding the relative merits of angleworms and spaghetti. Even worse and horrible to consider, suppose we feed our friend angleworms and tell him it's spaghetti. That, I'm afraid, would work too.

The **self-fulfilling prophecy** is an extension of the concept of definition of the situation. A self-fulfilling prophecy occurs when a *false* definition of a situation evokes a new behavior that makes the originally false conception come *true*. A student is worried that he will flunk an exam. This makes him so nervous and anxious that he is unable to study effectively, and he flunks the exam. Or the police believe Gorples to be very criminal types. Consequently, more police patrols and surveillance are concentrated in those sections of the city where the Gorples live. Increased police surveillance leads to greater visibility and reporting of crime, so the crime rate of Gorples increases. As Robert Merton explains, "Confident error generates its own spurious confirmation."[13] If we look carefully, we can see this phenomenon frequently in everyday situations.

Psychologist Robert Rosenthal performed several interesting experiments that illustrate what we are discussing here.[14] In one experiment, to which we referred in the second research question at the beginning of this chapter, six students were given five rats each. They were told that their rats were very bright ones, bred especially for running a maze. Six other students were given five rats each and were told that their rats were genetically inferior and probably would be poor at running through mazes. Then both groups of students attempted to train their rats to run the maze. What the students didn't know was that there were no "bright" rats and no "dull" rats; they were all the same with equal capacities for learning. What happened? Well, right from the beginning of the training the "bright" rats performed better. The "dull" rats made little progress and sometimes would not even budge from the starting position in the maze. And there were differences in more than just performance. The students found the "bright" rats to be brighter, more pleasant, and more likable than the "dull" rats. The self-fulfilling prophecy explains what happened: The students with "bright" rats helped, coaxed, and encouraged their animals to do what they expected them to, and the "dull" rats got little encouragement. Or perhaps because the students expected "brights" to do well and "dulls" to do poorly, that is what they saw, regardless of what was actually happening.

In Chapter One we discussed another of Rosenthal's experiments. As introduced in **Question 7,** you remember that Rosenthal and Lenore Jacobson (R&J) tested the intelligence of school children. R&J then told teachers that certain of their students were very sharp ("academic spurters") and could be expected to do quite well in school. As with the rats in the earlier study, the children were all alike. The "spurters" were *not* of higher intelligence, and in fact R&J had chosen them at random. The children were tested again later in the year, and the teachers evaluated the students. You can guess what happened. The students in the first and second grades who were labeled "academic spurters" were described as happier than other students, more curious, more interesting, more appealing, and better adjusted. They also showed greater gains in intelligence than did the other children. The teachers acted in terms of what they thought to be true (definition of the situation) and behaved so that what was expected to occur, did occur (self-fulfilling prophecy).[15] This illustrates our point once again: Our reality is subjective and socially structured.

In the preceding paragraph we introduced the process of **labeling,** which is an important one in sociology. Labeling refers to the tendency of people to stamp, typecast, or categorize others. Often these categories are negative, and understanding the concept of labeling is important for precisely the reasons we have just been discussing. If one is labeled "bright," "dull," "academic spurter," "stupid," "ex-con," "deviant," "lesbian," "mentally ill," or "sex offender," it is bound to have an effect on that individual. The label might be accurate or it might be false. But as we have just discovered, accuracy is of little consequence because reality is subjective and people believe what they are told. Labeling leads to different treatment by others (who respond to the label rather than to the individual), to effects on the individual's self-concept, and perhaps to a self-fulfilling prophecy. We will return to labeling in more detail in Chapter Fourteen.

## DEBATES AND ISSUES IN SOCIALIZATION

I might have implied that there is general agreement about the topics and issues we have discussed so far. I wish it were so, but the process of becoming a socialized person is a complicated one, and here, as in other areas of human behavior, there are great differences of opinion. One of the debates concerns the relative importance of biology and heredity on the one hand and of the social environment on the other. Sociologists see socialization as a *social* process. One's biology may set limits, but its overall effect in the development of the human being is relatively small. We believe that the key factor is the social context—the interaction between oneself and one's family, friends, and peers—and the social influence of one's religion, social class, race, gender, age, occupation, education, and so on. The diagram in Figure 2.1 illustrates this view. The social context is created by input from family, groups, friends, and numerous other sources (many more than we can show here). The effect of the social context on behavior is great; the effect of biology (B) is small. Now the argument starts. Is the effect of B really so small, or should it be written much larger in comparison with the social context? Many feel that the key is biology, that the B should be written very large because human behavior, like animal behavior, is basically genetically determined. These researchers find biological explanations for all sorts of behavior, from parental and sexual behavior to mental illness, aggression, shyness, and even alcoholism, to mention only a few. Sociologists grant that in certain types of cases—for example, a person born with a birth defect, a child with Down's syndrome, a very tall or very short person, a very old person—the B must be written larger; in such cases one's biology can have a greater influence on behavior. These cases are exceptional however, and even here, the effect of the social context is great.

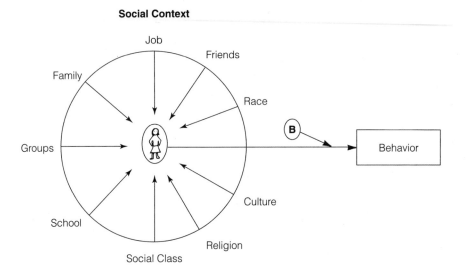

**FIGURE 2.1** *Effects of social context and biology (B) on behavior.*

Another issue concerns the importance of the early years of life. Sociologists believe that these years from birth to age six or seven are important but that the socialization process is lifelong; we continually change through interaction in the social environment. Psychologists tend to place much more emphasis on the early years. Many learning theorists believe that intelligence and learning capacity are determined in the first year or two of life. Isolation from or impaired interaction with people during these years, as in Anna's case discussed at the beginning of this chapter, causes irreparable damage. Examples of isolated children, and even studies with monkeys, are cited to show the effects of deprivation. But the evidence is ambiguous. Some children who are isolated never recover, and others do, as occurred with Isabelle. In a part of Holland there is a local custom of isolating a child in a room outside the house for the first ten months; the child is tightly bound and is given no toys and offered minimal interaction. At the age of one the child appears to be retarded, but by the age of five he or she is fully recovered. Jerome Kagan describes a village he visited in Guatemala in which the children are kept isolated in dark huts. They have no toys, and although parents may hold their children, they don't talk or interact with them.[16] At the age of one and a half, the children are retarded by our standards. Yet eleven-year-olds in this same village appear normal—they are alert, active, and intelligent. This leads Kagan to believe that too much emphasis has been placed on the first two years of life. He suggests that because children learn and develop at different rates, perhaps schools should rethink their policy of rank-ordering and labeling students in the early grades. We should keep an open mind on the importance of the early years and remember that socialization is a lifelong process.

## THEORY AND RESEARCH: A REVIEW

In Chapter One we stated that scientists use theory and research in their search for understanding. In Chapter Two we have seen some examples of their attempts. This chapter focuses on theories about the concepts of socialization, self, and personality. The ideas of Berger and Luckmann, Freud, Cooley, and Mead represent theoretical positions or viewpoints about how these basic concepts can best be understood. In this chapter at least, the *symbolic interactionist approach* seems to be the most useful theoretical view. Research discoveries can then help support or reject a particular position. Davis's *case histories* of isolation (Anna and Isabelle) tell us something about inadequate social interaction and its effect on the socialization process. Niederhoffer's *participant-observation study* of police officers provides research support for theoretical ideas about secondary socialization. Rosenthal's interesting *experiments* with students and their rats, and with teachers and their students, illustrate and support the ideas of definition of the situation and self-fulfilling prophecy.

Of course, there can be problems in the search for understanding. Some theories, such as Freud's id, superego, and ego and Mead's generalized other, are difficult to test. Sometimes research efforts yield conflicting results or are open to different interpretations. For example, why did Isabelle and Anna turn out so differently, and why are the repeats of Rosenthal's studies not consistent? The viewpoint of scientists, however, is that theory and research work together in a way that is both self-correcting and cumulative. Conflicting results prove the value of continued studies, and those theories that today seem difficult to test will submit to new methodological approaches in the future.

## SUMMARY

Our study of sociology starts with the assumption that humans are *social* beings. The obvious question is, How do they become that way? In this chapter we have discussed some of the processes that sociologists consider essential if we are to understand how one changes from a blob of matter into a complex, interacting social being.

The process through which this transformation occurs is called socialization. During socialization, the self is developed. The self is the individual's set of images about who or what he or she is. The development of self is a social process; it arises out of our interpretations of others' reactions to us. Several theories were examined that describe in detail how the self is developed. Cooley's explanation involves the looking-glass self; Mead describes self-growth through the generalized other and through assuming the role of the other. In Mead's view, language is a complex method of symbolic communication that is essential in socialization and in development of the self.

Although the individual is a combination of biological, hereditary, and social factors, sociologists focus their studies on interaction in the social environment—the stage or setting in which socialization and self-development take place. Determining the effects of the social environment on the individual, which include the interpretation and internalization of others' reactions to us, is also a very subjective process. There are substantial variations in how individuals perceive, interpret, and react to their social environment. Suppose 20 people observe an accident at a busy intersection, and then each person writes a description of what happened. How many different descriptions might we get? Probably at least 10 or 15, maybe even 20. Or observe how different students react to a good or bad grade on an exam; again, there is great variation. The effect of the social environment, then, is a very subjective and interpretive phenomenon; no two people react in exactly the same way. Differences multiply because future perceptions and interpretations are based on past perceptions and interpretations. In like manner, socialization and development of the self are building processes that take place throughout the life span.

The four readings that follow will help illustrate some of the concepts we have discussed in this chapter. In an excerpt from her book *The Story of My Life,* Helen Keller, who was deaf and blind from early childhood, recalls meeting her new teacher and beginning her education. This reading illustrates how important language and symbolic communication are in the processes of socialization and self-development. The second reading is an excerpt from Malcolm X's autobiography in which he describes how some events that occurred during his early teens seemed to have a great and lasting impact on him. In the third reading, a sad story about the death of a boy, Jean Mizer illustrates socialization, agents of socialization, and the self-fulfilling prophecy. The fourth reading reflects on the nature versus nurture, biology versus social environment debate. Dennis Overbye suggests that if a gene explains aggression (as has been suggested), perhaps we should also look for a gene that explains racism, hatred of gays, and the greed of white-collar criminals.

## TERMS FOR STUDY

anticipatory socialization (40)

definition of the situation (47)

ego (42)

generalized other (44)

I (44)

id (42)

labeling (49)

looking-glass self (43)

me (44)

personality (41)

primary socialization (38)

resocialization (40)

secondary socialization (38)

self (41)

self-fulfilling prophecy (48)

significant others (43)

social interaction (35)

socialization (35)

sociobiology (35)

superego (42)

For a discussion of Research Question 1, see page 37.
For a discussion of Research Question 2, see page 48.

**INFOTRAC COLLEGE EDITION**
**Search Word Summary**

To learn more about the topics from this chapter, you can use the following words to conduct an electronic search on InfoTrac College Edition, an online library of journals. Here you will find a multitude of articles from various sources and perspectives:

**www.infotrac-college.com/wadsworth/access.html**

advertising                                     Malcolm X

psychoanalysis                                prejudice

linguistics                                        Sigmund Freud

personality                                       military education

## Reading 2.1

# THE STORY OF MY LIFE

*Helen Keller*

*Socialization is defined as the process by which the organism becomes a social being—the learning of habits, skills, and other requirements for effective participation in social groups. Socialization is a continuing process that occurs through interaction with other humans—typically parents and other family members, teachers, and friends.*

*Helen Keller was deaf and blind from early childhood. In this excerpt from one of her books, she describes the period when her new teacher first arrived and a new form of learning began to take place.*

The most important day I remember in all my life is the one on which my teacher, Anne Mansfield Sullivan, came to me. I am filled with wonder when I consider the immeasurable contrasts between the two lives which it connects. It was the third of March, 1887, three months before I was seven years old.

Have you ever been to sea in a dense fog, when it seemed as if a tangible white darkness shut you in, and the great ship, tense and anxious, groped her way toward the shore with plummet and sounding-line, and you waited with beating heart for something to happen? I was like that ship before my education began, only I was without compass or sounding-line, and had no way of knowing how near the harbour was. "Light! give me light!" was the wordless cry of my soul, and the light of love shone on me in that very hour.

I felt approaching footsteps. I stretched out my hand, as I supposed to my mother. Someone took it, and I was caught up and held close in the arms of her who had come to reveal all things to me, and, more than all things else, to love me.

The morning after my teacher came she led me into her room and gave me a doll. The little blind children at the Perkins Institution had sent it and Laura Bridgman had dressed it; but I did not know this until afterward. When I had played with it a little while, Miss Sullivan slowly spelled into my hand the word "d-o-l-l." I was at once interested in this finger play and tried to imitate it. When I finally succeeded in making the letters correctly I was flushed with childish plea-sure and pride. Running downstairs to my mother I held up my hand and made the letters for doll. I did not know that I was spelling a word or even that words existed; I was simply making my fingers go in monkey-like imitation. In the days that followed I learned to spell in this uncomprehending way a great many words, among them pin, hat, cup, and a few verbs like sit, stand, and walk. But my teacher had been with me several weeks before I understood that everything has a name.

One day, while I was playing with my new doll, Miss Sullivan put my big rag doll into my lap also, spelled "d-o-l-l" and tried to make me understand that "d-o-l-l" applied to both. Earlier in the day we had had a tussle over the words "m-u-g" and "w-a-t-e-r." Miss Sullivan had tried to impress it upon me that "m-u-g" is mug and that "w-a-t-e-r" is water, but I persisted in confounding the two. In despair she had dropped the subject for the time, only to renew it at the first opportunity. I became impatient at her repeated attempts and, seizing the new doll, I dashed it upon the floor. I was keenly delighted when I felt the fragments of the broken doll at my feet. Neither sorrow nor regret followed my passionate outburst. I had not loved the doll. In the still, dark world in which I lived there was no strong sentiment of tenderness. I felt my teacher sweep the fragments to one side of the hearth, and I had a sense of satisfaction that the cause of my discomfort was removed. She brought me my hat, and I knew I was going out into the warm sunshine. This thought, if a wordless sensation may be called a thought, made me hop and skip with pleasure.

We walked down the path to the well-house, attracted by the fragrance of the honeysuckle with which it was covered. Someone was drawing water and my teacher placed my hand under the spout. As the cool stream gushed over one hand she spelled into the other the word water, first slowly, then rapidly. I stood still, my whole attention fixed upon the motions of her fingers. Suddenly I felt a misty con-sciousness as of something forgotten—a thrill of returning thought; and somehow the mystery of language was revealed to me. I knew then that "w-a-t-e-r" meant the wonderful cool something that was flowing over my hand. That living word awak-ened my soul, gave it light, hope, joy, set it free! There were barriers still, it is true, but barriers that could in time be swept away.

I left the well-house eager to learn. Everything had a name, and each name gave birth to a new thought. As we returned to the house every object which I touched seemed to quiver with life. That was because I saw everything with the strange, new sight that had come to me. On entering the door I remembered the doll I had broken. I felt my way to the hearth and picked up the pieces. I tried vainly to put them together. Then my eyes filled with tears; for I realized what I had done, and for the first time I felt repentance and sorrow.

I learned a great many new words that day. I do not remember what they all were; but I do know that mother, father, sister, teacher were among them—words that were to make the world blossom for me, "like Aaron's rod, with flowers." It would have been difficult to find a happier child than I was as I lay in my crib at the close of that eventful day and lived over the joys it had brought me, and for the first time longed for a new day to come.

I recall many incidents of the summer of 1887 that followed my soul's sudden awakening. I did nothing but explore with my hands and learn the name of every object that I touched; and the more I handled things and learned their names and uses, the more joyous and confident grew my sense of kinship with the rest of the world.

When the time of daisies and buttercups came Miss Sullivan took me by the hand across the fields, where men were preparing the earth for the seed, to the banks of the Tennessee River, and there, sitting on the warm grass, I had my first lessons in the beneficence of nature. I learned how the sun and the rain make to grow out of the ground every tree that is pleasant to the sight and good for food, how birds build their nests and live and thrive from land to land, how the squirrel, the deer, the lion, and every other creature finds food and shelter. As my knowledge of things grew I felt more and more the delight of the world I was in. Long before I learned to do a sum in arithmetic or describe the shape of the earth, Miss Sullivan had taught me to find beauty in the fragrant woods, in every blade of grass, and in the curves and dimples of my baby sister's hand. She linked my earliest thoughts with nature, and made me feel that "birds and flowers and I were happy peers."

I had now the key to all language, and I was eager to learn to use it. Children who hear acquire language without any particular effort; the words that fall from others' lips they catch on the wing, as it were, delightedly, while the little deaf child must trap them by a slow and often painful process. But whatever the process, the result is wonderful. Gradually from naming an object we advance step by step until we have traversed the vast distance between our first stammered syllable and the sweep of thought in a line of Shakespeare.

At first, when my teacher told me about a new thing I asked very few questions. My ideas were vague, and my vocabulary was inadequate; but as my knowledge of things grew, and I learned more and more words, my field of inquiry broadened, and I would return again and again to the same subject, eager for further information. Sometimes a new word revived an image that some earlier experience had engraved on my brain.

I remember the morning that I first asked the meaning of the word, "love." This was before I knew many words. I had found a few early violets in the garden and brought them to my teacher. She tried to kiss me; but at that time I did not like to have any one kiss me except my mother. Miss Sullivan put her arm gently round me and spelled into my hand, "I love Helen."

"What is love?" I asked.

She drew me closer to her and said, "It is here," pointing to my heart, whose beats I was conscious of for the first time. Her words puzzled me very much because I did not then understand anything unless I touched it.

I smelt the violets in her hand and asked, half in words, half in signs, a question which meant, "Is love the sweetness of flowers?"

"No," said my teacher.

Again I thought. The warm sun was shining on us.

"Is this not love?" I asked, pointing in the direction from which the heat came. "Is this not love?"

It seemed to me that there could be nothing more beautiful than the sun, whose warmth makes all things grow. But Miss Sullivan shook her head, and I was greatly puzzled and disappointed. I thought it strange that my teacher could not show me love.

A day or two afterward I was stringing beads of different sizes in symmetrical groups—two large beads, three small ones, and so on. I had made many mistakes, and Miss Sullivan had pointed them out again and again with gentle patience. Finally I noticed a very obvious error in the sequence and for an instant I concentrated my attention on the lesson and tried to think how I should have arranged the beads. Miss Sullivan touched my forehead and spelled with decided emphasis, "Think."

In a flash I knew that the word was the name of the process that was going on in my head. This was my first conscious perception of an abstract idea.

For a long time I was still—I was not thinking of the beads in my lap, but trying to find a meaning for "love" in the light of this new idea. The sun had been under a cloud all day, and there had been brief showers; but suddenly the sun broke forth in all its southern splendour.

Again I asked my teacher, "Is this not love?"

"Love is something like the clouds that were in the sky before the sun came out," she replied. Then in simpler words than these, which at that time I could not have understood, she explained: "You cannot touch the clouds, you know; but you feel the rain and know how glad the flowers and the thirsty earth are to have it after a hot day. You cannot touch love either; but you feel the sweetness that it pours into everything. Without love you would not be happy or want to play."

The beautiful truth burst upon my mind—I felt that there were invisible lines stretched between my spirit and the spirits of others.

# QUESTIONS 2.1

1. Without language and communication, socialization cannot take place. Discuss.

2. What skills were developed by Helen Keller during the period discussed in this reading? Could she have learned these skills without the help of others? Is the same true for a person who doesn't have the handicaps that Helen Keller had?

## Reading 2.2

# MASCOT

*Malcolm X*

*Socialization involves many things. Part of it is learning that you are male or female, short or tall, black or white; a major part is learning how people react to you because you are male or female, short or tall, black or white. These reactions and interpretations become an important part of the self. Many people probably do not recall the stage in the socialization process when they became aware of their gender or race. To remember, one must be very perceptive, and the new learning must be so unique or unexpected that it creates a lasting impression. In this excerpt from his autobiography, Malcolm X recalls learning what it means to be black in America.*

They told me I was going to go to a reform school. I was still thirteen years old.

But first I was going to the detention home. It was in Mason, Michigan, about twelve miles from Lansing. The detention home was where all the "bad" boys and girls from Ingham County were held, on their way to reform school—waiting for their hearings.

The white state man was a Mr. Maynard Allen. He was nicer to me than most of the state welfare people had been. He even had consoling words for the Gohannas and Mrs. Adcock and Big Boy; all of them were crying. But I wasn't. With the few clothes I owned stuffed into a box, we rode in his car to Mason. He talked as he drove along, saying that my school marks showed that if I would just straighten up, I could make something of myself. He said that reform school had the wrong reputation; he talked about what the word "reform" meant—to change and become better. He said the school was really a place where boys like me could have time to see their mistakes and start a new life and become somebody everyone would be proud of. And he told me that the lady in charge of the detention home, a Mrs. Swerlin, and her husband were very good people.

They were good people. Mrs. Swerlin was bigger than her husband, I remember, a big, buxom, robust, laughing woman, and Mr. Swerlin was thin, with black hair, and a black mustache and a red face, quiet and polite, even to me.

They liked me right away, too. Mrs. Swerlin showed me to my room, my own room—the first in my life. It was in one of those huge dormitory-like buildings where kids in detention were kept in those days—and still are in most places. I discovered next, with surprise, that I was allowed to eat with the Swerlins. It was the first time I'd eaten with white people—at least with grown white people—since the Seventh Day Adventist country meetings. It wasn't my own exclusive privilege, of course. Except for the very troublesome boys and girls at the detention home, who were kept locked up—those who had run away and been caught and brought back, or something like that—all of us ate with the Swerlins sitting at the head of the long tables.

They had a white cook-helper, I recall—Lucille Lathrop. (It amazes me how these names come back, from a time I haven't thought about for more than twenty years.) Lucille treated me well, too. Her husband's name was Duane Lathrop. He worked somewhere else, but he stayed there at the detention home on the weekends with Lucille.

I noticed again how white people smelled different from us, and how their food tasted different, not seasoned like Negro cooking. I began to sweep and mop and dust around in the Swerlins' house, as I had done with Big Boy at the Gohannas'.

They all liked my attitude, and it was out of their liking for me that I soon became accepted by them—as a mascot, I know now. They would talk about anything and everything with me standing right there hearing them, the same way people would talk freely in front of a pet canary. They would even talk about me, or about "niggers," as though I wasn't there, as if I wouldn't understand what the word meant. A hundred times a day, they used the word "nigger." I suppose that in their own minds, they meant no harm; in fact they probably meant well. It was the same with the cook, Lucille, and her husband, Duane. I remember one day when Mr. Swerlin, as nice as he was, came in from Lansing, where he had been through the Negro section, and said to Mrs. Swerlin right in front of me, "I just can't see how those niggers can be so happy and be so poor." He talked about how they lived in shacks, but had those big, shining cars out front.

And Mrs. Swerlin said, me standing right there, "Niggers are just that way. . . ." That scene always stayed with me.

It was the same with the other white people, most of them local politicians, when they would come visiting the Swerlins. One of their favorite parlor topics was "niggers." One of them was the judge who was in charge of me in Lansing. He was a close friend of the Swerlins. He would ask about me when he came, and they would call me in, and he would look me up and down, his expression approving, like he was examining a fine colt, or a pedigreed pup. I knew they must have told him how I acted and how I worked.

What I am trying to say is that it just never dawned upon them that I could understand, that I wasn't a pet, but a human being. They didn't give me credit for having the same sensitivity, intellect, and understanding that they would have been ready and willing to recognize in a white boy in my position. But it has historically been the case with white people, in their regard for black people, that even though we might be *with* them, we weren't considered *of* them. Even though they appeared to have opened the door, it was still closed. Thus they never did really see *me*.

This is the sort of kindly condescension which I try to clarify today, to these integration-hungry Negroes, about their "liberal" white friends, these so-called "good white people"—most of them anyway. I don't care how nice one is to you; the thing you must always remember is that almost never does he really see you as he sees himself, as he sees his own kind. He may stand with you through thin, but not thick; when the chips are down, you'll find that as fixed in him as his bone structure is his sometimes subconscious conviction that he's better than anybody black.

But I was no more than vaguely aware of anything like that in my detention-home years. I did my little chores around the house, and everything was fine. And each weekend, they didn't mind my catching a ride over to Lansing for the afternoon or evening. If I wasn't old enough, I sure was big enough by then, and nobody ever questioned my hanging out, even at night, in the streets of the Negro section.

I was growing up to be even bigger than Wilfred and Philbert, who had begun to meet girls at the school dances, and other places, and introduced me to a few. But the ones who seemed to like me, I didn't go for—and vice versa. I couldn't dance a lick, anyway, and I couldn't see squandering my few dimes on girls. So mostly I pleasured myself these Saturday nights by gawking around the Negro bars and restaurants. The jukeboxes were wailing Erskine Hawkins' "Tuxedo Junction," Slim and Slam's "Flatfoot Floogie," things like that. Sometimes, big bands from New York, out touring the one-night stands in the sticks, would play for big dances in Lansing. Everybody with legs would come out to see any performer who bore the magic name "New York." Which is how I first heard Lucky Thompson and Milt Jackson, both of whom I later got to know well in Harlem.

Many youngsters from the detention home, when their dates came up, went off to the reform school. But when mine came up—two or three times—it was always ignored. I saw new youngsters arrive and leave. I was glad and grateful. I knew it was Mrs. Swerlin's doing. I didn't want to leave.

She finally told me one day that I was going to be entered in Mason Junior High School. It was the only school in town. No ward of the detention home had ever gone to school there, at least while still a ward. So I entered their seventh grade. The only other Negroes there were some of the Lyons children, younger than I was, in the lower grades. The Lyons and I, as it happened, were the town's only Negroes. They were, as Negroes, very much respected. Mr. Lyons was a smart, hardworking man, and Mrs. Lyons was a very good woman. She and my mother, I had heard my mother say, were two of the four West Indians in that whole section of Michigan.

Some of the white kids at school, I found, were even friendlier than some of those in Lansing had been. Though some, including the teachers, called me "nigger," it was easy to see that they didn't mean any more harm by it than the Swerlins. As the "nigger" of my class, I was in fact extremely popular—I suppose partly because I was kind of a novelty. I was in demand, I had top priority. But I also benefited from the special prestige of having the seal of approval from that Very Important Woman about the town of Mason, Mrs. Swerlin. Nobody in Mason would have dreamed of getting on the wrong side of her. It became hard for me to get through a school day without someone after me to join this or head up that—the debating society, the junior high basketball team, or some other extracurricular activity. I never turned them down.

And I hadn't been in the school long when Mrs. Swerlin, knowing I could use spending money of my own, got me a job after school washing the dishes in a local restaurant. My boss there was the father of a white classmate whom I spent a lot of time with. His family lived over the restaurant. It was fine working there. Every Friday night when I got paid, I'd feel at least ten feet tall. I forget how much I made, but it seemed like a lot. It was the first time I'd ever had any money to speak of, all my own, in my whole life. As soon as I could afford it, I bought a green suit and some shoes, and at school I'd buy treats for the others in my class—at least as much as any of them did for me.

English and history were the subjects I liked most. My English teacher, I recall—a Mr. Ostrowski—was always giving advice about how to become something in life. The one thing I didn't like about history class was that the teacher, Mr. Williams, was a great one for "nigger" jokes. One day during my first week at school, I walked into the room and he started singing to the class, as a joke, "Way down yonder in the cotton field, some folks say that a nigger won't steal." Very funny. I liked history, but I never thereafter had much liking for Mr. Williams. Later, I remember, we came to the textbook section on Negro history. It was exactly one paragraph long. Mr. Williams laughed through it practically in a single breath, reading aloud how the Negroes had been slaves and then were freed, and how they were usually lazy and dumb and shiftless. He added, I remember, an anthropological footnote on his own, telling us between laughs how Negroes' feet were "so big that when they walk, they don't leave tracks, they leave a hole in the ground."

. . . Then, in the second semester of the seventh grade, I was elected class president. It surprised me even more than other people. But I can see now why the class might have done it. My grades were among the highest in the school. I was unique in my class, like a pink poodle. And I was proud; I'm not going to say I wasn't. In fact, by then, I didn't really have much feeling about being a Negro, because I was trying so hard, in every way I could, to be white. Which is why I am spending much of my life today telling the American black man that he's wasting his time straining to "integrate." I know from personal experience. I tried hard enough. . . .

That summer of 1940, in Lansing, I caught the Greyhound bus for Boston with my cardboard suitcase, and wearing my green suit. If someone had hung a sign, "hick," around my neck, I couldn't have looked much more obvious. They didn't have the turnpikes then; the bus stopped at what seemed every corner and cowpatch. From my seat in—you guessed it—the back of the bus, I gawked out of the window at white man's America rolling past for what seemed a month, but must have been only a day and a half.

When we finally arrived, Ella met me at the terminal and took me home. The house was on Waumbeck Street in the Sugar Hill section of Roxbury, the Harlem of Boston. I met Ella's second husband, Frank, who was now a soldier; and her brother Earl, the singer who called himself Jimmy Carleton; and Mary, who was very different from her older sister. It's funny how I seemed to think of Mary as Ella's sister, instead of her being, just as Ella is, my own half-sister. It's probably because Ella and I always were much closer as basic types; we're dominant people, and Mary has always been mild and quiet, almost shy.

Ella was busily involved in dozens of things. She belonged to I don't know how many different clubs; she was a leading light of local so-called "black society." I saw and met a hundred black people there whose big-city talk and ways left my mouth hanging open.

I couldn't have feigned indifference if I had tried to. People talked casually about Chicago, Detroit, New York. I didn't know the world contained as many Negroes as I saw thronging downtown Roxbury at night, especially on Saturdays. Neon lights, nightclubs, poolhalls, bars, the cars they drove! Restaurants made the streets smell—rich, greasy, down-home black cooking! Jukeboxes blared

Erskine Hawkins, Duke Ellington, Cootie Williams, dozens of others. If somebody had told me then that some day I'd know them all personally, I'd have found it hard to believe. The biggest bands, like these, played at the Roseland State Ballroom, on Boston's Massachusetts Avenue—one night for Negroes, the next night for whites.

I saw for the first time occasional black-white couples strolling around arm in arm. And on Sundays, when Ella, Mary, or somebody took me to church, I saw churches for black people such as I had never seen. They were many times finer than the white church I had attended back in Mason, Michigan. There, the white people just sat and worshipped with words; but the Boston Negroes, like all other Negroes I had ever seen at church, threw their souls and bodies wholly into worship.

Two or three times, I wrote letters to Wilfred intended for everybody back in Lansing. I said I'd try to describe it when I got back.

But I found I couldn't.

My restlessness with Mason—and for the first time in my life a restlessness with being around white people—began as soon as I got back home and entered eighth grade.

I continued to think constantly about all that I had seen in Boston, and about the way I had felt there. I know now that it was the sense of being a real part of a mass of my own kind for the first time.

The white people—classmates, the Swerlins, the people at the restaurant where I worked—noticed the change. They said, "You're acting so strange. You don't seem like yourself, Malcolm. What's the matter?"

I kept close to the top of the class, though. The top-most scholastic standing, I remember, kept shifting between me, a girl named Audrey Slaugh, and a boy named Jimmy Cotton.

It went on that way, as I became increasingly restless and disturbed through the first semester. And then one day, just about when those of us who had passed were about to move up to 8-A, from which we would enter high school the next year, something happened which was to become the first major turning point of my life.

Somehow, I happened to be alone in the classroom with Mr. Ostrowski, my English teacher. He was a tall, rather reddish white man and he had a thick mustache. I had gotten some of my best marks under him, and he had always made me feel that he liked me. He was, as I have mentioned, a natural-born "advisor," about what you ought to read, to do, or think—about any and everything. We used to make unkind jokes about him: why was he teaching in Mason instead of somewhere else, getting for himself some of the "success in life" that he kept telling us how to get?

I know that he probably meant well in what he happened to advise me that day. I doubt that he meant any harm. It was just in his nature as an American white man. I was one of his top students, one of the school's top students—but all he could see for me was the kind of future "in your place" that almost all white people see for black people.

He told me, "Malcolm, you ought to be thinking about a career. Have you been giving it thought?"

The truth is, I hadn't. I never had figured out why I told him, "Well, yes, sir, I've been thinking I'd like to be a lawyer." Lansing certainly had no Negro lawyers—or doctors either—in those days, to hold up an image I might have aspired to. All I really knew for certain was that a lawyer didn't wash dishes, as I was doing.

Mr. Ostrowski looked surprised, I remember, and leaned back in his chair and clasped his hands behind his head. He kind of half-smiled and said, "Malcolm, one of life's first needs is for us to be realistic. Don't misunderstand me, now. We all here like you, you know that. But you've got to be realistic about being a nigger. A lawyer—that's no realistic goal for a nigger. You need to think about something you *can* be. You're good with your hands—making things. Everybody admires your carpentry shop work. Why don't you plan on carpentry? People like you as a person—you'd get all kinds of work."

The more I thought afterwards about what he said, the more uneasy it made me. It just kept treading around in my mind.

What made it really begin to disturb me was Mr. Ostrowski's advice to others in my class—all of them white. Most of them told him they were planning to become farmers. But those who wanted to strike out on their own, to try something new, he had encouraged. Some, mostly girls, wanted to be teachers. A few wanted other professions, such as one boy who wanted to become a county agent; another, a veterinarian; and one girl wanted to be a nurse. They all reported that Mr. Ostrowski had encouraged what they had wanted. Yet nearly none of them had earned marks equal to mine.

It was a surprising thing that I had never thought of it that way before, but I realized that whatever I wasn't, I *was* smarter than nearly all of those white kids. But apparently I was still not intelligent enough, in their eyes, to become whatever *I* wanted to be.

It was then that I began to change—inside.

I drew away from white people. I came to class, and I answered when called upon. It became a physical strain simply to sit in Mr. Ostrowski's class.

Where "nigger" had slipped off my back before, whenever I heard it now, I stopped and looked at whoever said it. And they looked surprised that I did.

I quit hearing so much "nigger" and "What's wrong?"—which was the way I wanted it. Nobody, including the teachers, could decide what had come over me. I knew I was being discussed.

In a few more weeks, it was that way, too, at the restaurant where I worked washing dishes, and at the Swerlins'. . . .

---

# QUESTIONS 2.2

**1.** Malcolm X believes he was treated as a "mascot." What does he mean by this? Why was he treated this way? Discuss.

**2.** Outline the important aspects of socialization that occurred during the period described by Malcolm X. Who were the socialization agents in this period?

**3.** Review Chapter Two, and using examples from this reading, illustrate development of the self using Cooley's and Mead's theories.

## Reading 2.3

# CIPHER IN THE SNOW

*Jean E. Mizer*

---

*Socialization and development of self are interactive processes that are criti-
cally important in the growth and future of every individual. Seldom, however, are these processes so graphically and dramatically displayed as they are
in this reading. Jean Mizer was a school teacher in Idaho when she wrote this
memoir of a personal experience.*

It started with tragedy on a biting cold February morning. I was driving behind
the Milford Corners bus as I did most snowy mornings on my way to school. It
veered and stopped short at the hotel, which it had no business doing, and I was
annoyed as I had to come to an unexpected stop. A boy lurched out of the bus,
reeled, stumbled, and collapsed on the snowbank at the curb. The bus driver and
I reached him at the same moment. His thin, hollow face was white even against
the snow.

"He's dead," the driver whispered.

I didn't register for a minute. I glanced quickly at the scared young faces staring
down at us from the school bus. "A doctor! Quick! I'll phone from the hotel. . . ."

"No use. I tell you he's dead." The driver looked down at the boy's still form. "He
never even said he felt bad," he muttered, "just tapped me on the shoulder and
said, real quiet, 'I'm sorry. I have to get off at the hotel.' That's all. Polite and apolo-
gizing like."

At school, the giggling, shuffling morning noise quieted as the news went down
the halls. I passed a huddle of girls. "Who was it? Who dropped dead on the way to
school?" I heard one of them half-whisper.

"Don't know his name; some kid from Milford Corners," was the reply.

It was like that in the faculty room and the principal's office. "I'd appreciate your
going out to tell the parents," the principal told me. "They haven't a phone and, anyway, somebody from school should go there in person. I'll cover your classes."

"Why me?" I asked. "Wouldn't it be better if you did it?"

"I didn't know the boy," the principal admitted levelly. "And in last year's sophomore personalities column I note that you were listed as his favorite teacher."

I drove through the snow and cold down the bad canyon road to the Evans place
and thought about the boy, Cliff Evans. His favorite teacher! I thought. He hasn't
spoken two words to me in two years! I could see him in my mind's eye all right, sitting back there in the last seat in my afternoon literature class. He came in the room
by himself and left by himself. "Cliff Evans," I muttered to myself, "a boy who never
talked." I thought a minute. "A boy who never smiled. I never saw him smile once."

The big ranch kitchen was clean and warm. I blurted out my news somehow. Mrs.
Evans reached blindly toward a chair. "He never said anything about bein' ailing."

Reprinted from the November 1964 issue of *NEA Journal* by permission of the National
Education Association and the author.

His stepfather snorted. "He ain't said nothin' about anything since I moved in here."

Mrs. Evans pushed a pan to the back of the stove and began to untie her apron. "Now hold on," her husband snapped. "I got to have breakfast before I go to town. Nothin' we can do now anyway. If Cliff hadn't been so dumb, he'd have told us he didn't feel good."

After school I sat in the office and stared bleakly at the records spread out before me. I was to close the file and write the obituary for the school paper. The almost bare sheets mocked the effort. Cliff Evans, white, never legally adopted by stepfather, five young half-brothers and sisters. These meager strands of information and the list of D grades were all the records had to offer.

Cliff Evans had silently come in the school door in the mornings and gone out the school door in the evenings, and that was all. He had never belonged to a club. He had never played on a team. He had never held an office. As far as I could tell he had never done one happy, noisy kid thing. He had never been anybody at all.

How do you go about making a boy into a zero? The grade school records showed me. The first and second grade teachers' annotations read "sweet, shy child," "timid but eager." Then the third grade note had opened the attack. Some teacher had written in a good, firm hand, "Cliff won't talk. Uncooperative. Slow learner." The other academic sheep had followed with "dull"; "slow-witted"; "low I.Q." They became correct. The boy's I.Q. score in the ninth grade was listed at 83. But his I.Q. in the third grade had been 106. The score didn't go under 100 until the seventh grade. Even shy, timid, sweet children have resilience. It takes time to break them.

I stomped to the typewriter and wrote a savage report pointing out what education had done to Cliff Evans. I slapped a copy on the principal's desk and another in the sad, dog-eared file. I banged the typewriter and slammed the file and crashed the door shut, but I didn't feel much better. A little boy kept walking after me, a little boy with a peaked, pale face; a skinny body in faded jeans; and big eyes that had looked and searched for a long time and then had become veiled.

I could guess how many times he'd been chosen last to play sides in a game, how many whispered child conversations had excluded him, how many times he hadn't been asked. I could see and hear the faces and voices that said over and over, "You're dumb. You're dumb. You're nothing, Cliff Evans."

A child is a believing creature. Cliff undoubtedly believed them. Suddenly it seemed clear to me: When finally there was nothing left at all for Cliff Evans, he collapsed on a snowbank and went away. The doctor might list "heart failure" as the cause of death, but that wouldn't change my mind.

We couldn't find ten students in the school who had known Cliff well enough to attend the funeral as his friends. So the student body officers and a committee from the junior class went as a group to the church, being politely sad. I attended the services with them, and sat through it with a lump of cold lead in my chest and a big resolve growing through me.

I've never forgotten Cliff Evans nor that resolve. He has been my challenge year after year, class after class. I look up and down the rows carefully each September at the unfamiliar faces. I look for veiled eyes or bodies scrouged into a seat in an alien world. "Look, kids," I say silently, "I may not do anything else for you this year, but not one of you is going to come out of here a nobody. I'll work or fight to the bitter end doing battle with society and the school board, but I won't have one of you coming out of here thinking himself into a zero."

Most of the time—not always, but most of the time—I've succeeded.

---

## QUESTIONS 2.3

**1.** How does the self-fulfilling prophecy apply in this situation?

**2.** Who were the agents of socialization in this story? Why did they define Cliff Evans as they did?

---

### Reading 2.4

---

# BORN TO RAISE HELL?

*Dennis Overbye*

---

*Which is more important—social environment or biology? The nature versus nurture debate continues. Clearly, biological factors are an important part of who we are. For example, we see the biological stamp on such factors as skin and eye color and perhaps on susceptibility to certain diseases. However, sociologists believe that behavior is not biologically determined but is a product of one's social environment, past experiences and relationships, and the socialization process. Dennis Overbye gets involved in the debate and examines some of the implications of assuming that biology explains behavior.*

Some of us, it seems, were just born to be bad. Scientists say they are on the verge of pinning down genetic and biochemical abnormalities that predispose their bearers to violence. An article in the journal *Science* last summer carried the headline EVIDENCE FOUND FOR A POSSIBLE "AGGRESSION" GENE. Waiting in the wings are child-testing programs, drug manufacturers, insurance companies, civil rights advocates, defense attorneys and anxious citizens for whom the violent criminal has replaced the beady-eyed communist as the boogeyman. Crime thus joins homosexuality, smoking, divorce, schizophrenia, alcoholism, shyness, political liberalism, intelligence, religiosity, cancer and blue eyes among the many aspects of human life for which it is claimed that biology is destiny. Physicists have been pilloried for years for this kind of reductionism, but in biology it makes everybody happy: the scientists and pharmaceutical companies expand their domain; politicians have "progress" to point to; the smokers, divorcés and serial killers get to blame their problems on biology, and we get the satisfaction of knowing they are sick—not like us at all.

Admittedly, not even the most rabid sociobiologists contend that babies pop out of the womb with a thirst for bank robbing. Rather, they say, a constellation of influences leads to a life of crime, among them poverty, maleness, and a trait known as "impulsivity," presumably caused by bad brain chemistry, caused in turn by bad genes. What, you may ask, is impulsivity? The standard answer tends to involve people who can't control their emotions or who get into bar fights. A study conducted in Finland found that men so characterized tend to be deficient in the brain hormone serotonin—one of several chemical messengers that transmit signals between nerve cells. In another study researchers found that the men in a Dutch family with a history of male violence seemed to lack the ability to break down certain neurotransmitters, including serotonin, that build up in the brain during flight-or-fight situations. Couple this with persistent statistical surveys purporting to show that criminals tend to run in families and you have the logic behind the Violence Initiative dreamed up a couple of years ago by the Department of Health and Human Services, which included research to discover biological markers that could be used to distinguish violence-prone children as early as age five. Any doubts about the potential for abuse in such a program were erased when Frederick Goodwin, then director of the Alcohol, Drug Abuse and Mental Health Administration, lapsed into a comparison of inner-city youth to murderous oversexed monkeys during a speech about the initiative. "Maybe," he said, "it isn't just the careless use of the word when people call certain areas of certain cities jungles . . ." Amid the ensuing outcry, the National Institutes of Health canceled its financial support for a planned conference on the biology of violence. Now, however, bolstered by a cautious approval from the National Academy of Sciences, the conference is back on, and Goodwin is head of the National Institute of Mental Health.

Science marches on. Or does it? The whole affair is uncomfortably reminiscent—as the scientists admit—of the 1960s, when researchers theorized that carriers of an extra Y, or male, chromosome were predisposed to criminality, or of earlier attempts to read character from the bumps on people's skulls. As any doctor who ever testified for a tobacco company in one of those trials knows, the statistical association of two things, like, smoking and cancer, or guns and murders, does not necessarily imply cause and effect. We know too little about the biochemical cocktail that is the brain, and all too much about how stigmatized children live down to our expectations.

Being a great fan of science, I'm all in favor of more research, testing, poking, drugging, jailing, genetic therapy, amniocentesis, just as soon as someone can give me a scientific definition of impulsivity, one that provides, say, a cultural- and color-blind distinction between a spirited child and an impulsive one. If bar fights are the criteria, it's nice to think that a simple blood test might have spared us the antics of people like Billy Martin or George Steinbrenner.

My real complaint is that the violence initiative doesn't go nearly far enough. Some laboratory should be looking for the racism gene, or the homophobia gene. Goodwin was right; the inner city is a jungle. But so are the corporation, the newsroom, and the White House staff. The language of trial lawyers or bond traders in full testosterone fury is as blood-curdling as any mugger's. When it comes to social carnage, the convenience-store stickup can't compare with a leveraged buyout, trickling-down unemployment, depression, anger, alcoholism, divorce, domestic abuse, and addiction. I'd like to see white men with suspenders and cellular phones tested for the greed gene. The genomes of presidential candidates should be a matter of public record.

I'm a middle-aged white man as afraid as anyone else of being not quite alone on a dark New York City street, and I've stared into the stone-cold eyes of a mugger while he told me, "It's just you and me." When you hear those words, it's too late to fight the war on crime. The rush to define criminals as sick obscures an uncomfortable truth about our society, which is that crime and violence often pay handsomely. Just ask the conquistadores, the Menendez brothers, Oliver North or the comfortable and respected descendants of bootleggers and slaveholders. Ask the purveyors of the most violent television program in recent memory: the Gulf War.

---

## QUESTIONS 2.4

**1.** List a number of your own characteristics. Now try to assess their origin—are they (B)iological (I was born this way), or (E)nvironmental (this was a product of my life experiences)?

**2.** If there is an aggression gene, there must be a greed gene. Argue this statement pro and con.

**3.** The author suggests that the search to discover biological markers has the "potential for abuse." What might he have in mind?

---

## NOTES

1. This section is summarized from "Final Note on a Case of Extreme Isolation," by Kingsley Davis, *American Journal of Sociology* 52 (1947), pp. 432–437. Also see Bruno Bettelheim, "Feral Children and Autistic Children," *American Journal of Sociology* 64 (March 1959), pp. 455–467.

2. Edward C. Wilson's books include *Sociobiology: The New Synthesis* (Cambridge, Mass.: Harvard University Press, 1975) and *On Human Nature* (Cambridge, Mass.: Harvard University Press, 1978). His newer book devotes much more attention to human behavior than his earlier book did. Also see *Society* 15 (September/October 1978). The whole issue is on sociobiology. The quotation cited in the text comes from p. 10 of the *Society* issue.

3. Quite a bit of research has been done on this topic. For some of the most recent, see the two lead articles on confluence theory in the *American Sociological Review* 56 (April 1991), pp. 141–165.

4. See *Sociological Footprints*, 6th ed., by Leonard Cargan and Jeanne Ballantine (Belmont, Calif.: Wadsworth, 1994), pp. 53–68. Also see the works of Carol Gilligan and Jean Baker Miller, and Deborah Tannen's book *Talking from 9 to 5* (New York: William Morrow, 1994).

5. This discussion follows that contained in *The Social Construction of Reality*, by Peter Berger and Thomas Luckmann (Garden City, N.Y.: Doubleday, 1966), especially pp. 129–147.

6. See Arthur Niederhoffer, *Behind the Shield: The Police in Urban Society* (Garden City, N.Y.: Doubleday, 1967).

7. Numerous books and articles on prison life describe the language or argot of inmates. For example, see Gresham Sykes, *Society of Captives* (New York: Atheneum, 1958). The example I have used here comes from *The Felon*, by John Irwin (Englewood Cliffs, N.J.: Prentice-Hall, 1970), pp. 66–85. Also see Peter Letkemann, *Crime as Work* (Englewood Cliffs, N.J.: Prentice-Hall, 1973).

8. *The Brain Benders*, by Charles Brownfield (New York: Exposition Press, 1972), has an interesting description and analysis of brainwashing. Brownfield deals with the actual occurrence in wartime and with scientific experiments. The example cited here comes from p. 55.

9. Charles Horton Cooley's books are *Human Nature and the Social Order* (New York: Charles Scribner's, 1902) and *Social Organization* (New York: Charles Scribner's, 1909).

10. George Herbert Mead's writings include *Mind, Self, and Society* (Chicago: University of Chicago Press, 1934) and "The Genesis of the Self and Social Control," from *International Journal of Ethics* 35 (April 1925), pp. 251–273.

11. This discussion on the importance of language and the Sapir-Whorf school of thought draws in part from *Social Psychology,* rev. ed., by Alfred Lindesmith and Anselm Strauss (New York: Holt, 1956), chapter 8.

12. This section benefits from a summary of a number of research findings reported in *Newsweek,* February 19, 1996.

13. This discussion of definition of the situation and self-fulfilling prophecy follows that of Robert Merton in *Social Theory and Social Structure,* 2d ed. (New York: Free Press, 1957), pp. 421–436. The quote is from p. 128.

14. The study is first reported in "The Effect of Experimenter Bias on the Performance of the Albino Rat," by Robert Rosenthal and Kermit Fode, *Behavioral Science* 8 (July 1963), pp. 183–189. The discussion here is summarized from another report, "Teacher Expectations for the Disadvantaged," by Robert Rosenthal and Lenore Jacobson, *Scientific American* 218 (April 1968), pp. 19–23. Also see Rosenthal and Jacobson's book *Pygmalion in the Classroom: Teacher Expectation and Pupils' Intellectual Development* (New York: Holt, Rinehart & Winston, 1968).

15. We mentioned the importance of verification in science in Chapter One, and the Rosenthal study provides an interesting sidelight in this regard. When the study appeared in 1968, it drew wide interest. Then, a series of critical reviews appeared that took Rosenthal to task for the research methods he used in his study. Next, numerous attempts were made to repeat Rosenthal's study on other samples. Some researchers have been able to repeat his findings and conclusions, but many have not been able to do so, and the debate goes on about the original study. If you would like to get involved in the discussion, see Glen Mendels and James Flanders, "Teachers' Expectations and Pupil Performance," *American Educational Research Journal* 10 (Summer 1973), pp. 203–212, which describes a replication and cites a number of other attempts and critiques. It should be mentioned that the differences between "academic spurters" and others described in the text were distinct for the first and second graders but were mixed or nonexistent in grades three through six.

16. See "A Conversation With Jerome Kagan," *Saturday Review,* March 10, 1973.

## CHAPTER THREE

Elliott Erwitt/Magnum Photos Inc.

# Norms, Roles, Culture, Society

## TWO RESEARCH QUESTIONS

**1.** *For every 100 girls born worldwide, between 105 and 106 boys are born. But an official Chinese fertility survey in 1988 showed gender ratios of up to 120 boys per 100 girls among children born in China between 1980 and 1987. Why did the Chinese suddenly give birth to even more boys than girls? How can this be explained?*

**2.** *Some people are involved in many different jobs, activities, or positions in society; other people are more focused, are committed to one job or position. Would their approach to life, as indicated by the number of activities in which they are involved, have any effect on how long they live? Would the more focused live longer, would those with more varied lives live longer, or would their life spans be unrelated to style?*

IN THE PREVIOUS chapter we focused on how individuals develop into social beings. We discussed the socialization process, the development of the self and personality, and the importance of language. Our analysis was detailed, and it concentrated on the individual's growth from a helpless infant to a person capable of participating in complex social situations. The focus, again, was on the individual.

In this chapter, we move away from the individual to the *stage* or *setting* where socialization and development of the self take place. We call this stage the *social environment*. Here, individuals play parts, are governed by rules, and behave in predictable ways. We find the most important concepts in our study of the social environment to be *culture, society*, and *social structure*. Other concepts like norms, roles, and statuses help us better understand and explain what happens on this stage.

## THE MEANING OF CULTURE

The importance of culture in sociology is comparable to the importance of evolution in biology, of gravity in physics, and of disease in medicine. It is a central concept that allows us to organize and explain what we see. Much of human behavior can be understood and indeed predicted if we know a people's culture, their design for living.

For social scientists, the concept of culture is very important. By **culture** we mean that complex set of learned and shared beliefs, customs, skills, habits, traditions, and knowledge common to the members of a society. Culture is viewed as the social heritage of a society. According to anthropologist Clyde Kluckhohn, culture represents the distinctive way of life of a group of people, their complete design for living.[1]

There are many cultures in the world, some similar to each other and some very different from most others. A culture evolves through time, it is continually modified, and its complexity reflects its sources in human creativity. A culture is shared by the members of a society and is learned through the socialization process. The culture in which we live determines for us what we will want to eat, whom we will like or hate, what we will fear (snakes and mice, but probably not evil spirits), how we will express our emotions, how we will dress (blue jeans and bikinis, but not turbans or saris), what types of manners we will develop, and how we will celebrate New Year's Eve. I mentioned earlier that when I throw a volleyball to my son, he is likely to catch it with his hands and throw it back, whereas a French or English teenager is likely to head or kick the ball back to me. We can now see that these are *cultural* differences. In American culture, ball sports emphasize the use of hands—catching and throwing or hitting in basketball, baseball, volleyball, and handball, for example. In France and England and in many other cultures, the major sport is soccer, which emphasizes controlling the ball with one's feet and head.

When the details of a culture are examined, a distinction can be made between material and nonmaterial aspects. **Material culture** refers to the concrete things that a society creates and uses: screwdriver, house, classroom, desk, car, plane, telephone. **Nonmaterial culture** refers to the abstract creations of a society: customs, laws, ideas, values, beliefs. The nonmaterial culture would include beliefs about religion and courtship, ideas about democracy and communism, definitions of good manners, and rules for driving a car. One aspect of the nonmaterial culture that we don't usually think much about are the rules for conduct—what we are supposed to do in specific situations. We call these *norms*.

## NORMS

A sociology instructor is lecturing to a class of college freshmen and sophomores. At one point in her lecture, she asks a student to help her illustrate a point by standing up. The student is belligerent and upset at being interrupted in his reading of the school newspaper. The student replies that he would rather not help and that the teacher should ask someone else. The teacher asks the student once more to please help her with the experiment and to stand up. The student replies in a louder voice that he doesn't think much of this stupid class or the teacher and that he is not about to participate in some silly experiment. The now-angry teacher responds, "Either stand up or get out of this classroom!" The student replies, "I'm not standing up—and if you want me out of here you'll have to throw me out!" The two then glare at each other for what seems like a year.

If we were watching the scene and could take our eyes off the principal characters, we would notice that the rest of the class is behaving strangely. Nearly all appear to be very uncomfortable, and many are obviously embarrassed. They are twisting and squirming and are avoiding meeting the eyes of the instructor. At the beginning of the confrontation some may have laughed nervously; now all are quiet. They act as if they are trying to deny that what is happening is actually happening. . . . Why?

All societies have rules that specify what people should do in specific situations. Sociologists call these shared behavioral standards **norms.** Norms describe the accepted or required behavior for a person in a particular

situation. For example, when a person gets into a car, that person unconsciously begins to behave according to a whole set of norms relating to driving procedure. In America the driver knows that he or she should drive
on the right side of the road, pass on the left, signal before turning, and so
on. The norms vary from place to place, but they always exist. These
shared standards for behavior allow us to predict what other people will
do. Without even thinking about it or knowing the individual, I know how
the driver of a car approaching me will behave. The system of norms
allows us to predict what other people will do in specific situations and to
pattern our own behavior accordingly.

Where do norms come from? Like much of what we discuss in this
book, norms emerge through the process of social interaction. As people
interact, they come to understand that certain ways of behaving are proper
and acceptable and that other ways are improper and unacceptable. These
general opinions and beliefs are called **values.** Values represent the broad
perspectives people have. Observers of United States society believe you
can recognize the following predominant values in Americans: belief in
achievement and success, progress, individualism, equality and democracy,
patriotism, freedom, romantic love and monogamy, the importance of
activity and work, and the importance of telling the truth. Although norms
are based on values, the difference between the two concepts is important:
Values are more broadly based and represent general perspectives, whereas
norms represent specific rules for particular situations. Both norms and
values can be long-standing, born of custom and tradition, or they can be
of shorter duration.

Norms take many forms, from written laws to informal agreements
among group members as they try to complete a task or to manage emotion and interaction among members. Superstitions, customs, and myths
can act as norms: Many buildings don't have a 13th floor; people take care
not to walk under ladders or on cracks in the sidewalk; alcohol consumption is much higher around some holidays; and June is the month in which
to get married (which in turn probably influences the sale of rice). China's
population grew by about 15 million people in 1991. This is obviously
quite a lot, but it's not nearly as many as expected and 1.25 million *less*
than the increase in 1990. Amazingly, the birth rate in 1991 was even lower
than the government's planned quota. Why the decline? No major external
incident—for example, disease, war, or drastic change in the standard of
living—seems to be responsible. It appears that even China's severe family
planning laws were not the cause. Actually, the answer seems to be super-
stition! 1991 was the Year of Goat, and many rural families deliberately
avoided starting families in a year considered to be unlucky. 1992, however, was the highly auspicious Year of Monkey. A bright, favorable future
can be forecast for babies born in such a year. Although the data have not
yet been sufficiently analyzed, many experts predicted a significant
increase in births in 1992.

It disturbs us when the normative system breaks down and we can't predict
behavior. This may explain the strange behavior of the classroom students

# OBEDIENCE AND DISOBEDIENCE TO AUTHORITY

Social psychologist Stanley Milgram conducted a fascinating series of experiments on what happens when people are faced with conflicting norms. A subject is told by the experimenter to teach a learner a task, to test the learner, and to punish the learner if the learner makes mistakes. The subject punishes the learner by administering electric shocks. Shocks are given by a special machine controlled by the subject. Shocks are increased in intensity (marked from "Slight Shock" to "Danger: Severe Shock") with each wrong answer. The learner continually makes mistakes, continually gets shocked, and pleads with the subject to stop the shocking: "I can't stand the pain," "I have a heart condition," "Get me out of here. . . . I refuse to go on!" then shrieks of agony, and finally silence. The subject is reluctant to keep shocking but the experimenter tells the subject to continue. In reality, the experimenter and the learner are working together; the shocks, and the cries and screams are fake, but the subject doesn't know this. The subject has a problem, a norm conflict: Should the subject follow the orders of the official-looking white-coated experimenter, or heed the cries of the poor suffering learner? What would you do?

It turns out that the subjects were much more obedient to the experimenter than Milgram had imagined they would be. He had guessed that few would go beyond "Strong Shock," and yet in practice, many subjects were willing to give the most extreme shocks available. Milgram then added conditions to see if the degree of obedience would change. In one condition (Remote Feedback) the learner was in another room and couldn't be seen or heard by the subject, but as the shocks got higher the learner pounded on the wall. In a second condition (Voice Feedback) the learner couldn't be seen but could be heard. In a third condition (Proximity) the learner was in the same room, about two feet away from the subject. In a fourth condition (Touch-Proximity) the learner had to put his hand on a shock-plate, and if he refused, the subject had to force his hand on the plate. As you might guess, the closer the subject was to the learner, the more trouble the subject had going through with the shockings. But even in the last two conditions, 30 to 40 percent of the subjects remained obedient. Milgram also tried moving the authority figure, the experimenter, out of the room to see if that would have an effect. Obedience dropped off tremendously with the experimenter gone, and many of those subjects who remained obedient, cheated; they gave lower shocks than they were supposed to.

What does it all mean? Milgram suggests several reasons for why the subjects were as obedient as they were. For some subjects, perhaps the shocking provides a release for aggressive feelings. From a sociological viewpoint, the subjects became entangled in a complicated social situation from which they had trouble escaping. There were conflicting norms: Who should be heeded, the authority figure or the person in pain? Milgram somewhat pessimistically concludes, "A substantial proportion of people do what they are told to do, irrespective of the content of the act and without limitations of conscience, as long as they perceive that the command comes from a legitimate authority."

*Excerpt from "Some Conditions of Obedience and Disobedience to Authority,"* Human Relations, *February 1965, pp. 57–76.*

described in the hypothetical example that began this section. We have norms that govern the student-teacher relationship. These norms are so obvious and ingrained that we do not even think about them. Characteristically, norms do not become apparent until they are violated. According to norms concerning student-teacher relationships, the student will show respect for the teacher, will laugh at his or her jokes, and will generally obey the authority that the teacher represents; that is, students will respect any reasonable demands made by the teacher without resorting to mutiny. When, as in our previous example, behavior is contrary to the norms, the situation becomes uncomfortable for others. Instructors who have tried this "stand up" experiment on classes (with the aid of a willing villain) report that the experience is almost as hard on the two of them (instructor and villain), who know it's a fake, as it is for the rest of the class, who don't know that it has been set up.

Other experiments in norm violation have been devised.[2] For example, get somebody you don't know very well in a game of ticktacktoe. Invite the other person to make the first move. After he or she makes a mark, erase it, move it to another square, and make your own mark. Act as if nothing unusual has happened. Or select a person (not a family member or very close friend) with whom you will engage in an ordinary conversation. During the course of the conversation, without indicating that anything unusual is happening, bring your face closer to the person's until your noses are almost touching. The first experiment deals with norms concerning game playing, and the second with norms concerning spatial invasion, or the appropriate distance between people. In each case, the subject will react noticeably, and possibly unpredictably, because common norms that ordinarily one doesn't even think about are suddenly being violated.

Although it is rare, norms may break down. Perhaps during wars or disasters or other periods of great and sudden change, norms may cease to be effective at controlling people's behavior. When this state of "normlessness" is present, it can be very disorienting—one tends to lose all sense of stability, security, and orderlinesss, because you can no longer predict other people's behavior.

## Types of Norms

Norm strength varies greatly. One way of determining the relative strength of norms is by the sanctions that the norms carry. A **sanction** is the punishment one receives for violation of a norm, or the reward one receives for correct norm performance. Sanctions take a variety of forms: a look on someone's face, an A on an exam, a sharp word from a spouse or boss, a kind word from a parent, a traffic ticket from a police officer, a life sentence in a penitentiary. The norms that are most important to a society, that tend to be obeyed without question, and that have harsh sanctions if they are violated are called **mores.** In our society, norms dealing with taking another person's life, with eating human flesh, and with sexual activity with one's parents are examples of mores. Mores are often traditional norms that are a part of the customs of a society—"things have always been that way." When a society feels it necessary, perhaps because increasing violations mean that the informal sanctions aren't working, mores can be translated into written law. Whether written or unwritten, the emotional force of mores is strong; they still represent the "musts" of behavior.

We also have norms dealing with what we *should* do, rather than what we *must* do. These norms are less obligatory than mores, and the sanctions for violation are milder in degree. People may look at us rather strangely if we violate these norms, but they probably will not lock us up or banish us. These norms are called **folkways.** Practices such as shaking hands when meeting someone or norms regulating the type of clothes one wears would be considered folkways. A student who comes to class in a bathing suit and golf shoes would probably be allowed to stay in spite of the obvious violation of the folkways. At the same time, there would likely be ample private discussion of the student's character and intelligence.

A further distinction can be made between ideal norms and real norms. **Ideal norms** refer to what people agree *should* be done. **Real norms** refer to what they *actually* do. Ideal norms indicate that cheating on tests, extramarital sexual intercourse, and falsifying one's income-tax returns are wrong. What would observation of actual behavior—real norms—tell us? It would tell us that behavior doesn't always follow stated norms.

## Norm Conflict and Change

As we said earlier, norms develop from a set of beliefs and values people have about the way things should happen. Norms probably develop first at the group level when specific friendship groups, clubs, work groups, clans, organizations, or communities of people establish rules. Often these rules are informal and unwritten. Sometimes, if the people involved have sufficient power, their rules may become written down in the form of laws.

Conflict over norms is a common occurrence. There are several reasons for this. For one, different groups see things differently. Norms of the consumer may conflict with the norms of the manufacturer. Abortion rights and Right-to-Life groups find few norms on which they agree. Conflicting norms in the areas of drugs and of sexual conduct seem to be the rule rather than the exception. Narcotics laws in many areas severely punish possession and sale of certain drugs, and yet Californians recently passed a law legalizing the use of marijuana for medicinal purposes. Oregonians passed a bill legalizing assisted suicide, but some people disagree strongly and the federal government is threatening to step in and enact legislation that overrules the Oregon law.

Another reason for norm conflict is that norms constantly change, more rapidly in some groups than in others. Conflict between parents and teenagers (over language, clothes, behavior) is probably one example of this. Teenagers want their parents to get with it, while parents think to themselves, that isn't the way we did it when I was your age. Changing norms can be seen in my sociology department. Old-timers have been slow to adapt to the computer age, wonder why they need a password, and tell me that their typewriter never crashed! Younger faculty are so computer literate, it's scary—they are comfortable with the Internet, use their e-mail more than their phone, and know what http means.

Society's laws most often reflect the norms—they are usually "in synch." Sometimes they are not, however. Perhaps change in norms is so fast that change in laws can't keep up. (Again, laws reflecting norms dealing with drugs and sexual behavior might be examples.) Or, a group holding a minority viewpoint may somehow get enough clout to get a law passed supporting their view. Prohibition in the 1920s and 1930s comes to mind. One way a society handles the problem of out-of-date laws or laws reflecting a minority viewpoint is to not enforce such laws.

## CULTURAL DIVERSITY

When we compare different cultures we find similarities and differences. All cultures seem to have some form of family arrangement, and they all have some system for educating their young. I suppose that cultures are more similar than they are different. But the differences attract our attention and are often fascinating. Why are there differences? Several factors are probably at work. Cultural patterns can reflect responses to a harsh or benevolent physical environment (very wet or dry, agriculturally rich or poor, desert or mountain). Cultural patterns can be affected by accidents or variations in history—revolutions, wars, overpopulation, starvation, inventions, stage of development of neighbors, and so on. Finally, and perhaps most important, cultural patterns are a consequence of the almost random variation that emerges out of human creativity. As we look at a few examples of cultural diversity, try to figure out what produced these patterns.

Every four years, writers from England and Europe flock to the United States to observe a very peculiar ritual the like of which is unknown throughout the rest of the civilized world. These strange ceremonies are written up and read about by disbelieving audiences everywhere. The tribal celebrations being enacted are called locally the "American political conventions." Strange indeed.

In November 1995, the National Grid, the company that manages England's electric transmission system, announced that it was preparing for an anticipated major power surge. It seems that Princess Diana was to be interviewed on the BBC on the state of her marriage. A year earlier when Prince Charles was interviewed on the same subject, a 700-megawatt surge occurred after the broadcast as citizens all over the country turned on their tea kettles.

Anthropologists specialize in the study of cultural patterns in different societies, especially in primitive or preliterate societies, and in their studies they have made many discoveries that seem strange to Americans. They describe a society in which very fat women are highly regarded. Women in this society spend weeks in fattening sheds where they eat starchy, fatty foods and have their bodies greased to make them more attractive. On festival

## CHOICES

Diarrhea kills 3.1 million people annually, almost all of them children. Usha Bhagwani, the mother of two boys, lives in a one-room hovel in Thane, India. Her boys play barefoot in muddy fields, squatting and relieving themselves as the need arises, casual about the filth around them. The water that the village families drink comes from cracked pipes that run in a ditch filled with sewage. Sewage seeps into the water and produces diarrhea. The water has already killed two of her children, and Mrs. Bhagwani is worried about her 5- and 7-year-old boys. She has very limited money and she frets about which choice to make: Should she buy food so that the boys will get stronger? Or should she buy shoes so the boys will not get hookworms? Or should she send them to school? Or should she buy kerosene to boil the water?

—New York Times, *January 9, 1997*

day they are paraded before the king, who chooses the fattest and heaviest as his mate. Very peculiar people. But imagine for a moment that you have to explain to a member of that society the popularity in America of the various dieting and health spas where men and women spend great sums of money to lose weight.

Some Indian tribes living along the Amazon have an interesting reaction to childbirth. The woman breaks off from work in the fields and returns home for only two or three hours to give birth to the child. Meanwhile, her husband has been lying at home in a hammock, tossing about and groaning as if in great pain. Even after the birth when the woman has returned to the fields, the husband remains in bed with the baby to recuperate from his ordeal.

The first research question at the beginning of this chapter described a curious circumstance concerning the gender ratio in China—many more boy babies or many fewer girl babies were born during the 1980s than should have been the case. Demographers (those who study population patterns) are still working on this one, but the answer seems to be related to two factors. In 1979 the Chinese government, in attempts to control rapid population growth, introduced a strict policy limiting Chinese families to one child. The second factor, clearly a cultural one, was the desire of Chinese parents to have a boy. In that culture, a male baby is much preferred to a female baby. If the family was allowed to have only one child, it was very important that it be a boy. Birth statistics and the peculiar gender ratio implied that between 1980 and 1987 about 2.5 million female births were never reported. The experts first concluded that because boy babies were desired, girl babies were killed (female infanticide) or their births somehow hidden and not recorded. Perhaps that's what happened, but Swedish statistician Sten Johansson thinks there is another answer. He studied adoption records and found that the vast majority of children adopted during that time were girls. He concludes that more than 1 million girls were, in some way, left by their natural parents and adopted, some secretly and some openly, by foster parents. By the way, is Chinese culture alone in its preference for boy babies?

## Ethnocentrism and Cultural Relativism

All cultures differ to some extent, and yet because we are so used to our own, we often forget that basic fact. The tourist is reminded of this even in countries similar to our own when, for example, he or she encounters the English motorist driving on the left side of the road, the German driver approaching at a very high speed with lights flashing, Volkswagens and Porsches used as police cars, Spanish restaurants not opening for dinner until after 9:30 in the evening, French children drinking wine, or English police officers not carrying guns. The failure to anticipate and plan for cultural differences is common. Professor David Ricks has found that companies doing business in foreign countries make the same mistake. Chevrolet had trouble selling Novas in Latin American markets: *"No va"* in Spanish means "does not go." Pepsodent's promise of white teeth did not go over well in a part of Southeast Asia where black, discolored teeth are a sign of prestige. "Body by Fisher" translated to "Corpse by Fisher" in a Belgian ad. "Come alive with Pepsi" in a German translation became "Come alive out of the grave." Some firms forgot when labeling their products that green is the color of disease in Africa and white the color of death in Japan. And a firm had difficulty selling a refrigerator to the mostly Moslem Middle East when it used an ad picturing the refrigerator full of food, including a giant ham on the middle shelf. Some examples of odd notices in non-English-speaking countries include a Hong Kong dentist—"Teeth extracted by the latest Methodists"; a Greek tailor shop—"Order your summers suit. Because is big rush we will execute customers in strict rotation"; a Mexican hotel—"The manager has personally passed all the water served here."

Ethnocentrism describes a type of prejudice that says simply, my culture's ways are right and other cultures' ways, if they are not like mine, are wrong. Racism (a particular race is superior to others) and sexism (one sex is superior to another) are related to ethnocentrism but are more specific types of prejudice. Informal practices and formal policies (who gets hired, who gets paid most) can emerge to support racism, sexism, and ethnocentrism, as well as other types of prejudice. The ethnocentric person says that the familiar is good and the unfamiliar or foreign is bad. An ethnocentric person in the United States might maintain, among other things, that non-Christians are barbarians; that Inuit tribes practicing sexual hospitality are totally lacking in moral fiber; that anybody who eats dogmeat, horsemeat, or human flesh is not civilized; that democracy is the only way of government; and generally that we are doing other cultures a favor when we go in and Americanize them.

Ethnocentrism is, to some degree, difficult to avoid. Informally, while interacting with family and friends, and formally through a kind of indoctrination within the educational system, we are frequently taught that our ways are best and, at least by implication, the ways of others are less good. The mass media encourage ethnocentrism by treating the foreign and unfamiliar as easily recognized stereotypes. It is probably true that ethnocentrism is impossible to escape, for it is encouraged in one way or another by most societal institutions, such as the family, the church, the schools, and the government. This is at least partly due to the positive functions of ethnocentrism. Ethnocentrism probably leads to greater group solidarity, loyalty, and patriotism, and a certain degree of ethnocentrism may be essential for the survival of a culture. The effects of ethnocentrism are complicated: As we reinforce our belief in the goodness of our own ways, we make unfair and often derogatory judgments about the beliefs of others.

Even social scientists (who know better) sometimes run into difficulties when studying other cultures. Because most social scientists are white middle-class representatives of the dominant culture, they may have a tendency to describe minority cultures that are different—Amish, Inuit, delinquent gangs—from the viewpoint of their own value system rather than from the viewpoint of the people they are studying. As you come across descriptions of research in this book and elsewhere, see if the research seems to have been influenced by ethnocentrism. See if you can detect ethnocentrism in the choice of topic, in the way the research is done, or in the interpretation of results.

Related to ethnocentrism but opposite in meaning is the concept of **cultural relativism.** Cultural relativism suggests that each culture be judged from its own viewpoint without imposing outside standards of judgment. Behaviors, values, and beliefs are relative to the culture in which they appear. The cultural relativist believes that what is right in one society may be wrong in another and that what is considered civilized in one society may be seen as barbaric in another; however, the relativist believes basically that judgments should not be made about the "goodness" or "badness" of traits in cultures other than in one's own.

Recently, Robert Edgerton has raised some interesting questions that have broadened the ethnocentrism/cultural relativism debate. Edgerton feels that cultural relativism has been fashionable among social scientists and has led them to make some mistakes in analysis. Two assumptions that stem from cultural relativism bother him: (1) that primitive societies were more harmonious and better adapted to their environment than are larger more urbanized societies; and (2) that a society's long-standing beliefs and practices must play a positive role or these beliefs and practices would not have lasted.

Edgerton believes that it is possible to critically evaluate other societies without being ethnocentric. For example, he challenges the first assumption by citing research that shows that modern urban societies have done better than many primitive societies at feeding their populations, maintaining their health and quality of life (lower homicide rates than many primitive societies, much longer life expectancy than most primitive societies). He challenges the second assumption by citing such practices as cannibalism, torture, infanticide, feuding, witchcraft, painful female genital mutilation, ceremonial rape, and head hunting and wondering whether these practices serve beneficial or positive functions. Edgerton's point is that social scientists should be objective and critically evaluate—look at a society's beliefs and practices and determine if they are beneficial and adaptive (serve some useful purpose), or harmful and endanger people's health, happiness, or survival. We should recognize that some beliefs and behaviors serve human needs and social constraints better than others.[4]

## Subculture and Counterculture

Most societies, especially large, complex societies like the United States, have groups that, by their traits, beliefs, or interests, are somewhat separated and distinct from the rest of society. Such groups may share many of the characteristics of the dominant culture, but they have some of their own specific customs as well. If these groups have definite boundaries and if their differences from the rest of society have some permanence, they are called **subcultures.** Sociologists generally use the term *subculture* to refer to groups that stand out in that some of their values and customs are different from or even at odds with those of the rest of society. The sociologist asks such questions as, How do the subculture's values differ from those of the dominant culture? Is there conflict? Does the dominant culture attempt to change the subculture? What does the subculture do to maintain its separate identity?

Some religious groups, the Amish for example, seem to qualify as subcultures. A major problem for the Amish, as for many subcultures, is to maintain their identity, even their existence, in the face of a dominant culture frequently hostile to their beliefs. The opposition of the Amish to electricity, which means no lights at night on their horse-drawn carriages, and their opposition to any formal education beyond the eighth grade has brought well-publicized confrontations with the authorities in several midwestern states. Other examples of subcultures might include terrorist organizations, college students, Hare Krishnas, Chicanos, the Hutterites, professional baseball or football players, and sociologists.

The line of distinction between culture and subculture is not always clear. Some believe that over a period of time societies go through a melting-pot experience in which different nationality, racial, and interest groups become so mixed and merged together that their subcultural differences cease to exist. The result of the melting-pot experience would be a society representing a mix of the remnants of former subcultures.

There is an element of conflict in some subcultural behavior, and this has led to the use of the term **counterculture.** The central element in a counterculture is opposition to, or conflict with, certain norms and values of the dominant culture. Examples of countercultures might include youth gangs, motorcycle gangs, revolutionary groups, and terrorist organizations. *Subculture* is the more general term, *counterculture* the more specific. Countercultures are subcultures, but only those subcultures that

## AMISH GO WILD FOR SKATING

The Old Order Amish may not drive a car, ride a motorcycle or use a bicycle because of their religious beliefs. A horse and buggy is the typical mode of transportation. However, the *New York Times* reported in 1996 that, for young people at least, a break through had occurred: in-line skates. Hundreds of young Amish have taken up in-line skating, reported to be much faster than a buggy and providing greater freedom. Change is slow among the Amish and something new must fit in with their way of life. In-line skates are permissible because they are seen as a new version of roller skates, a cousin of the ice skate, and an improvement over the leg-powered scooter, all of which have long been used by the Amish.

include the element of opposition to the dominant culture could be called countercultures.

### Multiculturalism

Many societies, especially large complex societies like the United States, have a complicated mix of nationalities, cultures, subcultures, and counter-cultures. Do these separate groups mix and merge, ultimately losing their individual identity as in a melting pot? Or do they keep their distinctiveness and exist as separate groups or subcultures within the larger society, perhaps like a mosaic? More important, what *should* societies encourage and support—should the separate groups be expected to mix and merge, or should they be encouraged to keep their individuality?

Since the late 1980s, the United States has been confronting these issues in the debate over multiculturalism. **Multiculturalism** means acknowledging a society's cultural diversity by formally treating both genders and all racial and ethnic groups equally in the educational and political process. The focus is on public institutions, governmental policy, and perhaps most visibly, schools, colleges, and universities. Some examples of attempts to establish and encourage multiculturalism have had the following results: In Canada, concerns about protecting and promoting French distinctness has meant requirements for using the French language in Quebec. In the United States, the groups most often identified are African Americans, Asian Americans, Hispanic Americans, Native Americans, and women. In April 1993, after administrative resistance to establishing a separate Chicano studies department at the University of California at Los Angeles, a campus demonstration led to the establishment of the Cesar Chavez Center for Interdisciplinary Instruction in Chicana and Chicano Studies. The University of California at Berkeley now requires all undergraduates, whatever their ethnicity or major, to study at least three out of five cultural groups: Asians, Hispanic Americans, Native Americans, African Americans,

## MARGARET MEAD (1901–1978)

Margaret Mead earned her Ph.D. in Anthropology from Columbia University. Her work as a cultural anthropologist contributed greatly to the study of culture and socialization. While not denying the importance of biology and the natural environment, Mead demonstrated the central role of culture in studying human behavior and attitudes through her fieldwork in which she lived in many different, often primitive, societies. Her 1928 book, *Coming* *of Age in Somoa*, provoked a great and heated debate among sociocultural anthropologists regarding the proper methods and interpretation of field research. Mead also studied many contemporary problems including childcare, sexual behavior, adolescence, and American character and culture. In 1969, she was appointed full professor and head of the social sciences department at Fordham University.

and Europeans. The explicit goal: to move away from an "Anglocentric" curriculum toward one that validates other cultures as equally and essentially American. Stanford University replaced its Western Culture requirement with one called "Culture, Ideas, and Values," which added study of works of some non-European cultures and works by women, blacks, Hispanic Americans, Asians, and Native Americans. In late 1996, the Oakland, California, school board proposed that Ebonics (from ebony and phonics) be recognized as a distinct language. The school board hoped that recognition of and respect for "black english" would help improve the academic performance of black students in the Oakland school system.

Multiculturalism is controversial, ruffling feathers on all sides. Does it bring people together or is it divisive? Should society encourage the *melting pot* in which individual cultural differences eventually disappear, or the *mosaic* in which distinct cultural identities are maintained? Is there a core (language, common set of beliefs) to which all must adhere, or is the strength of a society gained through its diversity? Does multiculturalism help by encouraging identity and cultural awareness, or hinder by encouraging exclusiveness and separation? Culturally diverse societies like the United States that are committed, at least in principle, to equal representation of all must find an answer, a balance, that works.[5]

### Culture and Personality

Personality and temperament vary from one culture to another. In her book *Sex and Temperament*, Margaret Mead describes what she found when she visited three primitive societies in New Guinea. First, Mead describes the mountain-dwelling Arapesh, among whom both the men and women behave in a way that Americans would describe as maternal or feminine. Both parents devote their lives to raising the children. The men are gentle, and there is complete cooperation between both sexes at all times. Next,

Mead describes a cannibalistic tribe, the Mundugumor, living along a river. Here both the men and women behave in a way that Americans would describe as masculine. Men and women work in the fields together and are aggressive individualists. Finally, Mead describes the lake-dwelling Tchambuli. In this society, the men behave in a manner that Americans would define as feminine or maternal, and the women are masculine by our standards. The women spend the days fishing and weaving, and they have all the power. The men spend their time dancing in ceremonies, dressing and making themselves up, and engaging in artistic endeavors. The men gossip, quarrel, and become very jealous of each other over the affections of a woman. The women's attitude toward the men is one of kindly tolerance and appreciation; they watch the shows that the men put on.

We tend to believe that some patterns of behavior and temperament are automatically, necessarily related to gender. We might believe that to be male is to behave naturally in a certain way, and that female behavior is innate as well. The results of Mead's research suggest that the relationship between gender and the corresponding behavior and temperament is not necessarily biological, but is determined by the culture in which one lives—as are preferences for body shape and size, taste in food or art, belief in a supreme being, the nature of recreational activity, and many other characteristics.

David Riesman in *The Lonely Crowd* describes the effect of culture on personality in a somewhat different manner. His subject is American character, and he describes how character and personality change as other aspects within the culture change. According to Riesman, three types of social character have been dominant in American society. In earlier years the dominant type of character was *tradition-directed*. In a tradition-directed culture, behavior is carefully controlled. Routine orients and occupies the lives of everyone. Ritual, religion, and custom are dominant. New solutions are not sought, and change is very slow. Later, the *inner-directed* type of character appeared. Inner-directed people are taught early in life to have an inward focus, with emphasis on the self and its needs and gratifications. Other people are not of crucial importance. Individuals might be internally driven toward such ideals as power and wealth; they are encouraged to set their own goals and to be on their own. Their lives are concerned with self-mastery and accomplishment. Finally, and more recently, the *other-directed* type of character appeared. The chief interest for other-directed people is to be liked by other people. According to Riesman, other-directed people have built-in radar systems that search out the reactions and feelings of others so that they may adapt themselves to them. These people are more concerned with conformity, are shallower, friendlier, more unsure of themselves, and more demanding of approval from others. The peer group is all-important, as is the front that one puts up. Riesman believes that these character types result from other changes within the culture, such as population growth or change, changes in economic, industrial, and agricultural techniques, and urbanization. This major point applies to all of us: The culture in which we live plays an important part in determining who we are and how we behave, including patterns of personality and character.

## THE HIDDEN DIMENSION

Cultures differ in many ways, even in use of space. Anthropologist Edward T. Hall studied this topic and found that private or personal space was very important to Germans. Yards tend to be well fenced and doors thick, substantial, and usually closed in contrast to Americans whose doors are flimsy and often open. Americans move their chairs around to get the best view, whereas in Germany, it is a violation of the norms to move your chair. The English speak softly so only the person to whom they are speaking can hear.

They see Americans as speaking loudly and not caring who hears them. To Americans, walls of a house are fixed, but to Japanese, the walls of their homes are movable and the rooms are multipurpose. Pushing and shoving in public places is acceptable in Middle Eastern cultures but seen as rude by Americans. Arabs don't mind being crowded by people (and breathing on others while talking), but they hate to be hemmed in by walls. Ideally, homes should have few partitions, very high ceilings, and an unobstructed view.

American males have one of the highest heart-disease rates in the world. By contrast, Japanese males have one of the lowest rates. Obviously a cultural factor is working here, but which cultural factor is it? For years scientists thought the explanation to be diet. The Japanese diet of fish and rice is much lower in cholesterol than is the American diet of meat and dairy products. However, a nine-year study completed in 1975 of 4,000 Japanese Americans living in this country found another cultural factor to be the culprit.[6] The key turned out to be *lifestyle*. The Japanese living in the United States who maintained their traditional lifestyle—downplaying individual competition and accepting their place in family and society—had low heart-attack rates *regardless* of what they ate, how much they smoked, their blood pressure, or their weight. Those Japanese who adopted the American lifestyle (or personality) and became impatient, aggressive, hard-driving, competitive go-getters were five times as likely to have heart attacks as those who maintained Japanese ways. The Japanese culture is apparently better able to protect the individual against the effects of pressure and stress. Unhappily, then, one's culture dictates not only how one will live, but also how and why one will die.

## SOCIETY AND SOCIAL STRUCTURE

We learned in Chapter One that sociologists study the origin, development, organization, and functioning of human society. In its most general meaning, the term *society* refers to human association and the existence of social relationships. This is too general for our use, however, because it suggests that any set of people interacting could be called a society. A more precise definition states that a **society** is a continuing number of people living in a specific area who are relatively organized, self-sufficient, and independent

## THE AMERICANIZATION OF GEORGE

America is coming between me and my 12-year-old son. Actually, it started earlier, when George was hardly 10. Since then, I have helplessly watched this incursion, often with dismay and alarm.

Over two years ago we escaped the horrors of Lebanon, and chose quiet Princeton over the vibrancy of Harvard. . . . But little did I know that he was to face, at such an early age, the more subtle "terror" of American peer pressure, tantalizing communications media, and the unsettling dissonance of conflicting norms and expectations.

The family system in Lebanon is, on the whole, intimate, warm, and affectionate. A child there grows up in a nurturing atmosphere of extended kinship networks sustained by filial piety and mutual obligations.

The Lebanese, much like adjacent Mediterranean cultures, are very tactile. Touching, kissing, hugging, and the outward display of emotion—regardless of gender—are generously and spontaneously expressed. At least children of George's age indulge in these emotive expressions with little self-consciousness or feelings of shame or guilt.

I feel resentful that George should be disarmed of such harmless but reassuring expressions. I first noticed this transformation (or deformation) upon returning from a brief trip a few months after we had settled in Princeton.

Normally, even after the regular daily return home from work, George would interrupt his play and rush across the driveway to greet me; often, he would literally hurl himself into my open arms.

On that day, however, just as he was about to heed his normal impulse as he rushed across the driveway, he suddenly "froze" in mid-passage, looked in the direction of his watchful playmates, and with obvious hesitation and embarrassment, calmly walked over to greet me with a cold handshake and a casual "Hi, Dad." Bit by bit, even this gesture has been abandoned.

Such "frozen" moments have recurred and spilled over to other daily encounters with members of the family, and in particular acquaintances from Lebanon. I could see him fret as relatives and friends he has not seen for two years try, in vain, to solicit a hug or a kiss on the forehead. The reluctant denial has been transformed into a boast, that he is now an "American boy."

His "Americanization" was most forcefully conveyed by a recent incident on the tennis court. We were struggling in a doubles game against two other, more seasoned partners who normally beat us. After a long and heated game we won the set, partly because of two exquisite shots by George. He was ecstatic. As he rushed across to share his exuberance with me Lebanese style, he "froze" once again and treated me to a tamed version of the American "high five."

*Samir Khalaf,* The Christian Science Monitor, *September 22, 1987. Reprinted by permission of the author.*

and share a common culture. *Continuing* means that there is some permanence to the society. The number of people can vary greatly: American society numbers more than 260 million people, whereas other societies number fewer than 1,000 people. By *organized* we mean that there is some systematization and structure to the patterns of social interaction. A society occupies an area; it has boundaries that separate it from other societies. These boundaries—an ocean, a river, a mountain, or even a line drawn on a map—are not only *geographical* but *social* as well; the vast majority of the social interactions of a society take place within a particular area. Finally, each society, because of its particular cultural beliefs, habits, and traditions, has certain unique characteristics that distinguish it from other societies.

As you can see, several terms in our definition involve matters of degree. There is often much interaction among societies, and probably no society is totally independent of others. Although interpretation will always be necessary, this definition should give us a reasonably good idea of what is meant by society. However, other uses of the term don't fit our definition. Special-interest groups such as professional organizations (Society for the Study of Social Problems), social clubs (Second Street Skateboard Society), and other groups are *not* societies in the sociological sense. And whatever those people have whose pictures appear in the "society" pages of the newspaper, isn't a society. None of these groups alone constitutes a society; all of them are *part* of a society.

How do sociologists study human society? Let's suggest for a moment that any given society has a form, a composition, a structure. Imagine society as a building with many parts—wood, steel, bricks, and concrete—all somehow attached together in a coherent structure. If you look at the end product you'll see a five-story building, but thousands of elements actually make up the whole. Yet rather than a group of random parts, the building is a network, a structure. Now keep that image of society as a building: The many parts are people, and they are held together by the ways they relate to each other, their relationships to each other. The people of society, just like the steel and concrete of the building, are held together in very specific ways.

Just as we can take the five-story building apart to examine its elements, so can we take apart and examine society's elements. As we look at the ways people relate to each other, we find that these ways fit consistent patterns. The same important elements seem to appear again and again. **Social structure** refers to the network of ways people relate to each other in society. In their study of human society, sociologists concentrate on the important elements of social structure: status, role, group, and institution. Status and role will be discussed in this chapter; group and institution will be the focus of later chapters.

## STATUS AND ROLE

When students and an instructor walk into class the first day of the term, they know without thinking what to expect of each other and how each will behave. They know these things even though they have not seen each other before. The students know that the instructor will stand in front of the class behind a lectern, probably call roll, assign reading, and dismiss them early the first day. The instructor knows that, unless the class is required, students will be shopping around. They will be trying to decide whether to take this class, and their decision will be based on course content, the viewpoint and personality of the instructor, the amount and type of work required, and how the instructor is known to grade.

We know these things about each other partly because of the system of norms discussed previously. The concepts of *status* and *role* are closely related to norms, and they play a major part in the situation just described. People typically use *status* to refer to one's rank in society—"she's a neurosurgeon, which means she has high status and a Mercedes Benz. . . ." This usage is not incorrect, but when sociologists are discussing status and role we use the term *status* in a slightly different way. By **status** we mean a position in society or in a group. There are innumerable positions one can occupy: teacher, student, police officer, president, football player, father, wife, convict. Furthermore, each of us can occupy several positions at once: teacher, handball player, father, husband, and so on. By **role** we mean the behavior of one who occupies a particular status. As Robert Bierstedt puts it, a role is what an individual *does* in the status he or she occupies; statuses are occupied, roles are played.[7]

A set of norms surrounds each status and role. These norms, called **role requirements,** describe the behavior expected of people holding a particular position in society. Recalling our earlier example, the behavior of the student who refused to stand up was disturbing because it was unpredictable. He was occupying the status or position of student, but the role he played—his behavior—was contrary to the expected behavior of a person in that status. His behavior was outside the limits set by the norms, or role requirements.

Within the boundaries set by the role requirements, there is often extensive variation in how a role is played. On a football team, status would refer to the positions, role to the behavior of the incumbent of the position. One status would be quarterback. Role requirements of quarterbacks are generally to call the plays, to direct the team, and to try to move the ball down the field. But now look at the actual performance of several quarterbacks. One passes frequently, another seldom passes but often runs with the ball, and a third does neither but usually blocks. Compare four or five of your instructors in their role behavior. Although all occupy the status of college professor, no doubt their behavior varies markedly. One paces the floor; another stays behind the lectern while lecturing. One demands class discussion; the next dislikes having lectures interrupted. One has beautifully organized and prepared lectures; another has a disorganized, stream-of-consciousness presentation put together on the way to class. Or compare the behavior of the last three presidents of the United States. Again we see marked role differences within a given status. These differences in behavior obviously occur because people holding the same status define the role differently. This should sound familiar because it's essentially the idea we had in mind in discussing definition of the situation in Chapter Two. As a result of a particular pattern of socialization, each individual defines status and role in a particular way. Each individual brings to the situation a specific personality and a set of skills, interests, and abilities. Therefore, although each status carries with it certain role requirements, there is still variation and flexibility in actual behavior.

Role—behavior that is suited to a particular status—varies not only because of the style of a particular quarterback, college professor, or president; roles must also be seen in an *interaction* setting. While behaving, people are always socially interacting with others, and consequently their behavior adjusts to and is modified by the responses of others. Socialization and self-development occur through the process of social interaction. Roles we play are shaped by others' reactions to us. For example, after comments and complaints from an unhappy class, the unprepared, stream-of-consciousness professor mentioned earlier may modify his or her performance. It is a common occurrence to go into a situation prepared for one sort of role only to find that, in the process of the interaction, another sort of behavior is necessary.

Sociologist Erving Goffman analyzed people's behavior by likening it to being performers in a play. Each of us plays numerous roles in our everyday life, and in each of these roles we try to present our self in a particular and convincing way. There are several aspects to our performance, according to Goffman. There is the "front stage" where we play out a role for an audience. The teacher's front stage behavior takes place in a classroom in front of students, the coach on a basketball court or a football field in front of players, the doctor in an examining room in front of patients. There is a "setting," which involves clothes, furniture, props, and other background items that help make our front stage performance convincing. The teacher has a blackboard or an overhead projector; the coach has a whistle, clipboard, and particular set of clothes depending on whether it's practice day or game day; the doctor has a stethoscope, diplomas on the wall, and a receptionist to keep people out. There is also a "back stage" for after the performance is over. Now we are no longer trying to convince an audience—we can relax. Back stage behavior often contradicts the front stage image—the audience would be disappointed if they knew. The teacher gets home and can't answer the simplest question his child asks, or the coach who is such an authority figure on the field is a wimp with her friends, or the doctor doesn't look as God-like when in sloppy clothes and having trouble removing a splinter. Goffman's analysis is helpful in reminding us how we try to "manage" social interaction, and how we try to project the image of ourselves that we want others to believe.[8]

The second research question at the beginning of this chapter compared people who had many different jobs, activities, or positions in society with those who were more focused on one. We can now see that we were talking about roles—about people who are involved in many roles compared with those who concentrate more on one. Would the number of roles one is involved in have any effect on life span? A recent study says yes. A number of women were randomly selected and interviewed in 1956 and again in 1986. Those who actively "played" many roles, such as worker, church member, friend, neighbor, relative, and club or organization member, were compared with those who focused more on one role. The study found that those involved in many roles lived significantly longer. Of the various roles studied, active participation in clubs or organizations seemed to have the greatest effect on longevity. The scientists involved in the study concluded that role enhancement—being involved in multiple roles—seems to enhance health. This supports earlier studies that have shown that integration into society by "playing" multiple roles reduces the likelihood of psychological distress among both men and women.[9]

### Achieved and Ascribed Status

How do we happen to occupy the statuses that we do? Some, probably most, statuses are earned or achieved in some way, and hence these are called **achieved statuses**. The astronaut, police officer, college professor, and truck driver represent achieved statuses. Some statuses are automatically conferred on us with no effort or choice on our part. These are **ascribed statuses.** One's gender, race, and nationality are ascribed (although occasionally some changes can be made). Sometimes it is difficult to tell whether a status is ascribed or achieved. Consider the student who feels forced to go to college because of the wishes of his or her parents: Is the status of student ascribed or achieved? Or how about the statuses a child inherits from his or her parents, such as political and religious affiliations: Are they ascribed or achieved?

Statuses are stratified, or ranked, at several levels. Some statuses are of high rank and bring much prestige to the occupant. The doctor, board director of a large corporation, college president, author, artist, scientist, and movie star represent statuses that have high value and prestige in our society. An evaluation of position is usually determined by the requirements one must have to fill that status: extensive education, wealth, beauty, skill, or some other extraordinary characteristic. Sometimes this ranking is based on a societal tradition that automatically ranks certain characteristics above others, such as a particular gender, race, religion, or aristocratic affiliation. Sometimes having high ascribed status makes it easier to obtain desired achieved statuses. For example, a child born in a middle-class family (ascribed status) will have a better chance of becoming a doctor or scientist (achieved status) than will a lower-class child. Sometimes having high ascribed status is limiting, however. People who are members of royalty may find that their freedom is severely restricted. Living up to their ascribed statuses can mean declining to pursue more attractive achieved statuses.

### Role Strain and Role Conflict

Problems may occur when a person must play several roles simultaneously or when one role requires a person to perform in several different ways. These situations are called role conflict and role strain, and they can lead to personal stress and discomfort. **Role strain** refers to the situation in which there are differing and conflicting expectations regarding one's status or position. A student may experience role strain when he compares the expectations of his parents (to study, get As, prepare for a vocation) with the expectations of his fraternity brothers (to be social, to be active in fraternity affairs, to be athletic). Police officers, who are trained to arrest people who have committed crimes, probably feel role strain when they are ordered by superiors who do not want to make a bad situation worse to stand by and watch looting take place during a riot. A typical situation on college campuses leads to role strain for young professors: Their students expect them to be good teachers, but the school tells them that keeping their jobs and getting promoted will depend on how many articles and books they publish. Doing one takes valuable time from the other—what to do?

**Role conflict** occurs when a person occupies several statuses or positions that have contradictory role requirements. Here there is no confusion or disagreement about the requirements of a single role, as there was with role strain. The requirements of the roles are clearly understood; the problem is that the requirements of two or more roles are contradictory. Imagine the dilemma of a police officer invited by friends to a party where marijuana is being smoked. Police officers are trained to respond to violations of the law with the authority of their position, both on and off duty. But the officer is a normal citizen who is expected by friends—and who wants—to behave like everyone else at the party and have fun. The requirements of the two roles are clearly contradictory. Doctors seldom treat members of their own families because of the role conflict that can occur. Similarly, the football coach whose son is trying out for the team experiences conflict between the contradictory requirements of two different roles: coach and father.

## THEORY AND RESEARCH: A REVIEW

This chapter primarily introduces and illustrates a series of concepts: society, social structure, norm, status, role, and culture. Of the three theoretical perspectives introduced in Chapter One, the discussions in this chapter are most influenced by *functional analysis*. Specifically, the discussion of society and social structure and the sections on norms, status, and role, and much of the rest of the chapter, revolve around the ideas of stability and unity while focusing on the structure and the integration of the parts of society. All these elements are characteristic of functional analysis. Studies of the differing interests that lead to norm conflict (car owners versus manufacturers) and of the characteristics of countercultures turn more in the direction of *conflict theory*.

Of the research efforts described in this chapter, the comparative study of Japanese and American incidence of heart disease was *a nonreactive survey using records*. The example of the recalcitrant student who wouldn't stand up and Milgram's study on administering shocks are *experiments*. Milgram's study is interesting in that, by varying selected conditions in his experiment (for example, distance between subject and learner, and presence or absence of experimenter), he introduced certain *controls* that sharpened his findings. Mead's research on primitive societies is an example of *participant observation*.

## SUMMARY

In the first chapter in this section on socialization and culture, we focused on how individuals develop into social beings. In this chapter, we have turned our attention from individuals and the processes of socialization and self-development to the stage or setting where these processes take place. This setting is called the *social environment*. Within this environment, individuals exist in specific societies and cultures; they occupy numerous positions, are governed by rules, and behave in a variety of ways that are sometimes appropriate, sometimes inappropriate, but seldom unusual. The membership of individuals in a given society and culture, whose patterns and customs developed long before them and will probably long outlive them, affects and explains much of their behavior.

*Culture* is made up of the learned, shared patterns of behavior and knowledge common to a society, and includes both material and nonmaterial aspects. *Norms* are the rules for behavior. Because of these rules, we behave appropriately and behavior becomes predictable. Norm breakdown occasionally occurs, and when it does, it can produce crises both for society and for individuals. Norms vary in strength: "Shoulds" are called *folkways*; "musts" are called *mores*. There are *group norms* and *societal norms*, and sometimes these sets of norms conflict with each other. When this happens, those people influenced by the conflicting sets of norms experience problems. *Subcultures* refers to groups that share many of the traits of the dominant culture but have some unique customs and traits as well. Our cultural and subcultural affiliations are crucially important in determining who we are and what we do. The concept of *ethnocentrism* helps us understand the familiar tendency to assume that the world everywhere is the same as it is here and that if by some chance it's not, it should be. On the other hand, *cultural relativism* is an attitude that judges each culture from its own viewpoint.

*Society* is defined as a continuing number of people who live in a specific area; who are relatively organized, self-sufficient, and independent; and who share a common culture. A *status* is a position in society, and *role* describes the behavior of one who occupies a status. Norms define the boundaries for role requirements, but within these boundaries, the performance of roles will vary. Statuses, which can be either achieved or ascribed, are ranked in value or prestige. When a person occupies several statuses with contradictory role requirements, role conflict can occur. Role strain can result from one who tries to play a role that includes conflicting expectations.

Two readings follow that deal with norms, culture, and subculture. In the first reading, Terry Williams describes a teenage drug ring in New York City. In the second reading, the classic "One Hundred Percent American," Ralph Linton explores whether new elements of the American culture are introduced from within through invention or whether they are adapted from other cultures through the process of diffusion.

---

# TERMS FOR STUDY

| | |
|---|---|
| achieved status (89) | norms (71) |
| ascribed status (89) | real norms (75) |
| counterculture (80) | role (87) |
| cultural relativism (79) | role conflict (90) |
| culture (70) | role requirements (87) |
| ethnocentrism (78) | role strain (89) |
| folkways (75) | sanction (74) |
| ideal norms (75) | social structure (86) |
| material culture (71) | society (84) |
| mores (74) | status (87) |
| multiculturalism (81) | subculture (80) |
| nonmaterial culture (71) | values (72) |

For a discussion of Research Question 1, see page 77.
For a discussion of Research Question 2, see page 88.

---

 **INFOTRAC COLLEGE EDITION**
**Search Word Summary**

To learn more about the topics from this chapter, you can use the following words to conduct an electronic search on InfoTrac College Edition, an online library of journals. Here you will find a multitude of articles from various sources and perspectives:

**www.infotrac-college.com/wadsworth/access.html**

| | |
|---|---|
| ethnocentrism | Margaret Mead |
| Amish | drug dealers |
| Lebanese | Chinese culture |
| social status | Pro-Life Movement |

## Reading 3.1

# THE COCAINE KIDS

*Terry Williams*

*Subcultures and countercultures are groups that, by their traits, beliefs, interests, or behavior, are somewhat separated and distinct from the rest of society. Sometimes the line of distinction between culture and subculture is not clear. The teenagers and their behavior described in this excerpt from sociologist Terry Williams's book,* The Cocaine Kids, *seem clearly at odds with mainstream American culture. They make their living by dealing powder and crack cocaine in the Washington Heights area of New York City. On the other hand, the cocaine kids seem to share many, maybe most, of society's norms with the rest of us. Are they a subculture, a counterculture, or a slightly different segment of the dominant culture?*

Max was fourteen and already considered a "comer" in the cocaine business when I met him. We were introduced by a Dominican friend who knew of my interest in New York's underground cocaine culture and the teenagers who survive outside the regular economy. I assumed we would talk and then go our separate ways: He trusted my friend but he was shy; there was certainly no reason for him to talk with me about anything, and I was not about to press the issue. But there was something special about Max, and he became my friend and guide for nearly five years. I think we got along because I was an outsider and he had a story to tell, and he chose me to tell it to.

I believe Max trusted me for two reasons. First, I insisted on telling him the truth; second, I never revealed anything he said to me in confidence. My discretion was important not simply because of the danger he faced if his business became known to the wrong people, but because he is a private person and had been betrayed many times: by his own brother, who took cocaine and never returned the money; by girlfriends who used him because he had money, cocaine, and jewelry; and by others he thought he knew well. Trusting a complete stranger was risky, and Max at times tested me, telling me some things just to see if I would repeat them to those in the crew. I never did. He would often say, "Don't tell this to anybody" or ask, "Did you say anything to anybody about what I said to you?" Telling the truth was as important for me as it was for him. I had a great deal to lose if he thought I was deceitful.

Over several years, Max assembled the crew called here the Cocaine Kids. He introduced me to them—Chillie, Masterrap, Charlie, Hector, Jake, and Kitty—at *"la oficina,"* the office. This is an apartment, rented by Chillie (though not in his own name), where the Kids cut and mix cocaine, pick and pack crack; it is also the base from which they sell unpackaged cocaine to individual buyers.

## The Crew

Everybody sits around as Max prepares the crack. Max is a master at mixing. He uses his own recipe and is familiar with the effects of the drug in various combinations.

Jake snorts from a large bag of cocaine resting on the glass table, telling Chillie about a woman he met at Jump-Offs, an after-hours club. "*La jeva no era muy grande, pero enía lo de atrás* [The girl was not too big, but she had a big behind]." He kisses his fingers *"Hombre, 'mano."*

Max's recipe calls for an "eighth" of cocaine (1/8 kilo, or 125 grams), 60 grams of bicarbonate of soda (ordinary baking soda) and 40 grams of "comeback," an adulterant that has allowed Max to double his profits from crack: this chemical can be cooked with base and, when the base is dried, it smells, tastes, and looks very much like cocaine; all that is used "comes back." At $200 an ounce in 1984, it cost far less than the real thing.

He fills the Pyrex pot with tap water, and sets it on the stove to boil. After 20 minutes, he places the material in cold water to coagulate into crack, and members of the crew come forward to cut the hardened chunks with razor blades and pack the chips into red-topped capsules.

## Hector

Hector is skinny and freckle-faced. He is only three years older than Max, but he moves with the gait of an old man. His hands are rough, his bloodshot eyes dart from object to object with a twitchy nervousness. Though he was once a major dealer, and in many ways Max's mentor, he now looks to his little brother for support during hard times, and today is one of those times. He will only admit he made a mistake when he is high and "the cocaine is talking."

## Jake

Jake is rotund, and always wearing faded jeans, dirty unlaced sneakers, a soiled T-shirt and a bummy-looking leather jacket. He is the odd man in. He looks older than his seventeen years and is sometimes shy. Jake has been sniffing until it's time to pick and pack the white chips. He says cocaine gives him courage to face the unpredictable street.

A tireless worker, honest and loyal, Jake would never hurt Max. "I never lived in New York until my mother brought me here," he says, in a tone that is almost apologetic. He met Max by accident in the street. "He told me to go to this spot with him. He said he needed some back [backup or help] and he didn't have anybody. I went with him and made the move OK. After that we go back to see Chillie—Chillie used to have this spot in the Bronx then.

After this happened we come to Max's house and he asked me to go to work for him. I knew him because my sister knew his wife. Then we find out we're kin because of my aunt. I like working for him because it's easy money. I just watch myself and then I don't worry about too much."

## Chillie

Chillie is the boss at *la oficina*, which means that he supervises the work of Masterrap and Charlie. He and Max are the same age, and have worked together three years. He has dark, wavy hair and a sneaky smile that rarely surfaces; when it does it gives him a handsome but sinister look. He lets everybody know that he, not Max, should be controlling the cocaine business because he takes in more money than anyone in the crew except Max. "I made over a million dollars selling this stuff. If the connect [connection; the importer] knew what I was doing, he would want to see me. Max knows I do the best business out here. I don't want except a little money and a little respect." But Max won't introduce him to the Colombian supplier.

## Masterrap

Masterrap is slick, articulate and cool. He is quick to inform you that he is a ladies' man. "Rap is my name, females my game." When Chillie is busy or out of the office, he takes over; he is the second man behind the scale. While he does not appear ambitious, he is the first in the crew to see the bigger picture and is well aware that time is against everyone in the cocaine trade. He does not overindulge like the others do.

Masterrap has his heart set on a musical career and has written many "rap" songs that he hopes one day to record. "Coke is just a way for me to make some money and do some of the things I would otherwise not have the chance of doing in the real world. Coke ain't real. All this stuff and the things we do ain't real. If I told you half the things that go on in this place you wouldn't believe me. I wanna tell you my life story one day, and after you put it down I wanna see it and maybe then I'll believe this is really happening and not a dream."

## Charlie

Charlie is the only African-American on the crew. He is a bodyguard at the office; he and Chillie were high school friends, and now they are partners. He has taken three martial arts courses and learned how to shoot a gun after his uncle—a New York City corrections officer—took him to a rifle range.

Charlie is eighteen and looking to be the next man behind the scale.

## Kitty

The only woman in the crew, Kitty is five foot six, mulatto in complexion with high cheekbones and an engaging smile. She is depressed today; says she is tired of the cocaine-dealing hassle. "I really just want to go back to some school a few times a week. It would be better if I went Monday, Tuesday and Wednesday because then I would have time for my kid"—her son Armando, two years old.

(The kids often say they want to get out of the business when they are depressed—or when business is poor, after a bust, when there are family problems or lovers' quarrels. But when things are going well, they are eager to be out on the street trying to make a dollar.)

## Splib

Splib, Kitty's husband, is the only person present who deals as an independent, though he sometimes functions as a member of Max's crew. At nineteen, he is the oldest here. He is wise, handsome, and above all else a survivor. He also takes great pleasure in his ability to con and manipulate people.

"I never worry about money," he snaps. "I can always make money." Excited by his own story, he takes a folded dollar bill from his shirt pocket, and opens it to reveal what he announces to be the "purest cocaine in the world." Bending the edge of a matchbook cover into a vee, he takes two quick snorts. Refolding the bill with one hand, he is now ready to go into the street.

Splib, like most of the crew, is Dominican, but his ability to speak both Spanish and African-American slang with facility, and to mingle in both worlds is a valuable asset to the operation. He is aware of this, and high-handed about it: "The Indians [Colombians] have so much coke they can't off [sell] it without finding new markets. Blacks have proved they can organize and sell the shit, but the Indians don't know how to deal with Black cats. They don't understand their world or the way they do business. And they know it."

The crack is packed in vials, the powder allotted. Max tells Jake when to return for more. The money is to be dropped off at another location. Chillie and Kitty get their consignments; Charlie and Masterrap take one last snort before they depart. Everybody is ready to deal.

## The Neighborhood

Washington Heights stretches north on the west side of New York City from about 154th Street to 190th. In 1957, when Max's parents moved into the Heights, they were one of the first Dominican families to do so. Broadway, which runs north and south, was a dividing line: few Latinos and no African-Americans were allowed to rent apartments east of Broadway; few whites lived to the west.

Certainly it is true today that, although many different ethnic groups call Washington Heights home, Dominicans shape the character of this vibrant community. On a summer afternoon, the streets are teeming with Latino noises, smells and talk. Men—restaurant workers, street hustlers, store clerks, maintenance workers—gather to play dominoes or cards or shoot dice in the shade of a barbershop awning or a faded sign above *la grocería*. There is Astroni's, always serving breakfast, lunch and dinner: young dealers and customers go there to make phone calls; police officers drive up in the small hours for coffee take-out.

Saturday morning around 9:00 is a quiet time for the neighborhood drug trade, but vendors of other street merchandise are busy spreading their wares on the sidewalk and most of the regular businesses—Julio's Head Shop, Charlie's Metro Bar, the Greek-owned coffee shop, and many eateries—are always open. In the Monarch Bar, you can tell the dealers by the beepers clipped to their belts and by the way they handle money: they don't simply take out one bill to pay, but display the entire wad, counting off the bills in rapid strokes.

Along 156th Street, the house numbers signal the entranceways to many of the cocaine and crack houses in the area. Dealers like Jake stand cocksure in the doorways with their hands across their chests, luring customers—they are only a short walk from the subway stop—into the hallways. "Gypsy cab" drivers who have stopped to drink beer and snort a little *perico* talk near the barbershop; some are playing a favorite local card game, *viente-y-uno,* twenty-one. With money stacked high and every player eyeing the cash, they slam cards down with a cry loud enough to stop any passerby.

Illegal transactions are a leading activity here, and hustling is the name of the game. One sign of this is the large number of secondary operations catering to the drug trade: candy stores that sell drug paraphernalia and head shops that sell little else, such as Perran's, across the street from one of New York's most active copping zones, its display window filled with an assortment of lactose, dextrose, mannitol (all used to adulterate cocaine) and a host of water pipes for smoking crack or marijuana. Perran's is open 24 hours, and, like many businesses in the area, is thriving.

Less visible are bootleggers who sell a local version of corn whiskey popular among cocaine users, the cocaine bars frequented by local dealers and users day and night, and the after-hours spots, a community institution, that also survive on the cash spent by cocaine habitues.

Cocaine also occupies a central place in the lifeways of this community in another, invisible sense: Salsa, Latin jazz or the Spanish language have never received wide acceptance in the United States, but cocaine surely has. As coca and cocaine are, after all, Latin in origin, there is some nationalistic pride mixed with the Dominican and Latin entrepreneurial drive here.

## A Day in the Office

*La oficina* is located right in the middle of this bustling, mixed Washington Heights neighborhood. It has a large steel door specifically designed to prevent the police from knocking it down before the kids could dispose of the drugs. Chillie has hired the fourteen-year-old son of the building superintendent as a "catcher"—he is on call to retrieve any cocaine thrown out the window during a bust. The stock or "stash" of cocaine is kept in a bag stitched with beads worn by adherents of *Santería* (a set of religious practices involving Roman Catholic and African elements), and the boy's parents will not touch it because they fear it contains evil spirits. Chillie pays the boy in cash and cocaine.

The office is a small, one-bedroom apartment. A newcomer who enters the living room will see only a sofa, two chairs, a stereo, a TV, a tiny stool and some plants. Business hours are usually 1:00 PM to 5:00 AM, six days a week. The three office workers are Charlie, the armed door guard; Masterrap, who acts as host and receives requests from buyers; and Chillie, the man behind the scale, who sets prices, arranges to barter goods and services, gives credit and makes day-to-day decisions regarding sales.

At *la oficina,* unlike many "coke houses," the scales, packaging material and the drug itself are not immediately visible. Chillie prefers to deal from the thickly-carpeted bedroom, with a full-sized platform bed, a desk, and a large walk-in closet filled with candles, coconuts, beads in water-filled jars, coins in large bowls, cigars, and a silver plate holding food and money. The desk and telephone, the center of operations, sit near one window, facing the door. In the middle of the desktop are aluminum foil packets evenly cut to wrap the cocaine, and a triple-beam scale. No cocaine is visible until the buyer has shown his money.

Once in a while, the buyers are friends, and Charlie or Masterrap will negotiate with Chillie on their behalf. Chillie pays a bonus for every buyer they bring in, ranging from a few dollars on gram or half-gram sales to $100 for a one-ounce sale.

The phone rings constantly. Chillie answers from his room, and in a few minutes the doorbell rings. The first buyer today is a young Spanish woman who speaks halting English. She calls for Chillie to come out and meet her; they embrace and kiss. He asks how much she wants to buy and they go into his room. The slide on the scale makes tiny noises as it is moved along the ribbed bar. Ten minutes later she departs, after (he tells me) giving him $60 for a gram.

The doorbell rings again. Masterrap gets up and hollers "Back" to Charlie, who walks over and stands behind him. Masterrap peers into the peep hole; "it's cool," he announces, and Charlie relaxes and goes to sit in the kitchen. Two teenage Dominicans come in, wearing sneakers, leather jackets, and gold chains, and brand new blue jeans. Masterrap goes into a rap about music with them, then goes in to get them a taste from Chillie. They snort a bit, "take a freeze" (place a pinch of cocaine on the tongue), chat another minute or two, then go in to see Chillie.

After they leave, Masterrap and I talk about a movie we've both just seen. Charlie is in the kitchen, eating and yelling commands at a dog who is tied down. The dog, an akita ("They used to guard the emperors of Japan"), has a long curled tail, a strong face, and, though attentive, does not appear vicious or at all concerned with the goings-on. "They be mean dogs if you train them right," Charlie insists, lifting a piece of bread so the dog will jump for it. "Hey, they are better than them pit bulls out there," he asserts as if looking for an argument. When the dog moves about, Charlie shouts, *"calmate, calmate,"* and it sits, head held high. Some time later, Chillie shot the dog because it ate two ounces of cocaine mistakenly left on the table.

In all, I see fifteen buyers come and go that afternoon. Each one tastes the cocaine before purchasing; none stays more than twenty minutes. The first buyer is the lone woman. Masterrap explains, "You know we have plenty of females coming in here not just to cop but to hang out. Chillie don't go for that all the time. We gotta limit it. If things are slow, we let them stay longer or we might call a freak [a girl without inhibitions] to come over."

Buyers would say how much they wanted to purchase, and, after learning the price, would ask for discounts. Most sales were in the $60 to $80 range (1985). After the last customer, Chillie came out of the bathroom crowing, "I am The Deal-Maker."

## Risks of the Trade

Cocaine selling requires mobility, careful planning, and swift action. Max has used motorcycles, limousine services, messenger services and private cars to deliver and distribute cocaine. "I use the motorbike because I can get around fast and I can throw the stuff away if the cops follow me. If I got to take a big package, I never use my car, because the rule is, 'if you use it you lose it.' I don't use messengers any more because they might split with a package if they find out what's in it."

It also requires consistent caution; there is a sting of danger to every transaction, and sellers become expert at concealing the drug until they are sure they are dealing with a genuine user, not a police officer. Kitty says she always carries the coke in her bra when she is on the street; in a limo, she stashes it under the seat: "the windows are smoked so nobody can see in there and plus cops don't bother to stop limos anyway."

When she responds to a call from a prostitute whose customer wants the drug she has to worry about the "trick" being a cop, so she asks the women not to beep her until "the guy has already propositioned himself. Cops can't do that—that's entrapment. But a lot of these guys don't want sex, so it gets kind of tricky."

Those who work from a fixed location like an apartment are in a particularly dangerous position, as they are literally trapped, and must develop a strategy to avoid arrests and robberies. Chillie knows, "I'm the one that's gonna take the fall if this place is busted. The steel door is only so much protection. I gotta be able to get this stuff out of here in a hurry; I use Peppi [the super's son] to catch some of the stuff, but he ain't around all the time. I know the cops go to the back stairs sometimes when they bust people so that may not work for very long.

"But the apartment is not in my name, and I'm moving in a way that the whole thing is pre-packed: I'm gonna put all the cocaine into foil before the customer comes in here. I'm gonna make up all the twenties, the fifties [$20 and $50 packets], the whole thing so there won't be no hassles and no sitting around sniffing and shit. That way I won't have to have this scale no more."

(This effort did not survive in the market. Customers complained that the aluminum foil packets caused the drug to melt, and thought the amount was short; they began to demand their money back. Max, hearing of this, found waterproof packets, called snowseals, to replace the foil.)

(But many customers were still upset that they could no longer "taste" the cocaine, long a part of the sales ritual—and especially upset because Chillie still allowed some of his friends to sniff before they bought. Eventually, the complaints mushroomed to the point where he had to go back to using the scale and the old method of doing business.)

All the Kids snort cocaine regularly. This is accepted, but the use of crack is generally frowned upon: those who snort are thought to have more control and discipline than those who smoke crack or freebase. Most dealers see crack smokers as obsessive consumers who cannot take care of business; crack users, they say, tend to become agitated, quickly lose control and concentration, and take one dose after another at the expense of everything else. Snorters, however, can use the drug and still take care of business.

Yet the Kids who snort do so on almost any pretext: Chillie will sniff and ask Masterrap and Charlie to join him whenever he makes a sale of more than $100. One day they decided to celebrate because the New York Mets, their favorite baseball team, won a game. They called Max, began snorting and playing music, and called girls to join them. On less joyous occasions, cocaine serves a therapeutic purpose, as an antidote to stress, disappointment, and the problems of everyday life.

## QUESTIONS 3.1

**1.** What type of study is this? What are the strengths and weaknesses of this type of research?

**2.** Norms exist everywhere. Give examples of some of the cocaine kids' norms.

**3.** Behavioral differences are obvious, but what values do the cocaine kids share with mainstream American culture?

**4.** What do the cocaine kids represent—a subculture, a counterculture, or a segment of the predominant American culture? Make a case for each.

**5.** Take the subculture approach for a moment—of how many different subcultures could the cocaine kids be said to be members?

## Reading 3.2

# ONE HUNDRED PERCENT AMERICAN

*Ralph Linton*

*The content of a culture—skills, customs, material objects, and nonmaterial ideas—comes from several sources. Some aspects of culture appear spontaneously and are developed from within the culture through the process of invention. Much more, however, is borrowed from other cultures through the usually unconscious process of diffusion. Anthropologist Ralph Linton wrote this passage some years ago. It is often quoted and still remains a classic at demonstrating the importance of diffusion. Much of what we thought was "one hundred percent American" actually came long ago from societies we never heard of.*

There can be no question about the average American's Americanism or his desire to preserve this precious heritage at all costs. Nevertheless, some insidious

From *The American Mercury*, April 1937, pp. 427–429. Reprinted by permission of the publisher.

foreign ideas have already wormed their way into his civilization without his realizing what was going on. Thus dawn finds the unsuspecting patriot garbed in pajamas, a garment of East Indian origin; and lying in a bed built on a pattern which originated in either Persia or Asia Minor. He is muffled to the ears in un-American materials: cotton, first domesticated in India; linen, domesticated in the Near East; wool from an animal native to Asia Minor; or silk, whose uses were first discovered by the Chinese. All the substances have been transformed into cloth by a method invented in Southwestern Asia. If the weather is cold enough he may even be sleeping under an eiderdown quilt invented in Scandinavia.

On awakening he glances at the clock, a medieval European invention, uses one potent Latin word in abbreviated form, rises in haste, and goes to the bathroom. Here, if he stops to think about it, he must feel himself in the presence of a great American institution; he will have heard stories of both the quality and frequency of foreign plumbing and will know that in no other country does the average man perform his ablutions in the midst of such splendor. But the insidious foreign influence pursues him even here. Glass was invented by the ancient Egyptians, the use of glazed tiles for floors and walls in the Near East, porcelain in China, and the art of enameling on metal by Mediterranean artisans of the Bronze Age. Even his bathtub and toilet are but slightly modified copies of Roman originals. The only purely American contribution to the ensemble is the steam radiator.

In his bathroom the American washes with soap invented by the ancient Gauls. Next he cleans his teeth, a subversive European practice which did not invade America until the latter part of the eighteenth century. He then shaves, a masochistic rite first developed by the heathen priests of ancient Egypt and Sumer. The process is made less of a penance by the fact that his razor is of steel, an iron-carbon alloy discovered in either India or Turkestan. Lastly, he dries himself on a Turkish towel.

Returning to his bedroom, the unconscious victim of un-American practices removes his clothes from a chair, invented in the Near East, and proceeds to dress. He puts on close-fitting tailored garments whose form derives from the skin clothing of the ancient nomads of the Asiatic steppes and fastens them with buttons whose prototypes appeared in Europe at the close of the Stone Age. This costume is appropriate enough for outdoor exercise in a cold climate, but is quite unsuited to American summers, steam-heated houses, and Pullmans. Nevertheless, foreign ideas and habits hold the unfortunate man in thrall even when common sense tells him that the authentically American costume of gee string and moccasins would be far more comfortable. He puts on his feet stiff coverings made from hide prepared by a process invented in ancient Egypt and cut to a pattern which can be traced to ancient Greece, and makes sure they are properly polished, also a Greek idea. Lastly, he ties about his neck a strip of bright-colored cloth which is a vestigial survival of the shoulder shawls worn by seventeenth-century Croats. He gives himself a final appraisal in the mirror, an old Mediterranean invention, and goes downstairs to breakfast.

Here a whole new series of foreign things confront him. His food and drink are placed before him in pottery vessels, the popular name of which—china—is sufficient evidence of their origin. His fork is a medieval Italian invention and his spoon a copy of a Roman original. He will usually begin the meal with coffee, an Abyssinian plant first discovered by the Arabs. The American is quite likely to need it to dispel the morning-after effects of overindulgence in fermented drinks, invented in the

Near East, or distilled ones, invented by the alchemists of medieval Europe. Whereas the Arabs took their coffee straight, he will probably sweeten it with sugar, discovered in India, and dilute it with cream, both the domestication of cattle and the technique of milking having originated in Asia Minor.

If our patriot is old-fashioned enough to adhere to the so-called American breakfast, his coffee will be accompanied by an orange, domesticated in the Mediterranean region, a cantaloupe domesticated in Persia, or grapes, domesticated in Asia Minor. He will follow this with a bowl of cereal made from grain domesticated in the Near East and prepared by methods also invented there. From this he will go on to waffles, a Scandinavian invention, with plenty of butter, originally a Near Eastern cosmetic. As a side dish he may have the egg of a bird domesticated in Southeastern Asia or strips of the flesh of an animal domesticated in the same region, which have been salted and smoked by a process invented in Northern Europe.

Breakfast over, he places upon his head a molded piece of felt, invented by the nomads of Eastern Asia, and, if it looks like rain, puts on outer shoes of rubber, discovered by the ancient Mexicans, and takes an umbrella, invented in India. He then sprints for his train—the train, not the sprinting, being an English invention. At the station he pauses for a moment to buy a newspaper, paying for it with coins invented in ancient Lydia. Once on board he settles back to inhale the fumes of a cigarette invented in Mexico, or a cigar invented in Brazil. Meanwhile, he reads the news of the day, imprinted in characters invented by the ancient Semites by a process invented in Germany upon a material invented in China. As he scans the latest editorial pointing out the dire results to our institutions of accepting foreign ideas, he will not fail to thank a Hebrew God in an Indo-European language that he is a one hundred percent (decimal system invented by the Greeks) American (from Americus Vespucci, Italian geographer).

---

## QUESTIONS 3.2

**1.** Although it is true that many material objects have come to us through diffusion, most of the nonmaterial aspects of American culture are original and the products of invention. Discuss.

**2.** This article was written in the 1930s; the same sort of statements could not be made today. Discuss. To prove the point, draw up a list of aspects of American culture developed through invention.

**3.** Analyze Linton's article, using the terms *ethnocentrism* and *cultural relativism*.

---

## NOTES

1. Clyde Kluckhohn, "The Study of Culture," in *The Policy Sciences,* edited by Daniel Lerner and Harold D. Lasswell (Stanford, Calif.: Stanford University Press, 1951), p. 86.

2. These and other examples are contained in *Studies in Ethnomethodology,* by Harold Garfinkel (Englewood Cliffs, N.J.: Prentice-Hall, 1967).

3. The terms *folkways* and *mores* were introduced by William Graham Sumner in *Folkways* (Boston: Ginn, 1907).

4. See *Sick Societies,* by Robert Edgerton (New York: Macmillan, 1992).

5. See *Multiculturalism,* edited by Amy Gutman (Princeton, N.J.: Princeton University Press, 1994); *Race and Ethnic Relations,* 2d ed., by Martin Marger (Belmont, Calif.: Wadsworth, 1991), for an analysis of Canada; and *Time,* December 2, 1993, p. 73.

6. This study was done at the School of Public Health at the University of California at Berkeley and was reported in newspapers and newsmagazines beginning in August 1975.

7. Robert Bierstedt, *The Social Order,* 4th ed. (New York: McGraw-Hill, 1974), chapter 9.

8. See *The Presentation of Self in Everyday Life,* by Erving Goffman (Garden City, N.Y.: Doubleday, 1959).

9. See "Social Integration and Longevity: An Event History Analysis of Women's Roles and Resilience," by Phyllis Moen, Donna Dempster-McClain, and Robin Williams, Jr., *American Sociological Review* 54 (August 1989), pp. 635–644.

# PART TWO

# Social Organization

In preceding chapters we have looked at basic units of analysis used by sociologists to understand, describe, and explain human behavior. In this part our view becomes more general. If we combine these basic units and move to a higher level of abstraction, we come to the concept of social organization. If, in a like manner, we were analyzing football, we could first look at the basic elements—the players, a ball, a marked-off field, and maybe some spectators. But our description is helped a great deal if we move to the next level and describe a football game; thus it is now necessary to look beyond the elements and deal with the combination of the setting, the rules, the shared expectations, and the interaction patterns of participants—in other words, the social organization that allows 22 players, 6 officials, and 20,000 observers to mutually participate in and get something out of the same event.

Just as there is an extensive degree of organization to a football game, so there is to society. The basic elements of a society are a number of people and an inhabitable geographical area. But an understanding of a society cannot be obtained until we study the patterns of interaction and organization that are characteristic of the people. We see that cities exist (why do they?), that highways link cities, that educational facilities are developed, that some individuals are more highly valued than others, and that governments run the cities. These things don't just happen by accident; they are evidence that there is organization to society. People cooperate, interact, and share expectations and mutual interests. There is a structure or system to society much as there *is to a building or to a machine or to a football game: Parts link together to form a complex whole. The* **social organization** *of society is frequently referred to as a social fabric: an integrated set of norms, roles, cultural values, and beliefs through which people interact with each other, individually and through groups. The terms* social organization *and* social structure *(which we discussed in Chapter Three) are very similar concepts and are often used interchangeably. Although* social structure *can be used to refer more to a stable network of elements through which people relate to each other, and* social organization *can be used to refer more to the continually changing ordering and coordination of human activities, it is probably not worthwhile to try to make a distinction between the terms.*

*Studying various aspects of the social organization and social structure of society is the central task of sociology. We undertake this analysis in a variety of ways. It is as if we were looking at a subject, society in this case, through a number of windows. Each window is of a different shape, size, thickness, and color of glass. So, although we are looking at the same subject, each approach gives a somewhat different viewpoint, emphasizing some aspects and ignoring others. The following chapters will analyze the social organization of society through several windows. Groups will be the topic of Chapter Four, and several types of groups will be discussed in detail. Chapters Five and Six will deal with types of social differentiation and social inequality. In Chapters Seven through Ten we will discuss institutions as the sociologist sees them. And Chapter Eleven will cover population and ecology.*

David Hurn/Magnum Photos Inc.

# *Groups*

**Why Study Groups?**

**Groups and Nongroups**

**Types of Groups**
*Primary and Secondary Groups* • *Small Groups*

**Formal Organizations**
*Types of Organizations* • *Formal Organizations—Consequences and Concerns* •
*"The Organization Man"* • *"McDonaldization"* •
*Can Formal Organizations Change?*

## TWO RESEARCH QUESTIONS

**1.** *Is alcohol use related to either social class (one's rank in society) or religiosity (one's commitment to and involvement in religion)? If it is related, in what direction? Do higher-class people drink more or less? Do more religious people drink more or less? Which of the two—social class or religiosity— would have the greater influence on alcohol use? Why?*

**2.** *What determines how a new company will be organized? Can we tell ahead of time whether it will be run formally with lots of rules, or informally with decisions made on the spur of the moment? Will the proportion of women involved in the start-up period of the company affect what the company looks like?*

A GROUP—a major unit of analysis for sociologists—is simply defined as a collection of people. Not all collections of people, however, are defined as groups. Are the people on a bus a group? How about red-haired people between the ages of 30 and 45? Or what about the students at a large university? At the same time, collectivities that are traditionally defined as groups can vary tremendously in some characteristics. Your family, my sociology department, a college football team or sorority, and the president's cabinet could probably be defined as groups, but they are quite different from each other in many dimensions: size, complexity, type of interaction, and division of labor. Much of the study of social organization, therefore, could center around several questions: What is a group? How is the group developed? What are the different types of groups?

## WHY STUDY GROUPS?

Sociologists study groups for several reasons. The division of labor and the sharing of interests and jobs that occur in groups make it possible to complete tasks that would be impossible for individuals to accomplish alone. Far more important for the sociologist, however, is the role the group plays in the socialization and development of the individual. To a great degree, we feel that a person *is* the sum of the groups to which he or she belongs. Transmission of culture and learning of values, attitudes, and ways of behaving and believing occur mainly in groups. As we have pointed out in Chapters Two and Three, the socialization and development of self and personality are interactive processes. Individuals interact with others and modify their own behavior according to their interpretations of the responses of others. These others are individuals: mother, father, teacher, or another significant other. These others are also groups. Most of the socialization process, both primary and secondary, takes place in *groups*. First the family, then the classroom group, the team or club, the social group, the professional organization, and numerous other voluntary and involuntary associations influence and shape the individual.

The group is more than just the sum of the people who belong to it. Perhaps a good example of this is to compare a family of five people with five people we pick at random on the street. Even if we set each "group" in front of the fire or TV or around the same dinner table and start them interacting, the differences between the two would be clear. We would shortly see that the real family group had developed a set of norms, roles, and relationships to each other that went far beyond the interactions of the five separate people. Because of the presence of this network of relationships and understandings that exists in groups, the impact or power of the group in influencing the individual is extensive.

To get a better idea of the importance of groups, imagine the effects on two people of belonging to the following contrasting groups. One person is raised in a family in which competitiveness is valued; family members are constantly competing with each other and with outsiders. The other is raised in a family in which competition is discouraged. One is raised in a family that pursues artistic endeavors, and the other in a family that prefers athletics. One is a Jehovah's Witness; the other is a member of a Unitarian church. One is a member of a rock group; the other is a member of a chamber-music group. One is on the high-school basketball team; the other is on the yearbook staff. One goes to the University of California at Berkeley; the other goes to West Point. One goes through basic training in the Marine Corps; the other joins the Peace Corps. One is raised in a modern Western family, the other in a primitive tribal family.

You're probably thinking, "Sure, the Berkeley student is a lot different from the West Point student; but they had some of those differences before, which is why one chose Berkeley and the other chose West Point." That's probably true, but those differences, too, can be explained by other earlier group memberships and experiences. Each set of group experiences prepares and inclines the person toward another set, then another set, and so on. Or, consider this example. Imagine a 46-year-old doctor at the top of his field who climbs mountains, builds sailboats, races sports cars, writes books, and is a gourmet cook. How did this superachiever get this way? Was he born this way? Certainly not, says the sociologist. Study his socialization patterns through numerous group contexts, and you will find the answer. It may not be easy, but the basic explanations are in the complicated interaction between group and individual.

## GROUPS AND NONGROUPS

Sociologists say that a **group** comprises a number of people who (1) have shared or patterned interaction and (2) feel bound together by a "consciousness of kind" or a "we" feeling. "Consciousness of kind," a phrase coined by Franklin Giddings, refers to the individual's awareness of important similarities between himself or herself and certain others. This concept also refers to the awareness that the individual and other group members have common loyalties, share at least some similar values, and see themselves as set apart from the rest of the world because of their memberships in this particular group. Groups vary tremendously in variety, size, and shape. A group can be as small as two people or almost infinitely large. Groups can be simple in structure or exceedingly complex; they may involve close, intimate relationships between members or more distant and infrequent personal contacts. In other words, the definition of group—patterned interaction and "we" feeling—can fit an enormous variety of situations: a family, a basketball team, a sociology class, IBM, or General Motors.

This definition of groups, imprecise as it may seem, allows us to distinguish groups from other types of collectivities of people, which we could call *nongroups*. One type of nongroup, which we will call an **aggregate (or aggregation), consists of a number of people clustered together in one place.** Examples of aggregates might be all the people in New York City, or the pedestrians at a busy intersection waiting for the light to change to "walk," or all the people in North America, or the passengers on a jet from New York to San Francisco. A second useful nongroup, called a **category, consists of a number of people who have a particular characteristic in common.** Examples of categories would be all females, or all red-haired people, or all pilots, or all teenagers, or all whites.

Although we have called them nongroups, aggregates and categories can be transformed into groups should they develop patterned interaction among members and consciousness of kind. For example, let's examine our aggregate of 10 people waiting at the intersection for the light to change to "walk." Then suppose it *doesn't* change; for 5, 10, even 15 minutes the light refuses to budge from "wait." The pedestrians, strangers until now, begin talking to each other about the impossible situation. Should they race across through traffic against the light? Where's the cop?—you can never find one when you really need one. Some interaction takes place, and a consciousness of kind develops: a group of good people being victimized by a lousy, mechanical light. Or, take the passengers on the jet from New York to San Francisco, another aggregate. Somewhere over Pennsylvania the pilot says to himself, "I'm sick and tired of flying to San Francisco all the time; I guess I'll go to the North Pole." The passengers, who did not know each other before, would begin to interact, possibly in an agitated manner, and by the time they reached the Arctic Circle there would probably be highly developed interaction.

Categories can also become groups. All the red-haired people between the ages of 30 and 45 would constitute a category, as would all carpenters of Irish ancestry. But suppose the middle-aged redheads decided to get together and put out a journal telling of their common problems and aspirations. Or suppose the Carpenters of Irish Ancestry decide it's time to start an organization (CIA?), have a convention, and elect officers. In each case we might have a category developing into a group. These sound farfetched, but sociologists study both groups and nongroups (categories and aggregates), and the lines between these collectivities are somewhat fluid and easily crossed.

The study of people and society is a large task. It is made more manageable if the whole is broken down into more basic parts—parts that are smaller and have some characteristics in common. We can divide society into *categories* (males, upper-class blacks, teenage white females), *aggregates* —(a

crowd, a city), and *groups* —(a family, a study group). In later chapters we
will look at types of categories and aggregates, but the focus of this chapter
will be groups. Three sections follow; first we will look at specific types of
groups and then at groups at the two extremes of size: small groups and formal organizations.

## TYPES OF GROUPS

*The* New York Times *has described one of the most unusual clubs in the*
*world made up of a small number of usually serious scientists stationed at the*
*South Pole. It is called the "300 Club." Those who want to join must wait*
*until the temperature is at least 100 degrees Fahrenheit below zero, then strip*
*completely nude and dash 100 yards across the ice to a marker designating*
*the South Pole and 100 yards back to the scientific hut. Anyone surviving*
*becomes a member of this very exclusive group.*

Social groups can be classified in many ways as the following categories illustrate. In some groups, membership is automatic and the participant has no choice; in others, the option is open and individuals may or
may not join as they wish. These two types are called involuntary and
voluntary groups. **Involuntary groups** might include the family one is
born into or the army platoon one is drafted into. **Voluntary groups**
would include any of a vast number that an individual can exercise some
choice in joining: lodges, fraternities, bridge clubs, student governments,
or political organizations.

**Reference groups** are groups that serve as models for our behavior—
groups we might or might not actually belong to but whose perspectives we
assume and mold our behavior after. A reference group might be made up
of people one associates with or knows personally, or it might be an abstract
collectivity of individuals who represent models for our behavior. Each individual has many reference groups. As a teacher, I would have certain reference groups, as a sociologist others, as a husband others, and as a handball
player still others.

The first research question at the beginning of this chapter asked if alcohol use was related to social class and religiosity, and if so, what the relationship was. A recent paper examined this in some detail using reference
groups. There is a negative relationship between alcohol use and religiosity—
that is, the more religious person drinks less; the less religious person drinks
more. Religiosity was defined as church attendance, strength of belief, membership in church organizations, and belief in life after death. The relationship between social class and alcohol use is less powerful but appears to be
positive; that is, upper-class people drink more than lower-class people. The
real emphasis of the paper, however, was on which of the two factors was
more important, religiosity or social class. If the two were in conflict, which
would win out? For example, what would be the alcohol use of a highly religious upper-class person or of a less religious working-class person? The
answer was clear—religiosity would win out—and the explanation made

sense from a reference-group viewpoint. A social class is a *category*, a number of people who have a particular characteristic in common but who otherwise don't know each other and are less likely to share common norms or similar values. However, religiosity, especially as measured here, is more likely to reflect membership in a *group*. The authors suggest that a group—people with whom one has sustained and significant interaction—reflects a "primary environment of opinion." When this is in conflict with a "secondary environment of opinion" such as a category or aggregation, the group context will have greater influence.[1]

One's **peer group** is made up of people of relatively the same age, interests, and social position with whom one has reasonably close association and contact. A peer group can consist of a class at school, a street gang, or an occupational group such as the members of a college sociology department or a group of lawyers in a law firm. Not all the members of a peer group are necessarily friends, but the peer group exercises a major role in the socialization process. During adolescence, a peer group might be *the* major socializing agent.

Groups whose members come predominantly from one social-class level are called **horizontal groups.** Examples of horizontal groups would include almost any organization formed along occupational lines: an association of doctors, carpenters, or actors. If a group includes members from a variety of social classes, it could be called a **vertical group.** Vertical groups are more difficult to find in American society because many divisions are made along social-class lines. A church congregation might constitute a vertical group, and in some cases an army platoon made up of draftees would include members from a variety of social classes. Groups are defined as open or closed according to the ease of gaining membership. A white fraternity is often a **closed group** as far as a black male is concerned, but the United States Army is probably a very **open group** for the same individual. Now, what type of group is the "300 Club"?

### Primary and Secondary Groups

A family and a draftee's army platoon are involuntary groups, but they are quite different sorts of groups. A basketball team and the members of a large college class are voluntary groups, but again they are very different. Groups vary along a number of dimensions, and one of the most important of these is to what degree each group is primary or secondary. A family and a basketball team are more primary than are an army platoon and a large college class. Primary groups play an important role in human development. Most of the socialization process—learning society's norms and roles, and development of the self—takes place in small primary groups. The **primary group** was first described by Charles H. Cooley as referring to groups in which contacts between members are intimate, personal, and face-to-face. A great part of the individual's total life experience is bound up in the group and is known to other group members. The primary type of relationship is one that involves deep and

## "FRIENDS AND NETWORKS"—TALLY'S CORNER

More than most social worlds, perhaps, the streetcorner world takes its shape and color from the structure and character of the face-to-face relationships of the people who live in it. Unlike other areas in our society, . . . resources in the streetcorner world are almost entirely given over to the construction and maintenance of personal relationships. On the streetcorner, each man has his own network of these personal relationships. . . . At the edges of this network are those persons with whom his relationship is effectively neutral, such as area residents whom he has "seen around" but does not know except to nod or say "hi" to as they pass on the street. In toward the center are those persons he knows and likes best, those with whom he is "uptight": his "walking buddies," "good" or "best" friends, girl friends, and sometimes real or putative kinsmen. It is with these men and women that he spends his waking, nonworking hours, drinking, dancing, engaging in sex, playing the fool or the

wise man, passing the time at the Carry-out or on the streetcorner. . . . So important a part of daily life are these relationships that it seems like no life at all without them. Old Mr. Jenkins climbed out of his sickbed to take up a seat on the Coca-Cola case at the Carry-out for a couple of hours. "I can't stay home and play dead," he explained, "I got to get out and see my friends."

Preston was Clarence's uncle. They lived within a block of each other and within two blocks of the Carry-out. Clarence worked on a construction job and later got Preston a job at the same place. Tally, Wee Tom, and Budder also worked at the same construction site. The five men regularly walked back from the job to the streetcorner together, usually sharing a bottle along the way. On Friday afternoons, they continued drinking together for an hour or so after returning to the streetcorner. Tally referred to the other four men as his "drinking buddies."

---

personal interaction and communication. This interaction is an end in itself: Primary groups often exist because of the value of the primary relationship rather than because of other specific goals or tasks. People conform in primary groups because of strong informal norms—for example, fear of being ostracized, scorned, or ridiculed—rather than because of any formal written rules.

A **secondary group,** on the other hand, is more impersonal. Interaction is more superficial and probably based on utilitarian goals; that is, the person is less important than a particular skill he or she may offer the group. Interaction and communication are based on the value of one's particular skill rather than on interest in one's general personal qualities. Groups vary in their degree of primariness or secondariness. Moving from primary to secondary, we might see these groups: a married couple, an extended family including parents and grandparents, a basketball team, a sorority, a professional organization or labor union, or the employees of a large corporation.

Tally had met Wee Tom on the job. Through Tally, Wee Tom joined them on the walk home, began to hang around the Carry-out, and finally moved into the neighborhood as well. Budder had been the last to join the group at the construction site. He had known Preston and Clarence all along, but not well. He first knew Tally as a neighbor. They came to be friends through Tally's visits to the girl who lived with Budder, his common-law wife, and his wife's children. When Tally took Budder onto the job with him, Budder became a co-worker and drinking buddy, too. Thus, in Tally's network, Wee Tom began as co-worker, moved up to drinking buddy, neighbor, and finally close friend; Budder from neighbor and friend to co-worker. Importantly, and irrespective of the direction in which the relationships developed, the confluence of the co-worker and especially the neighbor relationship with friendship deepened the friend relationship.

The most common form of the pseudo-kin relationship between two men is known as "going for brothers." This means, simply, that two men agree to present themselves as brothers to the outside world and to deal with one another on the same basis. Going for brothers appears as a special case of friendship in which the usual claims, obligations, expectations, and loyalties of the friend relationship are publicly declared to be at their maximum.

Sea Cat and Arthur went for brothers. Sea Cat's room was Arthur's home so far as he had one anywhere. It was there that he kept his few clothes and other belongings, and it was on Sea Cat's dresser that he placed the pictures of his girl friends (sent "with love" or "love and kisses"). Sea Cat and Arthur wore one another's clothes and, whenever possible or practical, were in one another's company. Even when not together, each usually had a good idea of where the other was or had been or when he would return. Generally, they seem to prefer going with women who were themselves friends; for a period of a month or so, they went out with two sisters.

—*From* Tally's Corner *by Elliot Liebow. Copyright 1967 by Little, Brown and Co., Inc. Reprinted by permission of the publisher.*

## Small Groups

The study of small groups has long been an interest of sociology and psychology. Small groups are used as vehicles for treatment: Group therapy has proven to be a useful mechanism for change in prisons, mental hospitals, and organizations dealing with alcoholism and drug usage. Sensitivity-training and encounter groups are used by individuals to enhance self-awareness and by large organizations to improve communication and interaction. Many educators believe the small class works better than the large class in educating students. Small groups combine a primary group type of atmosphere with a task orientation. The assumption is that achievement of the task—better teaching, self-understanding, improved communication and management skills—is facilitated by the primary group atmosphere.

Small group studies in group pressure and conformity have reached some interesting conclusions. A stationary light in a dark room appears to move—this is called the *autokinetic effect*. Muzafer Sherif asked subjects to judge how far the light moved. When small groups were tested, they quickly arrived at an agreement on the light's movement—a group norm. When the members were tested individually, they stuck to the earlier established group norm. When other subjects were tested individually *first* (no group norm available), their judgments of the light's movement were more variable and erratic. In other studies, Solomon Asch asked small groups of people to compare the lengths of lines. Some members of the group were stooges, or confederates of the researcher. The stooges were instructed to make obviously wrong selections: Two lines of different lengths were said to be the same. In studies involving eight stooges and one subject, the subject was faced with the choice of going along with the majority or making an independent judgment. Many avoided stress and simply went along with the majority.

Small group studies in leadership have shown that two types of leaders typically appear—the task leader and the social-emotional leader. The *task leader* has skills that are important for the group to get a particular job done. But the task leader is not warm and friendly, not very approachable. The *social-emotional* leader, on the other hand, is highly approachable—he or she is the one to whom other group members can complain, show affection, or in other ways express feelings. Other group roles include nice guys, ignored, and rejected. *Nice guys* are approachable—not so much as the social-emotional leader but enough that their membership and contributions are valued. Those *ignored* are members who are usually inactive; they are an unknown quantity to the rest of the group. Those *rejected* are members perceived as unapproachable because they have no skills or characteristics valued by the group or because of hostile or confusing behavior on their part.

Small group studies of deviant behavior show some interesting patterns. Sometimes deviant members of a group are rejected or ostracized, especially if they are blocking a position that is popular with the group. In other situations, however, behavior of a deviant group member actually increased group solidarity—the rest of the group tried to help out or cover for the member who "couldn't keep up." And rank has its privileges: Certain fraternity members were found to have committed a crime. When it came time to punish, fraternity brothers gave the higher-ranking guilty party (officer, upper classman, social-emotional leader) less punishment than they gave to the lower-ranking guilty party. It seems that the brothers felt that the higher-ranking member was more valuable to the group, and they had more to lose by rejecting him.[2]

Some of the effects and constraints that small groups have on behavior can be seen in the actions of people on juries. The jury usually quickly selects a person of high status (wealth or education or both) as foreman. People of high status are more active on the jury; they talk more, have more influence, and are seen as more competent. Low-status people tend to defer to the higher-status members. Men are more active on juries, but women are better jurors; they take more care in considering testimony. Jurors are very

## CARL ROGERS ON ENCOUNTER GROUPS

You are in a small group with people you don't know particularly well and the following is said to you: "I feel threatened by your silence." "You remind me of my mother, with whom I had a tough time." "I took an instant dislike to you the first moment I saw you." "To me you're like a breath of fresh air in the group." "I dislike you more every time you speak up."

These statements are called "expression of feelings," and they are one of the things that happen in small intense groups called *encounter groups*. Key characteristics of such groups include providing a climate of safety that allows freedom of expression, and reducing defensive reactions to other group members and to one's self. The climate of mutual trust leads to expression of real feelings (positive and negative),

and less defensiveness means the possibility of change becomes less threatening. It becomes easier to hear others, and feedback helps one to see how he or she appears to others. It is hoped that skills and insights discovered in the group will carry over to relationships with others outside the group.

The expectation for encounter groups is that they will increase self-knowledge, and this will improve communication with others and help solve personal and group problems. They have been used widely by such varied groups as corporate executives, couples hoping for happier marriages, and people with drug and alcohol problems.

—*See* Carl Rogers on Encounter Groups *by Carl Rogers. New York: Harper & Row, 1970.*

reluctant to come back without a verdict. The self-confidence of witnesses often has more effect on the jury than the logic or soundness of what they say. Juries often try the lawyer rather than the person he or she is representing. Finally, the mere fact of a person's being brought to trial makes him or her suspect in the eyes of the jury, in spite of the presumption of innocence in our court system.[3]

## FORMAL ORGANIZATIONS

*The new director of the Department of Motor Vehicles (DMV) ran head-on into a problem recently. He had DMV workers paint "no parking" in Chinese in no-parking areas around the DMV buildings. There were already signs in English and Spanish as part of a program to help people unfamiliar with English. The director was proud and the local newspaper published a picture of the new signs. But the next day they were gone, painted over. It seems that a grounds supervisor had also seen the picture in the newspaper, had checked around and couldn't find the necessary work order for having the signs painted, and had them painted over. The next day the director had the signs painted in Chinese again, but only after he had obtained the necessary work order. "It was his first real experience with bureaucracy, and he lost it to the paper pushers," reported a spokesperson.*

## MAX WEBER (1864–1920)

Max Weber earned his Ph.D. in Law at the University of Berlin. There was an unresolved tension in Weber's personal life and in his work between the bureaucratic mind, as represented by his father, and his mother's religiosity. Weber differed from many of the economic theorists of his time like Marx and Engels. Where they focused on the importance of the economy as the determining factor for societies, Weber believed that ideas or "ethics" were fairly autonomous forces capable of profoundly affecting the economic world. Weber had a deep interest in religious ideas and their effect on the economy. In *The Protestant Ethic and the Spirit of Capitalism* (1904–1905), he was concerned with Protestantism as a system of ideas and its impact on the rise of another system of ideas, the "spirit of capitalism," and ultimately on capitalist economic systems. Weber, along with Emile Durkheim, is generally regarded as the founder of modern sociology as a distinct social science, and in 1920 Weber helped found the German Sociological Society. We will encounter Weber frequently in this book: in this chapter on bureaucracy, in Chapters Five and Nine on socioeconomic status and authority, and in Chapter Eight on the connection between Protestantism and capitalism.

Large-scale formal organizations are an essential part of modern society. These organizations, often called **bureaucracies,** arise as societies' activities become increasingly planned rather than spontaneous. People discovered long ago that if several people got together and planned an activity such as building a car or educating a student, they often got the job done more rapidly and efficiently together than if each pursued the task in his or her own way, spontaneously and haphazardly. Arthur Stinchcombe defines a formal organization as any social arrangement in which the activities of some people are systematically planned by other people to achieve some special purpose. Those who plan the activities automatically have authority over the others. Formal organizations seem to arise in societies that have a money economy and that pursue complex tasks requiring the coordinated efforts of a number of people. As the size of the organization increases, administrative tasks multiply, and this encourages further bureaucratization. These organizations have an enduring quality: People leave, but like the building they worked in, their jobs and the organization remain, seemingly forever.[4]

Some years ago, sociologist Max Weber identified the characteristics of an abstract, or pure, form of bureaucracy:

1. A precise division of labor exists and therefore each individual in the organization is a specialist—an expert in performing a specific task.

2. There is a hierarchy of authority, a chain of command in which each individual is clearly in control of a particular set of people and clearly responsible to another.

**3.** An exhaustive and consistent system of rules has been designed to ensure uniformity in the performance of every task.

**4.** Social relationships between individuals in the organization, especially between superiors and their subordinates, are formal and impersonal.

**5.** The bureaucracy is technically highly efficient, much like a machine.

**6.** Employees are highly trained for their tasks; they are, in turn, protected from arbitrary dismissal: Their jobs constitute a career with possibilities for advancement according to seniority or achievement.

Weber's characteristics described how bureaucracies *are supposed* to operate. Actually, formal organizations don't always follow Weber's description on all or even most points. In any organization, much that is unplanned occurs. The chain of command is often circumvented. Shortcuts reduce bureaucratic red tape in a variety of ways. Relationships between individuals are often personal and informal. Organizations occasionally are grossly inefficient. In fact, to counteract the somewhat negative view that many people have of the huge bureaucracy, organizations often consciously attempt to look and act less formal, less bureaucratic. Nevertheless, Weber's description provides us with a useful abstraction of the general characteristics of large organizations.

The bureaucracy I'm most familiar with is a university. If Max Weber wandered through our institution, he would probably note the following:

*Division of labor, specialization:* There are groundskeepers, secretaries, computer operators, librarians, janitors, and professors. The professors teach specific courses in specific areas, and these courses seldom overlap. In the sociology department I teach criminology, and another person teaches courses on Middle Eastern societies. We are both sociologists, yet he knows little about my area and I know even less about his. We are all highly specialized.

*Hierarchy of authority:* There is an authority ladder in each section and a massive organizational chart for the college (which looks like a map of the Los Angeles freeway system and I'm certain is understood by no one). Hierarchy runs from students (at the bottom) to professors to department heads to associate deans to deans to associate vice presidents to vice presidents to the president. Each must answer to the one above.

*System of rules:* Classes must be added or dropped by a certain day. Grades must be turned in by professors in a certain way by a certain time. Lists of rules describe how students can complain about injustices and how faculty can deal with unruly students. A faculty member wishing to leave campus to go to a professional meeting in another city must fill out a pile of forms weeks in advance of the trip. People are promoted by a mathematical model so complex that it defies comprehension.

*Formal and impersonal relationships:* All are handled alike; people in the same circumstances must be handled the same—the rules say so. We are all cogs in a large machine, and the machine works better if it does not have to concern itself with exceptions. People I know well become members of committees (budget, promotion, grievance), and suddenly they become very formal and impersonal.

*A highly efficient bureaucracy:* We have an assembly-line process with which we mass-educate thousands of students. We take flocks of uneducated high-school seniors and after only 4 years of subjecting them to some 40 sophisticated courses of instruction, we produce educated people.

*Highly trained employees, protected from dismissal:* Most of the faculty have Ph.D.s or the equivalent in their area of specialization. After remaining with the organization for a set number of years, tenure (protection from dismissal) is granted.

But there is another side—the nonbureaucratic side of the bureaucracies, the informal organization in the formal organization. Our institution has that aspect as well. Specialization breaks down somewhat as instructors, because of changing enrollment patterns, move into new teaching areas. It is common knowledge that, contrary to the hierarchy of authority, the university is really run by the secretaries. They are here every day, they work longer hours, they have more knowledge about how the place really works, and they end up making or influencing many of the decisions. Informal small primary groups arise within the organization that counteract formality and impersonality. Often these cut across the authority structure as well. Several faculty lunch groups constantly meet to pass on information and to plan strategy. A phone call to the right person (usually a secretary) quickly cuts through the system of rules and bypasses miles of red tape. The typical organization is also so large that it cannot keep track of bits and pieces. Many professors do not turn in travel plans; they know that they could leave for months and never be missed. And efficiency? The car manufacturer proudly counts the number of units manufactured—3,000 cars today. What they don't tell us is that 2,500 broke down within a mile of the assembly line. Our institution graduates 2,000 people a year, but there is growing concern over the inability of many of the graduates to read and understand the diploma they receive.

I have perhaps exaggerated to make the point: Bureaucracies and large organizations have typical formal characteristics that develop because of their size and functions. They also have typical informal characteristics that develop because of the needs of the people who work in the organizations.

### Types of Organizations

People who study organizations often come to the conclusion that organizations can be categorized. Max Weber categorized organizations as shown in Table 4.1. **Charisma,** which is a basic factor in Weber's charismatic retinue, refers to a certain superior quality an individual has that sets him or her

**TABLE 4.1**  Weber's Types of Organizations

|  | Characteristics | Examples |
|---|---|---|
| Charismatic Retinue | Made up of the followers and disciples of a charismatic leader | Followers of civil rights leader Martin Luther King, Jr.; Reverend Moon's Unification Church; followers of a great scientist or artist |
| Feudal Administration | Separate from but responsible to a parent organization; has local autonomy to act in terms of own interests; is obligated to superiors, but there is no constant flow of orders or supervision | New-car dealerships; chain or franchised stores, agencies, or theaters |
| Modern Bureaucracy | Close to Weber's characteristics of bureaucracy; specialization, hierarchy of authority, system of rules, formality | Governmental agencies, large public utilities, military organizations |
| Modern Professional Organization | Bureaucratic but with more individual freedom, more decentralization of responsibility once the individual's competency is certified | Universities, large hospitals |

Arthur Stinchcombe, "Formal Organizations," in Neil Smelser, ed., *Sociology: An Introduction,* 2d ed. (New York: Wiley, 1973).

apart from others. This individual is viewed as superhuman and capable of exceptional acts—the sort of person you might follow to the ends of the earth if you were asked to.

### Formal Organizations—Consequences and Concerns

*Parkinson's Law: The Formula of 1,000*
*Any enterprise with more than 1,000 employees becomes a self-perpetuating empire, creating so much internal work that it no longer needs any contact with the outside world.*

A proliferation in society of large organizations and the bureaucratic values of efficiency and rationality will have consequences—some obvious, some less so. Large organizations enable the performance of tasks that could not otherwise be accomplished. Products are undoubtedly produced with greater efficiency. At the same time, **alienation** (a feeling of isolation, meaninglessness, and powerlessness) can result when workers see themselves as small cogs in a huge, impersonal machine. Stinchcombe states that in societies dominated by formal organizations, people tend to become specialists in particular activities. Educational institutions respond by encouraging students to concentrate on learning a specific technique or skill rather than attaining a broader, more general education. In societies dominated by formal organizations, a larger percentage of a person's social relationships is planned instead of spontaneous. Interaction is more superficial, and emotions play less of a part in social relationships of this type. People are judged more by what they *do* (actions) than by what they *are* (their individual human qualities). To put it another way, interaction of all types, on the job or off, is more secondary, less primary. Some analysts of modern society have commented with concern on the decline in the number of primary groups with which one can affiliate. They argue that groups and contacts are becoming increasingly secondary in modern industrialized countries like the United States.

Scholars have long been interested in the contrast between primitive and modern societies or, to put it another way, the effects that modernization and industrialization have on a society. Obviously, the size of the society and the degree of modernization and development will change the social relationships within the society. Several of the typologies that have been developed to describe this process are shown in Table 4.2.

Many who feel as Cooley did—that primary group relationships are of great importance in the socialization process and in the development of the self and personality—are concerned that the increasing secondariness or *Gesellschaft*-like nature of modern societies (Table 4.2) will have serious effects on the developmental processes. Some would no doubt point to high rates of divorce, rising crime and delinquency, increased mental illness, and other individual and group pathologies as consequences of restricted primary-group contacts. Others would argue that these conditions are merely the consequences of any modern, complex society. Moreover, the huge, complicated, multilayered bureaucracies that perform tasks for us today are highly efficient even though somewhat impersonal. To idealize the *Gemeinschaft* society, they would argue, is unrealistic and shows a lack of understanding of the positive aspects of the large organizations of today. They would conclude that there is not really more pathology, such as crime, marital breakup, and mental illness; our efficient bureaucracies just do a better job of ferreting it out and letting us know about it.

In the early 1900s, a German sociologist named Robert Michels described a curious and perhaps threatening characteristic of large organizations. He used the phrase "Iron Law of Oligarchy" to describe what happens. Michels felt that **oligarchy**—that is, rule by a few—is inevitable because it is difficult for a large number of people to arrive at a decision efficiently. So, experts emerge to run things, and they eventually become indispensable. Once they are in power, they spend more and more of their time and energy maintaining their positions, rather than working toward the goals of the organization. You would think that the rest of the workers would get upset at this, but on the contrary, according to Michels, they are very happy to have somebody to look after things. The elite, as a consequence of working to stay in power, becomes increasingly conservative, and its interests become predominant over those of the masses they are supposedly serving. Michels

**TABLE 4.2** Types of Societies

| Author | Smaller, Less-Industrialized Societies | Larger, More-Industrialized Societies |
|---|---|---|
| Robert Redfield | **Folk** Isolated villages, small, homogeneous, self-contained, subsistence activities | **Urban** Large, industrialized nations, highly developed division of labor |
| Émile Durkheim | **Mechanical Solidarity** Homogeneous, preindustrial, uniformity among people, lack of differentiation or specialization | **Organic Solidarity** Complex societies, greater specialization and division of labor, increasing individualism |
| Ferdinand Töennies | **Gemeinschaft** Primary, close-knit, relationships are personal and informational, people are committed to community | **Gesellschaft** Secondary, impersonal, utilitarian, rational thought rather than emotion, contractual arrangements, well-developed division of labor |

concludes that in large organizations, oligarchy is inevitable and abuse of power follows.[5]

### "The Organization Man"

Another critique of organizations appeared some 40 years ago when W. H. Whyte, Jr., wrote *The Organization Man*. He painted a rather gloomy picture of how large organizations were affecting people's lives. Americans have long been guided by what has been called the Protestant Ethic—the idea that pursuit of individual salvation through hard work, thrift, and competitive struggle is the heart of the American achievement. According to Whyte, a new ethic has emerged as a consequence of the proliferation of large organizations. Now the young person going out into the world believes that to make a living he or she must do what somebody else tells him or her to do. The idea of individual achievement and the encouragement of personal creativity is disappearing. Instead, the guiding values of the organization person include a belief in the group as source of creativity, in belongingness and togetherness, and in the value of fitting in. These phrases are part of the organizational ethic: "Be a compromiser." "Maintain group solidarity at all costs." "Don't rock the boat." "The unorthodox can be dangerous." The "organization man" is conforming and "well-rounded"; he looks at his company this way:

> *Be loyal to the company and the company will be loyal to you. After all, if you do a good job for the organization, it is only good sense for the organization to be good to you, because that will be best for everybody. There are a bunch of real people around here. Tell them what you think and they will respect you*

*for it. They don't want a man to fret and stew about his work. It won't happen*
*to me. A man who gets ulcers probably shouldn't be in business anyway.*[6]

The organizational ethic maintains that the goals of the organization and the goals of the individual are the same. People don't buck the system; they cooperate with it.

Where did the "organization man" come from? He was produced by a society that is now geared to the large organization. The educational system, which is increasingly influenced by business, is the first step. The university curriculum has become more practical, vocational, and job-oriented, with a declining emphasis on liberal arts and the humanities. And what has become of the "organization man" now? He lives an outgoing, friendly life of inconspicuous consumption in suburbia, where he joins clubs, has kids, and plays golf, and where all the houses, cars, appliances, lawns, and even people tend to look alike.

And if the individual messes up, can't hack it, gets disgusted, and drops out—whose fault is it? The organization's? Certainly not. It's the *individual's* fault; he or she doesn't fit in. Our worship of organizations bothers Whyte, and he objects to the emerging values and behaviors that result.

Is it really as bad as he says? It's difficult to calculate the effects of formal organizations accurately because they are so much a part of our lives; but remember, he wrote *The Organization Man* many years ago (in 1956), and we are certainly a far more bureaucratized society now.

### "McDonaldization"

As Robin Leidner reports, no one walks into a McDonald's and asks, "So, what's good today?" McDonald's success is based on uniformity and predictability. The food is supposed to taste the same every day everywhere in the world and to be served quickly, courteously, and with a smile. Company workers are trained (at Hamburger University) to mass-produce food, friendliness, and good cheer. The company estimates that 95 percent of Americans eat at a McDonald's at least once a year, and the company reported profits of $1.4 billion for the first nine months of 1999.[7]

Take the characteristics of large organizations that we have been discussing in this chapter and combine these with an increasing emphasis on automation, mechanization, and computerization—several observers of society think we have a serious problem. They believe that this apparent obsession with efficiency and predictability is affecting our life in general, life beyond the bureaucracy. Sociologist George Ritzer has called this tendency the "McDonaldization of Society." He believes the "iron cage" of McDonaldization is hard for us to escape in our everyday lives. On every side we see shopping malls, Jiffy-Lube, AAMCO, Toys R Us, Super Cuts, Midas Muffler, H & R Block, Home Depot, Barnes & Noble, Petstuff, Kentucky Fried Chicken, Jenny Craig, and Wal-Mart. From birth (more use of fetal testing and cesarean sections) through death (prearranged funerals), every act is organized, routinized, and made predictable.

Ritzer makes the point that McDonaldization doesn't "work." It is often inefficient (long lines at checkouts), it can make products cost more, and worst of all, it encourages insincerity, superficiality, and dehumanization. But how do we avoid it? Ritzer does have some tips. Some large organizations carve out "skunk works"—situations in which people are insulated from routine and conformity and allowed to be creative and innovative. Work at home whatever hours you like and ignore the chain of command, for example. He has some other suggestions: Always go to the local mom-and-pop store for a haircut, pair of glasses, or tax advice; pay cash; go to restaurants that use real plates and silverware; send all junk mail back to the post office; when dialing a business, always insist on talking to a person; when receiving a call from a computer, set the phone down (don't hang up) and let it ramble on; never go to a domed stadium or one with artificial grass; take no classes that have computer-graded tests; avoid movies that have roman numerals after their names.[8]

The second research question at the beginning of this chapter deals with how companies, especially new companies, are structured. Do they start out highly organized and bureaucratic like the organizations we have just discussed, or do they start out small and informal? A recent paper examined this issue by looking at the early days (the first six years) of young technology start-up companies in California's Silicon Valley. The authors found that the structure of the new companies was determined by several factors. One key variable appeared to be the company founders' view of what the appropriate company structure should be. Some founders believed in the "star" model which envisioned employees attached to the company because of the challenging work, who were selected because of their long-term potential and who were controlled by their commitment to excellence. The "commitment" model saw employees attached to the company because of a family-like feeling or emotional bond, who were selected because they "fit" the company and who were controlled by their peer group. The "bureaucracy" and "autocracy" models imagined employees who were attached to the company because of challenging work (or for the money), who were selected for a specific skill, and who were controlled by formal or direct rules.

What did the study find? Not surprisingly, the authors found that in companies in which founders favored the bureaucratic or autocratic models, managerial intensity, control of employees, and supervision was greater. Both the star and the commitment models were less administratively intense, with the commitment model being by far the least bureaucratic. The effect of the proportion of women in the company was more of a puzzle to the authors. One argument suggests that companies worry about greater job turnover among women, which might lead to greater supervision and control in companies with a higher proportion of women. Some feminists have argued on the other hand that bureaucracy is primarily a male contrivance and that female employees require less supervision and management. Whatever the reason, the difference was significant—start-up companies with proportionately more women during their first year were less bureaucratized than otherwise similar firms.[9]

### Can Formal Organizations Change?

In sometimes interesting ways, formal organizations modify their forms and structures and change their philosophies. For example, bureaucracies have long held a basic belief in specialization: The simpler the task, the better and more efficiently it can be performed. (See Weber's first characteristic of bureaucracy earlier in this chapter.) Break the job down into its basic parts. Make it simple, precise, and repetitive, and the worker can't possibly mess it up. Well, this sort of approach to tasks—three turns on the same nut with the same wrench 3,000 times a day—might be more efficient, but people don't like it. They get bored, job satisfaction goes down, and absenteeism goes up. One popular management approach is to reverse the cherished specialization ethic and to *enrich* jobs by making them more interesting, more varied, and not so repetitive. This view has been supported widely by industry and by business graduate schools. Jobs should be meaningful, self-fulfillment is a worthwhile goal, and work groups are important for improving job satisfaction. Workers will take more pride in the product if they can follow it all the way through, performing a variety of tasks toward its completion.

They might be right, and it certainly sounds like a more enlightened approach to the work situation. But a skeptic might inquire about management's reasons for this new approach. Companies like to feel they are progressive. And, of course, it's good for the image. Ads have portrayed the happy Swedish automobile workers who start working on a car and then carry the job through the whole construction process. You just *know* that's a better car. Industry can be humanistic after all. The Japanese corporate model has also been seen as more worker-friendly. According to the model, the Japanese company is more a family—the employee expects to work for the company for life, decisions are collectively made, employees are broadly trained in all aspects of what the company does, and the company is involved in the worker's whole life, on and off the job.

But the clincher for the organization is the economic payoff. Improved worker morale means less absenteeism, less employee turnover, less sabotage, and improved quality of product. The best of all possible worlds exists if organizations can improve their images and cut costs at the same time. Imagine the company's dilemma, however, if the new program increases worker morale, happiness, and self-fulfillment—but efficiency and output are unchanged or reduced.

Weber's general characteristics of bureaucracy should not blind us to the fact that individual differences in organizations, as in people, are vast. Some are more affected by and responsive to change than others. For example, educational institutions are probably more likely to react to demands for change than are military organizations. Bureaucracies may be responsive to change in one area but resist it in another. The auto industry changes styles yearly, but its responses to suggestions of environmentalists come more slowly. And finally, organizations traditionally have been very slow to change. Problems inherent in industrial bureaucracies have been discussed for more than 100 years, yet little has changed. Today we are more bureaucratized than ever; large formal organizations are the hand-maidens of mass society.

## THEORY AND RESEARCH: A REVIEW

All three of the major theoretical perspectives introduced in Chapter One are included in this chapter, but *functional analysis* again, as in Chapter Three, is the major viewpoint. The characteristic emphasis on structure and function is seen clearly in the discussion of groups, types of groups, small groups, and large organizations. (See, for example, Weber's characteristics of bureaucracy and the analysis of types of organizations in Table 4.1.) *Symbolic interaction* is often helpful in understanding what happens in small groups. *Conflict theorists* would enjoy interpreting the findings of the study of fraternity deviants and also the description of jury behavior. Of the research efforts described in the chapter, Sherif's and Asch's classic studies as well as the study of fraternity deviants were *experiments*. Perhaps the best way to find out how a group works is to join it. Consequently, much of the best research done on small groups and large organizations involves *participant observation*. For example, Robin Leidner worked at McDonald's before writing her book—she even attended Hamburger University!

## SUMMARY

The social organization, or *social fabric*, of a society is woven of cultural norms, values, beliefs, and patterns of behavior through which people inter-act with each other. The analysis and description of this social organization is a major area of study for sociologists. The analysis of the social organiza-tion of a society can be approached from a number of viewpoints. The view-point we have taken in this chapter introduces the concepts of group, category, and aggregation. A *group* exists when a number of people experi-ence shared or patterned interaction and feel bound together by a con-sciousness of a kind. A collectivity of people lacking these traits is called a *category* (if they have a particular characteristic in common) or an *aggregate* (if they are located in a specific area).

Sociologists categorize groups in a variety of ways. Of special importance are primary groups, small groups, secondary groups, and formal organiza-tions. Characteristics of past, present, and future bureaucracies were com-pared. By categorizing and analyzing specific types of groups, we can better understand and explain the variety, importance, and effect of the group affiliations individuals maintain. However, any one group will cut across types and carry several labels. I work for a large formal organization in which I am a member of a group—a sociology department—that could be described as a reference group, a peer group, a voluntary group, a horizon-tal group, a closed group (depending on the job market and how many professors are seeking the California climate), and sometimes a primary group—sometimes a secondary group.

The first of the readings that follow is an excerpt from Christopher Edwards's book, *Crazy for God,* in which he describes how group pressure can be used to manipulate the individual. The second reading, by Barbara Garson, shows how automation and computerization can affect the everyday work situation.

## TERMS FOR STUDY

| | |
|---|---|
| aggregate (109) | mechanical solidarity (121) |
| alienation (119) | oligarchy (120) |
| bureaucracy (115) | open group (111) |
| category (109) | organic solidarity (121) |
| charisma (118) | peer group (111) |
| closed group (111) | primary group (111) |
| folk society (121) | reference group (110) |
| Gemeinschaft (121) | secondary group (112) |
| Gesellschaft (121) | social organization (105) |
| group (108) | urban society (121) |
| horizontal group (111) | vertical group (111) |
| involuntary group (110) | voluntary group (110) |

For a discussion of Research Question 1, see page 110.
For a discussion of Research Question 2, see page 123.

## INFOTRAC COLLEGE EDITION
### Search Word Summary

To learn more about the topics from this chapter, you can use the following words to conduct an electronic search on InfoTrac College Edition, an online library of journals. Here you will find a multitude of articles from various sources and perspectives:

**www.infotrac-college.com/wadsworth/access.html**

| | |
|---|---|
| alcoholism | fast food |
| deviant behavior | autonomy |
| Max Weber | corporate culture |
| bureaucracy | Silicon Valley |

# Reading 4.1

# CRAZY FOR GOD

*Christopher Edwards*

*Groups are important for the ways they affect and shape the individual. We don't generally think of the coercive power of groups or of how they indoctrinate members, but that is also part of group life. Group pressure is a powerful and very effective instrument for producing individual change, as can be seen in this excerpt from Christopher Edwards's book,* Crazy for God. *After his graduation from Yale, Edwards traveled to Berkeley, where he met some young people who invited him to a training camp for a "fun weekend." This led to his joining the Reverend Moon's Unification Church. He spent seven months as a "Moonie" until he was kidnapped by his father and deprogrammed.*

After a quick lunch on the lawn, Family members jumped to their feet, took their guests by the hands, and led them to a nearby field to play dodge ball. As I kicked at the ground, sending up clouds of dust, I savored schoolboy pleasures revisited. I looked around at the usual sea of smiling faces. But now deep in the Family members' eyes I glimpsed a chilling single-mindedness, an intent that both bewildered and fascinated me. What lay behind this supposedly innocent game?

Each group divided up into one of two teams. Each team was appointed a captain who suggested a cheer and team chant. During the entire game our team chanted loudly, "Bomb with Love," "Blast with Love," as the soft, round balls volleyed back and forth. Again I felt lost and confused, angry, remote and helpless, for the game had started without an explanation of the rules. The guests were being moved around the field like robots on roller skates.

"Listen, Chris," Jacob called from the sidelines. "If you don't understand the rules, just chant or cheer as loudly as possible. The important thing is to do whatever a Family member tells you. Remember, unity is everything here."

I dutifully started shouting.

"Louder, louder, Chris. That's it. Just follow me." He placed me in a new position and I clapped mechanically. A few minutes later, he moved me again.

I noticed how aggressively the Family members played, how they constantly eyed their guests, coaxing them to chant out loud. As I clapped and shouted, I could feel my tension slipping away, my sense of involvement growing. In spite of myself, I felt a desire to merge into this Family, this group, this game, to become a part of this vibrant, loving circle. "Give in, Chris," urged a voice within me. "Just be a child and obey. It's fun. It's trusting. Isn't this the innocence, the purity of love, you've been searching for?"

The game ended before I had figured out the rules. . . .

From the book, *Crazy for God,* by Christopher Edwards. Copyright ©1979 by Christopher Edwards. Published by Prentice-Hall, Inc., Englewood Cliffs, N.J. 07632.

Davey read the work assignments for each group. I was assigned to the trail crew, a group that was clearing a path up to the hills so that Family members could conduct spiritual hikes at sunset. After being given our picks and axes in the barn we proceeded to the trail. As we started to work, Edie shouted:

"Now, remember, first-weekers can't talk to first-weekers. Second-weekers can't talk to second-weekers. Promise you'll obey that!"

I resented this kind of regimentation, but I kept my mouth shut.

Seeing how awkwardly I was handling my pick, Scott stepped over to show me how to use it. As we were chatting, Jim suddenly appeared at our side.

"Hey, you're both first-weekers! You aren't allowed to talk to each other. If you want to know something or need help, just ask an older brother or sister. That's what we're here for. We've all been first-weekers. We know it's tough."

Annoyed at the interruption, I muttered "Okay," and picked away silently. Apparently sensing my anger, Jim suggested:

"How 'bout a song?" Did he really think he could dissipate my negative feelings that easily?

Jim distributed songbooks to the "new members," as first-weekers were called. All afternoon, we took turns singing religious and patriotic solos as everybody else busily picked and shoveled away. The hours passed quickly. My initial resentment toward Jim gradually melted away in the mass effect. . . .

The fifty people crammed in the little trailer suddenly whistled and burst into simultaneous applause. Everyone was smiling, grinning from cheek to cheek, delighted by Heavenly Father's power. Marilyn sat down, beaming. We all dug into our squash, bean, and salad dinner as Moses, at the center of the circle, called on another person to give testimony.

I could feel myself getting high, high from this smiling group, this happy Family surrounding me. Whatever my doubts might be, it seemed that this loving circle must have an element of goodness. And if there is goodness, then mustn't it be true?

"Individual Entertainment" followed. Various members of the audience volunteered to sing selections out of the songbook on an impromptu stage. What innocent joy, this entertainment was, I thought, a great way for guests to release their inhibitions and prove themselves to older members. Mary stood up and sang "Go Down, Moses." Keith, as an older member, sang a more militant song of self-righteousness, "Marching on, Heavenly Soldiers." The crowd cheered frantically. Edie, who was keeping close tabs on me as usual, elbowed me.

"What, Edie?"

"Well, aren't you going to sing? You're the only second-weeker who hasn't sung."

"I am?"

"Sure. Now, go sing."

"No thanks. I don't care to."

"Go. Go do what you're told."

"No. C'mon, Edie, I don't want to."

Phil, overhearing the conversation, collapsed in prayer behind me to ask for the salvation of my soul. I could feel my face grow hot. No one had ever prayed for me before, let alone for God to forgive me for not singing. Maybe I just had a lot to learn about the way in which God works. . . .

"Don't be such a spoiled child, Chris," Edie persisted. "Now, come on, do what I tell you. It's for your own good. And remember, if you want to grow in the Family you have to follow center. Come on, are you afraid of what people will think of you? Are you really afraid?"

Me—afraid? I'd never thought of it that way. It looked like I'd have to prove to her that I wasn't and that I wanted to be part of this Family.

"What should I sing?"

"Sing something that indicates your rebirth as a heavenly child. Sing something you'd never sing in Old Life."

As she said Old Life, Edie wrinkled her nose. She paged through the songbook, paused to think, tugged my shirt. "Sing 'God Bless America.' Go on!"

No, I could never sing that song. My God, I had been so ashamed of my country, a country consumed by worship of the dollar and aggrandizement of the self. What would my leftist college friends think if they saw me now? But since that was the most difficult song to sing of all, I decided I'd have to do it to prove to myself that I could, that I could even become a heavenly child if I wished. Edie raised my hand, and Moses called on me. I stood up and walked over to the stage.

"God bless America—land that I love . . . ," I began in a choked voice. An image flashed across my mind. I was nine years old, at a baseball game with my brother and Dad. A scratchy record was playing the National Anthem as the Yankees paused from their pregame warm-up. I felt the same lump in my throat that I was feeling now.

As I continued, singing louder, my voice cracked with emotion. My hands trembled. Everyone was staring at me, trying to establish eye contact, looking completely transfixed in typical Family fashion. My voice trembled and I began to choke again as I spilled out the faintly remembered words.

"God bless America—my home sweet home!"

The audience rose in unison, breaking into frenetic applause. Two brothers picked me up, led me around the room, and dropped me off next to an ecstatic Edie. Tears streamed down my face, tears of joy or sadness, or God knows what in this chaotic caldron of unbridled emotion.

I was a little child once again, the child who cried when Tarzan slipped from his vine or Lois Lane got captured by crooks. The Family kept jostling me, elbowing me, shaking my hand. Nobody had ever loved me so much, nobody had really cared. . . . This was where I wanted to stay, where I could be loved and accepted. This was the place where God wanted me to be. . . .

Remembering how Jacob had handled me that first weekend, I set out to do the same with Bill, to establish a good "give and take," as the lectures called it. It was essential for him to trust me, to consider me his friend, as I led him through the group rituals, deliberately creating a new reality for him, the reality of the group. One of my aims was to "love-bomb" him, to appeal to his ego in every way possible. I wanted him to feel as positive about himself as possible, for he had to feel accepted enough to want to stay before we could remold him, reshape him into the heavenly child we wanted him to be. I couldn't love him for what he was, for according to the Principle he was a product of Satan's world; I loved him for what I wanted him to *become,* an obedient follower of Reverend Moon, the mysterious Messiah behind this movement.

After two hour-long lectures, I took Bill aside and asked him, "What do you think of the Farm, Bill?"

"It's really amazing. The group's sincerity and dedication are fantastic. But I have so many questions about the lectures and about the social service projects in Berkeley, which nobody will talk about in any detail. And I can't get accustomed to the group life in our circles here. You know, how everything comes from or goes to Edie. Is this true in all the groups?"

"Well, yes. We feel that elder members have a special relationship to God, so the elder members direct the activities and mediate the conversations."

"I find that hard to accept, Chris."

"But you get used to it. You will learn that everything works better that way. And you'll find that the older members have a great deal of wisdom."

"Maybe so, but Edie can't or won't answer all my questions."

"Don't worry. They'll all be answered in time. You have to have faith, brother. Just have faith."

I patted him on the back and hugged him.

Our group walked out to the field for the usual weekend dodge ball game. Edie was the team captain for the four groups that composed our team.

"Okay, brothers and sisters. We have to think up a name for ourselves. How about the Heavenly Hustlers?"

"How 'bout the Boonville Bombers?"

"How's that, everybody?" Edie cried.

"Great!"

"Great! Just great!"

"Great!"

"Great!"

As I said "Great," I could hear the emptiness in my voice; I could feel a haunting hollowness in my head. The word echoed over and over after it spilled automatically from my lips. Over the weeks I learned that this word was an instant Family turn-on, as well as a term of obedience. I had learned to say "great" whenever somebody asked me if everything was all right, when they spoke in a tone that indicated that he or she demanded acknowledgment of the God-given authority of their orders.

I had noticed older Family members use the cue phrase in answer to their central figures even in the moments of depression, hopelessness, and despair they experienced while under heavenly subjugation. Their faces looked as rigid as mine, and their cheeks must have hurt as much as mine often did, the muscular pain of constant smiling, hour after hour, day after day.

So it would be the Boonville Bombers vs. Father's Flying Forces. Edie quickly explained the rules of the game, too quickly for the newcomers to understand, leaving them dependent on their spiritual parents, as I had been the first week. I marveled at the beauty of the system.

"Just follow me. I'll help guide you," I explained to Bill. Each team formed three lines, the front line always throwing the ball and back lines moving forward as the front-liners were knocked out of the game.

I realized by now that this was a particularly violent game, one in which a person could unleash his aggressions, his frustrations, at being controlled and manipulated. The older Family members played it extremely well, and I found myself getting better at playing it each weekend, enjoying this rare moment of individual expression. As the game progressed, Bill looked toward me, helplessly, for instructions. I told him where to stand, how to clap and chant, how to dodge the ball. . . .

Bill was obviously enjoying the game now, happy to be part of a team. I knew that teaching teamwork, teaching mutual interdependence was the first step on the road to indoctrination. Edie had explained that you have to consider yourself to be part of the team before you can accept the ideology of the team. This is a most important point, for it means that group affiliation is a stronger force than ideology, which only justifies and reinforces the affiliation. . . .

"I'd like to be excused to go to the bathroom," Bill announced. Edie look startled.

"Can't you wait until the group is finished?" she asked sharply.

"No, I want to go now."

"You must obey the rules of the group, Bill. That was assumed when you came here."

"Sorry, but I have to go *now.*"

"Okay, Chris will go with you."

We trotted off together. As Bill entered the stall, he said "Gee, I don't know why you had to come along."

"It's part of the Heavenly life, brother. We have to be so close that we'd follow each other anywhere." I didn't tell him that the real reason I had followed him was the Family's belief that Satan would attack him if he was left alone. Satan lurked everywhere, even in those stalls. Reverend Moon himself had been quoted many times as saying: "Satan is everywhere, and you are vulnerable to his attack!"

As we left the brothers' bathroom, I reached for Bill's hand. This time he instinctively pulled back. Looking him straight in the eyes, I said sternly, "You want unconditional love, don't you?"

"Why, yes, but . . ."

"No buts. If you want unconditional love, you have to obey the rules. That's all there is to it. As the principles say, Truth is Truth."

Bill reluctantly took my hand, and we returned to our circle, where Edie was leading a sensitivity game designed to get the group accustomed to taking orders from their leader, their center person.

"Now, I want everybody to lie down in the circle and put your feet in the center of the circle. Wiggle your toes. That's it. Now, wiggle your toes to say hello to the partners across from you. Good."

I admired the Family's genius, for I realized that this seemingly innocent exercise was, in fact, part of a complex and powerful mechanism for gaining control. I was to learn over the next few months that it worked as follows: You were cajoled to give up control to a person for five minutes, and that person structured your environment for that time. Then you gave up control for another twenty minutes, following the wave of group singing. Then you listened to lectures, giving up your critical control, since control in the discussion groups was contingent upon accepting the ideology of the lectures. You actually begin to listen to the lectures *only* to gain an awareness of what the group leaders would do to you and how they would justify it, for at that point it becomes a matter of survival.

Bill seemed to be responding well. When Edie asked for his inspiration, he replied, "I'm really impressed by the people here. I can't believe there's a community like this, smack in the center of California, a community of such loving, growing people—a real family. It's just tremendous. It's . . ."

---

## QUESTIONS 4.1

**1.** Outline the techniques the group in this selection used to manipulate the individual. Do groups generally use these same sorts of techniques to persuade group members to conform? If yes, show how; if no, are you sure? And why don't they?

**2.** Why are these group pressure techniques so effective? Discuss for the primary group, reference group, socialization, resocialization, and so on.

**3.** What was the point of age-segregating (first-weekers and second-weekers) the prospective members?

---

## Reading 4.2

# THE ELECTRONIC SWEATSHOP

*Barbara Garson*

---

*Barbara Garson believes that "the underlying premise of modern automation is a profound distrust of thinking human beings." She feels that computers and automation can be useful, but when they define and control the whole work experience, they can turn work, which should be a pleasant and creative experience, into an "electronic sweatshop."*

*In this excerpt from her book, Garson describes jobs in three workplaces (a griddleman at McDonald's, a reservations agent for American Airlines, and two social workers at the NTW [Not the Worst] welfare center in Boston). She focuses on how jobs have been changed by the emphasis on automation and computers.*

### Jason Pratt

"They call us the Green Machine," says Jason Pratt, recently retired McDonald's griddleman, "'cause the crew had green uniforms then. And that's what it is, a machine. You don't have to know how to cook, you don't have to know how to think. There's a procedure for everything and you just follow the procedures."

"Like?" I asked. I was interviewing Jason in the Pizza Hut across from his old McDonald's.

"Like, uh," the wiry teenager searched for a way to describe the all-encompassing procedures. "O.K., we'll start you off on something simple. You're on the ten-in-one grill, ten patties in a pound. Your basic burger. The guy on the bin calls, 'Six hamburgers.' So you lay your six pieces of meat on the grill and set the timer." Before my eyes Jason conjures up the gleaming, mechanized McDonald's kitchen. "Beepbeep, beep-beep, beep-beep. That's the beeper to sear 'em. It goes off in twenty seconds. Sup, sup, sup, sup, sup, sup." He presses each of the six patties down on the sizzling grill with an imaginary silver disk. "Now you turn off the sear beeper, put the buns in the oven, set the oven timer and then the next beeper is to turn the meat. This one goes beep-beep-beep, beep-beep-beep. So you turn your patties, and then you drop your re-cons on the meat, t-con, t-con, t-con." Here Jason takes two imaginary handfuls of reconstituted onions out of water and sets them out, two blops at a time, on top of the six patties he's arranged in two neat rows on our grill. "Now the bun oven buzzes [there are over a half dozen different timers with distinct beeps and buzzes in a McDonald's kitchen]. This one turns itself off when you open the oven door so you just take out your crowns, line 'em up and give 'em each a squirt of mustard and a squirt of ketchup." With mustard in his right hand and ketchup in his left, Jason wields the dispensers like a pair of six-shooters up and down the lines of buns. Each dispenser has two triggers. One fires the premeasured squirt for ten-in-ones—the second is set for quarter-pounders.

"Now," says Jason, slowing down, "now you get to put on the pickles. Two if they're regular, three if they're small. That's the creative part. Then the lettuce, then you ask for a cheese count ('cheese on four please'). Finally the last beep goes off and you lay your burger on the crowns."

"On the *crown* of the buns?" I ask, unable to visualize. "On top?"

"Yeah, you dress 'em upside down. Put 'em in the box upside down too. They flip 'em over when they serve 'em."

"Oh, I think I see."

"Then scoop up the heels [the bun bottoms] which are on top of the bun warmer, take the heels with one hand and push the tray out from underneath and they land (plip) one on each burger, right on top of the re-cons, neat and perfect. [The official time allotted by Hamburger Central, the McDonald's headquarters in Oak Brook, Ill., is ninety seconds to prepare and serve a burger.] It's like I told you. The procedures make the burgers. You don't have to know a thing."

"I would never go back to McDonald's," says Jason. "Not even as a manager." Jason is enrolled at the local junior college. "I'd like to run a real restaurant someday, but I'm taking data processing to fall back on." He's had many part-time jobs, the highest-paid at a hospital ($4.00 an hour), but that didn't last, and now dishwashing (at the $3.35 minimum). "Same as McDonald's. But I would never go back there. You're a complete robot."

"It seems like you can improvise a little with the onions," I suggested. "They're not premeasured." Indeed, the reconstituted onion shreds grabbed out of a container by the unscientific-looking wet handful struck me as oddly out of character in the McDonald's kitchen.

"There's supposed to be twelve onion bits per patty," Jason informed me. "They spot check."

"Oh come on."

"You think I'm kiddin'. They lift your heels and they say, 'You got too many onions.' It's portion control."

"Is there any freedom anywhere in the process?" I asked.

"Lettuce. They'll leave you alone as long as it's neat."

"So lettuce is freedom; pickles is judgment?"

"Yeah but you don't have time to play around with your pickles. They're never gonna say just six pickles except on the disk. [Each store has video disks to train the crew for each of about twenty work stations, like fries, register, lobby, quarter-pounder grill.] What you'll hear in real life is 'twelve and six on a turn-lay.' The first number is your hamburgers, the second is your Big Macs. On a turn-lay means you lay the first twelve, then you put down the second batch after you turn the first. So you got twenty-four burgers on the grill, in shifts. It's what they call a pro-duction mode. And remember you also got your fillets, your McNuggets. . . ."

"Wait, slow down." By then I was losing track of the patties on our imaginary grill. "I don't understand this turn-lay thing."

"Don't worry, you don't have to understand. You follow the beepers, you follow the buzzers and you turn your meat as fast as you can. It's like I told you, to work at McDonald's you don't need a face, you don't need a brain. You need to have two hands and two legs and move 'em as fast as you can. That's the whole system. I wouldn't go back there again for anything."

### Kenny

Until recently, airline reservation agents have been high-paid, long-term employees. They were valued because they had to learn all the company's routes, fares and policies. Then they had to apply this knowledge while responding on the telephone to the thousands of turns that even the simplest conversation can take. It seemed that this would always involve a great deal of personal judgment. After all, a two-way conversation isn't made out of interchangeable parts.

But in a feat of standardization even more phenomenal than McDonald's fry vat computer, the airlines have found ways to break down human conversation into predictable modules that can be handled almost as routinely as a bolt or a burger.

This has already downgraded reservation work. And as I listened to airline reserva-tion agents I slowly realized that something more than their job was being simplified.

I met young people like Kenny at American Airlines, who explained the dozens of ways in which the two-minute reservation conversation was now timed and graded. Kenny, a new and enthusiastic employee, told me that he liked the work, despite its stresses, because he liked being "involved with people." This modern young man seemed to have no idea that involvement had once meant something far more spontaneous than what he now did.

I also met many reservation agents (most of them older) who understood how their own responses were being streamlined, and they resisted. They would have followed company instructions for assembling a car, I believe, but assembling a conversation was somehow different.

"There's AHU, that After Hang Up time. It's supposed to be fourteen seconds. It just came down to thirteen. But my average is five seconds AHU, because I do most of the work while the customer's still on the phone. There's your talk time, your availability, your occupancy—that's the percent of time you're plugged in, which is supposed to be 98 percent. That's not the same as your availability. There's bookings. You're supposed to book 26 percent of your calls, which is very low. I've averaged 37 percent in the five months I've been here."

Kenny was explaining the computer-collected statistics on which his first raise would be based. Raises currently ranged from 2 to 7 percent. The exact rate depended on how many of the numerical standards, for AHU, talk time, occupancy and so on, you met or exceeded.

"One of the big numbers up to now is conversion rate." I'd never heard of a conversion rate.

"Every call is either action or potential," Kenny explained.

"If someone calls up and says, 'Hi, I want to go to L.A. first class,' that's action. There's nothing I have to do but get the information.

"But the fare shopping! [He held his hands up in horror.] Where they want to know is it cheaper if I go in the middle of the week or at night or after the holiday."

"You mean like I want to visit my mother in Florida, so I listen to all the rates and then say, 'Thanks, I'll call back after I talk to my mother'?"

"Right. Now if someone were listening in, I would be graded on how I probed for the business."

"O.K.," I said, becoming the customer. "Thank you. I'll have to call my mother and see what time is most convenient for her."

Kenny played the reservation agent: "And do you think you'll be flying toward the beginning or the middle of the week?"

"Well, uh, I have to be back by the following weekend so I guess the beginning."

"Tuesday we have a flight leaving at ten and at four. Which would you prefer? (See, I'm trying to get you to make a tentative booking.) Ten?"

"But I don't really know yet. I'll call back after I . . ."

"So I give you a sales pitch. 'It costs nothing to save a seat. There's limited seating at the SuperSaver fare.' I could give them every sales pitch in the world and they still say [now he imitates me, repeating like a parrot], 'I have to call my mother, I have to call my mother, I have to call my mother.'

"If we're going round and round and you keep resisting, I move into what we call a close. 'I still have room on the flight leaving at ten on Tuesday. Let's hold the seat for you tentatively. What's your last name?' And if you say, 'My last name is Garson,' then I know I've booked it. I've converted a potential!"

"Congratulations." I shook his hand across the table.

"But if you still say, 'Dah dah dah dah,' then I could go back to a sales pitch. 'We're coming on the holidays; summer is a hard time to fly; it doesn't cost you anything to hold the booking. Now of the ten o'clock and the four o'clock, which do you prefer?'"

"You're closing again," I said.

"No," he explained, "'Which do you prefer' is called a probe. 'What's your last name' is a close. Now if I come to the close and they don't book I could start around again. I could go around and around for two hours. It's always a pull between keeping your talk time down and getting your conversion rate up. There's no actual rule on it. Generally, I say if you come to close twice and they don't book, let them go."

I stared at Kenny. "That's amazing." I knew, of course, that good salesmen had always analyzed every phase of a contact. But it had never occurred to me that my two-minute reservation conversation could be broken down into opening, sales pitch, probe and close. When Kenny is monitored live, by his supervisor, he'll also be graded for such subjective qualities as tone of voice and effective use of name, as in, "If you like, Mrs. Garson, I can book you on the . . ."

"How does the monitoring work?" I asked. "I mean the human monitoring, what does it feel like?"

"We don't know when we're being listened to. You know afterwards because normally you get calls one after another without any let up. Suddenly there's a break, you look around and you see there's calls coming in to the other agents but you're not getting any. Then you hear a voice in your ear, 'Hello.' It's the supervisor from a listening post. Mine is usually on another floor. There are listening posts all over the office.

"You never know when it will happen. I haven't been monitored yet this month so I figure some time this week. It's usually once a month."

"What do they say?"

"Let me try to remember how it was last month. She said, 'Hello, I've been listening to your calls for an hour and you've been doing everything fine. Keep it up.'"

"What would a criticism be like?"

"Usually it's, 'You didn't ask for the business.' 'You should have closed.'

"Sometimes they also listen from the corporate headquarters in Dallas. It's called trunk monitoring. But *they* never come in on the line. You never know about Dallas."

## Eddy Malloy

When I was in college, many people studied social work with the expectation of a civil service job in some do-good agency, the largest of which was Welfare. They knew they'd never get rich as social workers, but, like teaching, it was a respectable profession that paid them for helping others. Besides, it was a secure job—for we have the poor always with us.

The main reward of these jobs was a work style (lifestyle was a later invention) that allowed social workers to interview, visit, fill out forms (alas), make evaluations and decisions and generally function as free human beings while doing worthwhile work.

CPAs deal with numbers. Lawyers deal with contracts. Social workers deal with people. It was the right kind of work for my fellow Berkeley students who pinned IBM registration cards onto their chests to declare *Do not fold, spindle or mutilate.* They demanded to be treated as individuals, not numbers. So they chose to work with individuals, not numbers.

Today I don't meet many undergraduates going into social work. In part this is a matter of fashion. Poor people are out of style. Teaching similarly declines in prestige when children decline in prestige.

But by now social work isn't just a less-prestigious or lower-paid profession; it's hardly a profession at all.

Of course the poor are still with us. So are the welfare centers. But these centers are no longer staffed by autonomous professionals. They're staffed by increasingly regimented clerks. These automated welfare workers have no more chance to deal with people than do airline reservation agents.

How can the "helping professions" be clericalized? How can you eliminate the listening, talking, asking and advising from a job intended to help? It sounds impossible to make social work as physically and mentally constrained as an assembly line job, but that's exactly what's being done.

"You mean that for every form you fill out, you have to fill out another form saying you filled out a form?"

"You'll never be able to get this system straight," Eddy expressed his confidence in me. For the entire lunch hour he'd been trying to explain the point system that had evolved from the time studies.

It went roughly like this:

Time standards, in tenths of an hour, had been established for each function that an FAW [Financial Assistance Worker] like Eddy performs. For instance:

Issue food stamp I.D. card _____ .3
    [meaning three-tenths of an hour, or 18 minutes]
Authorize funeral and burial expenses _____ .7
Replace lost or stolen check _____ .4
Complete redetermination _____ 1.8

The full list contained over sixty functions and was frequently updated.

Each function, like "update medical report" encompassed many "actions," such as going downstairs to see what the client wants, pulling the file, filling out several forms, distributing the forms to supervisors or the Electrical Data Processing center (EDP), returning the case file. The original time studies had identified and assigned time standards for over 1000 actions.

But how did they keep track of it all?

When a worker completed a function—when, for instance, he handed his supervisor the forms necessary to record a change of address or a change of food stamp benefits—he filled out an additional form called a "control slip." At the end of the month the supervisor used these control slips to count the number of points (tenths of hours) each worker had earned.

It was remarkably like the piece-work system in a mink factory I had visited. There, as you finished cleaning each pelt you tore off a tag and tossed it into a cigar box for the supervisor to count. Here, as you finished adding a baby you handed in a control slip. (At the welfare center the control slip was sometimes called a "penny slip," just as it is in a garment shop.)

At the end of the month the supervisor counted the worker's points, his tenths of hours, and calculated how close he came to using 100 percent of his available time. Each worker received a monthly report card with a percentage grade. So far a worker wasn't disciplined unless he fell below 70 percent. That was the system after the time study but before full computerization.

"So these control slips mean you fill out a form for every form you fill out?" I asked incredulously.

"You'll never get it down straight," Eddy assured me. "They keep changing everything anyway."

### Daniel Sheridan

Several people had described Daniel Sheridan as the fairest and most efficient supervisor at NTW. "Straight arrow" was the phrase that kept coming up.

. . . When we were seated and served, I posed my first question as open-endedly as I could. "What do you think of the time standards and the point system?"

"I blame the union for the way it's operating," Daniel answered concisely.

"You mean because they're sabotaging it?"

"No, because they're not sabotaging it."

"What do you mean?"

"If they followed the rules as the department issued them, this system would have collapsed in three months."

"You want it to collapse?"

"If I were a worker and a union activist," Daniel said, "the first time I did 100 percent in the first three weeks of the month I'd stop work. And if they tried to make me do anything over 100 percent, I'd fill out an overtime form.

"The problem," he declared, "is that all the workers have developed systems of their own to get the points they need and still deliver a timely service. That's what keeps this place going."

---

## QUESTIONS 4.2

**1.** See how many of Max Weber's characteristics of bureaucracy you can find in this reading.

**2.** Find examples of the other side— the nonbureaucratic side, the informal organization within the formal organization.

**3.** Look at the criticisms of bureaucracy under "Formal Organizations— Consequences and Concerns" in this chapter and describe which ones you think fit here. Or, with respect to the jobs described in this reading, analyze the statement: "It may seem cold and impersonal, but it's the fastest and most efficient way to get things done."

**4.** Make a list of five other jobs (perhaps several at your college or university) and show how they have become automated and computerized. Do you think this makes them less pleasant and creative?

---

## NOTES

1. See "Religiosity, Social Class and Alcohol Use: An Application of Reference Theory," by Leslie Clark, Leonard Beeghley, and John Cochran, *Sociological Perspectives* 33, (Summer 1990), pp. 201–218.

2. See Clovis Shepherd, *Small Groups* (San Francisco: Chandler, 1964), pp. 27–36. Also see Michael Olmsted, *The Small Group* (New York: Random House, 1959), pp. 117–132; *The Small Group,* 2d ed., by Michael Olmsted and Paul Hare (New York: Random House, 1978); Muzafer Sherif, *The Psychology of Social Norms* (New York: Harper, 1936); Solomon Asch, "Effects of Group Pressure on the Modification and Distortion of Judgment," in *Groups, Leadership and Men,* edited by M. H. Guetzkow (Pittsburgh:

Carnegie, 1951); S. Asch, "Studies of Independence and Conformity: A Minority of One Against a Unanimous Majority," *Psychological Monographs* 70, no. 9 (1956), whole no. 416; and Lowell W. Gerson, "Punishment and Position: The Sanctioning of Deviants in Small Groups," *Case Western Reserve Journal of Sociology*, vol. 1 (1967), pp. 54–62.

3. For more detail on these and other studies of jury behavior, see Hans Toch, ed., *Legal and Criminal Psychology* (New York: Holt, Rinehart & Winston, 1961).

4. See Arthur Stinchcombe, "Formal Organizations" in *Sociology: An Introduction*, 2d ed., edited by Neil Smelser (New York: Wiley, 1973). Also see Peter Blau and Marshall Meyer, *Bureaucracy in Modern Society*, 2d ed. (New York: Random House, 1971) and H. H. Gerth and C. W. Mills, *From Max Weber: Essays in Sociology* (New York: Oxford University Press, 1958).

5. The major source for this section is *Elites and Masses* by Martin Marger (New York: Van Nostrand, 1981). Especially see pp. 70–73.

6. William H. Whyte, Jr., *The Organization Man* (Garden City, N.Y.: Doubleday, 1956), p. 143.

7. See *Fast Food, Fast Talk*, by Robin Leidner (Berkeley: University of California Press, 1993), especially pp. 45–46.

8. See *The McDonaldization of Society*, by George Ritzer (Thousand Oaks, Calif.: Pine Forge Press, 1996). Material in these paragraphs benefits especially from pp. 3, 196–201, and chapters 8 and 9.

9. See James Baron, Michael Hannan, and Diane Burton, "Building the Iron Cage: Determinants of Managerial Intensity in the Early Years of Organizations," *American Sociological Review* 64 (August 1999), pp. 527–547.

## CHAPTER FIVE

Leonard Freed/Magnum Photos, Inc.

# Social Inequality:
# Social Class

**Social Stratification**

**Stratification Systems**

**Parallel Stratification Systems**

**Socioeconomic Status and Social Class**

*Wealth • Power • Prestige • How Many Social Classes? •
Determining Social Class • Social Class and Behavior • Social Mobility*

**Poverty and Welfare**

*The Welfare System • World Poverty*

## TWO RESEARCH QUESTIONS

*1. Are wealthy women more likely than poor women to have childbirth by cesarean section? Or would it be the other way around, or no difference? Why?*

*2. Are women on welfare more likely to give birth to children than women not on welfare? How does length of time on welfare affect the likelihood of giving birth?*

SOCIOLOGISTS STUDY GROUPS and nongroups to understand the social organization of society. The previous chapter dealt with groups. In this chapter our attention turns to nongroups. Many sociologists spend much of their time specializing in the study of *categories* of people—people who have a particular characteristic in common but who do not constitute a group. Those who specialize in the study of older people, an age category, are called gerontologists. Others study race, ethnic, religious, social-class, and gender categories.

The process of defining, describing, and distinguishing among different categories of people is called **social differentiation.** People differ across a range of variables, and we find ourselves automatically in some of these categories, such as age, gender, and race. Other categories, such as social class and religion, have a greater degree of flexibility, and to a certain extent one's position within them can be changed. (Recall the discussion of ascribed and achieved status in Chapter Three.)

Social differentiation leads to **social inequality.** As we differentiate among people, we "rank" them at different levels—we make judgments of inequality. These are social definitions, and they can differ from one society to the next, but they have important social meaning. Whether you have brown or blue eyes is probably of little significance to your future. However, think of the social meaning attached to being black or white, male or female, old or young, rich or poor. Among the questions that sociologists are interested in are these: *What* distinctions are made, *why* are these distinctions made, and what are the *consequences* of these distinctions for the people involved?

To get a clearer idea of social differentiation and its consequences, let's imagine two boys, Billy and Tommy. Billy is the one-year-old son of black parents in rural Mississippi. His parents are tenant farmers, the family income is less than $3,000 a year, he has six brothers and sisters, and neither of his parents went beyond the eighth grade in school. Tommy is the one-year-old son of white parents in Scarsdale, New York. His father is a surgeon, his mother is an author, the family income is more than $150,000 a year, he has one brother, and both his parents are college graduates. Now map out each boy's future. What are the chances of each living to the age of six? Of graduating from high school? Of going to college? Of going to an Ivy League school? Of receiving regular medical and dental care? Of living to the age of 65? Of getting married while still in his teens? Of having many children? Of having an arrest record? Of going to prison? Of having his name in *Who's Who in America?* These are two healthy, happy, bouncing babies, but their life chances are vastly different.

In this chapter and the next, we will focus on several important categories—social class, race, ethnicity, age, and gender—all of which involve judgments of social inequality.

## SOCIAL STRATIFICATION

*Semper Anglia—A dignified English solicitor-widower with a considerable income had long dreamed of playing Sandringham, one of Great Britain's most exclusive golf courses, and one day he made up his mind to chance it when he was traveling in the area.*

*Entering the clubhouse, he asked at the desk if he might play the course. The club secretary inquired, "Member?" "No, sir." "Guest of a member?" "No, sir." "Sorry."*

*As he turned to leave, the lawyer spotted a slightly familiar figure seated in the lounge, reading the* London Times. *It was Lord Parham. He approached and, bowing low, said, "I beg your pardon, your Lordship, but my name is Higginbotham of the London solicitors Higginbotham, Willingby, and Barclay. I should like to crave your Lordship's indulgence. Might I play this beautiful course as your guest?"*

*His Lordship gave Higginbotham a long look, put down his paper and asked, "Church?" "Church of England, sir, as was my late wife." "Education?" "Eton, sir, and Oxford." "Sport?" "Rugby, sir, a spot of tennis and a number four on the crew that beat Cambridge." "Service?" "Brigadier, sir, Coldstream Guards, Victoria Cross, and Knight of the Garter." "Campaigns?" "Dunkirk, El Alamein, and Normandy, sir." "Languages?" "Private tutor in French, fluent German, and a bit of Greek."*

*His Lordship considered briefly, then nodded to the club secretary and said, "Nine holes."*

The study of **social stratification** arises from the recognition that people are ranked or evaluated at a number of levels. As the geologist finds layers when examining a cross section of the earth, so the sociologist finds layers in the social world. Societies today are stratified into many layers, but this was not always the case. *Hunting and gathering* societies, the oldest known, were very egalitarian. Members of these small (50 people or less) societies spent all their time finding food; there was little, if any, division of labor and little social differentiation. The *horticultural* societies that followed were larger (from 200 to 4 million people). They were able to cultivate domestic plants and thereby produce a surplus, which led to activities beyond pure survival. Jobs aside from food gathering emerged and accumulation of wealth began. Specialization, division of labor, and presence of a surplus produced the beginnings of social stratification.

This trend continued in *agrarian* societies. Improved tools—the plow replaced the digging sticks and hoes of the horticultural societies—meant more efficient and productive use of the land and greater surpluses. This meant that a smaller proportion of the population remained involved in food production while more people occupied themselves as makers of tools and weapons, and as miners, merchants, traders, and soldiers. The division of labor and the greater surpluses in agrarian societies brought pronounced social stratification and social-class differences. These trends have continued and probably accelerated in *industrialized* societies. The division of labor is greater and stratification is marked.

## STRATIFICATION SYSTEMS

Different types of stratification systems have developed over time and in different parts of the world. **Slavery** is a form of stratification (based on economics) in which some human beings are owned by others. For a system of slavery to exist, laws in the places where it exists must specify that persons can be defined as the property of other persons. People have become slaves through birth, as a consequence of committing a crime, after a military defeat, through indebtedness, or by capture and sale. A number of countries have practiced slavery including Greece, the United States, Brazil, South Africa, and Canada. Slavery was not a totally closed system—occasionally, movement out of slave status was possible.

The **caste** system is a rigid system of stratification that is closed—one is born into a particular caste and movement to another is not possible. India's caste system is based on religion and consists of four main castes (priests, warriors, merchants, common laborers) and many subcastes. In addition, a large number of people (some 20 percent of the population) are outside the caste system and are called "outcasts." Because they are seen as lowly and unclean, they are commonly referred to as "untouchables." Each caste has specific rules about how to behave, and relationships between castes are carefully defined. South Africa is in the process of trying to dismantle a caste system based on race called apartheid. In Japan, members of a pariah caste called the Burakumin are seen as descendants of a less human race than that of the Japanese nation as a whole. The Burakumin are stigmatized and segregated residentially and occupationally. In the United States, a caste system based on race developed after the abolition of slavery. There were strict restrictions on access to schools, rest rooms, restaurants, hotels, and drinking fountains.

**Estate** systems have usually been found in advanced agricultural societies in which land is the most important economic resource. A few people, perhaps less than five percent of the population, own much of the land, and they receive most of the income. Sometimes there are two estates (landowners, peasants) but more often there are three (aristocracy or land owners, clergy, and peasants). Although on rare occasions one could move from one rank to another (peasant to clergy, or skilled warrior to landowner), the estate system was closed and there was almost no movement between levels. Estate systems have existed in Egypt, Central and South America, Russia, England, and France. Remnants of estate systems can be seen today, where wealthy landowners, often unknown and anonymous people, have great wealth based on inherited property. For example, in England, the Royal Family owns a large amount of valuable land, and a Viscount Portman owns 100 acres of expensive land in London along Oxford Street that was granted to his family in 1533.[1]

The industrial revolution created a new stratification system called the **class** system. Industry replaced land ownership as the driving energy. A more complex society developed that required a more educated and highly skilled workforce, and an extensive division of labor. This system was more open—movement from one level to another was more feasible than it had been in caste or estate systems. This was because, at least in theory, rank was based on merit and achievement rather than birthright. Also, at least in theory, the great separation in material benefits between elites and masses that exists in caste and estate systems was reduced. Many changes accompanied the industrial revolution and development of a class system including the following: movement from farm to city; movement from small firms to larger, more concentrated firms; and growth of government, bureaucratization, science, and technology.

## PARALLEL STRATIFICATION SYSTEMS

Most societies have more than one system of stratification. These parallel stratification systems are based on different factors. The system we usually use when discussing social class is based on occupation, education, wealth, power, and prestige. We could refer to this as socioeconomic status. Another stratification system might be based on race: Individuals are ranked high or low purely on the basis of racial characteristics. We see this kind of stratification in the United States and to a greater extent in South Africa. Age, gender, and religion can be the bases for other stratification systems. In some societies, older people automatically have higher status than do younger people. Northern Ireland is religiously stratified—the battles between Protestants and Catholics are often in the news.

When there are a number of parallel stratification systems, an individual will be ranked in several different systems at the same time: by race, then by socioeconomic factors, then by age, and so on. And though the stratification systems are parallel, they are not necessarily equal. The status of an American black who is upper-class in wealth and occupation might be roughly equivalent to the status of a white who is middle-class in wealth and occupation. The discrepancy is explained by the existence of two stratification systems—one that ranks a person according to socioeconomic status, and a second according to race.

Our society, then, is stratified in numerous ways. In the remainder of this chapter and in the next chapter we will examine several ways by which people are categorized and stratified. The subject of this chapter will be socioeconomic stratification. The following chapter will deal with race, ethnic, gender, and age categories.

## GOLD—WHO'S GOT IT?

Who makes the most money in American society? Is it people in business, entertainment, or sports? Well, people at the top in any of these fields do well, as can be seen below. The figures show 1998 income (in *millions* of dollars) of the top five earners in each of the three areas as reported by *Forbes* magazine.

### Business

| Who | How much |
| --- | --- |
| M. Eisner, Disney | $589 |
| M. Karmazin, CBS | 202 |
| S. Case, American Online | 159 |
| S. Hilbert, Conseco | 125 |
| C. Barrett, Intel | 117 |

### Sports

| Who | How Much |
| --- | --- |
| M. Jordan, basketball | $69 |
| M. Schumacher, auto racing | 38 |
| S. Fedorov, hockey | 30 |
| T. Woods, golf | 27 |
| D. Earnhardt, auto racing | 24 |

### Entertainment

| Who | How much |
| --- | --- |
| J. Seinfeld | $267 |
| L. David | 242 |
| S. Speilberg | 175 |
| O. Winfrey | 125 |
| J. Cameron | 115 |

—From *Forbes*, March 22, and May 17, 1999.

## SOCIOECONOMIC STATUS AND SOCIAL CLASS

Status based on socioeconomic factors represents one of the major systems of stratification. Following the ideas of Max Weber, socioeconomic status is usually determined by wealth, power, and prestige.[2]

### Wealth

Generally, when comparing and evaluating people, we rank higher those who have the greater *wealth* and store of material possessions—type and size of house, area of residence, make and number of cars, quality of clothes, and so on. In a society that places value on wealth and material possessions, it becomes important to allow others to find out how well-off one really is, and so we have **status symbols.** If a person drives a Rolls Royce, Lincoln Continental, or exotic foreign car with a low-numbered license plate; lives in Beverly Hills or on Park Avenue; is mentioned in the society pages; and winters in Palm Springs and summers in Switzerland, then we have a pretty good idea of where he or she belongs on the social-class ladder. The game becomes complicated, however, when status symbols change or become generally available. Then it becomes more difficult to tell who ranks where. Only the Internal Revenue Service knows for sure.

## WEALTH

On a day in July 1997, Microsoft stock increased in value by about $10 a share, which meant that Chairman Bill Gates's personal wealth increased by almost $3 billion! To put this in perspective, $3 billion would finance more than a dozen trips to the moon or buy 17 new Boeing 747s. Or even better, imagine a person who makes $150,000 a year—it would take that lucky person 20,000 years to make what Gates did in one day.

In June 2000 it was reported that the assets of the world's four richest men (Bill Gates, Larry Ellison, Warren Buffett, and Paul Allen) amounted to $163 billion, a figure that substantially exceeded the gross national product of the 43 least-developed countries whose total population is about 600 million people.

Wealth is strongly correlated with education, income, and occupation, and when socioeconomic status is measured, these other factors are usually included. Income refers to current earning capacity, whereas wealth refers more to an accumulation of money and property over time. The relationship between income and education is shown in Figure 5.1.

### Power

People are also ranked according to the amount of power we think they have. **Power** refers to the ability of one party (either an individual or a group) to affect the behavior of another party. Individuals who run things, who make the important decisions at a city or national level, whether they are formally elected or informally appointed, are accorded high status by those who know of the power they exercise. Elected government officials, advisors to presidents, and consultants in high, mysterious-sounding positions are automatically given high status because we know that they are at or near the centers of power.

Power is a fascinating variable that stimulates numerous questions: Who has it? How did they get it? How is it used? How is power distributed in society? Some analysts of American society believe that power is concentrated in the hands of a few people who are not directly responsible to anyone else. Others believe that power is fairly equally spread across a number of interest groups who respond to the will of the people. (These issues are hotly debated, and we will return to them in more detail in Chapter Nine.)

Some theorists suggest that it is basically wealth, accompanied by the related variables of education and occupation, that establishes a person's position in society and that this wealth *leads* to power. Wealth is not shared equally in our society and never has been. In 1990 the IRS reported that in the most recent year available (1986), the wealthiest 1.6 percent of the United States population—those with assets of $500,000 or more—controlled 29 percent of the wealth of the country. In fact, their holdings exceeded

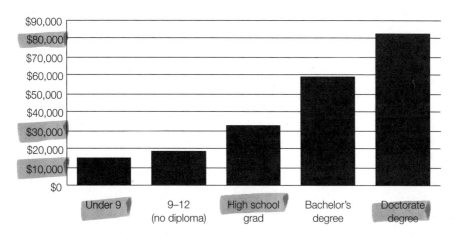

**FIGURE 5.1** *Median income and school years completed, 1996.*

United States Bureau of the Census, *Statistical Abstracts of the United States, 1998* (118th ed.), Washington, D.C.

**TABLE 5.1**  Proportion of Income Received by Each Fifth of the Population

| Population Group and 1996 Income | 1950 | 1985 | 1996 |
|---|---|---|---|
| Wealthiest fifth (more than $75,317) | 43% | 44% | 47% |
| Second fifth ($51,087–$75,316) | 23% | 24% | 23% |
| Third fifth ($34,316–$51,086) | 17% | 17% | 16% |
| Fourth fifth ($19,681–$34,315) | 12% | 11% | 10% |
| Poorest fifth (under $19,680) | 5% | 5% | 4% |

United States Bureau of the Census, *Statistical Abstracts of the United States, 1998* (118th ed.), Washington, D.C.

the entire gross national product of the country for that year! As the data in Table 5.1 indicate, the top fifth of the population controls far more than its share of national income, and the poorest fifth has only four percent of the national income. To put it another way, one-fifth of the population in the United States had nearly half of the annual income. The wealthiest 5 percent of the population in 1996 (those with an income of $128,000 or more) had 20 percent of the income.

### Prestige

**Prestige** means distinction or reputation and refers to how we subjectively evaluate others. It is usually connected to the position one holds in society. Some positions have high prestige (doctor, Supreme Court justice); others carry low prestige (shoeshiner, garbage collector). Prestige—distinction in the eyes of others—is also important although less concrete than wealth and power.

## STATUS SYMBOLS

If you live in Hong Kong, driving a Mercedes without an "appropriate" license plate is as awkward as wearing a tuxedo with sneakers. Chinese are known throughout Asia for their obsession with "lucky" numbers, and Hong Kong, where enormous amounts of money are paid to get the right license plate, is a remarkable example. Wealthy Chinese will pay thousands and sometimes millions of dollars for the right plate. A local business tycoon recently set a record by paying $3 million for the single-digit registration number nine, or "gaau," which sounds similar to the Cantonese word for longevity. Another unusual plate—AA4444—sold for $14,900. This is odd because four (a homonym of the word for death in Chinese) is usually considered unlucky, but apparently rarity outweighed bad luck. The government accommodates the situation by holding public auctions of special numbers—between 1973 and 1994 more than 200 auctions have been held, raising about $66 million.

Closer to home, everybody knows that the area code for New York City (Manhattan) is 212. By July 1999, because of numerous cell phones, faxes, and computers, phone numbers with that area code were scarce. The area code 212 became a status symbol, and the new area code (646) was not nearly as prestigious. 646? That could be Brooklyn or Staten Island or anyplace.

---

In 1947, Cecil North and Paul Hatt asked a national cross section of people to evaluate a series of occupations as "excellent," "good," "average," "somewhat below average," or "poor." This resulted in an occupational prestige scale that has been widely used by sociologists.[3] The study has been repeated and updated numerous times since 1947, and the occupational prestige scores have remained amazingly consistent. Some of the findings of one of these replications are shown in Table 5.2. Look at the occupations and prestige scores in Table 5.2 and see if you can figure out what leads people to rank occupations in particular ways. Is it pay, or power, or education, or government service, or what?

Kingsley Davis and Wilbert Moore believe that the prestige and status accorded one's position are determined by its scarcity or importance to society and by the amount of training or talent needed to obtain the position.[4] Logically, we would expect that one holding a position high in prestige would receive greater financial rewards than one would in a position of less prestige. This is usually true, but not always. A carpenter or plumber often makes more money than does a teacher, but the teacher, whose skill requires more training, has higher occupational prestige.

People in the situation just described—material wealth but low prestige, or vice versa—are victims of **status inconsistency.** They are status-inconsistent because the factors that determine their ranks in society are not consistent with each other. If we recall that there are many stratification systems—race, religion, age, gender, as well as socioeconomic factors—it is clear that there are many possibilities for status inconsistency. Examples of status-inconsistent

**TABLE 5.2** Occupational Prestige Scores

| Occupation | Prestige Score | Occupation | Prestige Score |
|---|---|---|---|
| Physician | 86 | Secretary | 46 |
| Lawyer | 75 | Telephone Operator | 40 |
| Computer Scientist | 74 | Prison Guard | 40 |
| College Professor | 74 | Farmer | 40 |
| Biologist | 73 | Apartment Manager | 39 |
| Architect | 73 | TV Repairman | 39 |
| Dentist | 72 | Barber | 36 |
| City Manager/Mayor | 70 | Piano Tuner | 35 |
| Registered Nurse | 66 | Plumber | 35 |
| High School Teacher | 66 | Baker | 35 |
| Accountant | 65 | Used Car Salesman | 34 |
| Professional Athlete | 65 | House Painter | 34 |
| Author | 63 | Housekeeper | 33 |
| Airline Pilot | 61 | Bus Driver | 32 |
| Police Officer | 60 | Cook | 31 |
| Statistician | 56 | Cashier | 29 |
| Ballet Dancer | 53 | Waiter/Waitress | 28 |
| Fireman | 53 | Taxicab Driver | 28 |
| Airplane Mechanic | 53 | Bartender | 25 |
| Electrician | 51 | Janitor | 22 |
| Funeral Director | 49 | Chambermaid | 20 |
| Mailman | 47 | Usher | 20 |

Keiko Nakoa and Judith Treas, "Computing Occupational Prestige Scores," *General Social Survey Methodological Report Number 70*, 1990.

people in the United States would include a college-educated carpenter, a wealthy business executive with an eighth-grade education, or a person of recent wealth but no prestige. The problem for the status-inconsistent person is that he or she behaves toward others in terms of his or her high rank (wealthy business executive), but others tend to behave toward him or her in terms of his or her low rank (eighth-grade education). Some sociologists report that status inconsistency can lead to other outcomes: stress, political liberalism, involvement in social movements, withdrawal, or psychosomatic illness.[5]

## How Many Social Classes?

A variety of factors, then, produce a layered society. The question of just exactly *how many* layers or social classes we have has long been of interest to sociologists. Karl Marx was an economic determinist. He believed that a person's position in the economic structure of society determined his or her lifestyle, values, beliefs, and behavior. In his writings in the mid-19th century, Marx described societies as being composed of two classes: the *bourgeoisie*, or capitalists, and the *proletariat*, or workers. The bourgeoisie controls the capital, the means of production. The proletariat owns only its own labor and, consequently, is continually exploited by the capitalists through

## KARL MARX (1818–1883)

Karl Marx earned his Ph.D. in Philosophy at the University of Berlin. Marx was viewed as a political radical and was a member of the international revolutionary movement. In 1848, Marx and his lifelong friend, Friedrich Engels, were commissioned by the Communist League to write a document on the principles of socialism that resulted in *The Communist Manifesto.*

Given his political belief in the creation of a communist society, it is not surprising that Marx devoted his attention to a critical analysis of capitalist society. Marx argued that problems of modern life can be traced to material sources—a consequence of capitalism—and that the solutions, therefore, can only be found in the overturning of capitalism by the collective actions of a large number of people. For Marx this meant practical activity, especially revolutionary activity. As Marx put it, "The philosophers have only *interpreted* the world in various ways; the point, however, is to *change* it." Because of his radical ideas, Marx was exiled from Germany and France. He sought refuge in London where he was safe from political persecution

Marx's ideas will be seen frequently in this book: in Chapter One (and most other chapters) on conflict theory, in this chapter on class structure, in Chapter Eight on religion, in Chapter Nine on economic and political institutions, in Chapter Eleven on population, and in Chapter Twelve on social change.

low wages and bad working conditions. Marx felt that once the workers realized their predicament and developed a class consciousness, there would be a continuing struggle between the two classes, leading to the overthrow of the bourgeoisie and eventually the emergence of a classless society. Marx's predictions were compromised somewhat by the emergence of a broad middle class (office workers, managers) in industrialized societies that had a greater influence than he had imagined. This class tended to insulate the capitalists from the workers. Workers haven't overthrown their oppressors as often as Marx predicted, and modern Marxists explain it this way: Workers have become pacified by labor union gains, by a higher standard of living in advanced capitalist nations, and by their own incorrect understanding (encouraged by the media) that their working-class status is appropriate and reasonable.

A modern Marxist, Erik O. Wright, has attempted to fine-tune and update Marx's theory by introducing a four-class model. Wright believes (like Marx) that class is determined by one's relation to the means of production, but he adds some categories to help explain the influence of the middle class. Wright's four classes are (1) *capitalists*, who own the means of production (factories, banks); (2) *managers*, who combine with capitalists to exploit workers by directing and supervising the work of others; (3) *petty bourgeoisie*, who are independent self-employed people who might own some small capital assets but employ few if any workers and do not exploit them; and (4) *workers*, who own only their own labor, which they sell to the capitalists. The managers are in an odd position: They are simultaneously workers in that they are working for others, and bourgeoisie in that they have power to exploit the labor of workers. By far the largest group numerically are the workers who make up approximately half of all people in the labor force.

Studies of social class in the United States have produced various results. Lloyd Warner studied a New England community and discovered six social classes. August Hollingshead, using a technique similar to Warner's in a mid-western town, found five social classes. Dennis Gilbert and Joseph Kahl have devised a system of six social classes that they believe provides a reasonable picture of class structure in the contemporary United States. They see class structure as growing out of the economic system and placement as based on source of income (either capitalist property, labor-force participation, or governmental transfers such as social security, welfare, and so on), type of occupation, and educational preparation. Table 5.3 summarizes their model.

Other students of stratification believe that it distorts the picture to speak of specific social classes, whether it might be two, three, five, six, nine, or whatever. It is more accurate, they feel, to speak of a stratification continuum consisting of many rankings with small gradations between them (much like a thermometer) and with no readily discernible social-class categories. Logically, this approach makes sense and might be the most accurate description of the way things really are. However, because people seem to believe in social classes, and because teachers and researchers often find it easier to make distinctions among several broad categories than among numerous slightly differing gradations, we usually view stratification by specific classes.[6]

### Determining Social Class

Suppose you want to know what social class you belong to. Sociologists determine rank in several ways. These methods include the subjective, reputational, and objective techniques. In the *subjective* method, individuals are asked what social class they think they belong to. Will the person be able to answer that? Many social scientists believe as Marx did that the world is divided along class lines and that a person's social-class identification is very much a part of his or her self-image; it is much of what he or she *is*. This view holds that social class is important to people, that these people therefore have a high degree of **class consciousness,** and that they should be able to place themselves accurately in the social-class spectrum.

The *reputational* method involves finding out what others think. People from the community are selected to act as prestige judges, and these judges in turn evaluate others in the community. Social class becomes what selected people say it is. This method is similar to the subjective technique, and again it hinges on community members' class consciousness.

The *objective* method differs from other methods in that individuals are evaluated on specific factors that sociologists assume are related to social class. For example, Warner examined four factors when he used the objective technique: the individual's occupation, his or her source of income, the

**TABLE 5.3**  Model of the American Class Structure

| Proportion of Households | Class | Education | Occupation of Family Head | Family Income, 1990 |
|---|---|---|---|---|
| 1% | Capitalist | Prestige university | Investors, heirs, executives | Over $750,000, mostly from assets |
| 14% | Upper middle | College, often with postgraduate study | Upper managers and professionals; medium business owners | $70,000 or more |
| 60% | Middle | At least high school; often some college or apprenticeship | Lower managers; semiprofessionals; nonretail sales; craftspeople, foremen | About $40,000 |
|  | Working | High school | Operatives; low-paid craftspeople; clerical workers; retail sales workers | About $25,000 |
| 25% | Working poor | Some high school | Service workers; laborers; low-paid operatives and clericals | Below $20,000 |
|  | Underclass | Some high school | Unemployed or part-time; many welfare recipients | Below $13,000 |

From Dennis Gilbert and Joseph Kahl, *The American Class Structure: A New Synthesis,* 4th ed. (Belmont, Calif.: Wadsworth, 1993), pp. 305–323.

type of house, and the area of the city in which he or she lives. By knowing these facts about a person, Warner held that he could objectively place him or her in a specific social class. Other objective factors that have been used include years of education, amount of income, type of possessions, and even type and quality of home furnishings.

Using any of these methods, especially the objective method, can give the impression that social class is a very clear-cut phenomenon and that each person can be adequately plugged in. This, of course, is an oversimplification. The categories are not precise and are not always related one to another. For example, in what social class do we put the status-inconsistent person whose occupational prestige is low, income is high, who lives in an apartment in a slum, and is a college graduate—and when we ask her, she says she doesn't believe in social class?

## Social Class and Behavior

When the Titanic sank in 1912, only 3 percent of the women in first class died, but 45 percent of the women in third class perished. It was supposed to be women and children into the lifeboats first, but the death rate was higher for children in third class than for men in first class.[7] We saw near the beginning of this chapter that the life chances of Tommy and Billy were affected drastically by their social class. It is clear that social class is related to numerous patterns of behavior.

Social-class differences in nutrition and health care lead to different disease patterns among the classes. A study comparing death rates in 1960 and 1986 found that although death rates had declined overall, poor and poorly educated people had higher death rates than people with higher incomes and better educations. The differences were greater in 1986 than they had been in 1960. Studies of social class and mental illness have found that psychoses, particularly schizophrenia and manic-depressive disorders, are more prevalent among lower-class than among middle-class people. Middle-class people, on the other hand, are more likely to be neurotic, or so diagnosed by psychiatrists, than are lower-class people. There are class differences in treatment as well. Middle-class people are likely to have psychotherapy, highly qualified practitioners, and more thorough treatment of mental illness. Lower-class people are more often treated by interns and less qualified practitioners, and they are more likely to be subjected to questionable procedures, such as shock treatments, lobotomies, and drugs.[8]

The first research question at the beginning of this chapter asked whether there might be a difference between rich and poor women in their rates of childbirth involving cesarean section. A University of California, Berkeley, study in 1989 reported that affluent women are nearly twice as likely as poor women to have cesarean sections. Several reasons are suggested. Cesarean sections cost more than natural childbirth, and wealthier people have the money, so physicians wanting to make more money might suggest the procedure. Or perhaps the wealthy are more likely to have cesarean sections if the birth looks difficult because physicians fear that these patients will sue them for malpractice if problems occur. Also, poor patients are more likely to give birth at large teaching hospitals staffed by young resident physicians who are closely supervised and limit cesarean sections to those cases in which they are truly necessary. Ironically, because cesarean sections are more dangerous than natural childbirth, poor women may be getting better care than richer women.[9]

Voting behavior is affected by many factors, and social class seems to be one of them. Traditionally, those toward the top of the social-class ladder (upper middle class, for example) are more conservative and most likely to vote Republican. Moving down the class ladder from middle to working to working poor (see Table 5.3), we find a greater tendency to vote Democratic. However, this relationship between social class and party voting is tenuous at best and can be changed by outside factors. For example, unemployment and inflation rates can lead disgruntled workers to change their voting patterns. The picture on voter participation is a bit clearer. Actually voting in an election seems to be correlated with social class in that those people with more education and more income are more likely to vote than are those with less. For example, in the 1996 presidential election, 28 percent of those with eight years of education or less voted whereas 73 percent of those with college degrees voted. It is interesting to note that this pattern is not true in other industrial democracies—there is a much higher rate of voter participation in general and by the lower classes in particular in other countries than in the United States.

Crime and deviant behavior show some variation by class. Homicide, assault, robbery, and burglary (commonly called "street" crimes) seem to be more prevalent toward the bottom of the social-class ladder, whereas embezzlement, income tax evasion, and various other types of "white-collar" crime are more prevalent among middle- and upper-middle-class people. Law enforcement likewise varies—a street crime is more likely to be reported to authorities and more vigorously pursued by the police than is a white-collar crime.[10]

Leisure-time activities are often class-related as well. Interest and participation in sports and other recreational activities are shown in Table 5.4. Upper-income households have greater participation in sports activities in general. Perhaps this is because they have more disposable income, or time, to spend on such activities. There are some differences between income levels. In some activities like backpacking, fishing, and exercise walking, high- and low-income households participate relatively equally. Higher-income households are more involved in aerobic exercising, bicycling, golf, skiing, swimming, and tennis. Can you give some reasons to explain the differences?

In how they raise their children, middle-class parents are probably more permissive, lower-class parents more rigid. Melvin Kohn reports that middle-class mothers value self-control, dependability, and consideration, whereas lower-class mothers value obedience and the ability in a boy to defend himself. Vanfossen suggests that the way children are raised reflects their parents' work situation. The middle-class child is raised in an atmosphere in which independence, achievement, curiosity, autonomy, and *self*-control are valued. The lower-class child is raised in an atmosphere in which the immediate and concrete rather than the new and unfamiliar are emphasized; the child is taught to focus on getting by, and he or she learns obedience, conformity, propriety, and control by *others*.[11]

Many of the current theories attempting to explain delinquent and criminal behavior are based on social class. Walter Miller, for example, argues that lower-class juveniles who become delinquents do so because of the lower-class value system. Values and beliefs important to lower-class youth include trouble, toughness, smartness, excitement, fate, and autonomy (resistance to being bossed or controlled). Miller believes that the greater presence of these values in the lower class makes it inevitable that many lower-class children will run afoul of the law because *laws* in our society emerge almost solely from the middle class.

Several other social-class-based theories see it differently. Albert Cohen, Richard Cloward, and Lloyd Ohlin believe that delinquency is a consequence of lower-class individuals' attempts to move into the middle class. Cohen believes that most won't make it, realize it, and, frustrated, do just the opposite of what the middle class likes. They get involved in malicious, nonutilitarian type of activities—vandalism, aggression, spreading graffiti.

**TABLE 5.4**   Participation in Sports Activities, 1996

| | Percentage Who Participate | |
|---|---|---|
| Activity | Household Income Under $15,000 | Household Income $75,000 and Over |
| Aerobic exercise | 7% ⟶ | 14% |
| Backpacking | 4% | 6% |
| Bicycling | 18% ⟶ | 28% |
| Exercise walking | 30% | 34% |
| Fishing | 19% | 21% |
| Golf | 4% ⟶ | 19% |
| Downhill skiing | 2% ⟶ | 12% |
| Swimming | 18% ⟶ | 33% |
| Tennis | 2% ⟶ | 9% |

United States Bureau of the Census, *Statistical Abstracts of the United States, 1998* (118th ed.), Washington, D.C.

Cloward and Ohlin likewise agree that lower-class individuals are motivated to move into the middle class, and many realize quickly that they are not going to make it, at least not legally. So, according to Cloward and Ohlin, they get involved in profitable criminal activity that might involve dealing drugs, stripping cars, or forming burglary rings. These activities will enable them to collect the "good things," the symbols of success and status associated the middle class. To summarize these ideas, Miller felt that lower-class delinquency was a consequence of being lower class in a society ruled by middle-class laws. Cohen, Cloward, and Ohlin felt that delinquency was a consequence of lower-class individuals trying to become middle class.

This detailed discussion of lower-class crime might leave you with the impression that crime is a lower-class phenomenon. This, of course, is not true: Types of crime (street crime, white-collar crime) vary by social class, and some crimes such as assault and robbery are much more easily detected and punished than are other crimes like fraud and income tax evasion. Also, middle-class people have the financial resources to deal with the law, to avoid arrest to begin with, and to obtain better legal defense should arrest occur. Typically, the middle-class juvenile delinquent is informally apprehended and returned to his or her parents, whereas the lower-class delinquent is formally arrested and sent to a juvenile hall. The actual extent of hidden middle-class crime is difficult to assess. Perhaps if all offenses could be known, the differences among social classes would be only in types of offenses, not in numbers.[12]

### Social Mobility

**Social mobility** refers to movement within the social-class structure. This movement can be up, down, or sideways. **Horizontal social mobility** refers to movement from one occupation to another within the same social class. If an architect becomes a minister or if a mail carrier becomes a carpenter, we would say that horizontal social mobility has occurred—occupational change but no change in social class. The term is also used by sociologists to describe geographical mobility, moving one's home from one place to another. Vertical social mobility is more relevant to our discussion here, however. **Vertical social mobility** refers to movement up or down the social-class ladder—movement from one rank to another.

Societies with unlimited possibilities for vertical social mobility are described as *open*. Those with no possibilities for mobility are called *closed*. In reality, all societies rank somewhere between the extremes of completely open and completely closed. For example, India's system of stratification—a caste system—is nearly closed. Lines between levels, or castes, are often firmly drawn; moving from the caste into which one is born can be difficult if not impossible. Some have suggested that we have a type of caste system in the United States relative to blacks and whites; that is, there is a firm line between black and white stratification systems, and although a black can rise within the black class system, he or she is either blocked from the white system or is automatically assigned low rank in the white system. In societies with stratification systems that are more open, mobility from one level to another might be both possible and frequent.

Examining mobility in a given country is a complicated procedure. Americans, for example, have traditionally considered their system to be completely open. Mobility and achievement are supposedly available to anyone who works hard and has led a clean life. We *know* that the rags-to-riches Horatio Alger story is true and that we can make it too. The truth is, however, that mobility is very much the same in all modern industrialized countries. The dramatic leap from rags to riches is a rare occurrence indeed. Mobility, when it occurs, usually happens in a series of small steps and possibly over several generations. Pure luck, of course, can increase one's opportunities, as can the needs of technology—the growth of the computer industry, for example. We find more mobility *within* classes (horizontal mobility) than *between* classes (vertical mobility). Even though there seems in general to be more upward vertical mobility than downward and many individuals are mobile, few move very far.

A study of the mobility of men and women in the United States reported an interesting finding: Women are more mobile, both upward and downward, through marriage than men are through occupations. Men are much more likely to inherit their fathers' statuses than are women. Because our society lacks mechanisms for women to directly inherit occupations, they are freer to marry into statuses both above and below their origins.[13]

To become upwardly mobile without relying completely on luck and wining the lottery, follow these general rules: Defer gratification, marry late, and have a small family if you must have one. If you live in a small town, leave. Vertical mobility is easier in medium-sized and large cities. If you are an immigrant to the United States, be Japanese, Jewish, or Scottish. These groups are more mobile than are other immigrant groups, who more often than not come in at the bottom of the social-class ladder and stay there. In modern societies in which skills and knowledge are increasingly important, education is essential for upward mobility. In the United States a college degree helps. For some not only the degree, but the right school—Harvard, Yale, Stanford, Princeton—is essential to mobility. Finally, if all else fails in your quest for upward mobility, marry someone who has followed these rules.

There are consequences to mobility. Sociologists have been studying the social-psychological effects of upward mobility. One result of uprooting oneself from the past and moving to a new level is that it becomes difficult to form satisfying relationships. One is alone, a **marginal person** between two worlds and a member of neither. Another view agrees that socially mobile people are isolates but argues that they were isolates to begin with. They were unhappy and isolated, and this led to attempts at mobility, so they are really no worse off. A third view of the situation, and the one that is most accurate according to current research on social mobility, is that socially mobile people have few problems because they are so anxious to get to the next level that they adopt the patterns of behavior of that level long before they ever get there; this behavior is called *anticipatory socialization*. Consequently, they are accepted and much more at home when they finally arrive.[14]

In a society that is upwardly oriented, downward mobility is hard to take. Studies cite relationships between downward mobility and both suicide and mental illness, especially schizophrenia. The same factors that are related to upward mobility are conversely related to downward mobility. Marrying early, having a large family, or failing to get an education can do it. Personal factors can help: business failure, sickness, or rejection of the ethic that says that climbing the social ladder is important.

## POVERTY AND WELFARE

Earlier in this chapter we learned that wealth is an important variable in determining social class and that wealth is not equally shared in American society. The top 20 percent of the population controls 47 percent of the national income, but the poorest 20 percent has only 4 percent of the national income (see Table 5.1). This emphasizes one of this country's major social issues: poverty. Poverty is a major social issue because of the links between it and crime, delinquency, self-esteem, mental illness, educational opportunities, physical health, life expectancy, infant mortality, and numerous other factors.

The United States government has developed a definition of poverty that is based on the amount of money needed to buy a nutritionally adequate diet, assuming that no more than one third of the family income is used for food. The poverty line varies according to family size, location, and the value of the dollar. In 1996, for example, the poverty line for a nonfarm family of four was $16,036. During that year, 36.5 million people (roughly 13.7 percent of the population) were below the poverty line. Large as that number is, it hardly represents the whole picture of poverty in this country. The poverty line refers to a so-called survival income, and many people who are above the poverty line don't have a normal or acceptable standard of living. They may live in substandard dwellings without electricity or running water, and although they might be able to survive, they know that most of the rest of the people in the country live far better than they do. It is difficult to arrive at an accurate estimate of how many poor and near-poor there are, but the bottom 20 percent of the population is the figure often used. Roughly 50 million people would fall into this category, and as we can see from Table 5.1, they were no better off in 1996 than they were in 1950. Each year they had only 4 to 5 percent of the country's total yearly income, although they represented 20 percent of the population.

Unemployment, inflation, and recession are part of the poverty picture. Although family income tends to increase each year because inflating dollars buy less, the "real" family income often *drops*. This forces still more Americans below the poverty level. In 1997, 6.7 million workers, or 4.9 percent of the labor force, were unemployed.

What are the characteristics of the poor? Statistics show the following groups to be overrepresented among the poor: minorities (African Americans and Hispanics), young people (15 to 24 years old), the illiterate and unemployable, and the disabled and handicapped. Poverty is higher in large families, in single-parent families, in rural areas, and in the South. Child poverty is high in the United States. A recent study by Syracuse researchers Timothy Smeeding and Lee Rainwater reports that the United States child poverty rate leads the developed world. They report that more than 21 percent of American children were defined as poor in 1991. The next closest country was Australia with 14 percent. Other countries included the United Kingdom and Italy, 10 percent; West Germany and France, 7 percent; Austria, 5 percent; and Switzerland and Denmark, 3 percent. Why does the United States lag so badly? Economists aren't sure but they suggest the following: (1) strong unions and minimum-wage laws keep wages higher for low-skilled European workers; (2) other countries have more generous safety nets like family or child allowances and national child and health care; and (3) 23 percent of United States families were headed by a single parent in 1990, which is much higher than Canada and most of Western Europe.[15]

How does a society that sees its country as a land of opportunity deal with the existence of poverty? In curious and interesting ways. One approach is to hide it. The poor are located in areas we don't see and consciously avoid. The sick and aged are almost by definition invisible. The subject is not popular and won't be found on prime-time television except for an occasional documentary. The poor have no economic or political clout. The poor are a large but essentially anonymous and hidden group; we try to deny their existence.

Another way of dealing with an unwelcome problem is to blame the victim. William Ryan begins his book *Blaming the Victim* with the story of actor Zero Mostel impersonating a pompous senator investigating the origins of World War II. The climax of the senator's investigation comes when he booms out with triumph and suspicion, "What was Pearl Harbor *doing* in the Pacific?" Exactly. Thus, according to this view, the poor are poor because they are lazy and unintelligent, they lack motivation, they tend to drink and have loose morals, they can't manage money, and they don't want to work anyway. Blaming the victim releases the nonvictims from responsibility for the victims' plight. Recent studies have found that many Americans (often, a majority of those asked) believe that (1) most people on welfare who can work don't try to find jobs, (2) many people on welfare are not honest about their need, and (3) many mothers on welfare have babies to increase their welfare payments. Actually, studies of those on welfare show that all of these beliefs are *incorrect.*

The second research question asked at the beginning of this chapter concerned the relationship between welfare and fertility—whether women on welfare are more likely to give birth than those who are not. A 1989 study reports that welfare recipients have a low fertility rate, considerably below that of women in the general population. The study found further that the longer a woman remains on welfare, the less likely she is to give birth. Interviews with women of childbearing age who were on welfare showed that the economic, social, and psychological pressures of being on welfare were not conducive to wanting more children. Having more children was clearly seen as making the situation worse, not better. Nearly all the women interviewed wanted to get off public assistance, and having more children was seen as a major hindrance to doing that.[16]

Why do we have these beliefs about the poor? Probably because Americans fervently believe in the work ethic. All able-bodied people should work, their income should come from work, and the more hard-working they are the better people they are. Work is good—even if it is low-paying, exhausting, and degrading temporary work. There is also a welfare ethic, which emphasizes that no one should accept money unless absolutely necessary. This leads to suspicions regarding welfare recipients, more extensive monitoring to find suspected welfare cheaters, and confirmation of the view that people who take charity are bad people to begin with.

Finally, we should recognize that poverty is *functional.* The existence of the poor and their exploitation benefit other Americans in many ways. The poor do society's dirty work—the dirty, dangerous, undignified, and menial jobs. They work for low wages, and as domestics, for example, they make life easier for the middle and upper classes. The existence of poverty creates all sorts of jobs and occupations to service the poor: social workers, police officers, ministers, pawnshop operators, and the peacetime army (which recruits heavily from the poor). The poor help the economy by purchasing goods that the more affluent don't want: day-old food, secondhand clothes, and deteriorating merchandise. In many ways, then, the existence of poverty is advantageous to the nonpoor segment of American society.[17]

### The Welfare System

The welfare system was devised ostensibly to assist the victims of poverty—to help them fare well in society. The idea of welfare has been around for a long time. Originally, churches were the primary agencies for charity and relief giving, but as the urban labor class grew, attempts to control the poor emerged. In 1351 under King Edward III of England, a law was passed that established maximum wages, required the unemployed to work for whomever demanded labor, restricted the travel of workers, and forbade charity to the able-bodied. In large part, these have become the basic features of welfare in the United States today.[18]

Public-program social-welfare expenditures in 1994 were more than $1,435 billion. These programs included aid for families with dependent children (AFDC); Medicaid; aid for the aged, the blind, and the disabled; food stamps; veterans aid; workers' compensation; black lung benefits; and so on. The list of programs is long, and the amount of money spent is substantial and increasing sharply.

Critics of the welfare system, however, believe that despite the huge amounts of money spent, the system actually *perpetuates* poverty. Those statutes that Edward III established more than 600 years ago helped him *control* the poor. Piven and Cloward argue that our current system has the same intent: to regulate and exploit the poor. During periods when high unemployment leads to public unrest, welfare programs are expanded to absorb and pacify the masses. At times of political and economic stability and low unemployment, welfare rolls are narrowed, forcing people into low-paying jobs in the labor market.[19]

Turner and Starnes suggest that we really have two systems operating, a wealthfare system and a welfare system. The **wealthfare system** enables the middle and especially the upper class to maintain their privileged position by allowing them to avoid taxation and by giving them and the businesses they own and operate direct cash payments and subsidies. The wealthfare system is an example of the exercise of power, and it works through the intricacies of the federal tax system.

The **welfare system** exists, according to Turner and Starnes, because it appears to be a humanitarian effort to keep people from starving; it is politically necessary because, due to their large numbers, the poor have some power and they may revolt; and this system is a nice balance to and keeps attention diverted from the far more expensive wealthfare system. The welfare system perpetuates poverty by keeping payments low, by forcing people to work in any available job, and by diffusing any collective power the poor may have.[20]

What is the answer? We must recognize that poverty is a problem we want to solve and try to redistribute wealth and resources, perhaps through a guaranteed annual income; possibly in these ways could inequality be reduced. But we cannot be hopeful about the prospect. If our analysis is accurate, the middle and upper classes benefit from the system as it exists. Poverty is unpleasant, but the alternative to it—massive changes in existing education, government, and economic institutions—and the threat that change suggests are more unpleasant for the affluent to consider, and power resides with privilege. Again we must consider power—who has it, how it is used, and for whose benefit.

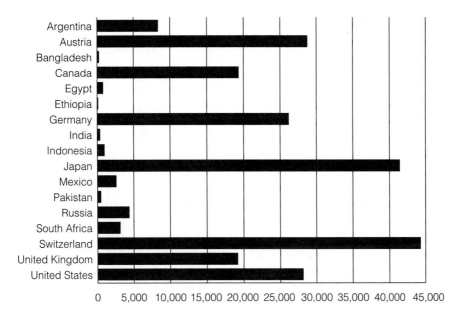

**FIGURE 5.2**   *Per capita gross national product (GNP) in selected countries (in dollars), 1995.*

United States Bureau of the Census, *Statistical Abstracts of the United States, 1998* (118th ed.), Washington, D.C.

## World Poverty

Pictures of starving people in Bangladesh, Ethiopia, and other countries should continually remind us that poverty is a worldwide problem. Wealth discrepancies between rich and poor countries are vast, and they represent a political powder keg and a disaster for humanity. More than half of the people on the planet (more than 3 billion people) have incomes of less than $2 a day. Figure 5.2 shows per capita gross national product figures for selected countries. Think of the amount of poverty and the plight of the poor in this country and then imagine the situation for a child born in a country in which the per capita income is a small fraction of what it is in the United States. What resources would such a country have to deal with poverty, starvation, or disaster?

Another ominous development is the gap between rich and poor *within* countries. The United Nations reported in July 1996 that the wealthy and the poor are living in increasingly separate worlds. The United States is slipping into a group of countries—Brazil, Britain, Guatemala—where the split is most pronounced, with the national per capita income four times or more higher than the average income of the poor. The gap between the rich and the poor in the United States is the widest in 50 years, and the distribution of income in America is now reported to be the most unequal among advanced countries. What are the potential political and human consequences if this continues?[21]

## THEORY AND RESEARCH: A REVIEW

This chapter represents an excellent opportunity to compare two of the theories that were introduced in Chapter One; in the area of social stratification, functional analysis and conflict theory come head to head, presenting quite contrasting analyses of this concept. And here, perhaps, the differences between the two theories become most evident. Proponents of functional analysis hold that societies are made up of complex and interconnected parts, each of which performs specific activities, or functions, to maintain the existence of the society as a whole. *Functionalists* like Parsons, Davis, and Moore suggest that stratification systems fit the pattern well. Their argument goes something like this: Societies that are at all complex must have an extensive division of labor to accomplish tasks and to assign workers to positions; this will guarantee smooth functioning and maintain stability. But the positions are not all equally important. Some (the physician, for example) require greater skill or responsibility than do others. To guarantee that the most qualified people, the ones with special talents, are encouraged to seek these jobs, greater rewards must be offered. Differential rewards result in the development of different classes of people. Every society must have a system that ensures that the most important positions will be filled by the most qualified people. Hence, stratification is a functional necessity.

*Conflict theorists* see several flaws in the argument. Functionalists say position and rank follow from merit—high status is determined by importance to society and by the amount of training and talent required to achieve it. However, no surgeon, nuclear physicist, or rocket scientist makes even a fraction of the money that is made by professional athletes like Michael Jordan or Mike Tyson or by movie stars. Income doesn't necessarily connect to importance to society. Further, much wealth is inherited, passed down from generation to generation, having much to do with luck and nothing at all to do with merit. Whereas functionalists suggest that stratification ensures the smooth and efficient operation of society, conflict theorists find that stratification is caused by the domination and exploitation of one group by another. Conflict theorists believe that society is best understood as a struggle (a conflict) for scarce resources such as food, land, wealth, or power. Some groups of people, perhaps because of inheritance or use of force, obtain more of these resources and are able to acquire a privileged position. People who have gained high positions will inevitably act in their own self-interest and will continually strive to maintain and increase their influence.

The key for conflict theorists is control of scarce goods and services; those who can gain this control will hold power and prestige. Some occupations are rewarded more than others, not because of the importance of the task (the functionalist argument), but because of the bargaining power of the group. For example, if an occupation, by requiring apprenticeship or complicated certification, limits membership to a very few, then the services of those few will be in greater demand. Perhaps physicians are highly paid not because they have some special skill but rather because admission to medical school is artificially limited. Or a group can enhance its position by creating demand where none existed. The fashion industry, with its designer jeans and running shoes, may be an example.

The functionalists see society as a smooth-running machine with the parts fitting nicely. The conflict theorists see instead an exploitive adversary system involving coercion and force, with dominant classes manipulating and controlling subordinate classes. It seems difficult to believe that these two groups of theorists are talking about the same thing. They are presenting contrary but valid interpretations of the same set of observations.

The data used in this chapter are descriptive and are for the most part census data obtained by *questionnaire* or *interview surveys*. The North-Hatt study and the sports-activities survey (Table 5.4) are also examples of *reactive surveys* that obtain information through mailed questionnaires. The data described in the section on social class and behavior were obtained in general through reactive surveys using questionnaires and interviews.

## SUMMARY

Sociologists study the social organization of society by studying groups and nongroups. The subject of the previous chapter was groups, and the characteristics and the variations in behavior in small and large groups were discussed. This chapter focused on social differentiation and social inequality. People are distinguished from each other by a number of categories: age, gender, race, social class, and religion. These are not groups as we discussed them earlier; they are nongroups, or categories, describing very large collectivities of people. A central point for us here is that when people are separated into categories, they are *ranked*. People regard some categories as better than others. Different stratification systems—slavery, caste, estate, and class—were discussed. In this chapter we focused on class or social stratification produced by socioeconomic factors.

The important variables in socioeconomic status are *wealth*, *power*, and *prestige*. Several interesting debates run through the discussion of social class. Who has the power in the United States? What does the American class structure look like? It depends on whom you ask. Some see a continuum with no specific classes; others see three, four, five, six, or even more reasonably well-defined classes.

We found that social class is related to numerous aspects of behavior: voting, mental illness, recreational patterns, homicide, child-rearing patterns, and delinquency. *Vertical social mobility* refers to movement up or down the social-class ladder. Analysis of social mobility in the United States indicates that Horatio Alger doesn't live here anymore; that is, few people move very far up or down the ladder.

Nearly 50 million people are poor or nearly poor in this country. Their poverty and how society deals with it are a major social issue for Americans.

Analysis of the welfare system suggests that although it was established to help the poor, it might actually perpetuate poverty.

In the next chapter, this analysis of social differentiation will continue as we direct our attention to other categories: race, ethnicity, gender, and age.

In the first of the readings that follow, Sarah Lyall illustrates a measure of social class that we don't often think about but that is actually quite commonly used. In the second reading, William Ryan examines several conflicting models in an attempt to understand what we really mean by equality. The third reading, by Kristin Luker, is a study of women's opposing views on abortion. This reading suggests that the women's social-class differences relate in part to their belief differences.

## TERMS FOR STUDY

| | |
|---|---|
| caste system (143) | social differentiation (141) |
| class consciousness (151) | social inequality (141) |
| class system (14) | social mobility (155) |
| estate system (143) | social stratification (142) |
| horizontal social mobility (155) | status inconsistency (148) |
| marginal person (157) | status symbol (145) |
| power (146) | vertical social mobility (156) |
| prestige (147) | wealthfare system (160) |
| slavery (143) | welfare system (160) |

For a discussion of Research Question 1, see page 153.
For a discussion of Research Question 2, see page 159.

 **INFOTRAC COLLEGE EDITION**
**Search Word Summary**

To learn more about the topics from this chapter, you can use the following words to conduct an electronic search on InfoTrac College Edition, an online library of journals. Here you will find a multitude of articles from various sources and perspectives:

**www.infotrac-college.com/wadsworth/access.html**

| | |
|---|---|
| social inequality | voting research |
| social class | class mobility |
| Northern Ireland | poverty |
| mental illness | |

## Reading 5.1

# IT'S NOT WHAT YOU SAY, IT'S HOW YOU SAY IT

*Sarah Lyall*

*Tony Blair is the prime minister of England and, as a politician, knows he must appeal to as many people as possible. This is tricky, however, because what is appealing to some may be offensive to others, and social class is treacherous ground indeed. Sarah Lyall, who wrote this article for the* New York Times, *describes an interesting attempt at bridging the class barrier.*

Tony Blair is a prime minister who prides himself on his common touch, so there was nothing particularly surprising about what he said recently on a British talk show.

Mr. Blair delighted his host, Des O'Connor, by telling a joke about his mother-in-law. He described how he hung up on Queen Elizabeth when an airplane pilot ordered him to turn off his cell phone. And he told a long story about how he was presented with a gift horse by the mayor of a French village. "I wasn't sure whether to ride it or eat it," Mr. Blair said.

Ha, ha (or not). But joking aside, where were the prime minister's T's? What happened to his H's? Why, when he tried to say, "they put on a little show for us with the mayor of the little village," did it come out as, "they pu' on a li'l show for us with they mayor of the li'l village?" And why did he say, referring to the horse, that " 'e's come back to England"?

Or, as Bruce Anderson said, writing in *The Daily Mail,* "Why did the prime minister take reverse elocution lessons?"

As it watched the show, a nation acutely aware of the nuances of accent noticed with a jolt that the prime minister had apparently changed his. The familiar Mr. Blair, with his soothing, almost preacher-like voice—a voice that sets him linguistically above the working classes and is a slightly modified example of what is known here as Received Pronunciation—had disappeared. In his place was a new Tony, a Tony speaking something called Estuary English, a hybrid accent that was first identified in 1984 and says something else altogether.

Taking some of its cues from traditional Cockney, such as the tendency to drop the "t" in words like "lot" and "little," Estuary English has become increasingly prevalent in the south of England and is generally used by modestly accented people seeking to appear more upscale, and by people with upscale accents, particularly teenagers in expensive schools, seeking to sound more street-wise.

Mr. Blair might well have picked up the accent from his young son, who attends the sort of posh London school where Estuary English flourishes in the mouths of students wanting street cred, said Paul Kerswill, a lecturer in linguistics at Reading University. Maybe Mr. Blair was looking for his own sort of street cred.

"I think he wants to be cool," said Mr. Kerswill, noting that Mr. Blair has made it clear that he enjoys playing the guitar in his spare time. "We probably got it from the Americans, the idea that you can be a politician and hip at the same time."

Finding itself a bit on the defensive about the episode, Downing Street acknowledged that the prime minister sounded different, but denied that he had actually put on a new accent.

"When you're on a show like that and you're telling an anecdote, your voice changes to reflect the tempo of the anecdote," a spokeswoman said. "We all change our voice to suit the thing that we're talking about. And the prime minister is very good at impersonations."

But in a country obsessed with accent, Mr. Blair's obvious shift downward was a sign of the times and an interesting insight into his style as prime minister. While other prime ministers have certainly taken their accents down a few notches when it suited them, their forays into regional or working-class accents have generally reflected their own backgrounds.

In a famous elocutionary makeover, Margaret Thatcher, a grocer's daughter from Grantham, in Leicestershire, used her iron will to conceal her humble verbal roots, losing her local dialect and developing a classically upper-class accent that she then fine-tuned downward again on the advice of her public-relations team.

"They said, 'You've got to remove some of the posh features,'" said John Honey, a professor of linguistics at the University of Botswana and the author of "Does Accent Matter?"

"She adopted that heavy breathing, intensive-care kind of voice," Mr. Honey said. "She dropped the pitch of her voice to sound more soothing and less strident, and she dropped the use of the impersonal 'one.'"

Mrs. Thatcher's real linguistic self rarely came out. But once, incandescent with fury in Parliament, she yelled "You're frit!," using dialect for "frightened." It was a slip that her Labor opponents would taunt her with for the rest of her career as prime minister.

While Mrs. Thatcher's successor, John Major, stuck to a bland if synthetic form of Received Pronunciation, Mr. Blair seems to be the first prime minister to have used Estuary English publicly, a state of affairs that caused dismay among some language purists, even those accustomed to hearing it spoken by other politicians.

"I was left screaming: 'Tony, you're a public-school, Oxbridge-educated lawyer,'" Brian Reade wrote in *The Daily Mirror.* "'Why are you patronizing us?'"

Anne Shelley, vice president of the Queen's English Society, said, "I was very disappointed with Tony Blair." Referring to the way he dropped his T's, she added, "His speech was slovenly and the glottal stop was the ugliest of the lot."

To several political reporters, the prime minister's sliding speech demonstrated nothing other than an alarming, Clintonesque desire to be all things to all people, to pretend he was "Reliable Ron from Romford," as Mr. Reade put it.

"The obvious explanation is vote-getting: trying to persuade his audience that he is really one of them," Mr. Anderson of *The Daily Mail* wrote. "Mr. Blair is a politician who adapts his personality and views to his surroundings to avoid disappointment in the opinion polls."

But veteran language-watchers said they were not too upset by his foray into Estuary English. They have previously noted anomalies in the prime minister's speech, like the American-style tendency to muddy hard vowels in words like "conservative," so that it comes out as "conserva-tuhve," and the tendency not to vocalize his L's, so that "arsenal" becomes "arsenaw."

"People in public life—particularly, perhaps, people in politics—need always to avoid the impression that they are in a world apart, in a sort of upper crust," said Lord Quirk, a linguist and member of the House of Lords. "On the other hand, they have to avoid giving the impression that they're talking down and joining in the yobbos, as it were. It is a very difficult mix to achieve."

## QUESTIONS 5.1

**1.** Apparently language is an important indicator of social class in England. Is the same true in the United States? Give examples.

**2.** What do "Received Pronunciation" and "Estuary English" mean?

**3.** Can language be a status symbol? Illustrate.

**4.** Why would Blair want to disguise his social class?

## Reading 5.2

# THE EQUALITY DILEMMA

*William Ryan*

*In this excerpt from his book* Equality, *William Ryan (a psychology professor at Boston College) examines two conflicting models of equality: Fair Shares and Fair Play. Ryan suggests that we face the following dilemma in American society: Although the Fair Play model is favored by most Americans, it is based on faulty assumptions; perhaps the Fair Shares model is more appropriate and closer to what is really meant by equality.*

Thinking about equality makes people fidgety. Insert the topic into a conversation and listen: voices rise, friends interrupt each other, utter conviction mingles with absolute confusion. The word slips out of our grasp as we try to define it. How can one say who is equal to whom and in what way? Are not some persons clearly superior to others, at least in some respects? Can superiority coexist with equality, or must the one demolish the other? Most important, are the existing inequalities such that they should be redressed? If they are, what precisely should be equalized? And by whom?

We feel constrained by the very word—to deny equality is almost to blaspheme. Yet, at the same time, something about the idea of equality is dimly sinful, subject to some obscure judgment looming above us. It is not really adequate to be "as good as." We should be "better than." And the striving for superiority fills much of the space in our lives, even filtering into radio commercials. Listen to a chorus of little boys singing, "My dog's better than your dog; my dog's better than yours." They are selling dog food, it turns out, and apparently successfully, which must mean that a lot of people want very much to have their dogs be better than your dogs. Their *dogs!*

This passing example suggests how far it is possible to carry the competition about who and whose is better. The young voices sing fearlessly. There is little danger that some fanatic will leap up and denounce the idea of one dog's being superior to another or quote some sacred text that asserts the equality of all dogs. In regard to people, however, we do hesitate to claim superiority or to imply inferiority. No commercials announce, "My son's better than your son" or "My wife's better than yours" or, most directly, "I'm better than you." That we remain reticent about flaunting such sentiments and yet devote ourselves to striving for superiority signals the clash of intensely contradictory beliefs about equality and inequality.

We all know, uneasily but without doubt, that our nation rests on a foundation of documents that contain mysterious assertions like "all men are created equal," and we comprehend that such phrases have become inseparable components of our license to nationhood. Our legitimacy as a particular society is rooted in them, and they bound and limit our behavior. We are thus obliged to agree that all men are created equal—whatever that might mean to us today.

But our lives are saturated with reminders about whose dog is better. In almost all our daily deeds, we silently pledge allegiance to inequality, insisting on the continual labeling of winners and losers, of Phi Beta Kappas and flunkouts, and on an order in which a few get much and the rest get little.

We re-create the ambiguity in the minds of our children as we teach them both sides of the contradiction. "No one is better than anyone else," we warn. "Don't act snotty and superior." At the same time we teach them that all are obliged to get ahead, to compete, to achieve. Everyone is equal? Yes. Everyone must try to be superior? Again, yes. The question reverberates. . . .

It should not surprise us, then, that the clause "all men are created equal" can be interpreted in quite different ways. Today, I would like to suggest, there are two major lines of interpretation: one, which I will call the "Fair Play" perspective, stresses the individual's right to pursue happiness and obtain resources; the other, which I will call the "Fair Shares" viewpoint, emphasizes the right of access to resources as a necessary condition for equal rights to life, liberty, and happiness.

Almost from the beginning, and most apparently during the past century or so, the Fair Play viewpoint has been dominant in America. This way of looking at the problem of equality stresses that each person should be equally free from all but the most minimal necessary interference with his right to "pursue happiness." It is frequently stressed that all are equally free to *pursue,* but have no guarantee of *attaining,* happiness.

The Fair Shares perspective, as compared with the Fair Play idea, concerns itself much more with equality of rights and of access, particularly the implicit rights to a reasonable share of society's resources, sufficient to sustain life at a decent standard of humanity and to preserve liberty and freedom from compulsion. Rather than focusing on the individual's pursuit of his own happiness, the advocate of Fair Shares is more committed to the principle that all members of the society obtain a reasonable portion of the goods that society produces. From this vantage point, the overzealous pursuit of private goals on the part of some individuals might even have to be bridled. From this it follows, too, that the proponent of Fair Shares has a different view of what constitutes fairness and justice, namely, an appropriate distribution throughout society of sufficient means for sustaining life and preserving liberty.

So the equality dilemma is built in to everyday life and thought in America; it comes with the territory. Rights, equality of rights—or at least interpretations of them—clash. The conflict between Fair Play and Fair Shares is real, deep, and serious, and it cannot be easily resolved. Some calculus of priorities must be established. Rules must be agreed upon. It is possible to imagine an almost endless number of such rules.

- Fair Shares until everyone has enough; Fair Play for the surplus

- Fair Shares in winter; Fair Play in summer

- Fair Play until the end of a specified "round," then "divvy up" Fair Shares, and start Fair Play all over again (like a series of Monopoly games)

- Fair Shares for white men; Fair Play for blacks and women

- Fair Play all the way, except that no one may actually be allowed to starve to death

The last rule is, I would argue, a perhaps bitter parody of the prevailing one in the United States. Equality of opportunity and the principle of meritocracy are the clearly dominant interpretation of "all men are created equal," mitigated by the principle (usually defined as charity rather than equality) that the weak, the helpless, the deficient will be more or less guaranteed a sufficient share to meet their minimal requirements for sustaining life.

The Fair Play concept is dominant in America partly because it puts forth two most compelling ideas: the time-honored principle of distributive justice and the cherished image of America as the land of opportunity. At least since Aristotle, the principle that rewards should accrue to each person in proportion of his worth or merit has seemed to many persons one that warrants intuitive acceptance. The more meritorious person—merit being some combination of ability and constructive effort—*deserves* a greater reward. From this perspective it is perfectly consistent to suppose that *unequal* shares could well be *fair* shares; moreover, within such a framework, it is very unlikely indeed that equal shares could be fair shares, since individuals are not equally meritorious.

The picture of America as the land of opportunity is also very appealing. The idea of a completely open society, where each person is entirely free to advance in his or her particular fashion, to become whatever he or she is inherently capable of becoming, with the sky the limit, is a universally inspiring one. This is a picture that makes most Americans proud.

But is it an accurate picture? Are these two connected ideas—unlimited opportunity and differential rewards fairly distributed according to differences in individual merit—congruent with the facts of life? The answer, of course, is yes and no. Yes, we see some vague congruence here and there—some evidence of upward mobility, some kinds of inequalities that can appear to be justifiable. But looking at the larger picture, we must answer with an unequivocal "No!" The fairness of unequal shares and the reality of equal opportunity are wishes and dreams, resting on a mushy, floating, purely imaginary foundation.

Income in the United States is concentrated in the hands of a few: one fifth of the population gets close to half of all the income, and the top five percent of this segment gets almost one fifth of it. The bottom three fifths of the population—that is, the majority of us—receive not much more than one third of all income. Giving a speech at a banquet, a friend of mine, James Breeden, described the distribution of income in terms of the dinners being served. It was a striking image, which I will try to reproduce here.

Imagine one hundred people at the banquet, seated at six tables. At the far right is a table set with English china and real silver, where five people sit comfortably. Next to them is another table, nicely set but nowhere near as fancy, where fifteen people sit. At each of the four remaining tables twenty people sit—the one on the far left has a stained paper tablecloth and plastic knives and forks. This arrangement is analogous to the spread of income groups—from the richest five percent at the right to the poorest twenty percent at the left.

Twenty waiters and waitresses come in, carrying one hundred delicious-looking dinners, just enough, one would suppose, for each of the one hundred guests. But, amazingly, four of the waiters bring twenty dinners to the five people at the fancy table on the right. There's hardly room for all the food. (If you go over and look a little closer, you will notice that two of the waiters are obsequiously fussing and trying to arrange ten dinners in front of just one of those five.) At the next-fanciest table, with the fifteen people, five waiters bring another twenty-five dinners. The twenty people at the third table get twenty-five dinners, fifteen go to the fourth table, and ten to the fifth. To the twenty people at the last table (the one with the paper tablecloth) a rude and clumsy waiter brings only five dinners. At the top table there are four dinners for each person; at the bottom table, four persons for each dinner. That's approximately the way income is distributed in America—fewer than half the people get even one dinner apiece.

When we move from income to wealth—from what you *get* to what you *own*—the degree of concentration makes the income distribution look almost fair by comparison. About one out of every four Americans owns *nothing*. Nothing! In fact, many of them *owe* more than they have. Their "wealth" is actually negative. The persons in the next quarter own about five percent of all personal assets. In other words, half of us own five percent, the other half own ninety-five percent. But it gets worse as you go up the scale. Those in the top six percent own half of all the wealth. Those in the top one percent own one-fourth of all the wealth.

These dramatically *unequal* shares are—it seems to me—clearly *unfair* —shares. Twenty million people are desperately poor, an additional forty million don't get enough income to meet the minimal requirements for a decent life, the great majority are just scraping by, a small minority are at least temporarily comfortable, and a tiny handful of persons live at levels of affluence and luxury that most persons cannot even imagine.

The central problem of inequality in America—the concentration of wealth and power in the hands of a tiny minority—cannot, then, be solved by any schemes that rest on the process of long division. We need, rather, to accustom ourselves to a different method of holding resources, namely, holding them in common, to be *shared* amongst us all—not divided up and parceled out, but shared. That is the basic principle of Fair Shares, and it is not at all foreign to our daily experience. To cite a banal example, we share the air we breathe, although some breathe in penthouses or sparsely settled suburbs and others in crowded slums. In a similar fashion, we share such resources as public parks and beaches, although, again, we cannot overlook the gross contrast between the size of vast private waterfront holdings and the tiny outlets to the oceans that are available to the public. No one in command of his senses would go to a public beach, count the number of people there, and suggest subdividing the beach into thirty-two-by-twenty-six-foot lots, one for each person. Such division would not only be unnecessary, it would ruin our enjoyment. If I were assigned to Lot No. 123, instead of enjoying the sun and going for a swim, I might sit and watch that sneaky little kid with the tin shovel to make sure he did not extend his sand castle onto my beach. So, we don't divide up the beach; we own it in common; it's *public;* and we just plain *share* it.

We use this mode of owning and sharing all the time and never give it a second thought. We share public schools, streets, libraries, sewers, and other public property and services, and we even think of them as being "free" (many libraries even have the word in their names). Nor do we need the "There's no such thing as a free lunch" folks reminding us that they're not really free; everyone is quite aware that taxes support them. We don't feel any need to divide up all the books in the library among all the citizens. And there's no sensible way of looking at the use of libraries in terms of "equal opportunity" as opposed to "equal results." Looking at the public library as a tiny example of what Fair Shares equality is all about, we note that it satisfies the principle of equal access if no one is *excluded* from the library on the irrelevant grounds of not owning enough or of having spent twelve years in school learning how not to read. And "equal results" is clearly quite meaningless. Some will withdraw many books; some, only a few; some will be so unwise as to never even use the facility.

The *idea* of sharing, then, which is the basic idea of equality, and the *practice* of sharing, which is the basic methodology of Fair Shares equality, are obviously quite familiar and acceptable to the American people in many areas of life. There are many institutions, activities, and services that the great majority believe should be located in the public sector, collectively owned and paid for, and equally accessible to everyone. We run into trouble when we start proposing the same system of ownership for the resources that the wealthy have corralled for themselves. It is then that the servants of the wealthy, the propagandists of Fair Play, get out their megaphones and yell at everybody that it's time to line up for the hundred-yard dash.

One can think of many similes and metaphors for life other than the footrace. Life is like a collection of craftsmen working together to construct a sturdy and beautiful building. Or, a bit more fancifully, it is like an orchestra—imagine a hundred members of a symphony orchestra racing to see who can finish first! When we experience a moving performance of the *Eroica,* how do we judge who the winners are? Is it the second violins, the horns, the conductor, Beethoven, the audience? Does it not make sense to say that, in this context, there are no losers, that all may be considered winners?

Most of the good things of life have either been provided free by God (nature, if you prefer) or have been produced by the combined efforts of many persons, sometimes of many generations. As all share in the making, so all should share in the use and the enjoyment. This may help convey a bit of what the Fair Shares idea of equality is all about.

---

## QUESTIONS 5.2

**1.** Distinguish between the Fair Shares and Fair Play views of equality. What are the problems or dilemmas with each?

**2.** Would functional theorists and conflict theorists respond differently to this selection? If so, how? If not, why not?

**3.** Is William Ryan a functionalist or a conflict theorist? Explain your choice.

**4.** Select the definition of equality that makes most sense to you and defend it. Briefly describe a program to make it work in the United States.

---

## Reading 5.3

# PRO-LIFE VERSUS PRO-CHOICE WOMEN— A STUDY IN CLASS CONTRASTS

*Kristin Luker*

*In this excerpt from her book* Abortion and the Politics of Motherhood, *Kristin Luker profiles the characteristics of women who are in favor of and opposed to abortion. She illustrates the point, a central one in sociology, that one's attitudes, values, and beliefs are a reflection of one's position in the social structure. In this case, peoples' social class and economic well-being affect how they respond to certain social issues.*

Ever since abortion became a divisive issue, many people have sensed something about the women actively involved on opposite sides. That something was difficult to articulate yet relevant to the bitterness of the controversy. Pro-choice women, it seemed, were *different* from pro-life women, and vice versa.

That perception, as it turns out, is accurate. A five-year study of a representative sample of these women activists demonstrates that the two groups are separated by far more than their disagreement over whether abortion is murder. By social class, by economic status, by educational attainment, they *are* different. And that is only the beginning. Their beliefs about abortion are simply the tip of the iceberg, reflecting attitudes that are also at odds on all the most central aspects of life: work, women's roles, marriage, parenthood, sex, and family.

Moreover, those attitudes were not arrived at by some intellectual process. If that were the case, there might be more hope of reversing the process and bridging the gap. Rather, their cherished beliefs flow out of realities in their own daily lives that are not easily changed.

This is why the debate is so passionate and uncompromising. The women who are most centrally involved in it have little basis for a dialogue and, in fact, few incentives to enter into one.

To explore what it is that makes the abortion debate so heated, a research team at the University of California at San Diego of which I was a part conducted interviews over a five-year period (1977–83) with a sample of nearly two hundred California women who spend at least five hours a week working for advocacy groups on that issue.

The typical pro-choice activist in our study was a forty-four-year-old married woman whose father was a college graduate. She married at age twenty-two or older, has one or two children and had some graduate or professional training after college. (Thirty-seven percent of all pro-choice women in this study have received at least some post-baccalaureate training.) She is married to a professional man, is employed and has a family income of more than $50,000 a year. She attends church rarely, if at all; indeed, religion is not particularly important to her.

The average pro-life activist is also forty-four and married. However, she wed at seventeen and has three or more children. (Sixteen percent of the pro-life women in the study have seven or more children.) Her father graduated from high school but not college, and only sixty percent of pro-life activists (compared with ninety percent of pro-choice activists) have a baccalaureate degree.

This typical pro-life activist is not employed, and is married to a small businessman or a lower-income white-collar worker; her family income is less than $30,000 a year. Her religion is one of the most important aspects of her life; she attends church at least once a week. She is probably a Catholic, but may be a convert to Catholicism. (Almost eighty percent of the pro-life activists in this study were Catholics at the time of the study, but only fifty-eight percent had been raised as Catholics.)

One small example points up the way these differences shape attitudes. A number of pro-life activists emphatically reject an expression that pro-choice women tend to use almost unthinkably—the expression "unwanted pregnancy." Pro-life women argue forcefully that a better term would be "surprise pregnancy," reflecting their view that although a pregnancy may be momentarily unwanted, the child that results from the pregnancy almost never is.

As our profile of the average pro-life activist makes clear, a woman who is not employed, who does not have an advanced degree, whose religion is important to her, and who has already committed herself wholeheartedly to marriage and a large family is well equipped to believe that an unanticipated pregnancy usually becomes a beloved child. Her life is so arranged that for her this belief is true, and her world view leads her to believe that everyone else can "make room for one more" as easily as she can, and that, therefore, abortion is cruel, wicked, and self-indulgent.

By the same token, our profile of the average pro-choice woman—a woman who is employed full-time, has an affluent lifestyle that depends in part on her contribution to the family income and expects to give her child a life at least as educationally, socially, and economically advantaged as her own—makes it evident that she draws on a different reality that makes her skeptical about the ability of the average person to transform unwanted pregnancies into well-loved and well-cared-for children.

Two prominent groups of women—blacks and working-class whites—are hardly represented at all among the ranks of abortion activists in our study. Many working-class women, often stereotyped as conservative on social issues, actually are pro-choice but seldom have the time or resources to be active. Blacks tend to be conservative on the abortion issue, but blacks often focus on issues perceived to be of more immediate concern to the black community, such as education.

Nevertheless, it is the activists who shape the debate, and it is the sharp cleavage among them that makes political compromise and dialogue so unlikely.

The gulf separating these two groups of women in our study is evident in the fact that few women on either side of the issue have any friends or even acquaintances who disagree with them about abortion. In fact, a number of pro-life women spontaneously declared during the interviews that they would end a friendship if they discovered that the friend did not share their views on abortion.

That is hardly surprising given the opposing views held by pro-choice and pro-life activists on so many issues. Consider, for example, attitudes about the roles of men and women. Pro-life activists believe that men and women are intrinsically different, and that this is both a cause and a product of different "natural" roles in life. As one said:

"Men and women were created differently, and were meant to complement each other, and when you get away from our proper roles as such, you start obscuring them. That's another part of the confusion going on now; people don't know where they stand, they don't know how to act, they don't know where they're coming from, so your psychiatrists' couches are filled with lost souls, with people who have gradually been led into confusion and don't even know it."

Men, the pro-life activists believe, are best suited to the public world of work and women to the private world of rearing children, managing homes, and caring for husbands.

Pro-choice activists reject this notion. They believe that men and women are fundamentally equal, by which they mean substantially similar, at least as regards rights and responsibilities. As a result, they see women's reproductive and family roles not as a natural niche, but as a potential barrier to full equality. So long as society is organized to maintain motherhood as an involuntary activity, they argue, "women's sphere" connotes a potentially low-status, unrewarded role to which women can be relegated at any time.

"I just feel that one of the main reasons women have been in a secondary position culturally is because of the natural way things happen. Women would bear children because they had no way to prevent it, except by having no sexual involvement. And that was not practical down through the years; so without knowing what to do to prevent it, women would continually have children. And then if they were the ones bearing the child, nursing the child, it just made sense for them to be the ones to rear the child. I think that was the natural order."

"When we advanced and found that we could control our reproduction, we could choose the size of our families, or whether we wanted families. But that changed the whole role for women in society. . . . It allowed us to be more than just the bearers of children, the homemakers. That's not to say that we shouldn't continue in that role. It's a good role, but it's not the only role for women."

For the pro-life people we talked to, the primary purpose of sexuality is procreation. It is not surprising, given this view, that the pro-life activists in this study are opposed to most contraceptives. Although they are careful to point out that the pro-life movement is officially neutral on the topic, most are confident that any law outlawing abortion would also outlaw The Pill and the I.U.D., a result they favor. A substantial number of pro-life activists use periodic abstinence, or natural family planning, as their only form of fertility control.

In contrast, pro-choice people generally focused on sexuality as an emotional outlet. They argued that the main purpose of sex is not to produce children but to afford pleasure, human contact and, perhaps most important, intimacy. But since such closeness requires trust, familiarity, and security, practice is required. As a result, contraception, which allows people to focus on the emotional aspects of sex without worrying about its procreative aspects, is a good thing, according to pro-choice activists.

A general theme in the interviews with pro-life activists (many of whom have large families, it will be recalled) is that there is an anti-child sentiment abroad in American society, as exemplified by the strong pressures to have only two children:

"My husband, being a scientist, gets a lot of questions. You know, having a large family, it's just for the poor, uneducated person, but if you have a doctor's degree and you have a large family, what's wrong with you?"

The values of pro-life activists about parenthood follow from their views on gender, sex, and contraception. Since the purpose of sex is procreation, they believe, married couples should be willing to have whatever number of children come, at whatever time they are conceived. Second, since motherhood is a natural role, one should not try to plan carefully for it through contraceptive use, and one need not prepare for it. Finally, since motherhood is the most satisfying and meaningful role for a woman, it is incomprehensible to pro-life people that a woman might want to postpone or avoid pregnancy in favor of something else, such as work, education, or worldly success.

Because pro-choice people see raising children as requiring financial resources, interpersonal and social skills, and emotional maturity, they often worry about how easy it is to have children. In their view, parenthood is far from natural; they feel that too many people stumble into it without appreciating what it takes:

"I would say that the tip of the iceberg is purposeful parenthood. I think life is too cheap. I think we're too easy-going. We assume that everyone will be a mother. . . .

Hell, it's a privilege; it's not special enough. The contraceptive agent affords us the opportunity to make motherhood really special."

Since pro-choice activists think that in the long run abortion will enhance the quality of parenting by making it optional, they see themselves as being on the *side* of children when they advocate abortion.

As one pro-choice woman said, "Many people don't want the child when they find out they're pregnant, but they resolve negative feelings and by the time the child is born they do want it. But if they don't want the child enough to seek an abortion, then probably they shouldn't have a child. I'm as concerned with the rights of the child as the rights of the woman."

Pro-life women have built their lives on the premise that reproduction is a *resource,* and they resist all those cultural values—small families, contraception, abortion, non-family roles for women, day care—which diminish the value of children or dilute the unique value of motherhood.

In the same way, pro-choice women resist those values which suggest that motherhood is a natural, primary, or inevitable role for a woman. Pro-choice activists believe that men and women are equal because, in their own lives, men and women have substantially the same kinds of experiences. The pro-choice women in this study have had approximately the same education as their husbands, and many of them have the same kinds of jobs—they are lawyers, accountants, pharmacists, ministers, and physicians. Even those who do not work in traditionally male occupations have salaried jobs and thus share common experiences.

Women who have "male" resources want to see motherhood recognized as discretionary. Women who have few of these resources, and limited opportunities, have incentives to see motherhood recognized as the most important role a woman can have.

---

## QUESTIONS 5.3

**1.** What type of research is this? Categorize using terms from Chapter One (reactive, nonreactive, and so on).

**2.** The point of this reading is that social class and economic well-being help mold beliefs. How and why did this happen? What is the logical connection between social-class position and beliefs about abortion?

**3.** What other aspects of these women's backgrounds aside from social class seem to have contributed to their viewpoints? What was the most important factor?

---

## NOTES

1. Sources for this section include *The Structure of Social Inequality* by Beth E. Vanfossen (Boston: Little, Brown, 1979), chapter 3; *Inequality and Stratification*, 2d ed., by Robert Rothman (Englewood Cliffs, N.J.: Prentice-Hall, 1993), chapter 3; and *Social Stratification and Inequality*, 2d ed., by Harold Kerbo (New York: McGraw-Hill, 1991), chapter 1.

2. H. H. Gerth and C. W. Mills, *From Max Weber: Essays in Sociology* (New York: Oxford University Press, 1958), chapter 7.

3. Cecil North and Paul Hatt, "Jobs and Occupations: A Popular Evaluation," *Opinion News* 9 (September 1947), pp. 3–13.

4. Kingsley Davis and Wilbert Moore, "Some Principles of Stratification," *American Sociological Review* 10 (April 1945), pp. 242–249.

5. See Gerhard Lenski, "Status Crystallization: A Non-Vertical Dimension," *American Sociological Review* 19 (August 1954), pp. 405–413; Gerhard Lenski, "Status Inconsistency and the Vote: A Four Nation Test," *American Journal of Sociology* 32 (April 1967), pp. 298–302; and "Status Consistency and Right-Wing Extremism," *American Sociological Review* 32 (February 1967), pp. 86–92.

6. Paragraphs on Marx and Wright benefit from Rothman, chapter 2, and Kerbo, chapter 5 (see note 1), and *Social Stratification*, 2d ed., by Daniel W. Rossides (Upper Saddle River, N.J.: Prentice-Hall, 1997), pp. 61, 139. Also see classic works in stratification including Lloyd Warner's study of Newburyport, Massachusetts, first described in Lloyd Warner and Paul S. Lunt, *The Social Life of a Modern Community* (New Haven, Conn.: Yale University Press, 1941). Robert S. Lynd and Helen M. Lynd described stratification in a midwestern city in *Middletown* (New York: Harcourt, 1959) and *Middletown in Transition* (New York: Harcourt, 1963). So did August Hollingshead in *Elmtown's Youth*, first published in 1949 and now available as *Elmtown's Youth and Elmtown Revisited* (New York: Wiley, 1975). Stratification along a continuum is discussed by John Cuber and William Kenkel in *Social Stratification in the United States* (New York: Appleton-Century-Crofts, 1954).

7. Walter Lord, *A Night to Remember* (New York: Holt, Rinehart & Winston, 1955).

8. The mortality data come from Gregory Pappas, Susan Queen, Wilbur Hadden, and Gail Fisher, "The Increasing Disparity in Mortality Between Socioeconomic Groups in the United States, 1960 and 1986," *The New England Journal of Medicine* 329 (July 8, 1993), pp. 103–109. Social class and mental illness data are from A. B. Hollingshead and F. C. Redlich, *Social Class and Mental Illness:* *A Community Study* (New York: Wiley, 1958).

9. See Jeffrey Gould, Becky Davey, and Randall Stafford, "Socioeconomic Differences in Rates of Cesarean Section," *The New England Journal of Medicine* 321 (July 27, 1989), pp. 233–239.

10. Marshall Clinard and Robert Meier discuss the data on homicide in *Sociology of Deviant Behavior*, 5th ed. (New York: Holt, Rinehart & Winston, 1979), p. 199. Information on voting comes from Kerbo, pp. 293–295 (see note 1).

11. See Melvin Kohn's book, *Class and Conformity* (Homewood, Ill.: Dorsey, 1969) and the discussion by Vanfossen in *The Structure of Social Inequality*, pp. 272–273 (see note 1).

12. The original works referred to in this section include William C. Kvaraceus and Walter B. Miller, *Delinquent Behavior: Culture and the Individual* (Washington, D.C.: National Education Association, 1959); Walter Miller, "Lower-Class Culture as a Generating Milieu of Gang Delinquency," *Journal of Social Issues* 14 (1958), pp. 5–19; Albert K. Cohen, *Delinquent Boys: The Culture of the Gang* (New York: Free Press, 1955); and Richard Cloward and Lloyd Ohlin, *Delinquency and Opportunity* (New York: Free Press, 1960). Also see "The Myth of Social Class and Criminality: An Empirical Assessment of the Empirical Evidence" by Charles Tittle, Wayne Villemez, and Douglas Smith, *American Sociological Review* 43 (October 1978), pp. 643–656.

13. Ivan Chase, "A Comparison of Men's and Women's Intergenerational Mobility in the United States," *American Sociological Review* 40 (August 1975), pp. 483–505.

14. See the discussion of social mobility and its effects in *Social Stratification in America* by Leonard Beeghley (Santa Monica, Calif.: Goodyear, 1978), pp. 311–318.

15. See "Doing Poorly: The Real Income of American Children in a Comparative Perspective," by Lee Rainwater and Timothy Smeeding, an unpublished paper from the Luxembourg Income Study. The poverty rate for this study was defined as the percent of children living in families with adjusted disposable incomes less than 50 percent of adjusted median income for

all persons. Years involved were 1984 through 1992.

16. The major sources I am using in this section on poverty are Beeghley, *Social Stratification in America,* chapter 8 (see note 14); Jonathan Turner and Charles Starnes, *Inequality: Privilege and Poverty in America* (Pacific Palisades, Calif.: Goodyear, 1976); Jerome Skolnick and Elliott Currie, *Crisis in American Institutions,* 5th ed. (Boston: Little, Brown, 1982); and William Ryan, *Blaming the Victim,* rev. ed. (New York: Random House, 1976). Facts on money and numbers come from *Statistical Abstracts of the United States, 1998 (118th* ed.), Washington, D.C. The study of welfare beliefs and facts is discussed in Beeghley, pp. 135–138. The interview study of the fertility rates of welfare women is from Mark Rank, "Fertility Among Women on Welfare: Incidence and Determinants," *American Sociological Review* 54 (April 1989), pp. 296–304.

17. Herbert Gans, "The Positive Functions of Poverty," *American Journal of Sociology* 78 (September 1972), pp. 275–289; and Beeghley, *Social Stratification in America,* p. 140.

18. Turner and Starnes, *Inequality: Privilege and Poverty in America,* p. 122.

19. Frances Piven and Richard Cloward, "The Relief of Welfare," in Skolnick and Currie, *Crisis in American Institutions,* 5th ed. (see note 16). Also see Beeghley, *Social Stratification in America,* pp. 139–140.

20. Turner and Starnes, *Inequality: Privilege and Poverty in America,* pp. 62–63, 93.

21. See Timothy Smeeding, "America's Income Inequality: Where Do We Stand?" *Challenge* (September–October, 1996), pp. 45–53.

## CHAPTER SIX

# Social Inequality:
# Race, Ethnicity, Gender, Age

### Minority Status

### Race and Ethnicity

*Prejudice and Discrimination* • *Blacks and African Americans* • *Native Americans* • *Hispanic Americans* • *Jews* • *Asian Americans*

### Gender

*Socialization and Gender Roles* • *Consequences of Gender-Role Typing*

### The Elderly

### Patterns of Interaction

### Minority-Group Reactions

## TWO RESEARCH QUESTIONS

*1. Race is viewed differently in the United States than it is in the Caribbean. How is it different? What happens when Caribbean Hispanics settle in the United States—do they adopt U.S. patterns or maintain their old beliefs?*

*2. In a group discussion involving men and women, who is more likely to interrupt whom, or is there any consistent pattern? Does the composition of the group affect interruption patterns?*

THE PREVIOUS CHAPTER introduced the concepts of social differentiation and social inequality. People fall into categories on the basis of characteristics such as age, gender, race, and social class. Further, the elements that make up these categories are ranked—people in them are not seen or treated equally. The study of differentiation and inequality is very helpful because we get a more precise picture of society. It also helps us understand the forces that bind people together and the characteristics of inequality.

In Chapter Five we discussed the inequality produced by socioeconomic factors. Socioeconomic status can be described as an achieved status because there is some flexibility in the ranking; one can perhaps move up or down the social-class ladder. In this chapter we look at categories that are more firmly established. They are automatically conferred, and there is little an individual can do to change them. We call them *ascribed statuses*.

## MINORITY STATUS

**Minority status** is an important ascribed status and can be helpful in explaining how people interact with one another. Minority status not only names the group or category to which one belongs but also describes one's social position—how one is likely to be treated by others.

We will find in this chapter that groups of minority status might be considered lower in social standing than the rest of society; they can be subject to domination and might be denied specific rights and privileges.

It is important to understand that minority status is a social condition, not necessarily a statistical one. A minority group could represent a majority of a society's population but hold minority status because of the way it is viewed and treated by other segments of society. For example, women in the United States are a numerical majority, but they have minority status because they are not treated on a level of equality with men. For years the black population in South Africa has held minority status even though it greatly outnumbers the white population, which has been dominant. In most instances, minority groups do represent a smaller proportion of the population—African Americans and Latinas and Latinos in the United States, for example—but it is important to recall the *social* nature of minority status.

Where do definitions of minority status come from? Who decides? A number of factors are probably at work. Some decisions about minority status arrive through custom and tradition. People who are successful are sometimes proprietary about their good fortune and look down on those who aspire to the same position. Some groups blend in more easily or appear to. If their skin color and language are the same and their cultural backgrounds are similar to those of the society they are entering, their differences will be less apparent, and they will be more likely to escape minority status. We also learned in the previous chapter about the importance of wealth and power. Assume that there is a limited supply of the good things in life; it follows, then, that those with wealth and power will want to protect their positions by denying access to wealth and power by others. By labeling and treating categories of people differently, as if of lower quality, these people are effectively removed as threats to established positions of privilege. The creation and maintenance of minority status can work much the same way as does the treatment of the poor (discussed in Chapter Five).

## RACE AND ETHNICITY

We have learned that people differ and are ranked by such socioeconomic factors as wealth, education, occupation, power, and prestige. People also differ and are ranked by race and ethnicity. A person's race and ethnic affiliation confer on him or her a status or position in society, and sometimes this is a minority status.

**Race** is a vague and ambiguous term that is generally defined as a group of people bound together by hereditary physical features. The difficulty with the notion of race is that practically no pure races exist. Substantial biological mixing has blurred boundaries to the point that it is difficult to define uniform hereditary physical features. Scientists attempting to define race have named anywhere from 3 racial categories to more than 30. Three definitions of race—biological, legal, and social—seem to emerge.[1] A *biological* definition is based on observable physical features such as skin color, hair texture, and eye color, or on differences in gene frequencies. A *legal* definition is that incorporated into the laws of states or nations. For example, the law in one state has defined a black as a person who has "one-eighth or more Negro blood," whereas in other states a black is a person with "any ascertainable trace of Negro blood." A *social* definition refers to what members of society consider the important distinctions about race. Brewton Berry suggested that in the United States the social definition of a black is anyone who identifies himself as black or who has any *known* trace of black ancestry. Clearly *race* is not a precise term, but it is nevertheless widely used.

A sociology class is shown slides of people who identify themselves as blacks but who, because of an absence of typical racial characteristics, don't *look* like blacks. The class (which is predominantly white) is asked to identify the race or nationality of the light-skinned, straight-haired people on the slides, and the guesses range across the globe. Finally the class is told, "They are all African Americans." The reaction is astonishment, of course, because "they don't look like blacks." This illustrates several things, among them the difficulty of defining race, the blurring of racial boundaries, and the importance of the social definition. Once the class is told that the people on the slides are identified as blacks (probably because of some African ancestry), the class believes it. No question about it, regardless of how they look.

When race is discussed, the question of racial differences inevitably comes up. That is, do races biologically differ in such matters as I.Q., achievement, or susceptibility to diseases? The haziness in definitions of race supplies part of the answer: If the boundaries of race are unclear, how can any statements about racial differences be made? Nevertheless they are, as witness the controversy over inherent racial I.Q. differences. Most social scientists conclude that there are no significant differences caused by race. There are numerous differences among people caused by cultural, social, and geographic factors, however, and these are often thought to be racial differences. For example, some gene frequencies and blood characteristics have developed in particular areas of the world. Population groups migrating from those areas, regardless of race, carry these characteristics to other parts of the world. Likewise, differences in attitude, achievement, and perception are explained by cultural differences: One culture emphasizes achievement, another tranquillity; one is aggressive, another encourages passivity; and so on. In short, social scientists argue from a cultural-determinist viewpoint and reject the idea of racially caused differences. Many people remain unconvinced, however, and the argument will undoubtedly continue.

**Ethnicity** and *ethnic group* are more useful terms than *race,* at least from a sociological viewpoint. Members of an ethnic group are bound together by cultural ties, which can have several origins. When groups migrate to a new area, their ties to a previous culture can remain strong. Religious beliefs can also provide a basis for ethnicity, as for the Jews, the Amish, and the Hutterites. The term *ethnicity* is useful because it appropriately brings to focus cultural similarities, which to the social scientist are more important and have more explanatory power than racial similarities.

The first research question at the beginning of this chapter concerned definitions of race in the Caribbean and in the United States. Race in the United States is an either-or concept—one is either black or white. In the Caribbean, however, race is a multicategory continuum, with many people falling in between black and white. Caribbean Hispanics are an interesting case in that although they are racially diverse—a mixture of black, white, Asian, and American Indian—they share a common ethnicity: their use of Spanish, their Spanish-colonial heritage, and their Catholic religion. In other words, their ethnic-group identity is more important than their racial identity.

Now, what happens when Caribbean Hispanics settle in the United States? Do they maintain their view of race and ethnicity, or do they change? A recent paper found that Caribbean Hispanics coming to the United States are torn between conforming to the American views of race and holding on to their own multiracial/Hispanic ethnic identity. The outcome is interesting but perhaps predictable. Caribbean Hispanics (especially Puerto Ricans)

## THE BELL CURVE DEBATE

A controversial book named *The Bell Curve*, published in 1994 by Charles Murray and Richard Herrnstein, made two important claims—society's rewards go to those with more intelligence, and intelligence is innate. Not only is intelligence determined by heredity but it varies by race. This started a long and rancorous debate over the relative importance of genetic and social factors. Researchers all over the country leaped to their data and a number of studies pointed out difficulties with *The Bell Curve's* assumptions. One of these, *Inequality by Design* by Claude Fischer et al., points out that IQ explains very little about income. If everyone had identical IQs, 90 to 95 percent of the inequality we see today would still be there. A Princeton study found that

going to school an extra year raised wages by about 10 percent, regardless of IQ. Experts recognize that most IQ tests measure reading and math ability, which is a product of society. These recent studies show that environmental factors—parents' education and income, type of family (intact, divorced, or single parent), neighborhood, and years of schooling—all have a major effect on life outcomes, independent of IQ. Bell Curve critics also conclude that poverty or lack of economic success are a consequence of a nation's social policies which neglect the disadvantaged.

—*See the article "New Work Refutes Conclusions of* The Bell Curve*," by Jonathan Marshall, San Francisco Chronicle, August 12, 1996.*

---

reject the bipolar racial ideas of the United States—they consider themselves to be racially intermediate between black and white. They continue to preserve an ethnic identity—"Spanish race"—that draws them together. However, census data suggest patterns of segregation similar to those in American society. Caribbean Hispanics are not segregated from white Hispanics, but they are highly segregated from both Hispanic and non-Hispanic blacks. They are also highly segregated from European Americans. It appears that people of mixed racial ancestry are accepted by white Hispanic Americans on the basis of shared ethnicity but are rejected by European Americans on the basis of race. The authors conclude that although both race and ethnicity are important in American life, race is especially so.[2]

### Prejudice and Discrimination

**Prejudice** has been defined as a fixed attitude—favorable or unfavorable—toward a person or thing, probably not based on actual experience. A prejudiced person ignores the individual and his or her particular qualities or characteristics and groups that individual with others who happen to have the same skin color (brown or yellow), or speak with the same accent (New England), or have the same type of name (Cohen, Greenberg), or come from the same part of the country (the South).

Prejudice tends to be generalized. People who are prejudiced against one group will probably be prejudiced against others. Some years ago E. L. Hartley asked college students what they thought of Canadians, Turks, Jews, atheists, Irish, Negroes, Wallonians, Pireneans, Danireans, and other groups. He found that people who were prejudiced against blacks were also likely to be prejudiced against Jews, Wallonians, and Danireans. Now there are about 15 million Jews in the world, but no Wallonians or Danireans. Hartley used these fictitious groups to see how people would react to groups they knew nothing about. He found that people tend to generalize their prejudices.

**Discrimination** is actual behavior unfavorable to a specific individual or group. When people are denied equal treatment, they are being discriminated against. Discrimination occurs when a person is denied a desired position or right because of irrelevant factors—for example, when skin color is used to determine eligibility to vote or when religious affiliation is used to determine one's place of residence.

A distinction can be made between individual and institutional discrimination. *Individual* discrimination refers to actions taken by individuals or groups to deny others equal treatment. Examples of individual discrimination might include a judge giving more severe sentences to African Americans, or a golf club limiting membership to whites, a bank "redlining" (refusing loans to) loan applicants based on their race or ethnicity, or a neighborhood using restrictive covenants to keep people not like themselves from living in their area. *Institutional* discrimination is less visible and intentional than individual discrimination. It refers to more general actions embedded in a society's norms and values that treat people unequally. These behaviors are legitimized as part of customary and routine behavior. South Africa for a long time had a system that formally guaranteed access to power, wealth, and prestige to whites at the expense of nonwhites. In the United States, inner-city schools typically have less money, and less adequate and more crowded facilities, meaning that the children who attend these schools (often the poor and minorities) will be disadvantaged compared with children who attend wealthier suburban schools with better facilities. The differences will carry through, as access to higher education, and ultimately, jobs will be affected.

A particular type of prejudice and discrimination is described by the term *racism* (and by the related terms *sexism* and *ageism*). **Racism** is the belief that people are divided into distinct hereditary groups that are innately different in their social behavior and mental capacities and that, because of these differences, people can be ranked as superior and inferior. This hereditary ranking then "justifies" differential treatment in access to society's resources such as jobs, living conditions, wealth, and power.[3] You can see that racism is involved in some of the definitions of race discussed earlier (see p. 181).

Prejudice and discrimination are usually associated but not always. The situation becomes complicated when people think one way but behave another way. Robert Merton attempted to describe what happens by suggesting four categories of people:[4]

*Unprejudiced nondiscriminator or all-weather liberal:* Confirmed and consistent liberal; not prejudiced; doesn't discriminate; believes in American creed of justice, freedom, equality of opportunity, and dignity of the individual.

*Unprejudiced discriminator or fair-weather liberal:* Thinks of expediency; not prejudiced but keeps quiet when bigots are about; will discriminate for fear that to do otherwise would "hurt business"; will make concessions to the intolerant.

*Prejudiced nondiscriminator or fair-weather illiberal:* Timid bigot; doesn't believe in the American creed but conforms to it when slightest pressure is applied; hates minorities but hires them to obey the law.

*Prejudiced discriminator or all-weather illiberal:* A bigot, pure and unashamed; doesn't believe in the American creed and doesn't hesitate to tell that to others; believes it is right and proper to discriminate, so does.

The origins of prejudice are explained in various ways. Some psychologists believe that a certain type of personality called an **authoritarian personality** is especially prone to prejudice.[5] The authoritarian personality is also ethnocentric, is rigidly conformist, and worships authority and strength. According to psychologists, the authoritarian personality can most often be traced to faulty emotional development born of harsh discipline and lack of affection and love from parents during childhood. Most sociologists believe that prejudice, like other attitudes and behaviors, is learned in interaction with others, mainly in the family, and that personality is not as important as is the social or cultural situation in which one interacts. We learn race prejudice in the same way we learn how to eat with a fork, to study for a test, or to drive a car. Also, some life situations affect how this learning takes place. For example, people who are downwardly mobile—moving from middle to lower class—show more intense feelings of prejudice toward blacks and Jews; and in times of rapid social change, prejudice and discrimination can become more intense and more generalized.

Sociologists also associate stereotyping with prejudice. In **stereotyping,** we apply a common label and set of characteristics to all people in a certain category, even though none or only some of the people in that category fit the description. The stereotyped male college professor is absentminded, smokes a pipe, is an extreme liberal, wears horn-rimmed glasses and a tweed jacket with elbow patches, and generally is not much in touch with the real world. This image is applied to everyone in the male college-professor category, even though most do not fit the part. Professors will probably survive their stereotype without damage because the characteristics of the stereotype are not negative. Besides, a person *chooses* as an adult to become a college teacher and, therefore, to enter a stereotyped category.

Someone of minority status, on the other hand, who is stereotyped from birth as ignorant, lazy, dirty, happy-go-lucky, morally primitive, emotionally unstable, and fit only for menial work can be psychologically damaged by the effects of the lifelong stereotype. When others react to us on the basis of such stereotypes, it strongly affects our self-concepts—the way we see ourselves. This again is the self-fulfilling prophecy, which we discussed several

---

## A RACIAL GAP—LUNG-CANCER SURGERY

A study in an October 1999 issue of the *New England Journal of Medicine* reported that black patients in the early stages of lung cancer are less likely to have surgery than are white patients with the same diagnosis, and as a result are more likely to die of the disease. Earlier studies have suggested that subtle racial biases lead doctors to treat minorities less intensively than white patients. One of the authors of the study concludes, "we need to understand better what's happening in that interaction between doctors and patients that leads to some patients having surgery less often than other patients."

---

times earlier. Remember what happened to the "bright" rats and "dull" rats? Or what happened when teachers thought some of their students were "academic spurters"?

A classic example of the effect of minority status on how people see themselves was provided by Professor Kenneth Clark when he asked 253 black children aged three to seven to choose between black and white dolls. Two thirds of the children preferred *white* dolls! This research, done in the 1940s, was cited by the Supreme Court in its 1954 school-desegregation decision. Those who hoped that recent gains in black achievements and renewed emphasis on black pride have improved images of self-worth were disappointed by several studies reported in 1987. Psychologist Darlene Powell-Hopson gave 155 children aged three to six a set of black and white Cabbage Patch Dolls and asked the children to give her the doll that "you want to be, you want to play with, is a nice color and would take home if you could." Sixty-five percent of the black children selected white dolls. From the results and from students' comments, Powell-Hopson concluded that black children are getting the message that it's preferable to be a member of another race. In a similar study by Sharon Gopaul-McNicol of 144 preschoolers in Trinidad, 74 percent of the black children chose white dolls. She suggests that although this Caribbean island has a black government and many successful blacks in business, they apparently are not enough to overcome the legacy of white supremacy passed on by 400 years of British rule and the influence of North American and European television.[6]

Several theories have been offered to explain prejudice and discrimination. One such theory explains conflict between unlike groups as scapegoating. If attention can be focused on some out-group, this can strengthen the boundaries and unity of the in-group. It is suggested that the Nazis' attacks on the Jews before and during World War II were so motivated. Marxist theory holds that economic competition is the best explanation for prejudice and discrimination. When access to desired goods or valued positions in society is limited, discrimination against specific categories of people helps ensure that others can more easily dominate and obtain their goals. By reducing the

**TABLE 6.1**     United States Population by Selected Ancestry Groups and Region (1990 unless otherwise noted)

| Ancestry Group/Region | Population |
|---|---|
| African American Descent (1997) | 33,947,000 |
| Asian American Descent | |
|     Chinese | 1,645,000 |
|     Filipino | 1,407,000 |
|     Japanese | 848,000 |
|     Korean | 799,000 |
|     Vietnamese | 615,000 |
|     Laotian | 149,000 |
|     Cambodian | 147,000 |
|     Others | 1,664,000 |
| European Descent | |
|     German | 57,947,000 |
|     Irish | 38,736,000 |
|     English | 33,771,000 |
|     Italian | 14,665,000 |
|     Scotch and Scotch-Irish | 11,012,000 |
|     French | 10,321,000 |
|     Polish | 9,366,000 |
|     Dutch | 6,227,000 |
|     Swedish | 4,681,000 |
|     Norwegian | 3,869,000 |
|     Others | 19,716,000 |
| Hispanic Descent (1997) | |
|     Mexican | 18,795,000 |
|     Central/South American | 4,292,000 |
|     Puerto Rican | 3,152,000 |
|     Cuban | 1,258,000 |
|     Others | 2,206,000 |
| Native American Descent | 1,938,000 |

United States Bureau of the Census, *Statistical Abstracts of the United States, 1998* (118th ed.), Washington, D.C.

status of others to second-class citizens and by eliminating some people from highly valued jobs, education, and access to wealth, we can guarantee that our own paths are free of obstacles. Conflict theorists believe that the conflict that arises when unlike groups meet can be best and most simply explained as resulting from a struggle for status among competing groups.

The United States is a mix of many different groups. Whether it is a melting pot, a mosaic, or a boiling cauldron is open to debate. But Table 6.1 clearly shows that there is great diversity in tracing ancestry. The data in Table 6.1 should be viewed as a very general estimate and not very precise. This is because not all ancestry groups are included, and, when asked to identify their ancestry, some people report being members of more than one ancestry group. Intermarriage and mixed-race children make a precise definition of race difficult, if not impossible. Indeed, in 1990, when asked their race, ten million Americans chose "other" rather than one of the 16 racial categories. Typically, one is asked to "select one" when checking racial categories. The 2000 census became more flexible by asking the respondent to check "one or more" of the 14 boxes representing 6 races and multiple subcategories that might apply.

In the following paragraphs we will examine some patterns of prejudice and discrimination as they relate to African Americans, Hispanic Americans, Asian Americans, Native Americans, Jews, women, and the elderly.

### Blacks and African Americans

There were 34 million blacks in the United States in 1997, or roughly 12 percent of the population. African Americans have been a part of American history from its beginning; the economy from the earliest days of our settlement was based on slavery, and we have yet to overcome its effects. Specific dates stand out:

1619 Twenty Africans purchased from a Dutch ship

1644 First slaves imported directly from Africa

1861–1865 Civil War with emancipation as a major issue

1865 Legal abolition of slavery by the Thirteenth Amendment to the Constitution

1896 *Plessy v. Ferguson* decision by the Supreme Court that supports "separate but equal" doctrine

1918–1960s Migration of African Americans from the rural South to jobs in the urban North

1954 *Brown v. Board of Education* decision by the Supreme Court stating that "separate facilities are inherently unequal"

1956 Montgomery bus boycott led by Martin Luther King, Jr.

1960–1965 Passive resistance, sit-ins, demonstrations

1965–1970 Urban riots, beginning of black militancy and "Black Power"

1970–1980s Prison riots, widespread disagreement over busing to achieve school desegregation, affirmative action, court suits over "reverse discrimination"

1990s Continued urban unrest, with focus on the criminal justice system (King case, Los Angeles riots, the Simpson trials), increased attention to hate crimes, and renewed political efforts to overturn affirmative action practices

The Kerner (riot commission) Report in 1968 stated that "our nation is moving toward two societies, one black, one white—separate and unequal." Some examples of how separate and unequal status affects other conditions of life are shown in Table 6.2.

**Jobs, Income, Unemployment**    The data in Tables 6.2 and 6.4 and Figure 6.1 show job- and income-related comparisons between blacks and whites. Median income of white families in 1996 was $18,000 higher than for black families. The jobless rate for blacks in 1997 was double that of whites.

**TABLE 6.2**   Comparison by Race on Selected Characteristics

|  |  | Whites | Blacks |
|---|---|---|---|
| Birth rate per 1,000 people (1997) |  | 14 | 18 |
| Life Expectancy (1996) |  |  |  |
|  | Males | 74 | 66 |
|  | Females | 80 | 74 |
| Infant deaths per 1,000 live births (1996) |  | 6 | 14 |
| Unmarried mothers percentage of all births (1996) |  | 26 | 70 |
| Homicide victims per 100,000 people (1995) |  |  |  |
|  | Males | 8 | 56 |
|  | Females | 3 | 11 |
| Suicide victims per 100,000 people (1996) |  |  |  |
|  | Males | 21 | 11 |
|  | Females | 5 | 2 |
| Percentage unemployed (1997) |  | 4 | 10 |
| Percentage below poverty line (1997) |  | 11 | 28 |
| Percentage of families earning under $10,000 (1997) |  | 6 | 19 |
| Percentage of families earning over $50,000 (1997) |  | 44 | 23 |
| Percentage completing four years of college (1997) |  | 25 | 13 |

United States Bureau of the Census, *Statistical Abstracts of the United States, 1998* (118th ed.), Washington, D.C.

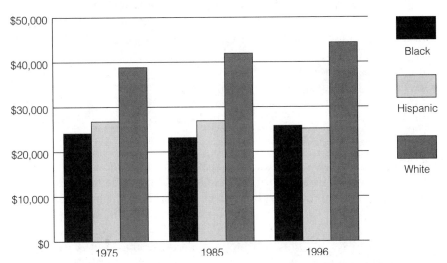

**FIGURE 6.1**   *Median family income (1996 dollars) by race and Hispanic origin.*

United States Bureau of the Census, *Statistical Abstracts of the United States, 1998* (118th edition), Washington, D.C.

Fourteen percent of white teenagers were unemployed; for black teenagers the figure was 32 percent. In 1997, although blacks were 12 percent of the population, they were 4 percent of the nation's physicians, 3 percent of the dentists, 3 percent of the lawyers and judges, 7 percent of the college and

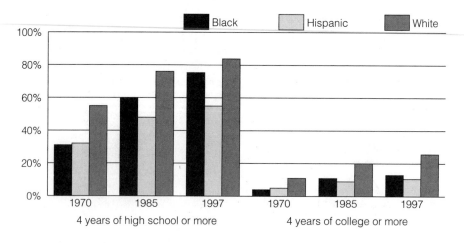

**FIGURE 6.2**   *Level of education, by race and Hispanic origin.*

United States Bureau of the Census, *Statistical Abstracts of the United States, 1998* (118th edition),
Washington, D.C.

university teachers, 28 percent of the postal clerks, 35 percent of the nursing
aides and orderlies, and 27 percent of the maids and housemen.

**Housing**   Finding adequate housing is difficult for blacks. First there were
restrictive laws, and then tacit agreements to keep blacks out of the better
districts in most cities and suburbs of the North, West, and South. Middle-
class whites leaving cities for suburbs have created *de facto* segregation,
turning inner cities into black ghettos. Blacks isolated in these areas find their
housing, their public facilities, and their schools automatically segregated
and made inadequate by the lowered tax base, the backwardness (and often
helplessness) of city government, and the exploitation or apathy of absentee
landlords and local businesspeople.

**Education**   Comparisons in educational attainment are shown in Figure 6.2.
The gaps between different groups are closing at both the high school and
college levels. The United States Supreme Court outlawed segregated schools
in *Brown v. Board of Education* in 1954, but integration of schools did not
proceed very rapidly. For some years now, the nation has struggled with ways
to achieve school integration in the face of housing segregation. One answer is
court-ordered busing of students from one area to another to maintain a racial
balance. The move in this direction was encouraged by a report by James S.
Coleman in 1966, which concluded that deprived students performed better
scholastically when their classmates came from backgrounds emphasizing
educational achievement. It was concluded that the educational chances of
blacks could be improved by moving them from poor school districts to
higher-quality (predominantly white, middle-class) school districts. Busing
has moved the cause of school integration along, but not without controversy
and some unanticipated consequences. Riots and demonstrations have

occurred. White parents have generally opposed busing, and many have responded by moving their children to private schools or by moving their family to areas not affected by busing, both of which might *increase* segregation.

Coleman complicated the issue with another report in 1975, which suggested that perhaps schools weren't so important as family background was in determining educational achievement; maybe there wasn't any point in busing after all. Then, another Coleman report in 1981 concluded that private schools do a better job than public schools even after family background factors are controlled—in other words, that schools *do* make a difference. Specific educational policies and techniques such as more homework, more-difficult subjects, patterns of discipline, and teacher interest in students can lead to higher achievement, regardless of family background. How can this be of help to minority families who cannot afford private schools? Coleman suggests tuition tax credits, which would probably increase the private-school enrollment of blacks and Hispanic Americans more than it would that of whites.[7] The equal education issue is a difficult one complicated by the interaction of several important factors—minority status, social class, family background, and educational policies—all playing against the backdrop of racism.

**Voting and Elected Officials**   Blacks were denied the vote for years in the South. However, implementation of the Civil Rights Act, calling for federal voting registrars in the South, seemed to change this form of discrimination. Voter-registration figures in 11 southern states in 1960 indicated that 61 percent of the eligible whites were registered, as compared with 29 percent of the eligible blacks. In 1996, however, 68 percent of the whites and 64 percent of blacks were registered. In 1970, there were fewer than 1,500 black office holders. In January 1993, 7,984 blacks were holding office in 45 states. The increase in the number of elected officials has been most evident in the South. In 1995, there were 40 blacks in the House of Representatives out of 435 individuals, and one black in the Senate (out of 100).

### Native Americans

In the middle- and late-19th century, Americans believed in the doctrine of Manifest Destiny, which held that it was the destiny of the United States to expand its territory and enhance its political, economic, and social influence over the whole of North America. And so, the original occupants of the land, the American Indians, were killed, captured, displaced, and placed on reservations as settlers moved west. It was too bad, as one Army general who supervised the removal of the Indians put it; they were a brave and proud people but they finally realized that it was their destiny to "give way to the insatiable progress of [the white] race. . . ."[8]

In his book *Custer Died for Your Sins*, Vine Deloria, Jr., discusses an interesting paradox in American society. Americans are told not to trust Communist countries, for they do not keep treaty commitments. We fought in Vietnam for years and thousands of people lost their lives ostensibly because the United States had to keep its commitments in Southeast Asia. The paradox is this: At the very time we were spending $100 billion keeping commitments in Vietnam, the United States government was also breaking the oldest Indian treaty, the Pickering Treaty of 1794 with the Seneca tribe of the Iroquois nation. In fact, as far as breaking treaties is concerned, it is doubtful that anybody can beat our record. Deloria reports that "America has yet to keep one Indian treaty or agreement despite the fact that the United States government signed over four hundred such treaties with Indian tribes."[9]

Desirable land has been systematically taken from Indians, who are then relocated on reservations in less desirable areas. Indians held 138 million acres in 1887; this was reduced to 55 million acres by 1970 and to 52 million by 1982. If formerly undesirable land becomes desirable because of the discovery of oil, or if it contains a river needed for power or irrigation, or if it is considered to be a prime area for development of any kind, further relocation and manipulation of Indian rights and lands are sure to follow.

The 1990 census reported that there were approximately 2 million American Indians in the United States. In general, living conditions on reservations have long been substandard. Median family income is low, and unemployment and the number of families below the poverty level is high (27 percent in 1990). Educational attainment and life expectancy are well below that of whites. Death statistics provide further evidence of the effects of minority status. Indians are less likely to die of heart disease and cancer than are other Americans, but they are about three times as likely to die of accidents. Indians have a high birth rate (about two and a half times that of whites), but Indian deaths in infancy and early childhood are about double that of other Americans. In 1990, 54 percent of all American Indian births were to unmarried mothers. Indian deaths by cirrhosis of the liver are more than double the national average, by homicide more than triple, and by tuberculosis four times the national average. A 1999 Department of Justice report found that American Indians experienced a rate of violent crime double that of the rest of the population. Aggravated assault and simple assault were especially high, and alcohol use was a major factor for Indians in committing violent offenses. The rate of alcohol-related offenses for Native Americans was more than double that found among non-Indians. Suicide rates are also very high, especially for Indians in their late teens and twenties. Indians suffer greatly the effects of isolation and alienation from society brought on by the general neglect of many generations. Racism and bigotry have an adverse effect on mental health, and the American Indian has been the victim of this abuse longer than has any other American.

### Hispanic Americans

Americans of Spanish and Latin origin have become this country's fastest growing minority. Census figures put the number of Hispanic Americans in 1997 at 29.7 million (roughly 10 percent of the population). The regions and countries that Hispanics come from are shown in Figure 6.3.

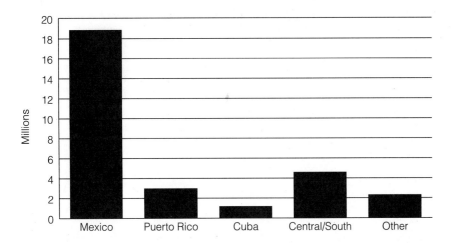

**FIGURE 6.3**   *Origins of Hispanic population, 1997.*

United States Bureau of the Census, *Statistical Abstracts of the United States, 1998* (118th edition), Washington, D.C.

Population projections suggest that the Hispanic population could outnumber blacks within the next decade.

Many Hispanic Americans face the conditions that seem to accompany minority status. Their median family income in 1996 was $26,179 (see Figure 6.1), and 26 percent of Hispanic families were below the poverty level. Hispanic Americans face health difficulties because, although their employment rates are high, they are severely underinsured (often being self-employed or working for small businesses). They might delay doctor visits because of lack of insurance and suspicion of the medical profession. Many Latinas and Latinos in the West are migrant farm workers laboring in the fields of California and Oregon. Farm labor is seasonal, uncertain, underpaid, and subject to difficult working conditions. Hispanic Americans are also underrepresented in the political process. In 1996, only 36 percent were registered to vote compared with over 60 percent of non-Hispanics. Hispanic Americans hold a low percentage of jobs in government. In 1994, 5,459 Hispanic Americans held office in 36 states. (More than 3,700 of these were in three states, Texas, New Mexico, and California.) In 1995, there were 17 Hispanics in the House of Representatives and no Hispanic senators.

The major unifying factors for Hispanic Americans are religion (Roman Catholicism) and language, but many diversities exist as well. Some of these are shown in Table 6.3. Reading Table 6.3, we see differences in education, unemployment, family status, and income. Cubans have higher levels of education, income, and intact families, whereas Latinos from Mexico have a high percentage of intact families but also high unemployment rates and a lower level of education. Most of the Latinos of Mexican origin have settled in the West and Southwest (more than 3 million in the Los Angeles metropolitan area alone). More than half of those from Cuba have settled in the

**TABLE 6.3** Characteristics of the Hispanic Population in the United States, 1997

| | Area of Origin | | | |
| --- | --- | --- | --- | --- |
| | Mexican | Puerto Rican | Cuban | Central/South American |
| Education | | | | |
| High School grad or more | 49% | 61% | 65% | 63% |
| Bachelor's degree or more | 7% | 11% | 20% | 15% |
| Labor Force | | | | |
| Unemployment Rate | 8% | 10% | 7% | 7% |
| Family Type | | | | |
| Married Couple | 72% | 54% | 77% | 65% |
| Female household, no spouse present | 20% | 39% | 17% | 27% |
| Family Income | | | | |
| Median Income | $25,347 | $23,646 | $35,616 | $29,960 |
| Families below poverty level | 28% | 33% | 13% | 19% |

United States Bureau of the Census, *Statistical Abstracts of the United States, 1998* (118th ed.), Washington, D.C.

Miami area, and more than 1 million Hispanics from Puerto Rico are living in the New York City area.

Perhaps the most serious problem for Hispanic Americans is in the realm of education. Figure 6.2 shows level of education (completion of high school and college) by race and Hispanic origin. As you can see, although Hispanics' level of education was slightly ahead of blacks in 1970, Hispanic Americans have not kept up with increases by whites and blacks since that time. The school dropout rate for Hispanic Americans is two or three times that of non-Hispanics. Poverty and the necessity to support one's family are clearly important factors, but another major issue is language. More than 17 million people in the United States speak Spanish at home, and 48 percent of these speak English less than "very well." Schools, however, tend to communicate with students and parents in English only. Exclusion of their language leads to disintegration of their family heritage and discourages participation in school affairs. Children in school have to become comfortable in a new language in addition to learning other material. They read and perform below grade level and ultimately may drop out of school early.

## Jews

There are approximately 14 million people of the Jewish faith in the world. There are about 6 million Jews in the United States, or 2.2 percent of our population in 1996. Religion is the foundation, but being Jewish also involves an ethnic and cultural identity. Unfortunately, persecution has been a consistent factor in Jewish cultural history. Jews were forced off their land and driven into other areas as early as the sixth century B.C. Especially sharp conflict between Christians and Jews emerged after the third century A.D.

Starting around the tenth century, Jews were segregated into ghettos, at first voluntarily, then on a compulsory basis. In 1555, the Pope decreed that Jews must be segregated strictly in their own quarter of the city surrounded by a high wall with gates closed at night. Jews were forced out of certain occupations and denied citizenship. They were used as scapegoats to distract attention from other issues and to promote national unity. This was carried to its most horrible extreme during World War II with the extermination of six million Jews in Germany.

The consequences of minority status for Jews in the United States do not seem to be as severe as they are for other groups we have considered. This is probably because Jews are less visible than are most minority groups and because they have been more upwardly mobile through their emphasis on education and participation in the professions. Jews, however, have faced certain types of prejudice and discrimination in the United States. For a time, certain jobs were not open to Jews. Colleges and universities, especially in the eastern United States, had quotas covering the percentage of Jews they would accept. In 1958, 33 percent of the real-estate agents in a large midwestern city reported they did not want to rent to Jews, and a study of 3,000 resorts across the country in the same year found that 22 percent discriminated against Jews. In the late 1960s and 1970s, such overt discrimination against Jews probably declined, although anti-Jewish prejudice remains. The FBI reported 1,109 anti-Jewish incidents and 1,182 anti-Jewish hate crimes in 1996. Seventy-nine percent of all religion-related hate crimes were anti-Jewish. Incidents and crimes include firebombing of a synagogue, the desecration of Jewish cemeteries, threats of violence, and the distribution of anti-Semitic leaflets.

### Asian Americans

There were more than 10 million Asian Americans in the United States in 1997, or roughly 3.7 percent of the population. Table 6.1 (p. 187) shows figures for countries based on 1990 and 1997 census data. Asian Americans are an extremely diverse population representing a wide variety of countries and cultures. As with other racial and ethnic categories (Hispanic, white, black), assuming uniformity and similarity in Asian Americans is a great oversimplification.

Two waves of Asian immigration have occurred. In the 19th century, Chinese, and in lesser numbers, Japanese, Koreans, and Filipinos, arrived. These were mainly unskilled laborers working in construction and agriculture. Difficult conditions in China, combined with the lure of the California gold strike in 1849 and the construction of the transcontinental railroads, encouraged Chinese immigration. Early Japanese immigrants likewise were typically single males working in low-paying and physically difficult jobs. White workers felt their jobs threatened by the newcomers, which led to the passage of various exclusionary laws in 1882, 1892, 1907, and 1924 to limit further immigration. Asian immigration was effectively cut off until the laws were revised in 1965. Widespread discrimination against Asian

**TABLE 6.4**  Population Characteristics in the United States, 1997

|  | Asian/Pacific Islander | Black | White |
|---|---|---|---|
| Education |  |  |  |
| High school grad or more | 85% | 75% | 83% |
| Bachelor's degree or more | 42% | 13% | 25% |
| Labor Force |  |  |  |
| Unemployment rate | 3% | 10% | 4% |
| Family Type |  |  |  |
| Married couple | 79% | 46% | 81% |
| Female household, no spouse present | 13% | 47% | 14% |
| Family Income |  |  |  |
| Median income | $49,105 | $26,522 | $44,756 |
| Families below poverty level | 13% | 26% | 9% |

United States Bureau of the Census, *Statistical Abstracts of the United States, 1998* (118th ed.), Washington, D.C.

Americans led to forcing immigrants into urban ghettos ("Chinatowns," for example) and later to the internment of Japanese Americans during World War II.

The second and current wave of immigration is different from the first in that the immigrants tend to be of higher social class, well educated, and occupationally skilled. Families rather than single males are immigrating, and their country of origin is much more diverse. The second wave of immigration is also much larger, which is leading to rapid population growth in the Asian American population. With the exception of the Japanese, most Asians in America now (more than 75 percent of the Chinese and Filipino populations and more than 90 percent of the Korean and Vietnamese population) are foreign born. In 1996, four Asian countries—the Philippines, India, Vietnam, and China—ranked third, fourth, fifth, and sixth in number of immigrants to this country. (Mexico and the former Soviet Union were first and second.) Another characteristic of the second wave of Asian immigration is the presence of refugees from Southeast Asia. Thousands of people from Vietnam, Laos, and Cambodia came to the United States following the fall of the South Vietnamese government in 1975.

Asian Americans have settled primarily in urban areas in the western part of the United States. More than 50 percent of Chinese, Vietnamese, Laotian, Hmong, and Cambodians and more than 70 percent of Filipinos and Japanese live in the West. Table 6.4 shows selected education, unemployment, family, and income characteristics across three racial categories. Keep in mind that the census category "Asian and Pacific Islander" includes a broad and diverse set of peoples, and grouping them all under one category loses much precision. Asian Americans consistently show high levels of education and income as can be seen in Table 6.4. A cultural focus on education is shown further in that Asian and Pacific Islanders were awarded 11.7 percent of the doctorate degrees in 1996 while representing less than 4 percent of the United States population.

Past cases of prejudice and discrimination toward Asian Americans have been severe—hostility toward Chinese and Japanese workers, discriminatory laws, urban segregation, internment of Japanese. In the late 20th century, the picture has changed somewhat. Asian countries have had remarkable economic successes. Asian Americans have done well compared with other minority groups. The prejudice and discrimination toward Asian Americans that occurs is perhaps a by-product of their success. Prestigious colleges and universities have been accused of limiting enrollment of Asian Americans. American businesspeople continually complain about what they believe are the unfair practices of Asian countries that work to undermine U.S. businesses. Resentment of blacks toward Korean store owners was a factor in the unrest that led to the riots in Los Angeles in 1992. Occasional instances of overt violence occur, but most instances of anti-Asian discrimination today are more subtle and indirect.[10]

## GENDER

Men and women seem to be treated differently in nearly all societies. This inequality, although universal, varies greatly from one society to the next. Primitive hunting and gathering societies seem to have had less inequality between males and females, whereas agrarian societies, in which the great universal patriarchal religions support women's inferiority, have more. In modern industrialized societies, although some differences exist because of local policies (regarding hiring practices, maternity leave, or child care, for example), women are very unequal to men.

Why is this so? Some say that biology is the answer. Men's greater physical strength, women's linkage to childbirth, and in-born differences in temperament and emotion—these mean that men will dominate situations. Not so, say other experts who, though granting biological differences between the sexes, argue that because cultures differ so greatly on male and female relationships, social factors must be crucial. This view suggests that differences in temperament and emotion (being aggressive, assertive, dominant, submissive) are traits present in some males and in some females. The presence of traits, whatever they are, depends on socialization, custom, and societal expectations, not biology.

Several theories outline how this societal training might have happened. For example, Marvin Harris suggests that in early times, warfare was common. Being bigger and stronger, men made better warriors. But they had to be encouraged somehow because who wants to fight and be killed. So the bravest warriors were rewarded with sexual access to women. Men were trained to fight, women were trained to be subservient to males.

Friedrich Engels who, like his colleague Karl Marx, was a 19th-century conflict theorist, was one of the first to focus on the social aspects of gender stratification. Engels believed that primitive societies showed relative gender equality, primarily because women participated equally in food production. As a culture's economy developed, however, and surpluses appeared, males wanted to pass these accumulated goods on to their offspring.

Inheritance along male lines and monogamous marriage followed, and women were excluded from economic work and relegated to household tasks. Engels saw "the first class oppression . . . [as] that of the female sex by the male."[11]

Whatever its original causes, gender inequality exists and leads to discriminatory behavior. If a society is committed, at least in theory, to equal treatment, then it is important to understand the processes by which inequality is encouraged. To get at this, we need to return to the socialization process.

### Socialization and Gender Roles

*Thirty pairs of parents were questioned within twenty-four hours after the birth of their first child and were asked to "describe your baby as you would to a close relative." Hospital information on the babies showed that the fifteen boy babies and the fifteen girl babies did not differ on such objective data as birth length, weight, irritability, etc. But the parents said that girl babies were softer, littler, more beautiful, prettier, more finely featured, cuter, and more inattentive than the boy babies. The fathers tended to label, or stereotype, the babies in this fashion more than the mothers. The authors of the study suggest that gender typing and gender-role socialization have already begun at birth.[12]*

How could this be? One day after birth and already babies are showing definite gender differences in physical appearance and temperament. The answer, of course, is that *it isn't so.* People assume that males and females are born with different abilities and temperaments. People assume it to be so and then behave as if it *were* so. We tend to act toward children one way if they are male, another way if they are female. We expect them to be a certain way, and they turn out that way. The self-fulfilling prophecy rides again.

To place the assumption about so-called inherited gender characteristics in correct perspective, recall Margaret Mead's description of gender and temperament among the Arapesh, Mundugumor, and Tchambuli. The behavior of the sexes in those societies varied a great deal. Compared with gender roles in America, roles were modified and reversed. The explanation is not that these societies are strange and peculiar, but more simply that the gender roles learned in those societies are different from those learned here.

Like other roles, gender roles are learned through the socialization process. It is possible that in our society the teaching of gender roles starts even earlier than the teaching of other roles. As we saw earlier, parents immediately start acting differently toward their children based on their gender. Look at their toys: Boys get tractors, trucks, tools, guns, and athletic equipment; girls get dolls, cooking sets, play perfume and cosmetic kits, and pretty clothes. Friends and grandparents, scandalized to hear that the mother concerned about gender typing bought her son a doll, were only partly mollified to find that at least it was a boy doll.

The schools continue the pattern. Textbooks from Dick and Jane on up portray boys in active, aggressive, so-called masculine roles and girls in passive, tender, so-called feminine roles. A study of teachers in nursery schools found that they spent more time with the boys in the class than with the girls. Boys were encouraged to work harder on academic subjects. They were given more rewards and more directions in how to do things. Boys were given instructions, then encouraged to complete the task themselves. If the girls did not quickly get the idea, the teacher would often intervene and do the task for them. There was one exception: The teachers did pay more attention to the girls on feminine gender-typed activities such as cooking. Even here girls got praise and assistance, whereas boys received detailed instructions. The boys were given more attention, and the environment was much more of a learning experience for them than it was for the girls. Studies have found the same thing: more attention to males than to females throughout the grades and a "let me do it for you" attitude toward females, even at the college level.[13]

Another group of researchers working in the Boston area found that gender-role differences were well developed in the majority of children by the age of five. The children knew which personality traits were "masculine" and which were "feminine." They knew which jobs were for men and which were for women. The experimenters developed a curriculum that attempted to make the children more flexible in their assumptions about the sexes. The outcome of the program was mixed. To the researchers' surprise, many of the fifth- and ninth-grade boys with whom they worked became *more* stereotyped in their views of women and more rigid and outspoken about what they thought to be the woman's place. The effects on the girls in the program were more positive, showing increased self-esteem and attitude change away from typical stereotypes.[14]

The researchers were hopeful about the school program they tested, but that attitude is not enough by itself. The socialization agents include family, peers, teachers, literature, entertainment, mass media, and the whole social tradition that socializes people to believe in the inherent psychological, temperamental, and attitudinal differences between the sexes. It should not surprise us at all when boys grow up to be physicians and mechanics and girls grow up to keep house, have babies, and take care of their families. They have been socialized into these roles from the very beginning.

### Consequences of Gender-Role Typing

The results of gender-role typing are many and varied, and not all the benefits are for males. The male is restricted in how he can show emotion: He is strong and silent, he does not show weakness, and he keeps his feelings under careful rein, at least outwardly. The female has far greater freedom to express emotion. The male is subject to much more stress and pressure to achieve and be successful. This is probably part of the reason why males have a shorter life expectancy, more heart disease, and higher rates of suicide and hospitalization for mental illness and related pathologies. Women have been much less involved in crime and deviant behavior than have men, and this too is related to gender-role differences. Some men would like to change roles and to be househusbands, staying home and cooking, working in the garden, and taking care of the kids. What are their chances in a society that sets up gender roles like ours does? The major consequence of gender-role typing, however, is that women are not able to participate fully in American society, a society in which they make up 51 percent of the members.

The second research question at the beginning of this chapter concerned interruptions in group discussions. Are men or women more likely to do the interrupting? A study conducted in the late 1980s discovered a sharp gender difference. The main observation of the research is that men discriminate in their attempts at interruption, whereas women do not. Men interrupt women far more often than they do other men. Women interrupt men and women equally often. Sometimes interruptions are positive or supportive, others are negative. The paper notes that women were three times as likely as men to yield the floor in the face of a negative interruption. The composition of the group also had some effects. For example, males in all-male groups tend to make supportive interruptions of the other males. As the number of women in the group increases, however, male interruptions are much less likely to be supportive. The authors find, perhaps not surprisingly, that everyday conversations both reflect and maintain a society's existing social inequalities.[15]

Evidence of the consequences of gender typing can be found in a variety of places. Income of male and female year-around, full-time workers is shown in Figure 6.4. As can be seen, men are paid more. The gap is closing, but slowly. In 1980, on average, women were paid 60 cents for every dollar paid to men. In 1994, women's pay had increased to 74 cents. Women are half of the workforce but generally hold lower-skilled and lower-paying jobs. Women are always well represented at the sales and clerical levels but less well represented elsewhere, although top female executives did make strides in 1995. For the first time, all the top 20 female executives in corporate America earned more than $1 million. The gap remains, however: in 1999, women were paid 75 cents for every dollar paid to men, and women held only 11.2 percent of the executive jobs in Fortune 500 companies.

Over the past 25 years, major changes have occurred regarding women's roles and accomplishments. Polls show that many support the aims and efforts of the movement to change women's status in society. More women are elected to public positions. In 1995, there were 47 women in the House of Representatives (of 435), 8 women in the United States Senate (of 100), and 2 women on the U.S. Supreme Court (of 9). In 1997, women were about 27 percent of the nation's lawyers and judges, 26 percent of the physicians, and 17 percent of the dentists. Women were 84 percent of the elementary-school teachers but only about 61 percent of the school administrators. In education, more women (36 percent in 1997) than men (32 percent) graduated from high school, bachelor's degrees were relatively evenly split between men (17%) and women (15%), but women received

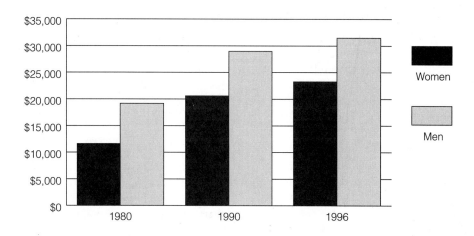

**FIGURE 6.4** *Median income of year around full-time workers, by gender.*
United States Bureau of the Census, *Statistical Abstracts of the United States, 1998* (118th ed.),
Washington, D.C.

40 percent and men received 60 percent of the doctorates awarded in 1996. Women's enrollment in business, medical, and law schools has increased substantially. Churches have ordained women and are creating opportunities for women to take leadership roles. Many police departments across the country now have women in jobs equal to those of male police officers. Women's participation in organized sports has increased. Title IX of the Education Amendments of 1972 prohibited sex discrimination in educational institutions that receive any federal funds, and this led to an expansion of opportunities between 1972 and 1978. Court cases in the 1990s have aided women's participation to the extent that 39 percent of NCAA athletes in 1997 were women. However, the percentage of college athletic budgets going to females is still much less than that going to males. In crime, men are and have always been far ahead of women. Even here, however, there is change. Arrests of males increased 2 percent between 1989 and 1998, whereas arrests of females increased 28 percent.

A final area of great concern is that of violence toward women. Violence toward women is not a new problem, but one that has finally gained the nation's attention. A few facts: Annually 100,000 rapes are reported to the police in this country, and it is estimated that two to six times that many actually occur—women are often reluctant to report rape in general, especially "date" rape and spousal rape. Ninety percent of the over 2,000 women murdered in 1997 were killed by men. Although approximately 600,000 assaults by intimates (friends, lovers, spouses) are officially reported to federal officials annually, the most conservative estimates indicate that two to four million women are battered each year. At least 170,000 of these incidents are serious enough to require hospitalization. Sexual harassment cases in businesses and in local and national government seem to be commonplace. The city where I live has a WEAVE (Women Escaping A Violent

## BLIND AUDITIONS

Zubin Mehta, former conductor of the New York Philharmonic, once said that "I just don't think women should be in an orchestra." And women weren't. Throughout the 1950s and 1960s, the Philharmonic was all male. Today, thanks to "blind" auditions, the situation is changing. A blind audition is one which disguises the identity of the candidate from the musical directors who might be gender or racially biased. The audition is conducted behind a screen and the musicians may even walk out on a rug or in socks to prevent listeners from hearing footsteps that might indicate gender. Before blind auditions started to become common (in the 1970s) women accounted for about 10 percent of new hires at major United States orchestras. Since the early 1980s, about half of new hires at the New York Philharmonic, 40 percent in San Francisco, and more than a third in Boston and Chicago have been women. Researchers estimate that using blind auditions explains anywhere from a quarter to nearly a half of the growing share of women in America's top orchestras between 1970 and 1990.

—*From "On Economics," by Jonathan Marshall*, San Francisco Chronicle, *February 10, 1997.*

Environment) chapter, a rape 24-hour crisis line, a rape–sexual assault center, a battered women's shelter, and a battered women's center.

The great success of the feminist movement has been in raising people's awareness—getting our attention—to the existence of gender stereotypes, of violence, and of unequal treatment. Education and change is occurring, and the progress in making people aware of issues and reducing discrimination is consistent if not dramatic. Changing traditional patterns of socialization for the young and resocializing adults are difficult at best.

## THE ELDERLY

Hope Bagger at the age of 70 was called for jury duty but then was told to go home because she was too old. She and many others with similar experiences have become active members of the Gray Panthers, an organization seeking to remedy the way society treats old people. In some parts of the world, old age is seen as a virtue—life in its very highest form. The elderly are revered and are well cared for by society as a whole. In this country, the elderly find that they are forced to retire, they have trouble getting other jobs, and they are exploited if they do get work. As one member of the Gray Panthers put it, the young seem to feel that old people descended from outer space. They forget that the elderly have a past and a future.

A baby born in the United States in 1900 had a life expectancy of 48 years; a child born today has a life expectancy of more than 76 years. There were 34 million people 65 and older in the United States in 1997 (12.7 percent of the population), and almost 15 million older than 75. These age groups are the fastest-growing population groups in the nation. By the year 2020, it is estimated that one-sixth of all Americans will be older than 65. Many of the elderly are poor and out of work, more than 50 percent are widowed, the suicide rate of older males is high, and age often compounds the effects of other minority statuses.

Aging affects people both physically and socially. In a general sense, physical aging begins in people around the age of 30 and continues throughout the life span. Certain body functions, muscle tone and strength, and the senses all reach their peaks around 30 and decline thereafter. Although many people tend to view it as such, aging is not a disease. People are more likely to die as they get older because the body becomes more vulnerable to outside influences, less adaptable, and more affected by stress and crisis.[16]

*Social* aging is more treacherous, yet potentially easier to deal with, than physical aging. Social aging refers to the attitudes we have about older people. We are indoctrinated with the idea that the old are different and inferior to the rest of us. We stereotype the aged, and they are victims of prejudice and discrimination. American society treats the elderly with the same sort of stereotyping, the same institutionalized discrimination, as that based on gender and race. The old are thought to be unintelligent, asexual, unemployable, mentally ill, and difficult to get along with. The fact that many if not most older people are *not* this way is beside the point. We continue to make them retire before they want to, to refuse to hire them, and to pay them less when they do get a job. And the self-fulfilling prophecy suggests to us that if people are viewed and treated a certain way long enough, they tend to become that way.

What do the elderly want? Our stereotypes tell us that they want a life of leisure (sitting out the last 20 years on the porch swing or in front of the tube); they want to grow old gracefully; they want a time of disengagement (to be able to relax now because the important things are done). On the contrary, to remain in good health old people need what everyone needs: work to do, money to live on, a place to live in, and other people to care for and to care for them. Society should recognize that the elderly are an important resource.[17]

As are many other victims of minority status, the elderly are becoming politically active. It probably starts with consciousness raising and moves to more militant stands on issues through collective action. Groups like the Gray Panthers, the National Retired Teachers Association, the National Council of Senior Citizens, and the American Association of Retired Persons (AARP) have formed a "gray lobby" of enormous strength in Washington. The AARP, founded in 1958 to provide insurance for retirees, now has some 27 million members and has become the nation's largest special-interest group. One of nine Americans belongs, and its members are flexing their political muscles.

It is important, then, that the elderly not be viewed as social obsoletes. They *do* have a past and a future. And remember that unlike many minority statuses, this is one that happens to *all* of us—if we're fortunate.

## PATTERNS OF INTERACTION

When populations that differ by race or ethnicity come together, a number of different reactions can occur, some peaceful, others not.[18] **Annihilation** is the elimination of one group by another. It can be intentional, as in the case of the Nazis' attempted extermination of a whole ethnic group—a practice called **genocide**—or it can be unintentional, as in the case of American Indians fallen prey to illnesses brought in by European settlers. **Expulsion** refers to the removal of a group from the territory in which it resides. The Japanese on the West Coast during World War II were removed from their homes and placed in detention camps, and American Indian tribes were forced from their homes onto reservations. **Segregation** refers to the setting apart of one group from another. It is not as extreme as expulsion, but the separation is nevertheless obvious. The segregation can be physical, such as a move to a separate area of the city; or it can be social, as when people living together are limited in how and with whom they can interact. Segregation from the minority's viewpoint can be either involuntary, as with blacks in America, or voluntary. Avoidance (voluntary self-segregation) occurs when a culture such as the Amish in America wants to maintain its identity and uniqueness against the influence of the dominant culture and therefore voluntarily separates itself.

Assimilation and amalgamation are other patterns of interaction that occur between unlike groups. **Assimilation** refers to the mixing and merging of unlike cultures so that two groups develop a common culture. When two cultures meet, each culture might adopt some of the other's traits so that the result is a true melting pot—the emergence of a new culture different from either of the old cultures. More often, however, assimilation means that the incoming or minority culture adapts to the dominant culture. **Amalgamation** refers to biological (rather than cultural) mixing. Amalgamation seems invariably to accompany the meeting of racial and ethnic groups, despite the best efforts of some to legislate against it. A **marginal person** is defined as one who lives between two antagonistic cultures, a product of each but a member of neither. Studies indicate that the consequences of marginality can be severe—personality disturbances or feelings of inferiority. Marginality is often defined as a product of amalgamation, but it can also result from cultural mixing (the southern rural migrant to the urban North, for example).

**Cultural pluralism** describes a pattern of interaction in which unlike cultures maintain their own identity and yet interact with each other relatively peacefully. Switzerland is usually cited as the best example of pluralism in that peoples of several nationalities and religions and even three or four different languages are able to get along peacefully. According to Brewton Berry, few countries have been able to successfully accomplish a system of pluralism. Some minorities in the United States have at various times voiced a preference for cultural pluralism, but the dominant culture seems to look either toward integration or toward discrimination and segregation.

## ELIJAH ANDERSON (1943– )

Elijah Anderson earned his Ph.D. in Sociology at Northwestern University in 1976. He has conducted extensive fieldwork on the plight of the urban poor and the disenfranchised. He writes about a complex and profoundly alienated subculture, something disturbingly alien to the experience of the mainstream culture most of us inhabit. "Good ethnography is part of a dialogue between classes," he explains. "It's opening up a line of communication between vastly different segments of society, trying to replace fear with understanding." In his 1999 book *Code* *of the Street,* Anderson describes the codes or temptations that exist in the inner city. The inner city that he describes is characterized by unemployment, teenage pregnancy, little interest in education, and much alienation and lack of hope. The inner-city codes stress achieving status and getting respect through verbal boasts, drug selling, sexual prowess, and violence. Anderson is a Professor of Sociology at the University of Pennsylvania. An excerpt from one of Anderson's books appears in Chapter Seven, pp. 257–261.

## MINORITY-GROUP REACTIONS

The patterns of interaction described here are usually dictated by the dominant group. What then of minority-group reactions to domination? Gordon Allport has outlined some types of reactions in his book *The Nature of Prejudice.*[19] The simplest response is to *deny* that one is a member of that minority group: Members of one race pass as members of another; people change their names to get rid of their ethnic identification. Similarly, **acculturation** refers to attempts by members of minority groups to assimilate, to blend in, and to take on as many characteristics of the dominant culture as possible. Hiding resentment behind *withdrawal* and passivity—"the mask of contentment"—is another reaction. **Secession** refers to the formal withdrawal of a group of people from a political, religious, or national group. The South seceded before the Civil War, and the citizens of Quebec have been considering secession from the rest of Canada. *Self-hatred* is another reaction, with the minority group identifying with the dominant group. Studies of Nazi concentration camps showed that after years of suffering, inmates began identifying with the guards. The prisoners wore bits of the guards' clothing, they became anti-Semitic, and in general they adopted the mentality of the oppressor. Likewise, if blacks value light skin more than dark, they might be accepting whites' evaluation of skin pigmentation. Self-hatred can lead to aggression against one's own group, as when a segment of the minority that thinks of itself as being of a higher class takes out its frustrations on a lower-ranked segment. Minorities sometimes show greater *prejudice* against other minorities and in other cases the opposite reaction—greater *sympathy*.

A *strengthening of in-group ties* is another reaction by minorities to domination. **Nativism** is a type of rejection of the dominant culture in which people attempt to improve their own existence by eliminating all foreign people, objects, and customs. This has been seen in the United States with slogans such as "Black Power" and "Don't buy where you can't work," and by the focus on the development of ethnic identity through emphasizing the minority's cultural and historical roots. Minorities might also react to domination by using *aggression.* The aggression can be individual and reflected in crime patterns, or it can be collective and culminate in urban disturbances or riots. In milder form, the aggression can appear in the literature and humor of the minority group as the group makes fun of or shows its bitterness toward the majority group. Humor is a common way of expressing feelings about other groups. In the following examples, who is putting down whom?

> An African American woman from Harlem wins the lottery and decides to buy a fur coat. She tries on a coat at Saks Fifth Avenue and it comes down to her ankles. As she looks at herself in the mirror, she turns to the saleslady and says, "Do you think this makes me look too Jewish?"

> A priest and a rabbi are each driving down a New York street. The rabbi stops at a light and the priest runs into him. An Irish cop comes up to the priest and asks, "Ah, Father, how fast might the rabbi have been going when he backed into you?"

> What do you have when you cross a WASP and a chimpanzee? A three-foot-tall, blond company president.

## THEORY AND RESEARCH: A REVIEW

*Conflict theorists* would have no difficulty explaining the sort of social differentiation we talked about in this chapter. Minority-status people are dominated and exploited by others because of the continuing struggle for scarce resources such as jobs, food, land, wealth, and power. Those who hold privileged positions attempt to ensure that they are safe by limiting access to these positions by whatever means possible. Racial, ethnic, gender, and age differences are easily identified and therefore become perfect "excluders." Color of skin, inability to speak the language well, or lack of citizenship become ready criteria for limiting access to jobs, education, or other valuable benefits. As in the discussion of stratification in the previous chapter, *functional analysis* would suggest that complex societies have numerous roles that must be performed. Some of these are more unpleasant than others, but all must be filled for society to run smoothly. Therefore, the stratification or ranking of society's members represents a good functional adaptation. By making distinctions between female and male, old and young, and recent arrivals and old timers, all slots are filled and all needed tasks from farm laborer and garbage collector to homemaker and company president are performed.

The data used in this chapter are descriptive and, for the most part, are census data obtained by *questionnaire* or *interview surveys.* Several studies cited in this chapter (Hartley's studies, Coleman's studies, the study of new-born babies) were also *reactive surveys.*

## SUMMARY

This is the second of two chapters on social differentiation and social inequality. Here we examined the characteristics of race, ethnic, gender, and age categories. We found in each case that elements within the categories are ranked and that inequality follows. One common component of these categories is that of *minority status.* The fact that categories are ranked and the consequences of this ranking were the central topics of this chapter. *Prejudice, discrimination,* and *stereotyping* are commonly used to reaffirm the status ranking we believe in. They help us keep others "in their place." We examined the types of reactions that occur when populations of different races or ethnic backgrounds meet, as well as the reactions of minority groups themselves to discrimination.

We examined a number of statistical examples of differential treatment—differences in education, jobs, income, opportunities in life, crime and deviant behavior, sickness, and others. Less easily displayed statistically but more important as a human concern is the effect of continued differential treatment on *how one sees oneself.* Some concepts studied earlier—socialization, role, self-concept, and self-fulfilling prophecy—help us understand the critical significance of minority status for the individual's development.

The first two readings that follow deal with the perceptions and judgments people make about others without knowing them—the foundation upon which stereotyping, prejudice, and discrimination are built. The first article is a first-person account by an African American man, who describes his urban adventures in being mistaken for someone he's not. The second article describes "the jeweler's dilemma"—in a dangerous world, whom do you let into your store? The third reading examines discrimination in the workplace and outlines ways to bypass or break through the "glass ceiling." The fourth reading is an excerpt from Dee Brown's *Bury My Heart at Wounded Knee,* which describes the consequences for Native Americans of the western expansion in the United States between 1860 and 1890. The final reading is the account of a person making his way from Mexico to the United States illegally and the border patrol's attempts to stop him.

---

## TERMS FOR STUDY

acculturation (205)

amalgamation (204)

annihilation (204)

assimilation (204)

authoritarian personality (185)

cultural pluralism (204)

discrimination (184)

ethnicity (182)

expulsion (204)

genocide (204)

| | |
|---|---|
| marginal person (204) | racism (184) |
| minority status (180) | secession (205) |
| nativism (206) | segregation (204) |
| prejudice (183) | stereotyping (185) |
| race (181) | |

For a discussion of Research Question 1, see page 182.
For a discussion of Research Question 2, see page 200.

---

## INFOTRAC COLLEGE EDITION
### Search Word Summary

To learn more about the topics from this chapter, you can use the following words to conduct an electronic search on InfoTrac College Edition, an online library of journals. Here you will find a multitude of articles from various sources and perspectives:

**www.infotrac-college.com/wadsworth/access.html**

| | |
|---|---|
| passive resistance | wages |
| Civil Rights Act | aging |
| health care | urban poor |
| immigration | |

**Reading 6.1**

---

# JUST WALK ON BY

*Brent Staples*

---

*Stereotyping, prejudice, and discrimination are based on attitudes and judgments we make about others without knowing them. We ignore the individual and generalize and categorize, often falsely. In this article, an African American man describes his experiences of being mistaken for someone he's not, continually, and beginning "to know the unwieldy inheritance I'd come into—the ability to alter public space in ugly ways."*

---

*Brent Staples has a Ph.D. in psychology, is an editorial writer for the* New York Times, *and is author of the 1994 memoir* Parallel Time: Growing Up in Black and White.

My first victim was a woman—white, well dressed, probably in her early twenties. I came upon her late one evening on a deserted street in Hyde Park, a relatively affluent neighborhood in an otherwise mean, impoverished section of Chicago. As I swung onto the avenue behind her, there seemed to be a discreet, uninflammatory distance between us. Not so. She cast back a worried glance. To her, the youngish black man—a broad six feet two inches with a beard and billowing hair, both hands shoved into the pockets of a bulky military jacket—seemed menacingly close. After a few more quick glimpses, she picked up her pace and was soon running in earnest. Within seconds she disappeared into a cross street.

That was more than a decade ago. I was twenty-two years old, a graduate student newly arrived at the University of Chicago. It was in the echo of that terrified woman's footfalls that I first began to know the unwieldy inheritance I'd come into—the ability to alter public space in ugly ways. It was clear that she thought herself the quarry of a mugger, a rapist, or worse. Suffering a bout of insomnia, however, I was stalking sleep, not defenseless wayfarers. As a softy who is scarcely able to take a knife to a raw chicken—let alone hold one to a person's throat—I was surprised, embarrassed, and dismayed all at once. Her flight made me feel like an accomplice in tyranny. It also made it clear that I was indistinguishable from the muggers who occasionally seeped into the area from the surrounding ghetto. That first encounter, and those that followed, signified that a vast, unnerving gulf lay between nighttime pedestrians—particularly women—and me. And I soon gathered that being perceived as danger-ous is a hazard in itself. I only needed to turn a corner into a dicey situation, or crowd some frightened, armed person in a foyer somewhere, or make an errant move after being pulled over by a policeman. Where fear and weapons meet—and they often do in urban America—there is always the possibility of death.

In that first year, my first away from my hometown, I was to become thoroughly famil-iar with the language of fear. At dark, shadowy intersections, I could cross in front of a car stopped at a traffic light and elicit the *thunk, thunk, thunk, thunk* of the driver—black, white, male, or female—hammering down the door locks. On less traveled streets after dark, I grew accustomed to but never comfortable with people crossing to the other side of the street rather than pass me. Then there were the standard unpleas-antries with policemen, doormen, bouncers, cabdrivers, and others whose business it is to screen out troublesome individuals *before* there is any nastiness.

I moved to New York nearly two years ago and I have remained an avid night walker. In central Manhattan, the near-constant crowd cover minimizes tense one-on-one street encounters. Elsewhere—in SoHo, for example, where sidewalks are narrow and tightly spaced buildings shut out the sky—things can get very taut indeed.

After dark, on the warrenlike streets of Brooklyn where I live, I often see women who fear the worst from me. They seem to have set their faces on neutral, and with their purse straps strung across their chests bandolier-style, they forge ahead as though bracing themselves against being tackled. I understand, of course, that the danger they perceive is not a hallucination. Women are particularly vulnerable to street violence, and young black males are drastically overrepresented among the perpetrators of that violence. Yet these truths are no solace against the kind of alienation that comes of being ever the suspect, a fearsome entity with whom pedestrians avoid making eye contact.

It is not altogether clear to me how I reached the ripe old age of twenty-two without being conscious of the lethality nighttime pedestrians attributed to me. Perhaps it was because in Chester, Pennsylvania, the small, angry industrial town where I came of age in the 1960s, I was scarcely noticeable against a backdrop of gang warfare, street knifings, and murders. I grew up one of the good boys, had perhaps a half-dozen fistfights. In retrospect, my shyness of combat has clear sources.

As a boy, I saw countless tough guys locked away; I have since buried several, too. They were babies, really—a teenage cousin, a brother of twenty-two, a childhood friend in his mid-twenties—all gone down in episodes of bravado played out in the streets. I came to doubt the virtues of intimidation early on. I chose, perhaps unconsciously, to remain a shadow—timid, but a survivor.

The fearsomeness mistakenly attributed to me in public places often has a perilous flavor. The most frightening of these confusions occurred in the late 1970s and early 1980s, when I worked as a journalist in Chicago. One day, rushing into the office of a magazine I was writing for with a deadline story in hand, I was mistaken for a burglar. The office manager called security and, with an ad hoc posse, pursued me through the labyrinthine halls, nearly to my editor's door. I had no way of proving who I was. I could only move briskly toward the company of someone who knew me.

Another time I was on assignment for a local paper and killing time before an interview. I entered a jewelry store on the city's affluent Near North Side. The proprietor excused herself and returned with an enormous red Doberman pinscher straining at the end of a leash. She stood, the dog extended toward me, silent to my questions, her eyes bulging nearly out of her head. I took a cursory look around, nodded, and bade her good night.

Relatively speaking, however, I never fared as badly as another black male journalist. He went to nearby Waukegan, Illinois, a couple of summers ago to work on a story about a murderer who was born there. Mistaking the reporter for the killer, police officers hauled him from his car at gunpoint and but for his press credentials would probably have tried to book him. Such episodes are not uncommon. Black men trade tales like this all the time.

Over the years, I learned to smother the rage I felt at so often being taken for a criminal. Not to do so would surely have led to madness. I now take precautions to make myself less threatening. I move about with care, particularly late in the evening. I give a wide berth to nervous people on subway platforms during the wee hours, particularly when I have exchanged business clothes for jeans. If I happen to be entering a building behind some people who appear skittish, I may walk by, letting them clear the lobby before I return, so as not to seem to be following them. I have been calm and extremely congenial on those rare occasions when I've been pulled over by the police.

And on late-evening constitutionals I employ what has proved to be an excellent tension-reducing measure: I whistle melodies from Beethoven and Vivaldi and the more popular classical composers. Even steely New Yorkers hunching toward nighttime destinations seem to relax, and occasionally they even join in the tune. Virtually everybody seems to sense that a mugger wouldn't be warbling bright, sunny selections from Vivaldi's *Four Seasons.* It is my equivalent of the cowbell that hikers wear when they know they are in bear country.

---

## QUESTIONS 6.1

**1.** Select at least five of the concepts discussed in this chapter and show how they are illustrated in this article.

**2.** Write a short essay on the effects of prejudice and discrimination on the individual, using concepts like socialization, self, looking-glass self, and significant and generalized other.

**3.** What sorts of patterns of interaction are illustrated in this article?

---

## Reading 6.2

---

# THE JEWELER'S DILEMMA

---

*A controversy broke out in Washington, D.C., in 1986 over the practice in certain jewelry stores of admitting customers only through a buzzer system. Some store owners used this system to exclude young black males on the grounds that these people were most likely to commit robbery. Some people argued that this was racism; others felt that the practice was justified.* The New Republic *magazine asked some of its readers to comment on this situation.*

### The Intelligent Bayesian

Men are not gods. Therefore men face challenges gods would not have to endure—ignorance and uncertainty. To make decisions, we need to have information about the world around us. The information we gather is not only imperfect, it is costly as well. So we learn to economize by guessing, prejudging, and using stereotypes.

Imagine you are challenged to a basketball game and must select five out of 20 people who appear to be equal in *every* respect except race and sex. There are five black and five white females, five black and five white males. You have no information about their basketball proficiency. There is a million-dollar prize for the contest. How would you choose a team? If you thought basketball skills were randomly distributed by race and sex, you would randomly select. Most people would perceive a strong associative relationship between basketball skills on the one hand, and race and sex on the other. Most would confine their choice to males, and their choice would be dominated by black males.

---

Can we say such a person is a sexist/racist? An alternative answer is that he is behaving like an intelligent Bayesian (Sir Thomas Bayes, the father of statistics). Inexpensively obtained information about race and sex is a proxy for information that costs more to obtain, namely, basketball proficiency.

There is a large class of human behavior that generally falls into the same testing procedure. Doctors can predict the probability of hypertension by knowing race, and osteoporosis by knowing sex. A white jeweler who does not open his door to young black males cannot be labeled a racist any more than a black taxi driver who refuses to pick up young black males at night. Black females and white females and white males commit holdups, but in this world of imperfect information cabdrivers and jewelers play the odds. To ask them to behave differently is to disarm them. The other side of the argument is that they are losing money if they err—i.e., turning away an honest customer.

People can be stupid Bayesians. In 1971 I owned a house in Washington's exclusive Chevy Chase suburb. My Saturday chore was to pick up motorists' litter along the side of my house. One time I was approached by a gentleman who stopped his car and said, "Sir, when you're finished working at this house, could you come over to my place and work for me?" I politely told the gentleman I would not have time, since the rest of the day would be spent working on my Ph.D. dissertation. The gentleman apologized profusely. (My wife once encountered another incompetent Bayesian. Awaiting a bus during a downpour, she was offered a ride to Chevy Chase Circle by a black woman who turned out to be a domestic servant. "Don't you just hate working for these white people out here?" her benefactor observed.) Mistaken identity is always at least a bit unpleasant, but it is not the same as racism.

WALTER E. WILLIAMS

Walter E. Williams is John M. Olin, Distinguished Professor of Economics at George Mason University.

## The Weight of Suspicion

Who would object to the refusal of a storekeeper, particularly of a store so obviously a target for crime as a jewelry store, to admit a stranger who looked "suspicious"? What is the purpose of a locked door and a buzzer system unless to keep out strangers who might wish to do harm? And since they are by definition strangers, how does anyone expect to distinguish between the innocuous and the malevolent strangers except by finding the latter "suspicious"?

I would find anyone suspicious who appeared different in certain ways from most of those who come into my shop. Teenagers, unless they were conventionally dressed, would be suspicious. People whose carriage and movements gave the impression of nervous tension would be suspicious. People whose voice on the intercom sounded tense, artificial, unclear would be suspicious. My decision to admit would be based on an intuitive calculation of the number of categories of suspicion and the intensity of my feeling, balanced by the knowledge that I must take adverse chances or encounter the greater risk of losing my store. And, of course, I add two decisive categories of suspicion, assuming one or more of the categories previously mentioned were positively present: sex and race.

To pretend that a roughly dressed, nervous, slurring, tense young black male has a democratic right to enter a jewelry store is to distort a rational process with an irrelevant abstraction. Does it salve my conscience to say that I exclude him because he's suspicious, not because I ascribe an innate criminal potentiality to males with black skins? Perhaps. My parents used to whisper the phrase "colored people" lest our Negro maid be embarrassed.

I too like to feel virtuously unprejudiced, and I have convinced myself that only so long as young black males commit an inordinate number of robberies will I continue to find decisive some suspicion-raising attributes I would overlook in others.

ROGER STARR

Roger Starr is an editorial writer at the *New York Times*.

## The Lazy Man's Substitute

Neither black nor white store owners are in business to display the virtues of admitting people of all colors, creeds, and fashions to their stores. They are in business to make money. I would want to take precautions to prevent robbery; I would look closely at people entering the store. The race of a potential customer would be one factor among many to be considered as I girded myself against thieves.

But in Washington and almost all other major cities, blacks do patronize jewelry stores. A jeweler in Beverly Hills who closed his or her door to heavily bejeweled Mr. T. would be foolishly closing his cash register. Unless I am a racist, race and age cannot be the sole deciding factors in calculating whom I will or will not let into my store. And I certainly would not close my door to, say, all young black men—not even to those who are casually dressed and behaving nervously. I would act cautiously in dealing with them, as I would with an antic, strangely dressed white man.

As a cabdriver I would apply the same considerations. Discrimination can be used judiciously. I would certainly exclude one class of people: those who struck me as dangerous. Nervous-looking people with bulges under their jackets would not be picked up; nor would those who looked obviously drunk or stoned. It all comes down to a subjective judgment of what dangerous people look like. This does not necessarily entail a racial judgment. Cabdrivers who don't pick up young black men as a rule are making a poorly informed decision. Racism is a lazy man's substitute for using good judgment.

. . . In this situation and all others, common sense is my constant guard. Common sense becomes racism when skin color becomes a formula for figuring out who is a danger to me.

JUAN WILLIAMS

Juan Williams is a reporter for the *Washington Post*.

## Black Man's Burden

While I am putting myself in the shoes of the jewelry store owner, I will also put myself in the skin of a black person. This makes the dilemma even more interesting.

In downtown Washington, where my jewelry store is located, I know that it is possible, but not likely, that a white man is going to rob me. To the extent that I worry about robbers, I mostly worry about blacks. Thus one effective method of keeping robbers out of my store will be to scrutinize blacks more carefully.

I don't much like to do this because I know how easily such occasional and sensible "discrimination" can turn into widespread and simple bigotry. For example, I read in the *Washington Post* recently that when I go out to look for a house or an apartment in the Washington area, I have a 50 percent chance of being steered away from white neighborhoods or not being shown the nicest apartment or not being told about all the amenities. Now, most of these real estate agents in Washington are white and most of them very courteous. You just never know when one of them may rip you off.

As a businessman, I like to be consistent in my dealings with the public. If I don't let a black guy in the door because he's black and he's wearing a baseball cap, maybe I should refuse to deal with a white real estate agent the next time I am apartment hunting. Or maybe I should refuse to be shown around by white real estate agents with Southern accents who voted for Ronald Reagan. But what would white people think if they knew that successful black men like myself were refusing to do business with them just because of their skin color?

The last thing black people need today is a lot of white people getting pissed off at them. So when facing a situation in which I might get ripped off, I think it's prudent to take a chance. The humiliation of getting ripped off by some speedy black kid who snatches one of my gold necklaces is comparable to the humiliation of getting discriminated against by some smiling white real estate agent. But, of course, the odds that black kid outside is going to rob me are nowhere near 50 percent. Someday I hope I won't have to wonder about white real estate agents. So I figure: unless he's really a dangerous-looking kid, let him in. A lot of white people won't understand that it's them, more than the kid, I'm giving a break. But maybe someday.

JEFFERSON MORLEY

---

Comments: To further complicate the situation, let's add some other information: (1) As the editors of *The New Republic* point out, what the jewelers are doing is illegal. If race is a factor in deciding whom you admit to your store, you're breaking the law. (2) Juan Williams (one of the letter writers) reports that the assumption that blacks commit more robberies is mistaken, "a convenient racist lie." He reports that young blacks' robbery rate is no higher than whites of the same age. (3) In 83 percent of robberies by blacks, their victims are black.

# QUESTIONS 6.2

1. "The behavior of the jewelers in denying admittance to certain people is blatant racism." Support or refute.

2. "Playing the odds in situations like this is only good and prudent business." Support or refute.

**3.** What does Walter Williams mean by "being an intelligent Bayesian"? Give examples in other types of social relationships.

**4.** How is the coach who selects five black basketball players to start the game while leaving several white players on the bench behaving differently from the jeweler who won't admit the black teenager to his store? Or is it the same? Discuss in terms of the meanings of stereotypes, prejudice, and discrimination.

---

## Reading 6.3

---

# EXILED IN GUYVILLE

*Lorraine Ali*

---

*Job discrimination takes several forms—not being able to get the job, getting the job but not being paid equally to others, or being denied promotion unfairly. All these exist for women. One component of the job situation for women is what has been called a "glass ceiling"—an invisible but very real barrier that keeps females from reaching high-level management positions in organizations. Does the glass ceiling exist, and if so what are some effective ways of bypassing or breaking through it? Los Angeles writer Lorraine Ali answers these questions as they relate to the music industry.*

"When I was a secretary, did I ever think I was going to become chairman of a major record label?" says Sylvia Rhone, the chairman and CEO of the recently linked Elektra Entertainment and East West Records. "I would say the answer would have to be no." Like most successful women in the pop music industry, Rhone, 42, started off answering phones and watching men climb the corporate ladder to success. Twenty years later, unlike most women, Rhone heads a corporate sphere that boasts nearly 200 acts and annual revenues estimated to be in the hundreds of millions.

Between the polar positions of secretary and CEO, women are working their way up the ranks of the music industry, breaking away from the terminal job of assistant and taking positions of responsibility previously reserved for men. Women can now be found signing new talent as A&R representatives, hiring and firing as vice presidents and boosting radio play as promotions people. But while considerable progress has been made during the past two decades, Rhone is still a rarity in the music business, where women fill only 3 percent of the senior positions available.

As of last year, only 25 percent of vice-president and department-head positions were held by women, yet when compared with the rest of the corporate world, the

From *Rolling Stone*, October 6, 1994. By Straight Arrow Publishers Company, L.P. 1994. All Rights Reserved. Reprinted by Permission.

music industry is actually six to eight times more likely to elevate women to this level. But as women gain higher positions in the industry, the money is slower to catch up. According to the most recent figures from the Bureau of Labor Statistics, women in the entertainment industry make roughly 74 percent of what similarly employed men make. That isn't much better than the national average, which indicates that women earn 71 cents to a man's every dollar.

With those kinds of numbers, victories have been bittersweet. While celebration would seem in order for a promotion like Rhone's, it also illustrates how hard women have had to push for such major advancements to become more than just cork-popping, isolated incidents.

"A glass ceiling definitely still exists, but I think it's being shattered every day," says Rhone, a Harlem, N.Y., native who grew up listening to the music of Aretha Franklin and Marvin Gaye.

"Those of us as minorities who break through the glass ceiling do get some scratches along the way," she says. "But I think there is definitely social change and headway being made. Some companies may hold out for a time, maintaining that archaic view of women as unsuitable executive material, but those old philosophies are changing drastically."

Such changes haven't come easy. Many women who entered the business in the '70s, including Warner Bros. East Coast A&R vice president Karin Berg, a former *Rolling Stone* writer who has brought bands like the B-52's and New Order to the label, remember when there was literally one other A&R woman and only a handful of female execs to take cues from. "Women just did publicity, and that's pretty much all they did," Berg says. "It was very unusual to find a woman as the head of a department."

Since then, cracking the exclusive mind-set and slanted policies of the record business' good-ol'-boys club has been a constant struggle. Virginia native Polly Anthony, 40, also started out as a secretary back in 1978 and is now general manager of the Sony 550 Music label. Artists she oversees at the year-old label include Celine Dion, Social Distortion and the The. "I think every woman who is successful in this business had to work a lot harder than men," Anthony says. "I think we've had to be patient, to turn the other cheek; we've had to prove ourselves time and time again. I think we've had to be incredibly tenacious."

Jenny Price, a 26-year-old Atlantic Records A&R rep, may represent a new generation of women entering the business, but her story isn't much different. "I've walked into meetings of all men where they acted like I wasn't even in the room," she says. "There was no conversation started with me; I didn't exist. So I had to step in and push extra hard to make way for myself, because there was nothing they were going to offer me as part of their group. It's just something I had to do."

Nancy Jeffries, one of pop's first A&R women, fell into the business end of rock almost accidentally when she took a job as a temp secretary at RCA in 1974. Jeffries, who had sung in the '60s psychedelic art band Insect Trust, would eventually end up discovering Evelyn "Champagne" King, Suzanne Vega and Lenny Kravitz. But regardless of her obvious talent, the New York native had to continually orchestrate her own plan of affirmative action in order to obtain well-deserved promotions or pay raises. Her MO: She would simply quit, only to be rehired with her demands met. "It seemed whenever I got a title anywhere, all the guys immediately had to have that title," says Jeffries, who is now senior VP of A&R at Elektra Records. "They would go crazy, pitch fits. It took over 10 years for the money to catch up, when I felt I was making the same amount as the guys."

Other women have taken greater steps to counter the injustices they have come up against in the industry. In November 1991, one month after Anita Hill testified before the Senate Judiciary Committee about Supreme Court nominee Clarence Thomas, Penny Muck filed a landmark sexual-harassment suit against Geffen Records. The 28-year-old secretary claimed her boss, Marko Babineau, then the general manager of Geffen's DGC label, had repeatedly harassed her, even to the point of masturbating on her desk. Muck also charged that previous complaints regarding Babineau made by fellow female staffers to Geffen execs had been over-looked. A year later, the case was settled for a reported $500,000. Released from his job at Geffen, Babineau went on to work for another major label.

Bypassing the corporate structure of rock altogether, many women have started companies of their own. "I feel less compromised," says Rosemary Carroll, co-founder of the law firm Codikow, Carroll and Regis.

"The more I got into the business," she says, "the more I saw the odds were stacked against, me, the more ambitious I got—if only just to show I could do it."

More and more women are going the route of Carroll and living off their own music-related businesses—from record labels to management and publicity firms. Most agree it's a more direct way to obtain their goals without the barriers or waiting periods created by sexist corporate policies.

Women new to the business like Price seem less optimistic about the current climate than women who have already achieved higher management positions. Price went from working in a record store to eventually assisting label kingpin Irving Azoff before becoming an A&R rep. Newcomers like her are moving up the ranks faster but still feel women are not being promoted as quickly as men. "A few years back I felt really disconcerted with the industry," Price says. "Seeing guys getting jobs out of nowhere and seeing myself and other women struggling as assistants was frus-trating. That's not over; I still know women who are getting slighted."

Bryn Bridenthal, a 48-year-old VP of media and artist relations at Geffen Records, sees it differently. "There was a time when I looked around, and it seemed the women who weren't getting what they wanted were citing discrimination, but in truth, they weren't focused or working hard enough," says Bridenthal, who special-izes in "difficult personalities and disaster," recently fielding the media's endless questions about Axl Rose's legal problems, Courtney Love's state of mind and the suicide of Kurt Cobain. "The record business is survival of the fittest. If you're going to wait around for someone to serve something up on a platter for you, you're gonna go hungry, because there are way too many people like me who will do anything to get what they want."

Competition in any corporate situation is fierce, and with high-level positions for women still limited, women are often thrown into a heavy-duty dog-eat-dog environ-ment. The question is, are women helping women in their wake or leaving them safely behind? "I think women were raised to compete against each other," says Anthony. "I think men were raised to play on teams together and have a healthy camaraderie and rivalry. We were raised with a different set of values when it came to relating to one another, as young women. Some of the strongest bonds in the world exist between two women, but there also needs to be a constant reminder to extend a hand out." Rhone never questioned the camaraderie between her and her female associates. "Those of us who took on roles of increasing responsibility knew we had a lot to prove," she says. "I always saw us as real trailblazers. We were linked to each other. If one of us failed, we all failed."

Publicity departments have long been referred to as the female ghetto of record companies, where a woman's career comes to a no-future halt. Bridenthal says that that is a misconception the record industry used to discredit the hard work of the female-dominated sector. When she founded the publicity department at Geffen in 1987—10 years after she entered the record business—Bridenthal sought to bust that "girly job" myth by infiltrating the record business with what she terms her kind of women. "What I meant by that was women who worked longer, harder, smarter and put work before anything else," she says, "who aren't here to get laid or get married but had what are traditionally thought of as more masculine goals."

The very need to assimilate to a man's frame of mind bothered Carroll, who once worked overtime in the masculinity department to counter being "blond" and "married to a rock star." "One thing I remember was that older male executives would treat me condescendingly, as if I were a glorified groupie," says Carroll. "That really surprised me. I went to Stanford Law School, I jumped through all the right hoops to be taken seriously. So to all of a sudden not be taken seriously surprised me. I went through a period of wearing those dumb pinstriped suits and putting my hair up in a bun, hoping that if I wore the right costume, people would let me play the role. Gradually they did, and I felt less pressured to play that role."

"I'm just starting to learn how to deal with men as an executive rather than as an assistant, and it's hard," says Price, who still finds her feminine side a risk. "Unfortunately the answer is sometimes not to act like a female, because femininity in this business is seen as a weakness."

Though femininity is often treated as counter-productive to good business, sometimes it also proves to be an asset in a world that's been male dominated since music was first pressed on vinyl. "It's less threatening to have a woman dealing with your creative aspects," says Jeffries. "It's much less confrontational and more cooperative. Women listen more. Men are more aggressive—more *me, me, me*."

Since the indie scene—a network of do-it-yourself record labels—was spawned by punk's rejection of traditional values, it has proved more accepting of women as decision makers and leaders and boasts more women starting their own labels and holding positions of responsibility.

Candice Pedersen started out as an intern at the independent label K Records, in Olympia, seven years ago and is now co-owner. "When I talk to indie labels, I'm always speaking to women who make decisions, but at majors, the women I speak to are usually publicists or sales people," says the 28-year-old, who works with the bands Mecca Normal and Halo Benders. "One of the reasons the indie scene exists is because people don't want to follow the rules. One of the ways you don't follow the rules is to choose the best person to work for you and not base the decision on preconceived ideas of which gender knows more about rock. Or, as women, you start up your own company."

A lot of ground has been broken by women in the music industry over the past 20 years, but the toughest advances have yet to come across the board. "To really judge our progress, we should look at where we are in five or 10 years," says Anthony. "The commitments are being made, women are being promoted and recognized, but let's see how many company presidents or people in truly senior management positions there are by then. The promise is there. Let's see how wisely we all play it out."

---

## QUESTIONS 6.3

**1.** What types of discrimination are illustrated in this article? Are these specific to the music industry or more generally present in most organizations?

**2.** What, according to this article, are some of the ways of dealing with a glass ceiling?

**3.** How would a conflict theorist explain what is happening in the reading? What explanation would a functionalist have?

**4.** According to this chapter, gender-role socialization is crucial. Give examples of this type of socialization from this reading.

---

## Reading 6.4

# WOUNDED KNEE

### *Dee Brown*

---

*In his book,* **Bury My Heart at Wounded Knee,** *Dee Brown writes about the critical period between 1860 and 1890, the 30 years during which the culture and civilization of the American Indian was destroyed. Brown tells the story in an unusual way—he uses the viewpoint and the words of the victims, the Indians. The story is a haunting and depressing one and includes descriptions of the long walk of the Navahos, the war to save the buffalo, the war for the Black Hills, the flight of the Nez Perce, and the story of Geronimo, the last of the Apache chiefs. The following excerpt from Brown's book describes the end of it, the "battle" at Wounded Knee Creek in South Dakota in late December 1890. The Indians' lands were gone then, their chiefs were dead or imprisoned, their own numbers decimated by 30 years of battles, by disease, by living on crowded reservations in unfamiliar places, and by heartbreak. In early December 1890 the great chief Sitting Bull was assassinated. Then . . .*

---

As soon as Big Foot learned that Sitting Bull had been killed, he started his people toward Pine Ridge, hoping that Red Cloud could protect them from the soldiers. En route, he fell ill of pneumonia, and when hemorrhaging began, he had to travel in a wagon. On December 28, as they neared Porcupine Creek, the Minneconjous sighted four troops of cavalry approaching. Big Foot immediately ordered a white flag run up over his wagon. About two o'clock in the afternoon he raised up from his blankets to greet Major Samuel Whitside, Seventh U. S. Cavalry. Big Foot's blankets were stained with blood from his lungs, and as he talked in a hoarse whisper with Whitside, red drops fell from his nose and froze in the bitter cold.

Whitside told Big Foot that he had orders to take him to a cavalry camp on Wounded Knee Creek. The Minneconjou chief replied that he was going in that direction; he was taking his people to Pine Ridge for safety.

Turning to his half-breed scout, John Shangreau, Major Whitside ordered him to begin disarming Big Foot's band.

"Look here, Major," Shangreau replied, "if you do that there is liable to be a fight here; and if there is, you will kill all those women and children and the men will get away from you."

Whitside insisted that his orders were to capture Big Foot's Indians and disarm and dismount them.

"We better take them to camp and then take their horses from them and their guns," Shangreau declared.

"All right," Whitside agreed. "You tell Big Foot to move down to camp at Wounded Knee."

The major glanced at the ailing chief, then gave an order for his Army ambulance to be brought forward. The ambulance would be warmer and would give Big Foot an easier ride than the jolting springless wagon. After the chief was transferred to the ambulance, Whitside formed a column for the march to Wounded Knee Creek. Two troops of cavalry took the lead, the ambulance and wagons following, the Indians herded into a compact group behind them, with the other two cavalry troops and a battery of two Hotchkiss guns bringing up the rear.

Twilight was falling when the column crawled over the last rise in the land and began descending the slope toward Chankpe Opi Wakpala, the creek called Wounded Knee. The wintry dusk and the tiny crystals of ice dancing in the dying light added a supernatural quality to the somber landscape. Somewhere along this frozen stream the heart of Crazy Horse lay in a secret place, and the Ghost Dancers believed that his disembodied spirit was waiting impatiently for the new earth that would surely come with the first green grass of spring.

At the cavalry tent camp on Wounded Knee Creek, the Indians were halted and children counted. There were 120 men and 230 women and children. Because of the gathering darkness, Major Whitside decided to wait until morning before disarming his prisoners. He assigned them a camping area immediately to the south of the military camp, issued them rations, and as there was a shortage of teepee covers, he furnished them several tents. Whitside ordered a stove placed in Big Foot's tent and sent a regimental surgeon to administer to the sick chief. To make certain that none of his prisoners escaped, the major stationed two troops of cavalry as sentinels around the Sioux teepees, and then posted his two Hotchkiss guns on top of a rise overlooking the camp. The barrels of these rifled guns, which could hurl explosive charges for more than two miles, were positioned to rake the length of the Indian lodges.

Later in the darkness of that December night the remainder of the Seventh Regiment marched in from the east and quietly bivouacked north of Major Whitside's troops. Colonel James W. Forsyth, commanding Custer's former regiment, now took charge of operations. He informed Whitside that he had received orders to take Big Foot's band to the Union Pacific Railroad for shipment to a military prison in Omaha.

After placing two more Hotchkiss guns on the slope beside the others, Forsyth and his officers settled down for the evening with a keg of whiskey to celebrate the capture of Big Foot.

The chief lay in his tent, too ill to sleep, barely able to breathe. Even with their protective Ghost Shirts and their belief in the prophecies of the new Messiah, his people were fearful of the pony soldiers camped all around them. Fourteen years before, on the Little Bighorn, some of these warriors had helped defeat some of these soldier chiefs—Moylan, Varnum, Wallace, Godfrey, Edgerly—and the Indians wondered if revenge could still be in their hearts.

"The following morning there was a bugle call," said Wasumaza, one of Big Foot's warriors who years afterward was to change his name to Dewey Beard. "Then I saw the soldiers mounting their horses and surrounding us. It was announced that all men should come to the center for a talk and that after the talk they were to move on to Pine Ridge agency. Big Foot was brought out of his teepee and sat in front of his tent and the older men were gathered around him and sitting right near him in the center."

After issuing hardtack for breakfast rations, Colonel Forsyth informed the Indians that they were now to be disarmed. "They called for guns and arms," White Lance said, "so all of us gave the guns and they were stacked up in the center." The soldier chiefs were not satisfied with the number of weapons surrendered, and so they sent details of troopers to search the teepees. "They would go right into the tents and come out with bundles and tear them open," Dog Chief said. "They brought our axes, knives, and tent stakes and piled them near the guns."

Still not satisfied, the soldier chiefs ordered the warriors to remove their blankets and submit to searches for weapons. The Indians' faces showed their anger, but only the medicine man, Yellow Bird, made any overt protest. He danced a few Ghost Dance steps, and chanted one of the holy songs, assuring the warriors that the soldiers' bullets could not penetrate their sacred garments. "The bullets will not go toward you," he chanted in Sioux. "The prairie is large and the bullets will not go toward you."

The troopers found only two rifles, one of them a new Winchester belonging to a young Minneconjou named Black Coyote. Black Coyote raised the Winchester above his head, shouting that he paid much money for the rifle and that it belonged to him. Some years afterward Dewey Beard recalled that Black Coyote was deaf. "If they had left him alone he was going to put his gun down where he should. They grabbed him and spun him in the east direction. He was still unconcerned even then. He hadn't his gun pointed at anyone. His intention was to put that gun down. They came on and grabbed the gun that he was going to put down. Right after they spun him around there was the report of a gun, was quite loud. I couldn't say that anyone was shot, but following that was a crash."

"It sounded much like the sound of tearing canvas, that was the crash," Rough Feather said. Afraid-of-the-Enemy described it as a "lightning crash."

Turning Hawk said that Black Coyote "was a crazy man, a young man of very bad influence and in fact a nobody." He said that Black Coyote fired his gun and that "immediately the soldiers returned fire and indiscriminate killing followed."

In the first seconds of violence, the firing of carbines was deafening, filling the air with powder smoke. Among the dying who lay sprawled on the frozen ground was Big Foot. Then there was a brief lull in the rattle of arms, with small groups of Indians and soldiers grappling at close quarters, using knives, clubs, pistols. As few of the Indians had arms, they soon had to flee, and then the big Hotchkiss guns on the hill opened upon them, firing almost a shell a second, raking the Indian camp, shredding the teepees with flying shrapnel, killing men, women, and children.

"We tried to run," Louise Weasel Bear said, "but they shot us like we were buffalo. I know there are some good white people, but the soldiers must be mean to shoot children and women. Indian soldiers would not do that to white children."

"I was running away from the place and followed those who were running away," said Hakiktawin, another of the young women. "My grandfather and grandmother and brother were killed as we crossed the ravine, and then I was shot on the right hip clear through and on my right wrist where I did not go any further as I was not able to walk, and after the soldier picked me up where a little girl came to me and crawled into the blanket."

When the madness ended, Big Foot and more than half of his people were dead or seriously wounded; 153 were known dead, but many of the wounded crawled away to die afterward. One estimate placed the final total of dead at very nearly 300 of the original 350 men, women, and children. The soldiers lost twenty-five dead and thirty-nine wounded, most of them struck by their own bullets or shrapnel.

After the wounded cavalrymen were started for the agency at Pine Ridge, a detail of soldiers went over the Wounded Knee battlefield, gathering up Indians who were still alive and loading them into wagons. As it was apparent by the end of the day that a blizzard was approaching, the dead Indians were left lying where they had fallen. (After the blizzard, when a burial party returned to Wounded Knee, they found the bodies, including Big Foot's, frozen into grotesque shapes.)

The wagonloads of wounded Sioux (four men and forty-seven women and children) reached Pine Ridge after dark. Because all available barracks were filled with soldiers, they were left lying in the open wagons in the bitter cold while an inept Army officer searched for shelter. Finally the Episcopal mission was opened, the benches taken out, and hay scattered over the rough flooring.

It was the fourth day after Christmas in the Year of Our Lord 1890. When the first torn and bleeding bodies were carried into the candlelit church, those who were conscious could see Christmas greenery hanging from the open rafters. Across the chancel front above the pulpit was strung a crudely lettered banner: PEACE ON EARTH, GOOD WILL TO MEN.

## QUESTIONS 6.4

**1.** Describe the treatment of the Indians in terms of the patterns of interaction described in the text. What is the pattern today?

**2.** Discuss the concept of Manifest Destiny: What might have caused it to appear? Were there other victims in

addition to the Indians? What are the consequences of this doctrine for its adherents as well as for its victims? Is this doctrine still with us?

**3.** Research the Indian situation in your area; what tribe lived there and what happened to them?

## Reading 6.5

# DANGEROUS CROSSING

*S. Lynne Walker*

*The Immigration and Naturalization Service estimated that there were five million undocumented immigrants in the United States in October 1996, two million in California alone. This population has increased 28 percent in four years. Popular resentment against illegal immigration is growing. Ballot measures are common in border states, and it has become politically popular to budget for higher-tech equipment and more border patrol agents to tighten border security. So far, these have not worked. There are several problems: When economic conditions in one area are seen as markedly better than in a neighboring area, there will be migration. And, there is always a demand for workers to do the "dirty work"—hard, tedious, boring, low-paid work—work that often only those at the bottom of the social-class ladder will do. The story told here is probably reasonably typical.*

The voice on the phone rattled off instructions: "Walk out of the bus station. Take a left. You'll see a bus that says 'Centro.' Tell the driver to let you off at Madero. Call again from the nearest pay phone."

Luis Muñoz, 21, hung up the phone on the wall of the Tijuana bus terminal and stuck his head out the door. He walked to the end of the long, vacant sidewalk in front of the bus station.

"Madero," he said as he pushed a peso into the driver's hand. "I want to get off at Madero."

The driver nodded and ground the granny gear, sending the one-time school bus lurching forward into the rainy dawn. Luis shifted uncomfortably and rubbed his neck, still sore from the 40-hour bus ride from Guanajuato to Tijuana.

El Guero was the name of the smuggler his brother said will help him get to Chicago: The White Man.

Luis jumped down from the bus into a puddle of water. He looked up and down the empty street. He lifted the receiver of a pay phone. Tentatively, he touched the buttons.

The same voice answered, offering new instructions:

"Look for a pharmacy with a blue awning. You'll see a man dressed in a sombrero and white jacket standing outside. He'll be whistling. Follow him. He'll take you to El Guero."

Luis turned around. There he was: a man in a white cowboy hat stood in the rain, whistling an indistinguishable, upbeat tune.

The man motioned and disappeared into the alley. Luis hesitated, then plunged into the garbage-filled passageway between two flophouses and followed him.

## Toward a Better Life

"When are we leaving? How long will it take us to get from San Diego to Chicago? Can we go in a plane?"

El Guero chuckled at Luis' impatient questions. They arrived at a safe house, a grimy two-story walk-up near Tijuana's tourist strip, Avenida Revolucion.

"Rest," he told Luis. "Have something to eat. We'll talk in a little while."

For the first time Luis had a few moments to reflect on the circumstances that led him from his home in Chichimequillas. Since he was 9 years old he had worked at a shoe factory in Leon, laboring from early morning to the middle of the night, six days a week, to produce 4,800 shoes a month and earn $1,450 a year.

But recently his mother told him it was time for him to follow in the footsteps of his father, his brothers and other men in the village and go to the United States to earn the money for a better life.

She had raised 12 children in a three-room adobe house with no running water. None of them had had an education, and many days they had gone without food.

She wanted more for them. Still, she wept when Luis left. "Every time one of you leaves," she told him, "a piece of me dies."

The door burst open. A scrawny young man named Jose stomped in.

Jose was El Guero's chief coyote, and he had just returned from downtown San Diego, 20 miles north, where he dropped off a group of migrants right under immigration officials' noses.

"When are we going? Where are we going?" Luis asked eagerly.

Jose just laughed.

"By 5 o'clock in the morning, you'll be in San Diego," he said.

The day passed slowly. One by one, more migrants drifted into the room. It was 3:30 p.m. when they were finally ready to leave.

The eight migrants climbed into a dilapidated van, men in back, so they couldn't be seen by police, women in front with Jose, who was cranking the key in the ignition until the motor grudgingly started.

## Captured

The van glided to a stop near the town of Tecate, 30 miles east of Tijuana.

"Hurry up," Jose, the coyote, shouted at the migrants. "If you don't hurry, the police will see you."

Luis jumped out, slid down a muddy hill and ducked under a barbed-wire fence.

"Hurry up!" he yelled. But it was too late.

Luis looked back and saw three men in black sliding down the hill toward him and the other migrants.

"Hold it right there," they ordered.

Luis and the others scrambled, but there were too many of the men in black to escape. The Mexican police caught Jose and a fellow coyote. The others simply gave up.

At the top of the hill, El Guero stood in the rain with a sheepish look on his face, a police officer at his side. He'd been captured for the first time in his smuggling career.

Twenty-four hours after being stopped on the hillside, Luis and the other migrants were released. Luis would never see El Guero again. As he and the others walked out the jailhouse door, the police asked them if they'd try to cross illegally again. The eight men and women admitted they would.

"Good luck," a police officer said as he ushered them out. "Hope it goes well for you."

## In Colonia Libertad

Luis stood in the chilly air at the edge of a gritty Tijuana neighborhood.

He was not alone. Almost 100 other men were waiting with him on the hillside. Like him, they had been brought there by coyotes to cross to the other side.

Near midnight a coyote named Javier emerged from the fog. He ordered the men to a nearby safe house, where they'd already spent a restless day.

"Nobody is going to get through tonight," he told them.

With the first streak of dawn, Luis awoke to see the miserable place where he spent the night. Colonia Libertad is a seamy community shoved up against Mexico's northern border. It is the neighborhood Mexicans call Liberty.

Every day, men like Luis, with nothing more than a change of clothes in a tattered plastic bag, stream past the houses in their determined march to find work in the United States.

"Do you have a blanket?" migrants ask the faces squinting behind tattered screen doors. "Can you spare a taco? Do you have a place where I can sleep?"

Those pleas changed the life of Carmen, a Mixtec Indian woman from Oaxaca who had worked in the United States, harvesting California strawberries for a decade.

First, she gave the men tacos. Then she started lending blankets. Then she gave them a place to sleep. One day, she decided to help them get across the border. That was the day she stopped being a law-abiding citizen and became a smuggler.

Now, she and her husband, Rafael, run a well-organized smuggling ring. They charge $450 for the trip from Tijuana to Los Angeles, while most smugglers charge almost $1,000. They also give the migrants a place to rest and a hot meal before their journey north.

Their job is becoming more risky, Rafael admitted. The Border Patrol has become tougher and better equipped in this presidential election year.

He and Carmen pay young men to lead migrants like Luis across the border. Every night, the coyotes wait on the hillside in Colonia Libertad, watching for the right moment to sneak another group of Mexicans into the United States.

## Confrontation

Luis hopped off and pressed against the 10-foot steel fence that divides Mexico from the United States, just before the spotlight of the Vietnam-era Hughes helicopter swept past, lighting up the border as if it were midday.

Luis never cursed back home. But he joined the other migrants in uttering profanities as he glared at the helicopter following the fence line from Tijuana toward Tecate.

Standing on the hillside, 17-year-old Javier, Carmen's youngest coyote, calculated the reward for his nightly work. Success meant he'd be paid $100 for every person he got through. When the flow was heavy, he earned more than $1,200 a week.

"I'd say we go right now," he told Luis, motioning north.

Suddenly, a sound of vehicles rumbled along the dirt roads. Bright lights flashed across the rugged terrain. The Broncos raced up the hill to the place where the men were gathered and screeched to a halt. The helicopter hovered overhead, illuminating the faces of the Border Patrol agents.

Luis ducked down. He peered through a hole in the fence at 12 tall, well-fed, muscular men standing shoulder to shoulder, a human line in the sand.

He'd never seen Border Patrol agents before. Now, the legendary men in green uniforms, the men whose exploits are exaggerated and repeated in every Mexican household, were standing no more than six feet away.

A balding, bespectacled man stepped forward from the line of green uniforms and looked at the eyes peeking at him through the fence.

"I understand that you have to make an effort to cross the border. But you are not going to cross here tonight," he said in fluent Spanish. "I'm sorry."

Quiet now, Luis and the other men on the hillside turned and walked back toward the safe house where they'd already spent two anxious days.

## In a Strange Land

The next night, Luis wedged himself under the forbidding steel fence that separates Mexico from the United States.

On the U.S. side, Javier gave the migrants final instructions before they dashed into San Diego.

"Don't look around for the *migra*," Javier barked. "Pay attention to where you're going. I don't want anyone falling down and busting their face."

Luis dashed headlong down the hill, his feet barely touching the soft earth. He ran for the first time on American soil, straight into a canyon.

Running beside him was the point man, a coyote named El Camaron, who teams with Javier. The others were behind him.

Luis ran faster than he ever ran on the soccer field back home.

The migrants sprinted past motion detectors laid by the Border Patrol, across a graded road and up the side of a steep hill.

On the barren hillside, migrants are easy targets for the Border Patrol's infrared night vision scopes. The heat-seeking devices set on 30-foot booms spot anything that moves—man, woman or beast. Mexicans call it the *ojo magico*—the magic eye.

But that night, the migrants were lucky. It was foggy and the ground was damp. The ojo magico had been turned off because in these conditions Border Patrol agents wouldn't stand a chance against the resolute migrants.

But there were still helicopters to fear, and Luis and the others ran with terror at their heels.

At the top, Luis found himself in a strange and foreign land. From the ridge, he could see the silhouettes of children's swing sets in the back yards, and televisions sets flickering in distant homes. A great city stretched out before him, glowing in radiant light.

He rested, savoring the moment. The others sat nearby, silent and awe-struck.

## Underground

As Luis and the others crept past a neighborhood in the San Diego suburb of San Ysidro, Javier raised his hand for them to stop, then waved them toward a storm sewer. The migrants' trek through canyons and over hills was finished. They were about to go underground.

Luis squeezed past a metal grate that partially covered the street gutter and stooped to step into a massive drainage pipe.

It was totally black inside. Try as he might, he couldn't see.

"Everybody hold hands," El Camaron said. "We've got to stay together. If you wander into one of the side tunnels, we'll never find you."

Luis grabbed the belt of the man in front and he reached for the hand of the man in back. The man behind him clutched the next hand, until they formed a human chain. Together, they moved ahead, feeling their way along the tunnel, inch by inch.

Luis's muscles ached from hunching over. They must have walked at least a mile.

At last, Luis stepped out into the orange glow of a neighborhood streetlight and breathed the cool night air.

The first leg of his trip to Chicago had come to an end. He was standing on U.S. soil, and no Border Patrol agents were in sight.

*[Luis's further adventures included avoiding border patrol check points on the California side of the border, picking up $1,000 his brothers had wired to him in Los Angeles, paying Rafael his $450 fee, buying new clothes, getting a fake green card ($45) and social security number, and flying to Chicago. There he immediately started a job his brothers had found for him washing dishes 11 hours a day, 6 days a week. He would take home $250 a week, nine times what he earned in Mexico. A fortune for him, and a house for his mother.]*

# QUESTIONS 6.5

**1.** "Illegal immigration is functional for society." Prepare an argument to support this statement.

**2.** Suppose you are a politician (state governor, senator)—prepare a reasonable and workable solution to the "immigration problem," keeping in mind the motivations of the parties on both sides as shown in this reading.

**3.** Examine this article using each of the three theories—conflict, functional analysis, symbolic interaction. Show what each would focus on and how they would analyze the material differently.

# NOTES

1. This distinction is suggested by Brewton Berry in *Race and Ethnic Relations,* 3d ed. (Boston: Houghton Mifflin, 1965), chapter 2.

2. See "Racial Identity Among Caribbean Hispanics: The Effect of Double Minority Status on Residential Segregation," by Nancy Denton and Douglas Massey, *American Sociological Review* 54 (October 1989), pp. 790–808.

3. The definition of racism is from Martin Marger, *Race and Ethnic Relations,* 2d ed. (Belmont, Calif.: Wadsworth, 1991), p. 27. For more on the mysterious Danireans and Wallonians, see E. L. Hartley, *Problems in Prejudice* (New York: Kings Crown, 1946).

4. Robert K. Merton, "Discrimination and the American Creed," in *Discrimination and National Welfare,* edited by R. M. MacIver (New York: Harper & Brothers, 1949), pp. 99–126. Also see Berry's summary in *Race and Ethnic Relations,* 3d ed., pp. 300–302 (see note 1).

5. Theodor W. Adorno, Else Frankel-Brunswick, D. J. Levinson, and R. N. Sanford, *The Authoritarian Personality* (New York: Harper & Row, 1950).

6. The research discussed in this paragraph was from papers presented at meetings of the American Psychological Association, 1987, as reported in *Time,* September 14, 1987, p. 74.

7. "The Meaning of the New Coleman Report," by Diane Ravitch, *Phi Delta Kappan,* June 1981, pp. 718–720.

8. General Carleton's views on Manifest Destiny and the Indians are quoted by Dee Brown in *Bury My Heart at Wounded Knee* (New York: Holt, Rinehart & Winston, Bantam Books, 1971), p. 31.

9. Vine Deloria, Jr., *Custer Died for Your Sins* (New York: Macmillan, 1969), pp. 28–29.

10. Several works have been very helpful for this section: Marger, *Race and Ethnic Relations,* 2d ed., chapter 9 (see note 3); and *Race Relations,* 5th ed., by Harry Kitano (Upper Saddle River, N.J.: Prentice-Hall, 1997).

11. Sources for this section are *Social Stratification,* 2d ed., by Daniel W. Rossides (Upper Saddle River, N.J.: Prentice-Hall, 1997), p. 446; and *Inequality and Stratification,* 2d ed., by Robert Rothman (Englewood Cliffs, N.J.: Prentice-Hall, 1993), pp. 27–28.

12. Jeffrey Rubin, Frank Provenzano, and Zella Luria, "The Eye of the Beholder: Parents' Views on Sex of New Borns," *American Journal of Orthopsychiatry* 44 (July 1974), pp. 512–519.

13. Lisa Serbin and Daniel O'Leary, "How Nursery Schools Teach Girls to Shut Up," *Psychology Today* 9 (December 1975), p. 57, and Myra and David Sadker, "Sexism in School," *Psychology Today* 19 (March 1985), pp. 54–57.

14. This research is summarized by Carol Tavris in "It's Tough to Nip Sexism in the Bud," *Psychology Today* 9 (December 1975), p. 58. Tavris states that the full report of this

research by Marcia Guttentag is available in *Undoing Sex Stereotypes* by Marcia Guttentag and Helen Bray (New York: McGraw-Hill, 1976).

15. See "Interruptions in Group Discussions: The Effects of Gender and Group Composition," by Lynn Smith-Lovin and Charles Brody, *American Sociological Review* 54 (June 1989), pp. 424–435.

16. Anita Harbert and Leon Ginsberg, *Human Services for Older Adults: Concepts & Skills* (Belmont, Calif.: Wadsworth, 1979), chapters 1 and 2.

17. Alexander Comfort, "Age Prejudice in America," *Social Policy,* November/December 1976, pp. 3–8.

18. These paragraphs on patterns of interaction are drawn from Berry, *Race and Ethnic Relations,* 3d ed., chapters 7–12 (cited in note 1).

19. See Gordon Allport, *The Nature of Prejudice* (Garden City, N.Y.: Doubleday, 1958), chapter 9.

# Institutions: Family

## Family

*Structural Variabilities: How Do Family Structures Differ?* • *Functional Uniformities: How Are Families Alike?*

## The Family: Contrast and Change

*The Family in China* • *The Family in America*

## Family Problems: The Conflict Approach

## The Family of the Future

## TWO RESEARCH QUESTIONS

*1. Do children repeat their parents' marital experiences (age at marriage, likelihood of being pregnant when married, chances of divorce, and chances of remarriage), or are children's patterns independent of their parents'?*

*2. What is the effect of having children on marital stability? Is the presence of children conducive to marital happiness? Which is more beneficial to marital happiness—young children or older (teenage) children?*

SO FAR WE have peered through several windows to discover the social organization of society. One view of social organization is through study of the nature and structure of groups. Analysis of the characteristics of categories and social differentiation provides another view. In this chapter, the study of social institutions will provide us with a third perspective.

One way of understanding social institutions as a concept is to describe the emergence of an institution. All societies have a set of constant and important central needs or problems that must be dealt with. A society, if it is to survive, must deal with such issues as reproduction of the species, socialization of the young, distribution of goods and services, care for the sick, and so on. Norms and roles emerge to organize and regulate behavior as society deems appropriate. And so a system of social relationships—a pattern of organization—develops. **Social institutions,** then, refer to this organized system of social relationships, common to all societies, that emerges to deal with these basic problems.[1]

For example, society's basic task of reproducing the species leads to the development of particular patterns of behavior. A society supports some of these behaviors (say, heterosexual contact) and discourages others (indiscriminate mating outside of marriage). This support reflects the development of norms: consistent views about the way things should and should not be done. The development of these norms and roles whereby the experimental becomes expected and the spontaneous becomes formalized is the process of **institutionalization.** As this process proceeds and the system of social relationships becomes more and more organized, a so-called institutional area emerges—in this case, the institution is the *family.*

The study of social organization makes a good deal more sense if we integrate several of the concepts we have studied. For example, if you merge the concepts of social institutions and groups, as shown in Table 7.1, you will discover that specific basic problems lead to the development of corresponding institutional areas. Next, notice that groups perform the functions that are important to the institutional area. Two important factors—the individual's interests and the institutional norms—come together in groups. These groups are of various sizes and complexities: from large, formal organizations to small, personal primary groups.

Let's follow an example across Table 7.1. People in most societies are concerned with life after death and seek an explanation for what is unknown or uncertain. Great power or faith might be invested in an idol, a god, a person,

**TABLE 7.1** Institutions in the United States

| Basic Problem or Issue Faced by Society | Institutional Area | Institutional Organization or Large Organizations | Institutional Units or Small Groups |
|---|---|---|---|
| Reproduction of the species | Family | Planned Parenthood; an organization of marriage counselors | My family |
| Socialization and training of the young | Education | University of California; Boston school system | A kindergarten class; a graduate seminar |
| Explanation for the unknown or uncertain | Religion | Catholic Church; the Council of Churches | Wednesday evening Bible-study group; a Sunday-school group |
| Distribution of goods and services | Economics | General Motors; IBM | Assembly-line workers installing brakes on Fords; the steno pool handling the Phillips account |
| Development of rules and regulations to govern people's behavior | Government | United States Senate; state assembly | Township subcommittee on dogcatching; committees for the preservation of historic landmarks |
| Use of free time; time away from work | Leisure | Professional Golfers Association; NCAA; Sacramento Recreation Association | A professional basketball team; a bowling team |
| Care and treatment of the sick | Health care | Massachusetts General Hospital; Kaiser health plan | Physicians and nurses on a cardiac ward; technicians in a medical lab |
| Collection and organization of empirical knowledge | Science | National Science Foundation; Salk Institute | Astronomers in an observatory studying the universe; a group of biologists experimenting with virus strains |

or an idea. At first experimentally and spontaneously, then more consistently, and finally in a formalized fashion, norms and roles emerge defining behaviors and beliefs felt to be appropriate in that society. This could be called the institutional area of religion. In America the institution of religion is organized around norms favoring Christianity. Large organizations such as the Catholic, Presbyterian, and Baptist churches and small groups such as the Wednesday evening Bible-study groups or Sunday-school groups make up the *locus* in which individual interests and institutional norms intersect. In these groups, institutional norms are played out. This same pattern is repeated in other institutional areas, as can be seen in the table.

In some institutional areas—particularly the family—smaller groups are all-important and assume the major, almost exclusive, role. In other institutional areas, especially economic and governmental ones, larger groups play a major role. The important fact, however, is that we can combine group and institution to understand social organization.

As you look at Table 7.1, you can see that institutional areas are not necessarily separate and isolated from each other. In fact, there is a good deal of overlap between them. Socialization of the young takes place in the family, and in educational and religious institutions as well. Groups from many institutional areas (science, health care, religion, education, leisure, government) fill economic functions: Scientists, physicians, clergy, teachers, professional basketball players, and lawyers all earn a living. So these institutional areas are overlapping circles, each intersecting with others in sometimes complicated ways.

Because the institutional areas are interrelated, we should expect that change in one institutional area will be related to change in another. Child-labor laws, reform movements, and the organization of labor have led to shorter workdays and workweeks and therefore much more free time, which in turn has given us the institutional area of leisure. Increased employment of women (economic institutional area) has affected the family. The increase of women's rights is having an impact on the family. Economic changes can have consequences for higher education: In a tight job market college students might become more concerned about employment after school so they might move to more "job-related" majors and away from the liberal arts and social sciences. Politicians running for office must contend with the effect of economic factors on voting behavior. When the Soviet Union launched its first Sputnik in 1957, many people in the United States became concerned that we had "fallen behind." One response was a renewed emphasis on science in schools: change in one area (education) to bring change in another (science). So it is clear that although we have introduced the institutional areas as separate entities, they are closely related and intersect at many points. The social organization of society is best seen as a social fabric—a network of norms, roles, and values through which people interact individually and in groups, and this point is evident again as we examine social institutions.

Although the same institutional areas are found in many societies, they might not *look* the same in every society. The mountain-dwelling Arapesh of New Guinea and the campus-dwelling college students of America both operate within the institutional area of the family, for example, which involves elaborate procedures for courtship, marriage, and the raising of children; yet the specific practices of the two societies are vastly different. Not only will the content be different, but the same institutional area in different societies can, at least to an extent, focus on different cultural concerns and serve different functions.

Institutions change, as do all aspects of human behavior. Change in institutions probably occurs more slowly than does the change in behavior and values of smaller elements like groups and individuals. Institutional change also varies from one society to another. To make a comparison of extremes, modern industrialized countries, for the most part, change more rapidly than do primitive preliterate societies. Even within one society, one institutional area might change slowly, another rapidly. We will discuss this topic of social change in more detail in Chapter Twelve.

In summary, social organization can be examined at several levels. We can understand people's behavior by focusing on the dynamics and characteristics of group behavior and on the particular types of groups to which people belong. We can also understand behavior by focusing on the attitudes, values, norms, procedures, and symbols that are part of certain institutional areas. Finally, the analysis is improved further by combining concepts and observing that groups operate in the context of the norms, roles, and values of an institutional area. To get a better idea of how the concept of institution is applied to specific areas of human behavior, the remainder of this chapter and the next three chapters will examine several institutional areas in more detail.

# FAMILY

Sociologists have always looked with great interest on the institution of the family, mainly because we tend to see the family as the most basic of all institutions, and perhaps also because family practices vary tremendously from one culture to another. These variations never cease to fascinate, and ethnocentrism plays a part in this fascination. Many people are convinced that their own way is the right and best way, and they are amazed to find other people behaving differently. This ethnocentrism is unfortunate. Perhaps these issues can be placed in better perspective if we examine cross-culturally some similarities and differences in family structure.

## Structural Variabilities: How Do Family Structures Differ?

Anthropologists, who study primitive societies all over the world, find that the structure of the family is almost infinitely variable. Imagine almost anything, and a society somewhere probably practices it (of course, the people there would be aghast if they knew the strange things *we* were doing).

It usually begins with some form of mating ritual: A couple can go on a first date; or a woman can inherit a husband; or a man can buy a spouse or two, or fall in love, or capture a mate. On the east coast of Greenland, the man goes to the woman's tent, grabs her by the hair, and drags her off to his tent.

In some cultures, parents are responsible for arranging marriages for their children. Chinese imperial law from around 1700 stated that a family elder who failed to find a husband for any young woman belonging to his household was condemning her to an unfulfilled life and was liable to receive publicly 80 blows with the bamboo.[2] In India, if a young woman's parents had been unable to find a husband for her as late as three years after she had reached puberty, then she could arrange a marriage for herself, but in so doing she would bring great disgrace on her family. Normally the marriage partner was chosen by the parents together. In China, however, the final decision on a prospective husband was typically made by the young woman's father; the mother, on the other hand, would select her son's future wife. Correct order had to be observed in marrying off children: Eldest children must be found mates first. This led to problems if the eldest child was unattractive or if younger children were especially attractive and precocious. This in turn forced desperate parents to trickery, switching one daughter for another at the last minute. Recall from the Bible that Jacob thought he had married Rachel but found out the next morning that the bride with whom he had spent the night was her elder sister Leah.

Beyond these prescribed, ritualistic mate-selection practices, there is also "love" as the basis for choosing a partner. This works in various ways: **Exchange theory** proposes the idea of a marketplace in mate selection. Mate seekers want to maximize their chances for a happy marriage and look for a good deal, a bargain if possible, in which their own assets and liabilities are compared with those of the potential partner. They continue the relationship on the assumption that they will get more out of it than it will cost. Anyone different from and especially below their own level of exchange is likely to be out of luck. **Complementary needs theory** suggests that people select partners who make up for, balance, or supply characteristics that they themselves don't have. Thus the talker looks for a listener, the eater for a cook, or the Mercedes-Benz fancier for a millionaire.

**Role and value theories** of mate selection suggest that people who share common values and common definitions about roles are more likely to select each other. If a woman is a big-city, sports-minded Episcopalian and a man is a small-town, sedentary Baptist (different values), or if she wants to stay home and cook and he pushes her to study medicine (differing role expectations), they will probably look elsewhere. **Process theory** suggests that mate selection involves a number of social and psychological processes. The field of potential mates is progressively narrowed down by physical attraction, religious and racial differences, role and value similarities, psychological influences, and so on. Bernard Murstein, an exchange theorist, believes that most of us settle for a partner rather than choose one; only people with many assets and few liabilities actually choose each other—the rest of us just settle.[3] How depressing . . .

At any rate, whomever you date, buy, inherit, are forced to accept, capture, or settle for is related, in part, to whether your society practices **endogamy** (marriage within a certain group) or **exogamy** (marriage outside that certain group). In the United States, as in most societies, both are practiced. We forbid marriage within the immediate family, and it is therefore necessary to marry an outsider: exogamy. At the same time interracial marriages are often discouraged, and some religious groups encourage their members to marry people of the same faith: endogamy. The number of spouses you can have at one time varies depending on the culture in which you live. **Monogamy** means having one spouse at a time. It is estimated that most people in the world practice monogamy. The United States is technically monogamous, but with our high rates of divorce and family instability, it might be more accurate to say that we practice a type of serial or musical-chairs monogamy.

**Polygamy** is marriage to more than one spouse and includes three types: polygyny, polyandry, and group marriage. **Polygyny** refers to the practice of one man having several wives at a time. G. P. Murdock studied 250 societies and found that polygyny was the preferred form of marriage in 77 percent of them, although, as mentioned above, most people in the world practice monogamy.[4] Even in polygynous societies, most people practice monogamy. Although both sexes favor polygyny in such societies, the wealthy are more able to afford it and are more likely to practice it. **Polyandry** refers to the very rare practice of one woman having several husbands and, when it occurs, is usually found in poor societies with a shortage of women. Occasionally such societies kill female babies to keep the population down. When the woman becomes pregnant, deciding which husband is the father is a problem. This is usually solved by some sort of ceremony, possibly by shooting arrows at a distant target and designating the owner of the closest arrow to be the father. **Group marriage** refers to the marriage of two or more men to two or more women at the same time. The successful, long-term practice of group marriage is extremely rare and, when it does occur, is usually limited to closed groups or communes.

Family practices vary in many other ways in societies across the world. Where do newlyweds live after marriage? They may be expected to live either with the husband's family or with the wife's family, or to alternate between the husband's and wife's families. In some societies (such as the United States) the couple is expected to live apart from both families. How is ancestry traced? In some societies family lineage is traced through males, in others through females, and in still others through both sides equally. As to who has the formal power and legal authority in the family and who makes the decisions, in **patriarchal societies** it is the male and in **matriarchal societies** the female. It is interesting to note that many of the various practices we have mentioned arise at least partially in reaction to the incest taboo. Almost all societies forbid or at least limit sexual contact between family members with restrictions that are called **incest taboos.** The particular lines of the taboo vary from one society to another. To help regulate the situation, various rules—residence, descent, and so on—were developed. It should be clear by now that almost any imaginable practice exists in some society somewhere. And all of it seems to work.

A number of other terms have emerged that help describe aspects of family structure. The family one is born into is called the **family of orientation.** The family of which one is a parent is called the **family of procreation.** The **nuclear family** includes the married couple and their children. The **extended family** includes more than two generations in close association or under the same roof, and the **compound family** includes multiple spouses—several wives or husbands at the same time.

In summary, we could describe the peculiar American family as a nuclear, monogamous, patriarchal, exogamous structure practicing marriage by romance.

## ARRANGED MARRIAGE IN NEW YORK

Throughout New York, in communities that still practice arranged marriages—Sikhs, Muslims, Hindus, and Hasidic Jews—young people say they feel good enough about the ancient practice to continue the tradition, albeit in a more "American" way.

In India, families routinely do matchmaking, but many young Sikhs have recently emigrated to the U.S. alone and are living in small isolated enclaves. Some have been relying on the Internet to find partners. Sites such as SuitableMatch.com and INDOlink.com run "matrimonial" ads for the entire Indian community. The ads are usually placed by parents. Inesha, a 15-year-old American-born girl, goes to a Queens school with few other Sikhs. She hopes to fall in love and choose her own mate, although one her parents would accept—perhaps a highly educated Sikh from the Punjab region. "If I can't find someone by the time I'm 22 or 23, I will go to my parents for help," she asserts.

In traditional Islamic communities, at about age 18, couples begin going on "Islamically acceptable" chaperoned dates, followed by a short engagement period of maybe three months before they are wed. "Allah knows what's in our hearts, so there's no need for a long engagement if you are with who you are meant to be with."

Nearly 3000 Patels, members of a prominent Hindu clan, gather from all of the U.S. in Atlanta for what has been called a "meat market." Seven hundred register as "single." The Patels meet for three days of socials, panels, and vegetarian-friendly meals in a high-speed attempt at finding new family members. Three hundred marriages per year are generated by this event and by follow-up mailings.

Professor S. N. Sridhar sees a new marriage model among Hindus: the child-initiated, parent-arranged marriage. "It was after my wife and I decided to get married that our parents ran background checks on the families, and then planned and hosted the wedding. It's a common modern Indian compromise."

"The Internet is bringing evil into the house," proclaimed a Hasidic father. Lubavitch is the only Hasidic sect that embraces the Net. Its late leader declared that all technology should be used to spread Hasidism among Jews. Traditionally, in Crown Heights, professional matchmakers organize meticulous index files containing photos, educational backgrounds, family information, and medical histories of marriage-aged prospects. Parents set up in-house, supervised dates. If all goes well, a pair ventures out on their own to a public place like Central Park. Couples date on average two weeks to three months before an engagement is announced. In other Hasidic sects, couples meet only once before they marry.

*—From "Sikh and Ye Shall Find,"*
Village Voice, *December 15, 1998, by Rebecca Segall and Lauren Reynolds.*

### Functional Uniformities: How Are Families Alike?

The institution of the family exists in all societies in some form. Functional analysis explains it this way: A family must serve a purpose, fill basic needs, and perform essential tasks. Those who have studied family institutions have discovered a number of such functions, some obvious and important and some less obvious. An obvious function of the family is to provide for continuation of the species. Societies, to exist, need people; the institution of the family functions to control reproduction. A second function of the family is to control sexual expression. The family institution attempts to deal with powerful sexual needs and desires by defining the who, when, where, why, and how of sexual activity.

A third function of the family is to care for and socialize children. The baby cannot survive without the care that the family provides. Socialization of the young by the family transmits the culture and prepares the child for participation in the adult world. Much of the child's later life is patterned on family models. It is possible that personal pathology and individual unhappiness will follow from inadequate family models or from a bad family atmosphere. The child might frequently be the defenseless target of the parents' own frustrations, anxieties, neuroses, and maladjustments.

The first research question at the beginning of this chapter was essentially a socialization question: Do children repeat their parents' marital experiences or do they take off in new directions? Probably some of both, but a recent paper discovered some striking similarities between mothers' and daughters' behaviors. The author used data drawn from interviews conducted with mothers and children in the Detroit area between 1962 and 1985. The researcher found that daughters of mothers who married young and who were pregnant at marriage entered into cohabitational and marital unions sooner and at a higher rate than other children. Daughters were also quite likely to repeat their mother's pattern regarding the aftermath of divorce. Daughters were far more likely to remarry if their mothers did than if their mothers did not. Even parents' marital experiences become part of the socialization process and are clearly important for their children.[5]

The family can also provide close affectional and emotional ties for the individual. In some societies, many groups and associations can perform this function, but in mass, industrialized secondary societies such as the United States, the family might be one of the few remaining institutions in which primary relationships are possible. The family also functions to provide placement or status ascription. The family we are born into provides us with a *place* in society. If a person is a Protestant, middle-class, midwestern American Democrat, it is more than likely to be because he or she inherited certain statuses, such as religion, social class, region, nationality, and political preference, from his or her family of orientation.

These are some functions of the family, and there are probably others. These functions are not necessarily completely separate from each other; for example, reproduction of the species and control of sexual expression seem to be related. And further, other institutions can serve some of the same functions the family institution serves. The institution of education certainly deals with socialization of the young and transmission of culture, as does the institution of religion.

## THE FAMILY: CONTRAST AND CHANGE

We have made the point that even though the family exists everywhere, there are many variations in its form. Also, the family continually changes, in some societies more rapidly than in others, and a variety of forces produce these changes. A brief look at another family system will help illustrate this point.

### The Family in China

Ross Eshleman has described several stages of change in the Chinese family.[6] China is a country with a population four or five times larger than the U.S. population and with a family system far different from ours. Until 1900 or so, the most prevalent type of family in China was the traditional family. The clan—the central characteristic of the traditional family—was made up of all people with a common surname who descended from a common ancestor. A clan could be as large as several thousand people and included gentry (intellectuals, landowners, government officials) and peasants who cultivated the land. The clan performed major functions such as lending money to members, establishing schools, settling disputes, collecting taxes, and maintaining ancestral burial grounds. Ancestors were worshipped (given offerings of goods and food) even after death.

Males were dominant in the Chinese family. A wife had to obey three people: her father before marriage, her husband after marriage, and her son after her husband died. Female infanticide was common, especially among the poor. Marriages were arranged through matchmakers; many women met their husbands for the first time on their wedding day. Women's duties were to have male children and aid in the work. Men could have additional wives, but women could have only one husband and were not allowed to remarry if widowed. Binding of women's feet was practiced; this reduced the foot in length by about three inches and resulted in permanent crippling. Footbinding was practiced more among the upper class than among the peasants because the latter needed freedom of movement to work in the fields. Suicide was seen as a socially acceptable way for women to deal with their problems.

Change began to modify the traditional Chinese family in the 1900s. These changes were influenced by events occurring in the rest of the world, especially in Russia. Karl Marx's collaborator Friedrich Engels felt that in a capitalist society the family becomes the chief source of female oppression. Private property leads to subjugation and slavery of women and elevation of the status of men. One of the consequences of the revolution in Russia was the passage of new marriage and divorce laws, the granting of abortion on demand, and the granting of legal equality to women.

China did not follow the Russian pattern precisely but did work to reform what were seen as the evils of the traditional family. A Communist Party conference in 1922 called for equal voting rights for women and an end to the traditional maltreatment of women. The

Marriage Law of 1950 represented a major attempt to restructure the traditional family. This law enacted a new democratic marriage system based on free choice of partners, monogamy, and equal rights for both sexes. It prohibited bigamy, child betrothal, and infanticide. It gave husband and wife equal status in the home, granted divorce if both wanted it, and allowed the divorced woman to keep property that belonged to her before her marriage.

So, what emerges as the Chinese family of today? As we might expect, characteristics of both the traditional family and the family changed by law blend together. Rural and urban differences are great; thus, the rural families keep more of the traditional practices than do the urban ones. Generally, although women have made progress toward equality, discrimination still exists in access to jobs and adequate pay. Most of the important jobs are held by men; and sons are valued over daughters. Parental arrangement of marriages still occurs, premarital sex and illegitimate births are infrequent, and divorce is not widespread. The strict family planning policy limiting families to one child (enacted in 1979) has produced strains. The policy is at odds with the traditional wish for sons and for large families, and has led to enforcement problems; those living in the countryside and the wealthy are avoiding punishment for violating the law, whereas city dwellers and the less wealthy are not. Several observers say the policy has also led to a generation of overindulged, spoiled children, called "little emperors." (See readings 7.2 and 11.2 for more on this subject.)

In conclusion, changes in the political system in China produced pressures for change in other institutions. Communist ideology and the economic goals of the state pushed for an altered family in which women were more equal and able to participate in production, whereas family or clan-held property was confiscated and clans became communes. The contemporary Chinese family is consequently vastly different from the traditional one, but it has many remnants of the traditional family and is described as strong and stable.

### The Family in America

If we look at the American family over the last 100 years or so we can see a number of changes. The family is smaller than it used to be; there are fewer children and fewer adults. Birth rates have been dropping in the United States for a long period of time, and therefore the 1970s had the lowest rates on record. The large-family model of the 19th century has given way to the smaller family of two or fewer children. Women's attitudes concerning the number of children they want have changed over the past 50 years. Larger families—two to four or more children—used to be favored, but now smaller families—one or two children—are the norm. The American family is also small in that it is a *nuclear* rather than an *extended* family. The extended family of a century ago had aunts, uncles, parents, grandparents, and children all living under the same roof. Compare that with the nuclear family of today and think of the differences in roles, division of labor, diffusion of responsibilities, child care, expression of affection and emotion, and individual freedom. The shift from the extended to the nuclear family brought tremendous changes in the roles of individual family members. The family of today is segregated: Each segment of that large extended family of 100 years ago now lives by itself as a separate family unit.

Although it is dangerous to generalize, today's family is probably more egalitarian and less patriarchal than it used to be. People are more likely to share authority and decision making. Women are more likely to be working outside the home—perhaps for economic reasons or perhaps because traditional values are changing. The increasingly real possibility of a meaningful career has given women alternatives to early marriage or even to marriage at all. Both males and females are increasingly choosing not to marry or to marry late, as we shall see later in this chapter. Men's roles are changing as well—some choose alternatives to customary patterns to become househusbands, and some companies provide paternity leave. Change is a struggle, however, as both men and women have trouble breaking loose from traditional roles and expectations. Many women who work find they must deal with two jobs: working outside the home for pay and then coming home to a "second shift"—cleaning, cooking, and caring for children. Many husbands resist helping out with these activities, which they view as "women's work," and this puts added strain on marriages.

The American family has moved from farm to city. Last century's farm family now lives in or near an urban area. And the family is much more geographically mobile than it used to be. Currently, 16 percent of Americans move from one house to another in a year, and 2 percent move from one state to another.

The family used to be the all-powerful institution, but gradually other institutions—religious, educational, governmental—have become involved in some of the functions that the family formerly handled exclusively. So far this has resulted not in the disappearance of the family, but in increased emphasis on and jealous guarding of the remaining functions that the family fills: "Sex education in the schools?? Certainly not! Keep it in the home where it belongs!" As mentioned before, perhaps one of the major functions of the family in mass society is to provide for the affectional and emotional needs of people. The family is nearly the last refuge of the true primary type of relationship, and if it is accurate to call this a basic social need, then this function of the family becomes exceedingly important as other functions decrease in importance.

The second research question at the beginning of this chapter concerned the relationship between children and marital stability. It is often suggested that the presence of children is one factor that might hold marriages together. Is this a "commonsense" suggestion, and is it really true? A recent study examined this question by looking at interviews with about 5,000 couples who were first married between 1968 and 1985. The researchers found some interesting patterns. The first birth has a major effect in *increasing* the stability of a marriage. (Exception: A child born before the parents were married decreases marital stability.) The presence of additional younger children (less than six years old) also seems conducive to helping a marriage stay together. However, the presence of children between 6 and 12 years of age appears to be a neutral factor—they neither increase nor decrease the chances that their parents' marriage will break up. Finally, older children (teenagers) are a problem for marital stability—their presence significantly *increases* the chances that parents will dissolve their marriage! The researchers suggest that the presence of young children may hold a marriage together because couples who plan to stay together are more likely to plan having children; their first child is a "signal" of their stability. Instability in families with teenagers perhaps reflects marriages in which the parents were unhappy for some time but waited to divorce until the children were older, or perhaps it reflects the "outside-world" problems that teenagers have and the resulting strains these problems place on their parents' marriages. Does having children increase marital stability? As the authors of the study say, "clearly, yes and no."[7]

Several interesting and important changes are taking place in the American family. Census bureau documents show some sharp contrasts between 1970 and 1998.[8] Today's families and households are smaller. There were 3.6 people per family in 1970, 3.2 in 1998. Households went from 3.1 people to 2.6. Twenty-one percent of all households had five or more persons in 1970 but only 10 percent were that large in 1998. Geographically, people are moving from the Northeast and Midwest to the South and West, and they have become more urbanized—households in metropolitan areas increased from 69 percent in 1970 to 80 percent in 1996. Several of the most important changes are highlighted in the following sections.

**Later Marriage**   The median age at first marriage (see Figure 7.1), which didn't change much between 1950 and 1970, climbed steadily between 1970 and the present. In 1997, the median age of first marriage for men was 26.8 years, higher than any previous figure going back 100 years. For women it was 25, also significantly higher than any previous figure.

**Staying Single and Living Alone**   One of the most remarkable changes in the family during the last 25 years is the shift from the traditional married couple with children to the number of people living alone. These changes are shown in Figure 7.2; 40 percent of all households were married couples with children in the early 1970s, but by 1998 this had dropped to 26 percent. Households with people living alone increased from 17 to 26 percent during the same period. Some of this is a consequence of later marriage. The proportion of adults staying single for a longer period of their life even if they ultimately marry is increasing. In 1998, 22 percent of the women and 29 percent of the men aged 30 to 34 had never married. Twenty-six million people older than 18 lived alone in 1998 (compared with 11 million in 1970). Most of the growth in this category came from divorced or never-married people, and the elderly.

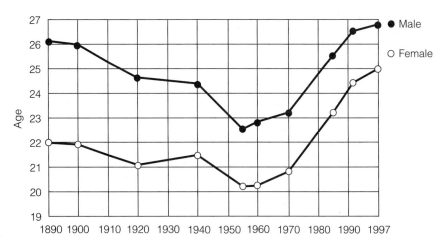

**FIGURE 7.1**   *Median age at first marriage.*

United States Bureau of the Census, "Marital Status and Living Arrangements: March 1997," *Current Population Reports,* Series P-20, No. 506.

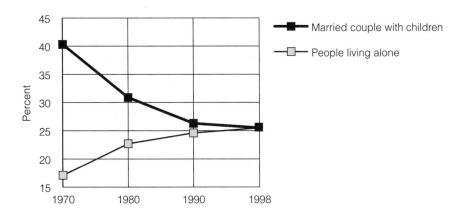

**FIGURE 7.2**   *Household composition in the United States.*

United States Bureau of the Census, *Statistical Abstracts of the United States, 1999* (119th ed.), Washington, D.C.

Thirty-one percent of people older than 65 (41 percent of women and 17 percent of men) lived alone.

**Other Living Arrangements**   Twenty-nine percent of all children younger than 18 lived with only one parent in 1998. Four million grandchildren younger than 18 lived in the home of their grandparents, and many (36 percent) had no parent in the home. The number of households occupied by two unrelated adults of the opposite sex more than doubled between 1980

(1.6 million) and 1998 (4.2 million). The number of unmarried-couple households is relatively small when compared with the number of married couples, but the proportions are changing.

**Divorce**     Marriage and divorce rates in the United States are shown in Figure 7.3. Marriage rates have fluctuated over the years with a downward trend since 1980. Divorce rates went up sharply, starting in the middle 1960s with a high in 1980 and have declined slightly since then. One of the many consequences of this higher divorce rate is that in 1998, 71 percent of children younger than 18 lived with two parents, compared with 85 percent in 1970. The following list gives some other facts about marriage and divorce in the 1990s.

**Census Bureau Facts About Marriage and Divorce in the 1990s**

- Women in their thirties and early forties have the highest divorce rate.

- First marriages that end in divorce typically last about six years.

- Women who divorce typically wait two and a half years before remarrying.

- Three out of ten divorced women who remarry divorce again.

- Four out of ten first marriages end in divorce.

- The proportion of women who never marry went from 5 to 10 percent.

- Only 75 percent of black women will ever marry compared with 91 percent of white women.

The American family of today, then, is many things. The "traditional family," in which the husband works and the wife stays home with the children, represents one quarter to one third of all families. Young people seem to be postponing marriage, perhaps because of unstable economic conditions, perhaps because a new variety in living arrangements enables people to enjoy primary group benefits without the necessity of a traditional marriage. People who do marry do so later. Some population analysts take this as a hopeful sign. Maybe Americans are being more careful about selecting mates. Taking more time and marrying at a later age may lead to more stable marriages in the long run, and perhaps today's high divorce rates will begin to decline. At any rate, if the marriage is not happy, it is likely to be dissolved quickly. There are now more single-parent families because of both divorce and single people being allowed to adopt children. The nuclear family (two parents plus children) is a major part of American family life, but it continues to shrink in size as the birth rate drops. The retired couple in their sixties and seventies represents another facet of the American family, one whose numbers are growing substantially as our population grows older and as the birth rate decreases.

Perhaps, then, we see two contrasting threads in the American family. One is an experimental, postponing-or-never-marrying, alternative-lifestyle thread; the other is the traditional nuclear family, progressing from young to middle-aged parents with their children to elderly retired couples. Time has

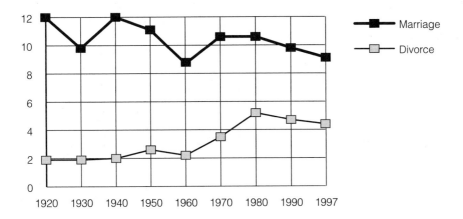

**FIGURE 7.3**    *Marriage and divorce rates (per 1,000 people) in the United States.*
United States Bureau of the Census, *Statistical Abstracts of the United States, 1999* (119th ed.), Washington, D.C.

changed the nuclear family: It is a smaller unit with more equal relationships between partners. Although the two threads contrast, they also intersect; many people find themselves first being experimental and later following a more traditional path as their own needs and definitions undergo change.

## FAMILY PROBLEMS: THE CONFLICT APPROACH

Several times in this book we have examined the contrasting viewpoints provided by functional theorists and conflict theorists. When the family is studied, one generally starts (as we did earlier in this chapter) with the functions that the family provides. The family in all societies does have numerous and important functions, but the conflict-theory approach deserves examination as well. Just as the family fulfills certain functions, it provides a setting for conflict. As we noted earlier, Engels felt that the capitalist family was the chief source of oppression of females and that the "battle between the sexes" is probably waged more often in the family arena than anywhere else. Conflict is a sign that all is not well, and in this section we will examine the evidence of problems and instability in the American family.

Divorce is a good place to start in looking at family problems. There were 1.16 million divorces in 1997. The divorce rate climbed sharply from the middle 1960s to 1980 and is at about that same level currently. Divorce is rising in many countries, but the United States continues to have one of the highest rates in the world. Why? Among the many explanations offered, the following seem most convincing: Divorce is becoming more and more accepted. It is not the stigma it once was, and people are much more inclined to go through it today. Laws have changed; many states have enacted no-fault divorce laws, making divorces much easier to obtain. Another explanation has to do with the emphasis on individuality and happiness. Parties to a marriage that isn't working are less likely to want to stick with it today than they used to be. The view is that if you're not happy, then get out and try again, or try something else. Both parties now know that they have a good chance to make meaningful lives for themselves in some way other than by staying in a bad marriage. And with fewer children, perhaps fewer marriages are kept together "for the sake of the children." Unmarried people living together also separate. Courts have been faced with difficult cases that were essentially requests to have the legal benefits of divorce, such as community property and alimony payments, extended to people who had never married.

A recent study of husbands and wives attempted to find out what factors in a marriage were most likely to cause either party to think about divorce. Some interesting relationships were discovered. The wife's thoughts of divorce were *increased* if she had work experience, if she had a youngest child between the ages of 6 and 11, and if she felt that housework should be divided equally. Her thoughts of divorce were *decreased* according to her age at marriage (the older she was when she got married, the less likely she was to think of divorce), the length of the marriage, and her husband's increasing contribution to the housework. The husband's thoughts of divorce were *increased* if his wife worked outside the home, if there were religious differences, and if his wife believed that housework should be divided equally. His thoughts of divorce were *decreased* by the presence of a child under six, by the length of the marriage, and by being older than his wife. These findings are explained in part by the fact that women's economic roles are changing faster than are men's. This has increased her freedom and independence and has modified some of her expectations of marriage. Some husbands are having difficulty adjusting to these changes.[9]

An even more obvious example of conflict within the family is family violence. Nearly one fifth of all murders in the United States involve a person killing a member of his or her own family. Spousal abuse and child abuse are much more widespread than was earlier thought, and might be the most underreported of all crimes. Research by Murray Straus found that in 16 percent of the families studied, an act of violence was committed by one partner against the other within the past year.[10] If this is projected over the life of the marriage, violent acts occurred in 28 percent of the marriages. The researchers believe that this estimate is low and that the actual rate of family violence is closer to 50 to 60 percent of all couples. Generalizing from this study, it is estimated that 1.8 million wives are beaten by their husbands every year, and surprisingly, more than 2 million husbands are beaten by their wives.

Child abuse, both physical and sexual, has become an increasingly visible issue. The study by Straus and his coworkers, cited in the previous paragraph, found violence toward children to be widespread in American society. Based on their interviews with 2,100 families, the researchers found that within the past year more than half of the parents had slapped or spanked their children, 13 percent had hit their children with an object, and 3 percent had kicked, bit, or punched their children. The researchers estimated that some 46,000 children in the United States had a knife or gun

# MURRAY A. STRAUS (1926–  )

Murray Straus earned his Ph.D. in Sociology from the University of Wisconsin. He focused on the sociology of the family, and much of his research has dealt with the topic of violence within the family. In books such as *Beating the Devil Out of Them: Corporal Punishment in American Families* (1994) and *Intimate Violence: The Definitive Study of the Causes and Consequences of Abuse in the American Family* (1988), Straus has attempted to discover the extensiveness and impact of violence on the family. He has also placed emphasis on the controversial aspect of female spouses abusing their husbands. In *Behind Closed Doors: Violence in the American Family* (1980) Straus states, "The number of wives who throw things at their husbands is almost twice as large as the number of husbands who throw things at their wives. The rate of kicking and hitting with an object is also higher for wives than for husbands." Throughout his research Straus has attempted to find the personal and societal factors that make violence possible, and the policy solutions that should be implemented to eradicate it. Murray Straus is a professor of sociology and the codirector of the Family Research Laboratory at the University of New Hampshire.

used against them during the last year! They also found that mothers are more severe, more abusive, and more often violent with their children than are fathers. Recent headlines have reported sexual abuse of children in families, in foster homes, and in day-care centers. Children have been required to testify in court against adults, sometimes against their own parents, which has led to current controversy on two fronts: On the one hand, the child experiences the double trauma of being a victim and of then having to testify about it; on the other hand, questions have arisen about the accuracy of some of the children's testimonies and about the competence of children in general as witnesses.

Other conditions of concern in American society have roots that can be traced to the family. Seventeen percent of the arrests for violent crime and 33 percent of the arrests for property crime in 1998 were of people under the age of 18. Suicide rates among teenagers are increasing. Drug abuse and alcohol abuse by both young people and adults are serious problems. Illegitimate births increased by 50 percent during the 1970s, and by 1997, approximately 32 percent of American babies were born to unmarried mothers.

Explanations for at least some of what is happening to the family are not difficult to find. Society changes, and its institutions change as well. Fewer bonds tie family members together today, and the consequence is seen in divorce and in young people leaving home earlier. Both parents are more likely to be working, and so children might be supervised less closely than they used to be. Also, outside agencies such as day-care centers play an increasingly important role in "family" life. Less desire for having children, along with increased freedom and independence for women, has led to

The Soviet newspaper *Moskovskaya Pravda* reported in 1986 that only 37 percent of the marriages in the Soviet Union survive three years or more. Why? Drunkenness used to be the primary reason for divorce, but that seems to have changed. Housing facilities are so scarce that only one third of young married couples have an apartment of their own. Most have to live with parents or relatives and consequently have little privacy. Vacations together are difficult to arrange because of conflicting work schedules. Also, divorces in childless marriages are easy to obtain. How to slow down the divorce rate? Some say tighten divorce laws, whereas others suggests that solving the housing shortage is the answer.

increasingly diverse arrangements for satisfying affectional and emotional needs. We will likely see more changes and redefinitions in the family as other aspects of society continue to change.

## THE FAMILY OF THE FUTURE

The institution of the family, which changed slowly for 200 years, now appears to be changing rapidly. Some guesses about the future given current conditions look like this: Some experts predict a continuing decline in the number of traditional nuclear families consisting of two parents and one or more children. Perhaps single-parent families will continue to increase because later marriage and never-marry choices will continue to increase. Families of the future are likely to have both partners working, meaning that if they have children, child care outside the home will become a greater issue. As somewhat of a countertrend, the number of parents doing homeschooling—taking their children out of schools and educating them at home—is increasing. With the onset of the computer age, working out of the home will become an option for many. Perhaps the increased focus on child abuse and spousal abuse will help us take steps toward becoming a less violent society. Scientists have accomplished human embryo transplants and test-tube babies, and they have cloned animals. It is daunting to try to predict the future consequences for the American family of what seems to be an accelerating rate of change and innovation.

## THEORY AND RESEARCH: A REVIEW

The study of social institutions has traditionally started from the viewpoint of *functional analysis*. If you look back at the discussion of social institutions at the beginning of this chapter, you will find institutions to be an "organized system of social relationships" that "emerges to deal with certain basic problems." Societies need to have specific basic functions performed. An institutional structure arises around attempts to supply these functions.

Change is slow and the feeling of an organized and integrated system develops. Social institutions—family, organized religion, education, economics, government, and so on—show functional analysis at its clearest. This chapter focuses on structures developed and functions provided by the institution of the family.

Recently in the study of social institutions, *conflict theory* has become more prominent. In part, this is a reflection of obvious conflict—high divorce rates, crime, and violence between family members—existing in or emerging from the family. The emphasis on conflict theory is also a reflection of the increasingly popular view that scarcity, competition, exploitation, and conflict exist in all social situations. Notice the various aspects of the conflict perspective in dating, the Chinese family, and family problems discussed earlier in the chapter. The family is perhaps best seen as a small group held together by affectional ties and pulled apart by the inevitable conflicts that exist in any group of people. *Symbolic interaction* would focus on the patterns of interaction among members—how people interpret, define, and react to the behavior (symbols, gestures) of those around them. Proponents would be interested in such questions as these: How is discipline handled? How do children manipulate parents and vice versa? Who exercises power and in what ways? What effect does the placement of the child in the family (firstborn, middle child, and so on) have, and why does this happen? How do aging, changing roles, social class, and other life situations affect patterns of interaction?

A number of ethnographies are referred to in this chapter in discussing family practices in other societies; most of these have been done by anthropologists, and some involve *participant observation*. The census data describing the family in the United States today were obtained through *questionnaire* or *interview surveys*. The study examining the relationship between presence of children and marital stability, and the study of children repeating their parents' marital experiences, were *reactive surveys*, as was Straus's study of family violence. One of the major concerns in the family-violence study was the *sample*. The sociologists selected a national probability sample of 2,100 families so that they could generalize their findings to families throughout the United States. As always happens, there were several drawbacks or problems. The researchers chose to look at intact families, so single-parent families are not represented in the sample. Parent-child relations with children younger than three years of age were not included, even though very young children present an especially stressful situation. And the researchers, even after as many as four visits to the family's house and letters and financial incentives, were able to reach just 65 percent of the sample. Were the families they couldn't contact different from most others in patterns of violence? The researchers hope not, but it's impossible to know.

## SUMMARY

The major topics of the previous chapters in this section on social organization were *groups, categories,* and *social differentiation.* These concepts permit the study of several different types of collectivities of people. In this chapter we moved to another, somewhat more abstract level and focused on social institutions. Sociologists define an institution as a system of social relationships—an organized way of doing things that meets certain basic needs of society. All societies have basic needs, such as reproduction of the species and socialization of the young, for example. Consequently, many of the same institutions—family, religion, education, government, economics—appear in all societies, although the content of a given institution might vary from one society to another. As societies develop and change, other institutions, such as science and leisure, may emerge.

Sociological research and study are more apt to be concerned with specific institutions than with the concept itself. The major part of this chapter has presented an analysis of the institution of the family. We examined cross-cultural variations in courtship and the family to make the following point: Although the institution of the family exists everywhere, it is not the *same* everywhere. Constants or uniformities were discussed, and the functions that the family seems to fulfill were outlined. We took a brief look at the family in China to illustrate not only the contrast between that family system and ours, but also the idea that *change* is a constant in any institution. Finally, we examined the family in America from several perspectives: current trends and changes, problems, and predictions about the family of the future.

A major point of this chapter is that although a social institution (in this case, the family) exists in all societies, it varies tremendously from one society to the next. The group of three readings that follows describes some unique characteristics of the contemporary family in Japan, in China, and in the inner-city black community in the United States.

---

## TERMS FOR STUDY

| | |
|---|---|
| complementary needs theory (235) | matriarchal society (236) |
| compound family (236) | monogamy (235) |
| endogamy (235) | nuclear family (236) |
| exchange theory (235) | patriarchal society (236) |
| exogamy (235) | polyandry (236) |
| extended family (236) | polygamy (236) |
| family of orientation (236) | polygyny (236) |
| family of procreation (236) | process theory (235) |
| group marriage (236) | role and value theories (235) |
| incest taboo (236) | social institutions (231) |
| institutionalization (231) | |

For a discussion of Research Question 1, see page 238.
For a discussion of Research Question 2, see page 241.

⚓ **INFOTRAC COLLEGE EDITION**
**Search Word Summary**

To learn more about the topics from this chapter, you can use the following words to conduct an electronic search on InfoTrac College Edition, an online library of journals. Here you will find a multitude of articles from various sources and perspectives:

**www.infotrac-college.com/wadsworth/access.html**

| | |
|---|---|
| Sikhs | divorce |
| Friedrich Engels | child abuse |
| Marriage Law | single parents |
| sex education | unemployment |

**Reading 7.1**

# THEY GET BY WITH A LOT OF HELP FROM THEIR *KYOIKU MAMAS*

*Carol Simons*

*The institution of the family fulfills many functions, but perhaps its major role is to care for and socialize children. Socialization of the young by the family transmits the culture and prepares the child for future participation in the adult world. Much of the child's later life is patterned on family models. This reading and the two following examine particular aspects of three contrasting family systems—the family in Japan, in China, and in inner-city America.*

*Carol Simons is a former associate editor for* Smithsonian *and lives in Tokyo. This reading is an excerpt from an article that appeared in* The Smithsonian.

Two-year-old Hiromasa Itoh doesn't know it yet, but he's preparing for one of the most important milestones of his life, the examination for entry into first grade. Already he has learned to march correctly around the classroom in time with the piano and follow the green tape stuck to the floor—ignoring the red, blue and yellow tapes that lead in different directions. With the other 14 children in his class at a central Tokyo nursery school, he obeys the "cleaning-up music" and sings the

From *Smithsonian*, March 1987. Reprinted with permission.

good-bye song. His mother, observing through a one-way glass window, says that it's all in preparation for an entrance examination in two or three years, when Hiromasa will try for admission to one of Tokyo's prestigious private schools.

Forty-five minutes south of the capital city by train, in the small suburb of Myorenji, near Yokohama, 13-year-old Naoko Masuo returns from school, slips quietly into her family's two-story house and settles into her homework. She is wearing a plaid skirt and blue blazer, the uniform of the Sho-ei Girls School, where she is a seventh-grader. "I made it," her smile seems to say. For three years, when she was in fourth through sixth grades in public school, Naoko's schedule was high-pressure: she would rush home from school, study for a short time and then leave again to attend *juku,* or cram school, three hours a day three times a week. Her goal was to enter a good private school, and the exam would be tough.

Her brother, Toshihiro, passed a similar exam with flying colors several years ago and entered one of the elite national schools in Tokyo. The summer before the exam, he went to *juku* eight hours a day. Now, as a high school graduate, he is attending prep school—preparing for university entrance exams that he will take in March.

Little Hiromasa, Naoko, and Toshihiro are all on the Japanese road to success. And alongside them, in what must surely be one of the world's greatest traffic jams, are thousands of the nation's children, each one trying to pass exams, enter good schools and attain the good jobs that mark the end of a race well run.

But such children are by no means running as independents. They are guided and coached, trained and fed every step of the way by their mothers, who have had sharp eyes on the finish line right from the start.

No one doubts that behind every high-scoring Japanese student—and they are among the highest scoring in the world—there stands a mother, supportive, aggressive and completely involved in her child's education. She studies, she packs lunches, she waits for hours in lines to register her child for exams and waits again in the hallways for hours while he takes them. She denies herself TV so her child can study in quiet and she stirs noodles at 11 for the scholar's snack. She shuttles youngsters from exercise class to rhythm class to calligraphy and piano, to swimming and martial arts. She helps every day with homework, hires tutors and works part-time to pay for *juku.* Sometimes she enrolls in "mother's class" so she can help with the drills at home.

So accepted is this role that it has spawned its own label, *kyoiku mama* (education mother). This title is not worn openly. Many Japanese mothers are embarrassed, or modest, and simply say, "I do my best." But that best is a lot, because to Japanese women, motherhood is a profession, demanding and prestigious, with education of the child the number-one responsibility. Cutthroat competition in postwar Japan has made her job harder than ever. And while many critics tend to play down the idea of the perpetually pushy mother, there are those who say that a good proportion of the credit for Japan's economic miracle can be laid at her feet.

"Much of a mother's sense of personal accomplishment is tied to the educational achievements of her children, and she expends great effort helping them," states *Japanese Education Today,* a major report issued in January by the U.S. Department of Education. "In addition, there is considerable peer pressure on the mother. The community's perception of a woman's success as a mother depends in large part on how well her children do in school."

Naoko's and Toshihiro's mother, Mieko Masuo, fully fills the role of the education mother, although she'd be the last to take credit for her children's accomplishments. This 46-year-old homemaker with a B.A. in psychology is a whiz at making her family tick. She's the last one to go to bed at night ("I wait until my son has finished his homework. Then I check the gas and also for fire. My mother stayed up and my husband's mother, and it's the custom for me, too") and the first one up in the morning, at 6. She prepares a traditional breakfast for the family, including *miso* soup, rice, egg, vegetable and fish. At the same time, she cooks lunch for her husband and Naoko, which she packs up in a lunch box, or *o-bento*. She displays the *o-bento* that Naoko will carry to school. In the pink plastic box, looking like a culinary jigsaw puzzle, are fried chicken, boiled eggs, rice, lotus roots, mint leaves, tomatoes, carrots, fruit salad and chopsticks. No pb&j sandwiches in brown bags for this family.

"Every morning, every week, every year, I cook rice and make *o-bento*," Mrs. Masuo says with a laugh, winking at Naoko. "I wouldn't want to give her a *tenuki o-bento*." Naturally. Everyone knows that a "sloppy lunch box" indicates an uncaring mother.

But Mrs. Masuo doesn't live in the kitchen. She never misses a school mother's meeting. She knows all the teachers well, has researched their backgrounds and how successful their previous students have been in passing exams. She carefully chose her children's schools and *juku,* and has spent hours accompanying them to classes. "It's a pity our children have to study so much," she apologizes. "But it's necessary." She says that someday she'd like to get a part-time job—perhaps when exams are over. "But at the moment I must help my children. So I provide psychological help and *o-bento help.*" Then she laughs.

Like most Japanese mothers, Mrs. Itoh spends most of her time with her son and her six-month-old daughter, Emi. Baby-sitters and play groups are not part of her life. She has dinner with the children well before her husband comes home from work. She takes them to the park, to swimming lessons and music, much of the time carrying her baby in a pack on her back. Indeed, a young mother with an infant in a sling and a toddler by the hand walking along a subway platform or a city street is a sight that evokes the very essence of motherhood to most Japanese.

This physical tie between mother and child is only a small part of the strong social relationship that binds members of the family together in mutual dependency and obligation. It's the mother's job to foster this relationship. From the beginning, the child is rarely left alone, sleeps with the parents, is governed with affectionate permissiveness and learns through low-key signals what is expected and what to expect in return.

Many American children are also raised with affection and physical contact, but the idea is to create independent youngsters. Discipline begins early. Children have bedrooms separate from their parents. They spend time playing alone or staying with strangers and learn early that the individual is responsible for his own actions. An American mother, in disciplining, is more likely to scold or demand; a Japanese mother is apt to show displeasure with a mild rebuke, an approach that prompted one American six-year-old to tell his own mother: "If I had to be gotten mad at by someone, I wish it would be by a Japanese."

Even a casual observer is struck by the strong yet tender mother-child connection. A Japanese senior high school teacher said that many wives, including his own, sleep in a room with their children and not their husbands. "Is it the same in America?" he asked. At a dinner party, a businessman made his wife's excuses:

"I'm sorry she couldn't come tonight. My son has an exam tomorrow." Even if the excuse was not true, the use of it says a lot.

The relationship of dependency and obligation fostered in the child by the mother extends to family, school, company, and country and is the essence of Japanese society. The child is taught early that he must do well or people will laugh at him— and laugh at his mother as well. "Most Japanese mothers feel ashamed if their children do not do well at school," said one mother. "It is our responsibility to see that the child fulfills his responsibility." Bad behavior may bring shame, but good behavior has its own rewards. One woman described a friend by saying: "Her son studied very hard in order to get into a good high school and he got in. She is very clever."

This attitude is precisely what gives education mothers such as Mrs. Masuo and her *o-bento* philosophy such esteem and why they take such pride in their role, even if they don't admit it. Their goal is clear: success in entrance exams, good school, a good college and a good job. (For daughters the goal has a twist: good schools lead to good husbands.)

Today, more and more educators and parents are questioning the high-pressure system that gives rise to such popular sayings as "Sleep four hours, pass; sleep five hours, fail." Educationists speak of lost childhoods, kids never getting a chance to play, "eating facts" to pass exams, and the production of students who memorize answers but can't create ideas. They cite the cruelty of students who take pleasure when their classmates fail, increasing delinquency, and high incidence of bullying in the schools.

Not surprisingly, Japanese mothers have been among the major critics, perhaps because they bear much of the brunt and witness the effects of the pressure on their children. "My son kept getting headaches and then he didn't want to go to school," said one mother. "So I stopped the *juku*." Recently, such mothers have gained an ally in Prime Minister Yasuhiro Nakasone, whose government has been seeking ways to depressurize the education system. Nevertheless, many doubt that his efforts will have any effect in a society dedicated to hard work and competition.

## Reading 7.2

# THE LITTLE EMPERORS, PART 1

*Daniela Deane*

*In an attempt to control its exploding population, China passed a one child per family policy in 1979. How has this policy affected the Chinese family? Daniela Deane explored this question while living in Hong Kong in an article she wrote for the* Los Angeles Times Magazine.

Xu Ming sits on the worn sofa with his short, chubby arms and legs splayed, forced open by fat and the layers of padded clothing worn in northern China to ward off the relentless chill. To reach the floor, the tubby 8-year-old rocks back and forth on his big bottom, inching forward slowly, eventually ending upright. Xu Ming finds it hard to move.

"He got fat when he was about 3," says his father, Xu Jianguo, holding the boy's bloated, dimpled hand. "We were living with my parents and they were very good to him. He's the only grandson. It's a tradition in China that boys are very loved. They love him very much, and so they feed him a lot. They give him everything he wants."

Xu Ming weighs 135 pounds, about twice what he should at his age. He's one of hundreds of children who have sought help in the past few years at the Beijing Children's Hospital, which recently began the first American-style fat farm for obese children in what was once the land of skin and bones.

"We used to get a lot of cases of malnutrition," says Dr. Ni Guichen, director of endocrinology at the hospital and founder of the weight-reduction classes. "But in the last 10 years, the problem has become obese children. The number of fat children in China is growing very fast. The main reason is the one-child policy," she says, speaking in a drab waiting room. "Because parents can only have one child, the families take extra good care of that one child, which means feeding him too much."

Bulging waistlines are one result of China's tough campaign to curb its population. The one-child campaign, a strict national directive that seeks to limit each Chinese couple to a single son or daughter, has other dramatic consequences: millions of abortions, fewer girls, and a generation of spoiled children.

The 10-day weight-reduction sessions—a combination of exercise, nutritional guidance and psychological counseling—are very popular. Hundreds of children—some so fat they can hardly walk—are turned away for each class.

According to Ni, about 5 percent of children in China's cities are obese, with two obese boys for every overweight girl, the traditional preference toward boys being reflected in the amount of attention lavished on the child. "Part of the course is also centered on the parents. We try to teach them how to bring their children up properly, not just by spoiling them," Ni says.

Ming's father is proud that his son, after two sessions at the fat farm, has managed to halve his intake of jiaozi, the stodgy meat-filled dumplings that are Ming's particular weakness, from 30 to 15 at a sitting. "Even if he's not full, that's all he gets," he says. "In the beginning, it was very difficult. He would put his arms around our necks and beg us for more food. We couldn't bear it, so we'd give him a little more."

Ming lost a few pounds but hasn't been able to keep the weight off. He's a bit slimmer now, but only because he's taller. "I want to lose weight," says Ming, who spends his afternoons snacking at his grandparents' house and his evenings plopped in front of the television set at home. "The kids make fun of me, they call me a fat pig. I hate the nicknames. In sports class, I can't do what the teacher says. I can run a little bit, but after a while I have to sit down. The teacher puts me at the front of the class where all the other kids can see me. They all laugh and make fun of me."

The many fat children visible on China's city streets are just the most obvious example of 13 years of the country's one-child policy. In the vast countryside, the policy has meant shadowy lives as second-class citizens for thousands of girls or, worse, death. It has made abortion a way of life and a couple's sexual intimacy the government's concern. Even women's menstrual cycles are monitored. Under the directive, couples literally have to line up for permission to procreate. Second children are sometimes possible, but only on payment of a heavy fine.

The policy is an unparalleled intrusion into the private lives of a nation's citizens, an experiment on a scale never attempted elsewhere in the world. But no expert will argue that China—by far the world's most populous country with 1.16 billion people—could continue without strict curbs on its population.

Dinner time for one 5-year-old girl consists of granddad chasing her through the house, bowl and spoon in hand, barking like a dog or mewing like a cat. If he performs authentically enough, she rewards him by accepting a mouthful of food. No problem, insists granddad, "it's good exercise for her."

An 11-year-old boy never gets up to go to the toilet during the night. That's because his mother, summoned by a shout, gets up instead and positions a bottle under the covers for him. "We wouldn't want him to have to get up in the night," his mother says.

Another mother wanted her 16-year-old to eat some fruit, but the teenager was engrossed in a video game. Not wanting him to get his fingers sticky or daring to interrupt, she peeled several grapes and popped one after another into his mouth. "Not so fast," he snapped. "Can't you see I have to spit out the seeds?"

Stories like these are routinely published in China's newspapers, evidence that the government-imposed birth-control policy has produced an emerging generation of spoiled, lazy, selfish, self-centered and overweight children. There are about 40 million only children in China. Dubbed the country's "Little Emperors," their behavior toward their elders is likened to that of the young emperor Pu Yi, who heaped indignities on his eunuch servants while making them cater to his whims, as chronicled in Bernardo Bertolucci's film "The Last Emperor."

Many studies on China's only children have been done. One such study confirmed that only children generally are not well liked. The study, conducted by a team of Chinese psychologists, asked a group of 360 Chinese children, half who have siblings and half who don't, to rate each other's behavior. The only children were, without fail, the least popular, regardless of age or social background. Peers rated them more uncooperative and selfish than children with brothers and sisters. They bragged more, were less helpful in group activities and more apt to follow their own selfish interests. And they wouldn't share their toys.

The Chinese lay a lot of blame on what they call the "4-2-1" syndrome—four doting grandparents, two overindulgent parents, all pinning their hopes and ambitions on one child.

Besides stuffing them with food, Chinese parents have very high expectations of their one *bao bei,* or treasured object. Some have their still-in-strollers babies tested for IQ levels. Others try to teach toddlers Tang Dynasty poetry. Many shell out months of their hard-earned salaries for music lessons and instruments for children who have no talent or interest in playing. They fill their kids' lives with lessons in piano, English, gymnastics, and typing.

The one-child parents, most of them from traditionally large Chinese families, grew up during the chaotic, 10-year Cultural Revolution, when many of the country's cultural treasures were destroyed and schools were closed for long periods of time. Because many of that generation spent years toiling in the fields rather than studying, they demand—and put all their hopes into—academic achievement for their children.

"We've already invested a lot of money in his intellectual development," Wang Zhouzhi told me in her Spartan home in a tiny village of Changping county outside Beijing, discussing her son, Chenqian, an only child. "I don't care how much money we spend on him. We've bought him an organ and we push him hard. Unfortunately, he's only a mediocre student," she says, looking toward the 10-year-old boy. Chenqian, dressed in a child-sized Chinese army uniform, ate 10 pieces of candy during the half-hour interview and repeatedly fired off his toy pistol, all without a word of reproach from his mother.

Would Chenqian have liked a sibling to play with? "No," he answers loudly, firing a rapid, jarring succession of shots. His mother breaks in: "If he had a little brother or sister, he wouldn't get everything he wants. Of course he doesn't want one. With only one child, I give my full care and concern to him."

But how will these children, now entering their teenage years and moving quickly toward adulthood, become the collectivist-minded citizens China's hard-line Communist leadership demands? Some think they never will. Ironically, it may be just these overindulged children who will change Chinese society. After growing up doing as they wished, ruling their immediate families, they're not likely to obey a central government that tells them to fall in line. This new generation of egotists, who haven't been taught to take even their parents into consideration, simply may not be able to think of the society as a whole—the basic principle of communism.

## Reading 7.3

# FAMILY AND PEER GROUP IN NORTHTON

*Elijah Anderson*

*Unemployment, racism, drug use, and lack of hope for the future are a major part of inner-city neighborhoods. How do these factors affect the family, and how are they influenced by the family? Elijah Anderson has been involved in field research in black underclass neighborhoods in Philadelphia and Chicago for more than twenty years. In this excerpt from his book,* Streetwise: Race, Class, and Change in an Urban Community, *Anderson examines some of the characteristics of inner-city families in a neighborhood he calls Northton. Anderson is a sociology professor at the University of Pennsylvania.*

The young woman who becomes addicted to "rock cocaine" or crack often comes from an unstable family. The economy and drugs are often implicated in this instability. In sharp contrast, the strong ghetto family, often with both a husband and a wife, but sometimes only a strong-willed mother and her children who have the help of close relatives, seems to instill in girls a hopeful sense that they can reproduce this strong family or one even "better." Such units, when they exist, are generally regarded as advantaged. The father usually works at a regular job and has a sense that his values have paid off. Both parents, or close kin, strive to instill in the children the work ethic, common decency, and social and moral responsibility.

Young girls growing up in such situations, strongly encouraged by their mothers and fathers and other kin, are sometimes highly motivated to avoid habits and situations that would undermine their movement toward "the good life." They are encouraged to look beyond their immediate circumstances; they often have dreams of marriage, a family, and a home. The strict control their parents exert, combined with their own hopes for the future, makes them watch themselves and work hard to achieve a life better than that lived by so many in the ghetto. Because of such interests and concerns, these young women tend not to follow the social track leading to drug addiction, though even they can become victims of crack, since because of its highly addictive nature one behavioral lapse or seeking an "experience" can fundamentally change a person's life.

By contrast, young girls and boys emerging from homes that lack a strong, intact family unit, usually headed by a single mother working or on welfare and trying desperately to make ends meet, become especially vulnerable to drug addiction and unwed pregnancies. The process leading up to such events usually begins in early childhood with full participation in play groups on the streets. Children who become deeply engaged in the drug culture often come from homes where they have relatively little adult attention, little moral training, and limited family encouragement to strive for a life much different from the one they are living.

Coming from a single-parent household, often headed by a nearly destitute mother, children may have almost no effective adult supervision. As some people in the community say, "they just grow up." Many become "street kids," left largely on their own, and by the time they are preteens some are becoming "street smart," beginning to experiment with sex and drugs. The local play group may become something of a family for youths who lack support at home.

The ghetto street culture can be glamorous and seductive to the adolescent, promising its followers the chance of being "hip" and popular with certain "cool" peers who hang out on the streets or near the neighborhood school. Often such teenagers lack interest in school, and in time they may drop out in favor of spending time with their street-oriented peers.

To be sure, most adolescents dabble in the street culture, including experimenting with sex and drugs; for status and esteem, they must learn to interpret and manipulate its emblems. Those who are most successful may become invested in such pursuits to the point of becoming overwhelmed and socially defined by them. Such definitions contribute to the formation of a street identity, which then gives such youths further interest in behaving in accordance with the peer-group's norms—or even inventing them.

The morally strict and financially stable intact nuclear families, on the decline in the underclass neighborhood, with their strong emotional and social ties and their aspirations for their children, must engage in sometimes fierce competition with the peer group. With a variety of social supports, including extended kin networks and strong religious affiliations, such families can often withstand the lure of the street culture, but even they may succumb and lose control of their offspring, sometimes permanently. The much less viable family headed by an impoverished young woman who has her hands full working, socializing, and mothering stands little chance in the struggle for a child's allegiance and loyalty and often does not prevail.

Though stable families possess an outlook that considers upward mobility a real possibility and try to instill this outlook in their children, such value transfers are difficult to accomplish. One important reason is the street culture's very strong and insistent attraction. But also important is the fact that the wider culture and its institutions are perceived, quite accurately at times, as unreceptive and unyielding to the efforts of ghetto youths.

Resolving this situation depends to some degree on luck. Youths from the most socially "promising" domestic arrangements sometimes become fixated and invested in the ghetto street culture, but they may have a chance to grow out of it. . . . Those from the poorest families, with the least control over social resources, appear to be most vulnerable to the street's seductive draw.

Vulnerability to the street may be manifested in a variety of ways, from adolescent pregnancy to street crime to serious drug abuse. One may approach "hipness" through good looks and grooming, nice clothes, dancing ability, sexual activity, crime, and drug use—which may be more or less important depending upon one's particular crowd or peer group.

. . . The economic noose restricting ghetto life encourages both men and women to try to extract maximum personal benefit from sexual relationships. The dreams of a middle-class life-style nurtured by young inner-city women are thwarted by the harsh socioeconomic realities of the ghetto. Young men without job prospects cling to the support offered by their peer groups and their mothers and shy away from lasting relationships with girlfriends. Thus girls and boys alike scramble to take what they can from each other, trusting only their own ability to trick the other into giving them something that will establish their version of the good life—the best life they can put together in their environment.

We should remember that the people we are talking about are very young—mainly from fifteen years old to their early twenties. Their bodies are grown, but they are emotionally immature. These girls and boys often have no very clear notion of the long-term consequences of their behavior, and they have few trustworthy role models to instruct them.

Although middle-class youths and poor youths may have much in common sexually, their level of practical education differs. The ignorance of inner-city girls about their bodies astonishes the middle-class observer. Many have only a vague notion of where babies come from, and they generally know nothing about birth control until after they have their first child—and sometimes not then. Parents in this culture are extremely reticent about discussing sex and birth control with their children. Many mothers are ashamed to "talk about it" or feel they are in no position to do so, since they behaved the same way as their daughters when they were young. Education thus becomes a community health problem, but most girls come in contact with community health services only when they become pregnant—sometimes many months into their pregnancies.

Many women in the black underclass emerge from a fundamentalist religious orientation and hold a "pro-life" philosophy. Abortion is therefore usually not an option. . . . New life is sometimes characterized as a "heavenly gift," an infant is sacred to the young woman, and the extended family always seems to make do somehow with another baby. A birth is usually met with great praise, regardless of its circumstances, and the child is genuinely valued. Such ready social approval works against many efforts to avoid illegitimate births.

In fact, in cold economic terms, a baby can be an asset, which is without doubt an important factor behind exploitative sex and out-of-wedlock babies. Public assistance is one of the few reliable sources of money, and, for many, drugs are another. The most desperate people thus feed on one another. Babies and sex may be used for income; women receive money from welfare for having babies, and men sometimes act as prostitutes to pry the money from them.

The lack of gainful employment not only keeps the entire community in a pit of poverty, but also deprives young men of the traditional American way of proving their manhood—supporting a family. They must thus prove themselves in other ways. Casual sex with as many women as possible, impregnating one or more, and getting them to "have your baby" brings a boy the ultimate in esteem from his peers and makes him a man. "Casual" sex is therefore fraught with social significance for the boy who has little or no hope of achieving financial stability and hence cannot see himself taking care of a family.

The meshing of these forces can be clearly seen. Adolescents, trapped in poverty, ignorant of the long-term consequences of their behavior but aware of the immediate benefits, engage in a mating game. The girl has her dream of a family and a home, of a good man who will provide for her and her children. The boy, knowing he cannot be that family man because he has few job prospects, yet needing to have sex to achieve manhood in the eyes of his peer group, pretends to be the "good" man and so convinces her to give him sex and perhaps a baby. He may then abandon her, and she realizes he was not the good man after all, but a nothing out to exploit her. The boy has gotten what he wanted, but the girl learns that she has gotten something too. The baby may bring her a certain amount of praise, a steady welfare check, and a measure of independence. Her family often helps out as best they can. As she becomes older and wiser, she can use her income to turn the tables, attracting her original man or other men.

. . . In Northton there are an overwhelming number of poor, female-headed families, a significant proportion of them on welfare. As these families grow older, the children meet the street culture and must come to terms with it. Some dabble in it, experiencing enough to try to pass as hip among their friends; few want to be known as "square" or "lame." For those only loosely anchored to conventional institutions (extended families, churches, or community organizations of one kind or another), the street culture calls loudly and insistently.

Most young boys can no longer earn enough to support a family, in the regular economy or the underground economy or both. Yet they engage in sex for both biological and peer-group reasons and often father children out of wedlock. Over time such behavior becomes incorporated into sex codes that reflect the reality of the situation. The young men often adjust to their financial situation by remaining with their mothers, retaining their "freedom" and all the comforts of home; they may sometimes play the role of man of the house and aid the household through meager financial contributions. But the youth is unwilling to "play house" with the mother of his child, an adaptation his peer group turns into a virtue. The girls are often headed for adolescent pregnancy, but they are doubly unable to provide for their children, with no skills, no provision for child care, and no vision of anything else.

## QUESTIONS 7.1–7.3

**1.** Take several concepts from earlier chapters (for example, socialization, self-concept, status, generalized other, and so on) and apply them across the three family styles described in the readings.

**2.** Connect family structure to behavior—how does parental conduct seem to affect children's behavior in each case?

**3.** How has the structure of society changed the structure of the family in each case?

**4.** Contrast male and female roles in the three types of families. Make hypotheses in those instances in which you cannot find specific examples.

**5.** How do these three family structures compare with the "typical American middle-class family"? List similarities and differences.

## NOTES

1. This discussion benefits from the discussion of social institutions by S. N. Eisenstadt in the *International Encyclopedia of the Social Sciences* (New York: Crowell, Collier & Macmillan, 1968), vol. 14, pp. 409–429.

2. The discussion in this paragraph is taken in part from David and Vera Mace, *Marriage: East and West* (Garden City, N.Y.: Doubleday, 1960), pp. 141–142.

3. These two paragraphs on courtship theory benefit from the discussion by Marcia and Thomas Lasswell in *Marriage and the Family* (Lexington, Mass.: D. C. Heath, 1982), chapter 6.

4. George P. Murdock, *Social Structure* (New York: Macmillan, 1949), p. 28.

5. See "Influence of the Marital History of Parents on the Marital and Cohabitational Experiences of Children," by Arland Thornton, *American Journal of Sociology* 96 (January 1991), pp. 868–894.

6. J. Ross Eshleman, *The Family: An Introduction*, 3d ed. (Boston: Allyn & Bacon, 1981), chapter 6.

7. See "Children and Marital Disruption," by Linda Waite and Lee Lillard, *American Journal of Sociology* 96 (January 1991), pp. 930–953.

8. Numerous Census Bureau documents were used for this section on marriage and divorce. Especially see "Marital Status and Living Arrangements: March 1997," *Current Population Reports*, Series P-20, no. 506, and *Statistical Abstracts of the United States*, 1999.

9. See "Considering Divorce: An Expansion of Becker's Theory of Marital Instability," by Joan Huber and Glenna Spitze, *American Journal of Sociology* 86 (July 1980), pp. 75–89.

10. This research was done by Murray A. Straus, Richard J. Gelles, and Suzanne K. Steinmetz and is reported in their book, *Behind Closed Doors: Violence in the American Family* (Garden City, N.Y.: Doubleday, 1980), chapter 2.

## CHAPTER EIGHT

Michael S. Yamashita/CORBIS

# Institutions: Religion

Religion: The Functional Approach

Religious Organizations

Religion in America

## TWO RESEARCH QUESTIONS

**1.** *Is religion related to health? Would belonging to a religious community make one more or less healthy, or are the variables unrelated?*

**2.** *Is there a relationship between religion and suicide? Specifically, will suicide rates vary depending on the type of church one belongs to? If so, why?*

> *. . . I will give a short account of the psychology of myself when my hansom cab ran into the side of a motor omnibus, and I hope hurt it.*
> *. . . In those few moments, while my cab was tearing towards the traffic of the Strand . . . I did really have, in that short and shrieking period, a rapid succession of a number of fundamental points of view. I had, so to speak, about five religions in almost as many seconds. My first religion was pure Paganism, which among sincere men is more shortly described as extreme fear. Then there succeeded a state of mind which is quite real, but for which no proper name has ever been found. The ancients called it Stoicism, and I think it must be what some German lunatics mean (if they mean anything) when they talk about Pessimism. It was an empty and open acceptance of the thing that happens—as if one had got beyond the value of it. And then, curiously enough, came a very strong contrary feeling—that things mattered very much indeed, and yet that they were something more than tragic. It was a feeling, not that life was unimportant, but that life was much too important ever to be anything but life. I hope that this was Christianity. At any rate, it occurred at the moment when we went crash into the omnibus.[1]*

IN OUR DISCUSSION of institutions in Chapter Seven, we suggested that all societies have specific constant and important central needs or problems that must be addressed. Norms and roles emerge, and a pattern of organization or social institution develops. In the previous chapter we looked at the family, and in this chapter we will examine the institution of religion.

Some eternal questions face all of us: What happens after death? What is the meaning of life? Given the misfortunes of the present, can there be hope for the future? As people struggle with these concerns, they develop a set of beliefs, norms, and roles that represent their attempts to answer these questions. We refer to this increasingly organized set of norms and roles that develops as the institution of religion.

The central concerns from which the institution of religion develops are usually abstract—much more so than those we discussed in the previous chapter on the family. One consequence is that definitions of and explanations for the institution of religion are more varied than are those for some other institutions. For example, Andrew Greeley suggests that religion has at its origins the human propensity to hope; that in the face of life's disappointments, hope needs to be renewed; that a variety of experiences may renew hope; and that religion is a collectively created phenomenon that effectively renews hope. This approach focuses on what religion *does* (renews hope), whereas some other approaches focus on what religion *is* ("a set of sacred beliefs and symbols").[2]

Religion, like the family, has appeared in all societies. If one compares different societies, however, one finds tremendous variation in religious practices. To get an idea of the variation, look at Tables 8.1 and 8.2. Table 8.1 shows a breakdown of the religious population of the world. Table 8.2 shows recent membership figures of religious groups in the United States. Not all of America's churches are included in Table 8.2; only those with memberships of at least 500,000 are listed. This table (8.2) shows church *members*, and these represent about 60 percent of the United States population. Forty percent of the population doesn't claim any church membership. As you view the figures in these tables, keep in mind also that religious bodies count membership in different ways. For example, some churches count all baptized people, including infants, whereas others count only those people who have attained full membership, which usually includes only individuals over 13 years of age. Thus, these figures should be taken as no more than general estimates.

All countries have at least one religion, although not all people are religious. Ethnocentrism relates to our discussion of religion, as it does to our discussion of the family: The lack of acceptance of others' ways of doing things, the hostility to difference, is probably even more pronounced with regard to religion. Whereas people with peculiar practices (meaning those different from ours) in marriage, family, clothing, or eating are seen as alien, primitive, or ignorant, people with religious beliefs other than our own are frequently seen as evil, pagan, or guided by the devil. Great rivalries and conflicts have occurred between religious groups, for religious faith and conviction can be a powerful force. Knowing that "it's God's will" or that "God is on our side" has justified all kinds of behavior.

## RELIGION: THE FUNCTIONAL APPROACH

Sociologists who study religion are not concerned with the possible truth or falsity of religious beliefs. They are interested in what people believe and especially in what believing does for people. Although there is great variation among religions, a discussion of some common definitions and descriptions might help our understanding of the institution of religion. **Religion** can be described as a unified system of beliefs, feelings, and behaviors related to things defined as sacred. Societies define as sacred, for example, objects (a tree, the moon, a cross, a book), animals, ideas (science, communism, democracy), and people (Buddha, Christ, saints). When things become sacred, they are endowed with a special quality or power and are treated with awe, reverence, and respect. Beliefs about the sacred entities develop, and then people often lend a supernatural quality to their beliefs. A set of appropriate feelings—reverence, happiness, sadness, fear, terror, ecstasy—is established. Finally, behaviors or rituals are developed consistent with the beliefs and feelings: the confessional, the rosary, communion, dietary laws, or the content of a particular worship service.[3]

**TABLE 8.1**  Religious Population of the World, 1998

| Faith | Population (millions) | Percent | Faith | Population (millions) | Percent |
|---|---|---|---|---|---|
| Christian | 1,943 | 33 | Christian | | |
| Muslim | 1,164 | 20 | Catholic | 1,027 | 51 |
| Hindu | 762 | 13 | Protestant | 316 | 16 |
| Buddhist | 354 | 6 | Orthodox | 214 | 11 |
| Chinese folk religions | 379 | 6 | Anglican | 64 | 3 |
| New religionists | 100 | 2 | Other | 374 | 19 |
| Ethnic religionists | 249 | 4 | | | |
| Jewish | 14 | 0.2 | | | |
| Sikhs | 22 | 0.4 | | | |
| Spiritists | 12 | 0.2 | | | |
| Confucian | 6 | 0.1 | | | |
| Shinto | 3 | 0.05 | | | |
| Nonreligious | 760 | 13 | | | |
| Atheist | 150 | 3 | | | |

United States Bureau of the Census, *Statistical Abstracts of the United States, 1999* (119th ed.), Washington, D.C.

**TABLE 8.2**  Membership in Selected Religious Bodies in the United States, 1996

| Religious Group | Members (thousands) |
|---|---|
| Adventist | 808 |
| Baptist | 36,613 |
| Christian churches | 2,000 |
| Churches of Christ | 1,651 |
| Eastern Orthodox | 5,302 |
| Episcopal | 2,505 |
| Jehovah's Witnesses | 946 |
| Jewish | 4,300 |
| Latter-Day Saints | 4,166 |
| Lutheran | 8,350 |
| Methodist | 13,533 |
| Muslim | 5,100 |
| Pentecostal | 10,606 |
| Presbyterian | 4,193 |
| Roman Catholic | 60,191 |

*The World Almanac and Book of Facts,* 1997 (Mahwah, N.J.: Funk & Wagnalls, 1997).

Traditionally, institutions are examined by the functions they serve. What functions does religion offer for society? First, religion provides a way for people to deal with the unknown; it supplies some measure of certainty in an uncertain world. There are a number of things we cannot explain. There are many other things that can be explained and understood but are still difficult to accept, such as the certainty of death or the loss of loved ones. Religion enables people to adjust to these situations by providing a sacred, supernatural being or object to explain the unknowable; religion

## ÉMILE DURKHEIM (1858–1917)

Émile Durkheim earned a Ph.D. in Philosophy at the Ecole Normale Superieure in France. In his book *Rules of Sociological Method*, Durkheim argued that it is the special task of sociology to study what he called "social facts." He defined social facts as forces and structures that are *external* to the individual. He opposed the utilitarian tradition in British social thought that explained social phenomena by focusing on the actions and motives of individuals. Durkheim didn't think the individualism of the utilitarian approach could provide the basis on which to build a stable society.

Durkheim first taught at the University of Bordeaux and then at the Sorbonne in Paris where he was a professor of Science in Education and Sociology. He appears to have been a masterful lecturer who held his audience so much in thrall that one of his students could write, "Those who wished to escape his influence had to flee from his courses; on those who attended he imposed, willy-nilly, his mastery." Durkheim helped establish the autonomy of sociology as a discipline and is widely acknowledged as a "founding father."

Durkheim appears often in this book. In this chapter, we look at his view of the functions of religion, and at his view of the relationship between religion and suicide. Earlier, in Chapter Four, we examined his typology of societies. In Chapter Twelve we look at his ideas on social change and anomie, and in Chapter Fourteen we review Durkheim's views on the positive functions of deviance, and the details of his classic study of suicide.

---

provides a belief system to use to deal with situations that are difficult to accept, such as the certainty of death or a life after death. Anthropologists have noted that sacred rituals and beliefs are usually found in situations that are most difficult to control, in which there is most uncertainty. Our own appeals for divine guidance and support usually come at times when uncertainty is present: when the jet accelerates for takeoff, or just before a particularly difficult test.

Religion also provides people with a perspective, a viewpoint, a way of looking at the world. It provides value orientations about how the world ought to be; it gives meaning to life. Max Weber held that people are motivated by ideologies or beliefs that stem in great part from religious origins and that these are more basic than are the economic values believed by Marx to be so important. In his classic work *The Protestant Ethic and the Spirit of Capitalism*, Weber's analysis of history led him to the conclusion that the beliefs surrounding Protestantism were central to the rise of capitalism. For example, Protestant values encourage the view that work is a calling: To work hard is virtuous, acquisition is supported, and sobriety is encouraged. Far from being a needless distraction, religion is the source, according to Weber, of many of a society's most basic values, perspectives, and orientations to the world.

Religion can serve an integrative function in that it promotes solidarity. People with similar beliefs and viewpoints are drawn together and are more unified than those without this common experience. The presence of religion does not guarantee solidarity within a society, however, as other factors are also important. Many societies—the United States, for example—have several religions, and these can conflict with each other. In this case, there can be a degree of solidarity among those of the same faith but an absence of solidarity in the society; religion can prove to be disintegrative for society even though it is integrative for the individual.

The first research question at the beginning of this chapter concerned a possible relationship between religion and health. It has been suggested that religion might affect health by regulating behavior—diet, smoking, and drinking—and by providing social support and enhancing social integration. Some time ago, sociologist Émile Durkheim emphasized the integrative and regulative functions that religion can provide for members. Whether these functions could be related to health was recently studied in an interesting way.

Two small, cohesive kibbutzim (collective settlements or communities) in the Negev region of Israel were chosen for study. They were similar in size (about 400 people) and age of members, but one kibbutz was religious and the other was nonreligious, or secular. A questionnaire given to community members included several measures of health behavior (like diet), health status (such as disabilities or sicknesses), illness behavior (for example, number of doctor visits and medication taken), and measures of religiosity (religious practice, private praying). The researchers found that the religious kibbutz practiced more health-promoting behavior and that the members of that community reported that they were more healthy and had less illness behavior than did the secular kibbutz. However, those who practiced individualistic religious activities, like private praying, were less healthy and reported more psychological distress. It seems to be the collective aspect of religion rather than the individual aspect of religiosity that is important. The authors suggest that a religious community provides additional regulation and integration (beyond that of other types of communities) that result in more healthy behavior and better psychological and physical well-being.[4]

Religion may function as an agency of social control. We noted earlier that religion provides people with a value orientation, a perspective on life, a view of the world. From shared viewpoints, common agreements arise as to what people should and should not do. Because these norms emerge from a religious base, the church becomes involved in ensuring that they are observed. They might be informally encouraged or enforced, or they might be enacted into law. For example, many religions provide education for the young in schools run by the church. Some churches do not recognize marriages that take place outside the church. The clergy perform marriages and engage in marriage counseling. A little more indirectly, through pressure groups and lobbies, religions are involved in ensuring the passage or rejection of laws felt to be crucial; laws dealing with liquor, pornography, abortion, and birth control are recent examples in the United States. In countries in which there is less separation of church and state, religious organizations

A week's schedule from a church bulletin board:

THIS WEEK'S GROUP MEETINGS—7 P.M. IN THE SANCTUARY

| | |
|---|---|
| Monday | Say No to Drugs |
| Tuesday | The Puzzle of Dyslexia |
| Wednesday | Abused Spouses |
| Thursday | Codependency Recovery Group |
| Friday | Alcoholics Anonymous |
| Saturday | Food Closet and Aid for the Homeless |

NEXT SUNDAY 9 A.M. and 11 A.M. REVEREND CHAVOOR

*"Rejoicing about America's Future"*

and their representatives are more directly involved with government in determining right and wrong.

In a variety of ways, then, organized religious groups assert moral authority and attempt to provide guidance; they act as agents of social control. Problems can arise when values and viewpoints on issues change, as they have, for example, on legalized abortion and birth control in the United States. Formal church doctrines change much more slowly than do the values and attitudes of the people. The result is that the social-control function of the church is compromised when its members no longer hold its values; if this occurs on many issues, the authority of the church is likely to be seriously undermined.

A report published in 1976 provides a good example of how this can happen: In 1968, Pope Paul VI ruled against the use of artificial birth control by Catholics. Earlier in the 1960s the Second Vatican Council had made church reforms that were well accepted by Catholics in the United States. However, the Pope's 1968 ruling was, according to Andrew Greeley, who studied the consequences and authored the report, "as far as the American church goes, one of the worst catastrophes in religious history." The birth-control ruling did serious damage to the authority of the church and to the credibility of the Pope. Apparently as a consequence of the ruling, between 1963 and 1974 weekly attendance at Mass dropped from 71 to 50 percent, monthly confession dropped from 38 to 17 percent, families who favored a son becoming a priest dropped from 63 to 50 percent, parochial school enrollment dropped from 5.6 million to 3.5 million, and so on. The Pope's ruling was issued to restore faith in the church, but, in pushing a view no longer held by most members, the church actually badly weakened itself.[5]

How is religion in America best explained in a functional sense? Greeley suggests a combination of three factors: belonging, meaning, and comfort. Belonging refers to the idea originated by Durkheim that religion is the source of social interaction; it provides the cement that holds society together in the face of internal and external threats; it provides unity. Meaning refers to Weber's idea that religion provides a way of looking at the world—a set of definitions and ideas that guides one's thinking. Religion gives meaning to that which is difficult to interpret. It also influences the character and direction of other institutions (economic, for example) because people's views of the world are shaped by their religions. Finally, religion provides comfort, serenity, and reassurance. It supports existing institutions and provides social stability. Greeley believes that these three factors, especially the relationship between the meaning and belonging functions, are responsible for the vigorous religious behavior in the United States.[6]

The functions we have discussed—to explain the unknown, to provide a viewpoint (meaning), to afford solidarity (belonging), and to exercise social control—represent some of the more obvious functions of religion. These functions are not mutually exclusive, and there are others as well. It is interesting to note here while we are discussing functions that *magic* in primitive societies and *science* in more advanced societies serve many of the same functions as does religion. Without careful study, it is sometimes difficult for laypeople to distinguish among statements from magic, science, and religion.

## RELIGIOUS ORGANIZATIONS

So far we have examined the elements of religion (beliefs, feelings, and behavior related to sacred objects) and some of the functions religion can serve. Religion can also be studied from the viewpoint of the organizational structure of religious groups. Our knowledge of bureaucracies and large organizations can be applied to churches. One way of categorizing religious bodies would be by their ceremonies, their pattern of public worship. Greeley suggests the breakdown shown in Table 8.3. Religious ceremony is examined by how rational or emotional it is (Apollonian or Dionysian, respectively) and by the degree of formality (simple church or high church).

This breakdown provides four categories, each of which describes a fairly distinct type of religious experience. **Apollonian–simple-church ceremonies** appear often in American Protestant denominations. Services are "relatively plain and simple and at the same time sober, restrained, and dignified." Even the hymns are sedate and restrained. **Dionysian–simple-church ceremonies** involve "direct intervention of the spirit" and a simple and matter-of-fact approach to prayer, which at the same time encourages members to freely and openly express their emotions. **Apollonian–high-church services** are "dignified, rational, and restrained, while at the same time elaborate, artistic, and stylized." **Dionysian–high-church services** are "ecstatic, emotional, and nonrational, and at the same time elaborate, artistic, and stylized." The latter is a difficult combination, and this category of experience is not frequently found in churches in the United States. Greeley notices several trends: Along the simple-church–high-church dimension, Catholicism in this country seems to be moving in the direction of simple church, whereas some Protestant denominations are going the other way. Along the Apollonian-Dionysian dimension, however, the direction is clear.

**TABLE 8.3**  Types of Religious Ceremony

|  | Simple Church (simple, informal ceremony) | High Church (formal, complex ceremony) |
| --- | --- | --- |
| Apollonian (rational, sober, unemotional) | Traditional Protestant— Presbyterian, Congregational, Methodist | Latin mass, solemn high mass of Roman Catholic Church |
| Dionysian (emotional, nonrational, ecstatic) | Pentecostal, holy roller, snake handlers | Coptic mass, Javanese dance |

Adapted from Andrew Greeley, *The Denominational Society* (Glenview, Ill.: Scott, Foresman & Co., 1972).

The revolt against scientific rationalism is leading churches to become increasingly Dionysian—nonrational, emotional, and ecstatic.[7]

Terminology has also been developed to categorize churches by the size and scope of the organization. One distinction that is commonly made is between church and sect. A **church,** which is large and highly organized, represents and supports the *status quo;* it is considered a respectable organization, and membership is usually automatic—one is born into the church. A **sect,** on the other hand, is smaller and less organized, and membership is voluntary. Members of sects usually show greater depth and fervor in their religious commitment than do members of churches. The sect is more closely associated with the lower classes, the church with the middle and upper classes. The sect usually arises because of protest when church members become disaffected. They may feel that the church is too compromising or religiously too conservative, or in some other way is not responding to the wishes or convictions of its membership. The disaffected, deciding they would rather switch than fight, establish a sect based on purity of belief and on the individual religious needs of its membership.[8]

Milton Yinger sees church and sect as two types on a continuum of religious organizations. He adds several other types, including denomination and cult, as shown in Table 8.4. The **denomination** appeals to a somewhat smaller category of people—a racial, ethnic, or social-class grouping—than does the church. Yinger defines **cult** as a small, short-lived, often local group frequently built around a dominant leader. Cults and sects arise, it is argued, because people feel deprivation in some form—poverty, ill health, value conflict, anomie. This deprivation leads to the formation of new religious movements. Once formed, the movement might later die out, or it might become larger, more established, and conservative, and move to another category of organization—from cult to sect to denomination, for example.

## RELIGION IN AMERICA

To a sociologist looking at religion in the United States, particular aspects stand out. In America, as in other countries, there is *differential involvement* in religious activities. If we take church attendance as a measure of religious involvement, we find that women are more active than men, Catholics more than Protestants, college-educated more than those with a high-school

**TABLE 8.4** Religious Organizations

| Type of Organization | Characteristics | Examples |
|---|---|---|
| Church | Large; universal appeal; very respectable; members are born into it; supports the status quo and existing nonreligious (secular) institutions; formal structure | Roman Catholic Church of the Middle Ages |
| Denomination | Less universal appeal but organized around class, racial, or regional boundaries; conventional and respectable; less emotionalism and fervor than is shown in the sect; mainly middle class; stable, organized, and at peace with secular institutions | Presbyterian Church, Lutheran Church |
| Sect | Smaller; arises out of protest; membership is voluntary; uncompromising on religious doctrine; more emotional; less formal organization; more withdrawn from society and at war with secular institutions; it alone "has truth"; mainly lower class | Pentecostal and evangelical movements, Jehovah's Witnesses |
| Cult | Smallest; less organized; somewhat formless and often short-term; frequently centered around a charismatic leader; concerned mainly with problems of the individual (loneliness, alienation); more exotic ritual; may focus on a religious theme (salvation) or on a more secular theme (self-improvement, self-awareness) | Snake worshipers, Branch Davidians, urban storefront churches |

Adapted from J. Milton Yinger, *Religion, Society and the Individual* (New York: Macmillan, 1957), ch. 6.

education or less, people older than 30 more than those younger than 30, and people in the East most and those in the West least. Middle- and upper-class people attend and participate in church activities more than those of the lower class. Lower-class people tend to show greater intensity and emotion in religious feeling. Religious groups also seem to be stratified according to social class. Episcopalians, Congregationalists, and Jews tend to be of higher social class than Presbyterians and Methodists, who in turn tend to rank higher than Baptists and Catholics. This ranking in great part reflects ethnic or national origin and recency of immigration into the United States. The religious and social-class picture is complicated by the apparent tendency of Americans to change their religious affiliation as they move up the social-class ladder.[9]

Great *religious diversity* characterizes this country. The United States has no official state-supported religion, and unlike some countries, America has attempted to maintain a separation between church and state. Broadly speaking, three religious bodies—Protestant, Catholic, and Jewish—dominate. However, there are numerous Protestant denominations, of which the largest are Baptist, Methodist, Lutheran, and Presbyterian, as well as a multitude of sects and cults. Some authorities have noted this diversity in American religious groups and have argued that, on the contrary, the differences are in name only. These authorities believe that there is a growing consensus on religious beliefs and that we are approaching a common religion in America that is eroding the traditional differences among Protestantism, Catholicism, and Judaism. Charles Glock and Rodney Stark tested this idea by asking several members of Protestant and Catholic churches about their beliefs. They concluded that great diversity does exist, not only between Protestants and Catholics but among various Protestant denominations as well. In fact, Glock and Stark believe that the differences among various Protestant denominations are so great that it is inappropriate to speak of "Protestants"—there is no such thing as a unified Protestant religion.

Protestantism, Catholicism, and Judaism in the United States are probably peculiarly American; often they differ markedly from their antecedents in other countries. A *Time* magazine poll of American Catholics conducted on the eve of a visit to the United States by Pope John Paul II seemed to support this. Often, American Catholics differed with the official church: American Catholics favored permitting women priests (52 percent), married priests (53 percent), and remarriage in the church for the divorced (76 percent). Only 24 percent viewed artificial birth control as wrong. Seventy-eight percent of those polled felt it was permissible for Catholics to make up their own minds on such moral issues as birth control and abortion. Often, the viewpoints of Catholics and Protestants were similar: 27 percent of Catholics and 34 percent of Protestants felt that women should have the right to abortion on demand; 43 percent of Catholics and 41 percent of Protestants agree that the Pope is out of touch with Catholics in the United States. The melting pot has modified traditional religions so that they all probably reflect the cultural values of the United States.

The second question at the beginning of this chapter dealt with the possible relationship between religion and suicide. Durkheim started the research on this issue in 1897, and much has been done since his time. A study on suicide is relevant to our discussion here on religious changes in the United States. This study, which examined suicide rates within different religious groups, found that religious affiliation was associated with suicide rates in complicated ways. Catholics, Jews, and evangelical Protestants had lower suicide rates, whereas Presbyterians, Methodists, and Lutherans had higher suicide rates. Religions that might be called "mainstream" and liberal or moderate had higher rates. The authors suggest that people belong to different "networks" in society. People's membership in groups or organizations and their affiliations are examples of networks. Some of these networks are tightly knit, more controlling, and provide more energy and support for members; others are more loosely structured and provide less support for day-to-day behavior. Some religions (Catholicism, evangelical Protestant denominations) are more tightly involved in their members' lives and provide additional social support and guidance in times of personal crises. Other religions (mainstream Protestant denominations) have responded to changes in society and have become looser and consequently less central to members' lives. They may provide less energy and support and be less able to help in times of crisis, perhaps leading to higher suicide rates.[10]

In the late 1970s, the White House was occupied by a president who was a born-again Christian, and numerous people on the national scene, from Watergate conspirators to athletic heroes, publicly became religious. This trend continued through the 1980s and into the 1990s, with religious groups like the Moral Majority exerting political pressure at high levels, and with the spread of electronic religion, in which radio and television sermons reach a wide audience. The United States appears to be going through a religious revival of sorts, with renewed interest in fundamentalist, pentecostal, and evangelical types of experiences. Evangelical movements have a high profile and are continually in the news. Television ministries ("televangelism"), a phenomenon of the cable and satellite technologies of the 1980s, are reaching a large audience, attracting much money, and exerting political influence. Some traditional churches have discovered to their dismay that many of their members have found themselves attracted to a different message, perhaps to more emotional and exotic religious conduct like faith healing, speaking in tongues, mass revival meetings, and so on. Membership trends in the mid-1990s showed declines in old-line Protestant churches (Methodist, Episcopal, Presbyterian) and increases in the conservative Protestant and evangelical churches. Switching from the religious tradition in which one was raised is more common, especially among liberal and moderate Protestants. It also seems to have become more acceptable and common to state no religious affiliation and to have no attachment to any organized church.

A passing parade of recent religion-related events includes the following: scandals involving financial and sexual improprieties of religious leaders; strong interest in Eastern religions and mysticism; members of Hare Krishna chanting in the streets; government investigations of Scientology; Reverend Moon's Unification Church holding mass rallies and mass weddings; religious groups being charged with brainwashing converts while parents attempt to kidnap their children from the group; a gun battle that left four ATF agents and six members of the Branch Davidians dead, followed in April 1993 by a fire in which sect leader David Koresh and 78 of his disciples died; increasing involvement of the Christian Right in political campaigns; religious groups rushing on-line on the Internet with home pages, bulletin boards, and theological news groups.

Religious behavior in general seems to be fragmenting and taking many different forms. Overall church attendance has not changed much between 1970 and the present—it is the *form* of religious expression that is undergoing change. Great diversity continues to characterize religion in America.

Religion in America is *losing functions:* Other institutions are taking over the traditional functions that religion has served. Education and socialization of the young are handled for the most part by educational institutions rather than by the church. More dramatic is the rise in the importance of science. Today we tend to put our faith in science rather than in religion when dealing with the unknown or uncertain. Scientific knowledge increasingly forces a redefining of religious beliefs. How do you describe heaven when rockets and spaceships are racing through the universe to the moon and beyond? What is the soul when hearts and possibly even brains can be transplanted

## CULTS IN JAPAN

*The Japan Study Group for Quitting Cults estimates that about 1 million people or close to 1 percent of Japan's population are members of a cult. One cult (Aum) made headlines in 1995 when they sprayed nerve gas in a Tokyo subway killing 12 people and injuring 5,000. What is the attraction of cults in Japan? The experts suggest the following: the end of state Shintoism, which left many without spiritual guidance; pressures in Japanese society toward rigid conformity; a prolonged economic downturn; and a highly competitive educational system. Catherine Makino describes some current cults.*

**Aum Shinri Kyo (Supreme Truth):** Founded in 1987, Aum's combination of Christianity and Buddhism once attracted a peak membership of about 20,000 followers worldwide.

Its spiritual leader is Shoko Asahara, the son of a tatami straw mat maker who has worked as an acupuncturist and yoga instructor.

In the past, Asahara has predicted major disasters will occur, including a nuclear war between the United States and Japan.

He is on trial, accused of masterminding the Tokyo subway nerve-gas attack; another cult official has been sentenced to death. Group membership has plummeted to several thousand. The cult also recently changed it name to Aleph, the first letter of the Hebrew alphabet.

**Yamagishi:** Miyozo Yamagishi, a chicken farmer, founded this group in 1953 to create a rural utopia. Its members live in communes, denounce personal possessions and are considered to be "one family."

Yamagishi's organic dairy products and vegetables are sold widely from the back of vans or at upscale department stores. Annual sales total $1.7 million.

The group claims to have 5,000 members in communes in 40 locations in Japan and seven outside the country, including Brazil, Thailand, Germany, Australia and the United States.

**Life Space:** Koji Takahashi, a former tax accountant, founded this group in

---

from one body to another? Today it seems that we look to science for the answers rather than to religion; the scientist is more of a folk hero than is the minister. We might predict that either religion will change to adapt to a scientific world or that religion will have ever decreasing influence. Some religious organizations, in a spirit of "if you can't lick 'em, join 'em," are attempting to develop new functions to replace those they have lost. Usually this means becoming more "this-worldly" and less "other-worldly." The new church looks like a social-work agency, and religious leaders frequently lead the way in speaking out on social issues. Clergy put their bodies on the line in the 1960s in civil rights demonstrations in the South. Individual representatives of religions spoke out on the draft and on the war in Southeast Asia among other things, and have been active in the farm-labor movement in California and in other parts of the Southwest. Religious organizations have provided food and shelter for youth "doing their own thing." It is frequently noted that the clergy in the pulpit sound more as though their college degrees were in psychology or sociology than in theology.

1983. At that time, he gave "self-enlightment" seminars for $5,000 a person. These programs, in which participants role-played as beggars or blind people to experience different perspectives, were an instant success.

The group is controversial, however, for its belief that the human body never dies. Takahashi says he can revive a dead person by touching palms on the corpse's head and body. To date, two bodies have been found mummified, and Takahashi was arrested last month in connection with one of the deaths.

As its peak, Life Space membership numbered 10,000.

**Sukyo Mahikari (True Light):** Founded in 1960 by Okada Kotama, this sect has been accused of brainwashing its members.

The core belief is that one's "primary soul" lies 10 centimeters (about 4 inches) behind the forehead and that people inherit bad karma which can only be purged in special sessions. Each member is given an amulet to prevent "spirit possession."

The cult is now led by Kotama's daughter Kishu, who is known as the "teaching master" and said to be one of Japan's richest women. The group claims to have 500,000 followers.

**Hono Hana Sampogyo (Flower of Discipline):** Members of this group, led by 54-year-old Hogen Fukunaga, believe they can cure illnesses by reading the soles of the feet. The cult is now under investigation for tax evasion, fraud and treating patients without a medical license. The group has an estimated 10,000 members.

**Kensho-Kai:** This Buddhist group was established in 1982 and became an authorized religious organization in 1996. Its members advocate revising Japan's constitution to establish Kensho-Kai as a national religion.

The group's recruiters typically focus on junior high and high school students, and some schools have asked authorities for help with students who solicit others to join the cult. It has about 6,000 members, including senior-ranking members of the armed forces.

—*From "Cult Nation,"* San Francisco Chronicle, *March 21, 2000, by Catherine Makino.*

---

These trends might not be widespread or even widely accepted in the very churches in which they are taking place. Some argue that these changes are certain guarantees of the decline of religion. After all, churches were formerly offering something different, a special way of looking at the world unattainable elsewhere. Now the church is no different from other secular organizations and therefore has lost any unique appeal it might have had. Others believe, however, that in a changing world, religion too must change or run the risk of becoming irrelevant to modern people.

Finally, the **secularization** of American religion bears discussion. By secular we mean nonspiritual, nonsacred, and this-worldly. Americans seem to be religious and nonreligious at the same time. Millions of Bibles are sold yearly, and polls indicate that more than 90 percent of Americans say they believe in God, yet 53 percent of the people asked could not name one of the four authors of the first four books of the New Testament, less than 50 percent could name at least four of the Ten Commandments, and 46 percent of those polled by Gallup in 1980 felt that religion was losing its influence.

Americans believe in prayer, in heaven and hell, and in life after death, yet church attendance has dropped from 49 percent in a given week in 1958 to 40 percent in 1971, and it remains at about that level today. A 1993 survey of 4,000 Americans found the following: 30 percent described themselves as totally secular, 29 percent described themselves as barely religious, 22 percent described themselves as modestly religious, and 19 percent said that they regularly practiced their religion.

In 1957, 97 percent of Americans identified with a religious denomination. Yet at about the same time a panel of outstanding Americans, when asked to rate the most significant events in history, gave first place to Columbus's discovery of America. Christ's birth or crucifixion came in 14th, tied with the discovery of the X ray and the Wright brothers' first flight.[11] Religion used to be of vital significance; it guided the pioneers' way of life. People read the Bible regularly at family gatherings. It was the law, and its teachings were carefully followed. But as society changes in other ways, so does religious observation. Today people join churches and say they are religious, but perhaps the meaning of religion and the reasons for believing have changed. Although many remain religious in the traditional sense, many others are religious for social and secular reasons—it looks right and it makes one respectable. Religion is used to obtain other desired results. Belonging to the right church will help the aspiring businessperson make the right contacts, meet the right people, and have the right friends. Athletic teams pray before games; even the United States Marine Corps holds special religious services before shooting the rifle on qualification day. We seem to feel that if we have religion—belong, or go to church—other good things will happen to us. The result is an increasing secularization of American religion. As Will Herberg puts it, contemporary American religion is *man-centered*. People, not God, are the beginning and end of the spiritual system of much of present-day American religiosity. The result is a religiousness without religion, a way of sociability or belonging rather than a way of orienting one's life to God.[12]

Making categorical statements about religion in America is risky business, however. In 1992, 75 percent of Americans polled were members of churches or synagogues, more than half said religion was very important in their personal lives, and the public consistently says that it has more confidence in the church and organized religion than in other key institutions. Although less than half of the American population attends church regularly (36 percent in 1996), this percentage is higher than in most other predominantly Protestant nations in the world. Interest in television ministries and evangelical churches is strong. Next to the secular strain of religion is another strain emphasizing its spiritual and emotional aspects. This seems to be the keynote of religion in America: great variation in belief and practice.

## THEORY AND RESEARCH: A REVIEW

As mentioned in the previous chapter, the study of social institutions usually starts with *functional analysis*. Institutions imply order, stability, and a unified system with integrated parts, the whole thing working together to make society run—an apt description of the functional approach. *Conflict theory* provides another perspective by focusing on the role religion plays in social conflict. Contrasting religious beliefs have often been related to political differences and to conflict—see, for example, the ongoing struggles in Northern Ireland and in the Middle East. Karl Marx was critical of religion because of people's tendency to use it in times of uncertainty or stress. Marx described religion as the opiate of the masses because he felt that it took their minds off their real problems in favor of the other-worldly answers of Christianity. The working class's real problem in Marx's view was its manipulation and exploitation by the wealthy industrialists—the bourgeoisie—in capitalist economic systems. Religion was a needless distraction that pacified the oppressed; it could only slow the workers' revolt and subsequent freedom from oppression.

Religion plays an important role in social conflict, according to conflict theorists, because the dominant religion in society usually represents the dominant social class. The dominant religion supports the status quo, offers justifications for existing class lines, and operates to make working-class and lower-class people satisfied with their condition. The conflicts among social classes can lead to the emergence of specific religious groups that represent specific class interests. One example is the People's Temple of the late Reverend Jim Jones, which reportedly drew much of its membership from the ranks of the disaffected and alienated in society. According to the conflict perspective, a consequence of the development of class-based religions is that often what appear to be religious differences between opposing religious groups (for example, Protestants and Catholics in Northern Ireland) actually turn out to be class differences. One religious group represents money and propertied people and the other represents the working class. Conflict, then, is a central element in the discussion of religion, according to this viewpoint.

General information about church membership and religious opinions and beliefs comes from *reactive surveys* (public opinion polls) and *nonreactive surveys* (analysis of membership records and historical data). The reactive studies that ask people about their religious opinions and beliefs suffer from the disadvantage that questionnaire studies typically face—response rate. The response rate is often below 50 percent. Critics of the research wonder if the viewpoints of those who returned the questionnaires might be different from the viewpoints of those who did not.

## SUMMARY

Religion has been the subject of this second chapter on social institutions. We have suggested that institutions emerge because of societies' important needs. In the case of religion, these needs involve renewing hope, giving meaning to life, explaining the unknown, and dealing with death. Religion can be involved in other functions such as socialization, social control, and the promotion of solidarity and unity. Religion also plays an important role in social conflict: Marx felt that religion distracted people from their real problems, making it less likely that these problems would ever be corrected.

Religious groups sometimes battle each other or support one social class or ethnic or racial group to the disadvantage of another.

After examining the functions of religion, we dealt with the diversity of religious groups by suggesting several ways to distinguish among various types of religious organizations. Finally, we suggested that religion in America is characterized by differential involvement in religious activities, much religious diversity, an apparent loss of functions, and some interesting trends on the topic of secularization.

Religious groups can be similar in their beliefs and ceremonies, or they can be vastly different. The dramatic contrast to familiar "mainstream" religions is shown in the two readings that follow. In the first reading, Michael Watterlond describes the Holy Ghost People of Appalachia, and in the second, Richard Lacayo tries to explain and place in perspective the behavior of the Heaven's Gate cult.

---

## TERMS FOR STUDY

| | |
|---|---|
| Apollonian ceremony (269) | Dionysian ceremony (269) |
| church (270) | religion (264) |
| cult (270) | sect (270) |
| denomination (270) | secularization (275) |

For a discussion of Research Question 1, see page 267.
For a discussion of Research Question 2, see page 272.

---

## INFOTRAC COLLEGE EDITION
### Search Word Summary

To learn more about the topics from this chapter, you can use the following words to conduct an electronic search on InfoTrac College Edition, an online library of journals. Here you will find a multitude of articles from various sources and perspectives:

**www.infotrac-college.com/wadsworth/access.html**

| | |
|---|---|
| Shinto | evangelists |
| Emile Durkheim | Protestantism |
| cults | Kibbutzim |
| abortion | |

## Reading 8.1

# THE HOLY GHOST PEOPLE

*Michael Watterlond*

*Writer Michael Watterlond's description of the Holy Ghost People is interesting on several counts. The practices of this religious group appear quite bizarre, something most of us haven't experienced. Comparing their activities with the formal ritual of the Catholic Church makes us wonder if both could possibly be part of the same social institution. And yet, if we study the Holy Ghost People carefully, we should be able to discern certain elements that this religious group has in common with the mainstream religions with which we are more familiar.*

For the Holy Ghost people of Appalachia, it is a fragile fabric that separates this world from the next.

They are serpent handlers.

They take up poisonous snakes in church—timber rattlers, copperheads, even cobras—and they drink strychnine. They handle fire, using coal oil torches or blow torches. And they speak "with new tongues." They lay hands on the sick, trusting that—God willing—the sick shall recover. And when necessary, they cast out devils in Jesus' name.

"We live in the world, but we are not of the world," they say repeatedly during their twice-weekly church services, as though this were the statement that most distinctly defines them from us.

That these people are of another plane is unquestionable. They exist wholly in the thin, almost dimensionless region of "hard doctrine," of "getting right with Jesus" and most importantly of what they offhandedly call "this thing" or "this."

"This" is the featureless article of language they use to encompass the fiery, dramatic, even deadly practices that set their religion, their lives, and their select social grouping apart from most of Christian culture.

They belong to a variety of independent, fundamentalist sects called loosely Jesus Only. And they belong specifically to churches that subscribe to a doctrine based on what they call "the signs" or "the signs following."

The basis for these doctrines is a stone-hard reading of the last few verses of the Gospel according to Mark and other passages in the New and Old Testaments as presented in the King James version of the Bible, an English translation published in 1611.

In the pertinent section of Mark, Jesus has already risen from the tomb and appeared to several characters, including his disciples. As he is about to ascend, he issues these final pronouncements:

Mark 16:17 "And these signs shall follow them that believe; In my name shall they cast out devils; they shall speak with new tongues;

Reprinted from the May issue of *Science 83*.

Mark 16:18 "they shall take up serpents; and if they drink any deadly thing, it shall not hurt them; they shall lay hands on the sick, and they shall recover."

The Book of Daniel, in which Shadrach, Meshach, and Abednego are cast into the fiery furnace, is the basis for fire handling.

The fundamentalist serpent handlers take these passages as absolute. That these verses in Mark do not appear in the earliest extant texts is of no importance to them at all. Only the King James Bible counts. It acts as a major initiator of personal action, behavior, and decision making in their lives. It defines them.

Mary Lee Daugherty, formerly a professor of religion at the University of Charleston, estimates that there are now about 1,000 members of serpent-handling sects in West Virginia. She says numbers have dwindled recently because high unemployment has forced many members to migrate to urban areas. Since no organization links these congregations, a total member count would be guesswork.

Though there are local laws against handling dangerous animals in public in many states where serpent-handling churches exist, the laws aren't always strictly enforced. There are congregations in most areas of Appalachia, stretching from West Virginia south through Kentucky, Tennessee, the Carolinas, and into Georgia. Migrations from hill country into some of the urban areas of the Midwest have led to serpent-handling churches in nonrural areas such as Columbus, Cleveland, Flint, Indianapolis, and Detroit. The Full Gospel Jesus Church of Columbus, under the leadership of Willie Sizemore, has purchased and renovated a building in one of the city's industrial areas.

George Hensley, who began the practice of snake handling in rural Tennessee around 1909, could not have predicted its spread to urban environments. Hensley, a Holiness circuit preacher, died of a snake bite near Atha, Florida, in 1955. He is not revered by these people and is not considered to have been a prophet of any sort. His death, however, does illustrate a point one frequently hears in Jesus Only congregations: The Bible says to take up serpents; it doesn't say they won't bite.

"Make sure it's God," a deacon warns. He holds the microphone close to his lips like an entertainer. His hefty coal miner's shoulders straighten as he whips the mike cord away from his feet. "Only God can do it. Not me. Not you."

The loud, mechanical sounding buzz from the white pine serpent box makes his point unmistakable. The box this afternoon contains four timber rattlers and a rosy, velvet-textured copperhead.

"Death is in there," he says. "Don't go in the box unless it's with you."

By "it," the deacon means "the anointing"—the protection and spiritual direction of God that is manifest in what believers report as physical and emotional sensations.

"It's different for everyone," according to Sizemore. "Some people get a cold feeling in their hands or in their stomachs. Some don't."

Investigators in the past have reported that members say the anointing has a different and, for them, recognizably distinct sensation for each sign. The anointing to handle serpents may appear as a tingling, chilled feeling in the hands. An anointing to drink strychnine may appear as a trembling in the gut. The reports vary just as the sensations vary.

The deacon's warning about acting on the signs only if the anointing is present is a typical part of services, and the saints, as the members call each other, enthusiastically applaud. As he turns from the pulpit, the underlying rhythmic drum beat that has throbbed subliminally in the background becomes distinct and powerful. It is joined by electric organ, electric guitar, tambourine, and a room full of the clapping hands, stomping feet, and hearty, frantic voices of 40 saints.

As the service gets underway, the saints stand, clapping with the music, and walk slowly toward the open area between the pulpit and the first row of pews. The movement forward and together is called a "press." There is some feeling that the spiritual power of the group is concentrated by this gathering together. Should a member be bitten by a snake or "get down on strychnine," the gathering of saints around the victim is also called a press. Members refer to this communal praying and support of the stricken person as a "good press."

The services themselves have an informal structure that begins with loud, insistent singing. Most songs are belted out in four-four time, and each lasts as long as 20 or 30 minutes. After a few songs there will be requests for prayers or healings and testament. The prayer requests may be for "sinner children" or for members of the church who are sick or injured. There will be more songs before the sermon. Services last from two to five hours depending on the time of year and work schedules in the mines.

The signs may show themselves at any time during the service but are most apt to be manifest in association with the driving rhythms of the music. The enthusiasm may verge on violence. One woman slumps to the floor, "slain in the spirit," they say. As she trembles in front of the pulpit, other saints close in on her screaming, "JESUS! JESUS! JESUS!" in her ears.

"I saw one man get up and run around the inside of the church more than 50 times," says sociologist Michael Carter of Warner Southern College in Lake Wales, Florida, an experienced observer of the sect.

Members wail and shake and lapse into the unintelligible, ecstatic "new tongues" of glossolalia. Each member has his or her own style of speaking. As one saint sails into a new oration of tongues, another clamps his hands to her head and speaks in tongues himself; the ecstasy spreads like contagion. "He's translating," Carter observes.

There are two camps of psychological explanation and description of such services. Some researchers theorize that the wild activities are attempts to transcend reality, while others believe the point is self-actualization. Possibly both are right. It takes only a brief conversation with participants to learn that their daily life is one ongoing seance. They see God's movement and directives in every action, object, or thought.

Jerking as though his head were being battered by some unseen opponent, one brother moves across the floor toward the canister of coal oil next to the pulpit. He lights it quickly and thrusts his hands into the fire as though he were washing himself in flame. Next to him a young woman has been dervishlike for several minutes. She continues spinning under the tent of her long brown hair for nearly half an hour. In the back, the children play tic-tac-toe, practice spelling exercises, or sleep, unconnected to the proceedings.

Investigation into psychological and sociological aspects of fire and serpent handling has been conducted sporadically since the 1940s. A paper presented last year by Carter and sociologist Kenneth Ambrose of Marshall University in

Huntington, West Virginia, at the Fifth Annual Appalachian Studies Conference, attempted to determine the satisfaction church members derive from such activities. They interviewed members of an urban congregation to compare the rewards of church versus non-church activity.

Results of the study indicate that taking part in the signs gives these people personal reward equaled in no other aspect of their lives. Participation in "signs of the spirit" were statistically evaluated by Ambrose and Carter to reveal the highest level of satisfaction for members when they spoke in tongues. Handling serpents and handling fire were also rated highly, while drinking strychnine—clearly the most deadly sign—provided more shallow levels of satisfaction.

"When you drink strychnine," members say, "you're already bit"—your fate has been decided. Strychnine, commonly a white or colorless powder, causes a warm feeling in the gut, tingling sensations, and muscle spasms. In poisonous doses—15 to 30 milligrams for a human—the spasms can be severe enough, as the serpent handlers say, to "snap muscles right off the bone" and stop the heart. However, it is possible to ingest a considerable amount of strychnine and not exhibit the symptoms of poisoning. The drug does not accumulate in the body because it is rapidly oxidized in the liver and excreted in urine. The likelihood of developing a tolerance to it is remote.

While members of serpent handling churches—as well as most Jesus Only churches—forgo worldly diversions such as films, television, and politics, they do hold down jobs. Sizemore, for example, is a factory worker. But they consider mingling with the world on the job part of their earthly burden.

Since most members have had to face stern criticism and even ostracism by friends and relatives—including husbands and wives at times—they are even more strongly pushed into this cluster of supportive friends at church, where they speak of themselves as the chosen people. The social and psychological bonds are reinforced more or less constantly by hugging, touching, kissing. This creates a rich, meaningful world for them—a sense of being special.

Also, the pastor rides herd spiritually on the congregation. During service he will approach a member he suspects (or "discerns," as they would say) is backsliding, or going to other churches. He will preach to that member eye-to-eye, only the microphone between them, and talk in generalities about worldliness, sin, or traipsing around.

"I don't hold with people going from one church to another," says Brother Sizemore.

It is sometimes a strenuous task for the outsider to pry himself loose from all this music and high spirits and remind himself that these people take poisonous reptiles out of the countryside, bring them into their churches and drape them over their bodies, wear them in bundles on their heads. While the churchgoers are enthusiastic and emotional and committed, they do not appear to be disturbed and certainly not suicidal.

"I'm just as afraid of serpents as anybody," one 22-year-old West Virginia man says. He has attended these churches since childhood and has left the church his mother attends in recent years because that congregation has stopped handling serpents. "I'm afraid when I am in the flesh," he explains, "but when it is the spirit, there's nothing. I'm just not afraid."

He pulls up his sleeve to illustrate the critical point he wants to make: that these serpents are real and that their venom is real.

"It bit me here," he says, pointing to his wrist. "And it swelled up so much that the skin just pulled apart up here."

"When Richard died," one member says, "his whole arm split open from the shoulder down to the elbow."

Richard was Richard Williams. He died in 1974 after being viciously bitten by a huge eastern diamondback. Like most members who suffer bites, Williams refused medical aid and waited for fate to reveal itself.

The people of this church still talk about Richard Williams' anointing and about the serpent-handling feats he performed. Pictures on the walls of the church show him with his face in a mound of snakes; lying with his head resting on them like pillows, stuffing them into his shirt next to his skin. In a photo of Williams and the snake that killed him, it appears as if he is holding the felled limb of a tree rather than a snake.

"What really killed him was that the serpent got his vein when it hit the second time," one member says.

"You just can't tell," Willie Sizemore says, "you just have to make sure you have the anointing."

The folk myth that individuals who suffer repeated snakebites develop an immunity to the venom is viewed skeptically by Sherman A. Minton, Indiana University School of Medicine microbiologist and toxicologist. Minton says that such immunity is developed rarely and only when regular, gradually increased doses of the venom are administered. He points out that many people have allergies to venom, making successive bites more painful and causing greater swelling, asthma, and other symptoms.

As one snake expert puts it, however, the chances of dying by snakebite in the United States are comparable to the probability of being struck by lightning. One 10-year study showed that 8,000 venomous bites occur each year resulting in an average of only 14 deaths.

There are many theories about why snakes strike the saints infrequently. It is possible that given many warm-bodied targets pressing closely around it, a snake lapses into a sort of negative panic, a hysteria that makes it unable to single out one target.

Some observers have reported that church members' hands feel cold to the touch after handling fire or snakes. "I have felt their hands after serpent handling or fire handling," says Ambrose, who has observed these services for 15 years. "Their hands are definitely cold, even after handling fire." This would correspond with research in trance states involved in other religious cultures. It would also account for the vagueness of memory, almost sensory amnesia, that researchers have reported in serpent handlers as well as fire handlers. In his doctoral dissertation at Princeton University in 1974, Steven Kane reported that in North Carolina serpent handling congregations, young women were known to embrace hot stovepipes without injury or memory of the event. It has also been suggested by some observers that cold hands on the body of the snake would camouflage the touch and prevent it from feeling the handler.

Retired sociologist Nathan Gerrard, formerly of the University of Charleston, observed serpent handlers for seven years in the 1960s. By administering portions of a psychological test to measure deviate personalities, Gerrard concluded that serpent handlers had healthier attitudes about death, suffered less from pessimistic hypochondria, and generally seemed better adjusted in certain ways than "conventional" churchgoers he used as a control group.

However, one of Gerrard's conclusions at the time was also that serpent handling represented a sop to desperation. He called them the "stationary working class" and attributed their stern doctrine to highland ignorance and poverty—a fatalistic creed that offered death and salvation as the only way out of West Virginia. Since that time, however, Jesus Only churches have spread out of Appalachia as economic hardships pushed followers into urban centers perched on the edge of the Midwest.

Also, the members of the Full Gospel Jesus Church of Micco, a creek-bank town about 10 miles from Logan, West Virginia, are distinctly not the archetypal mountain folks. They come to church in well-polished, late-model automobiles and dress like middle-class people in most parts of the country.

While it may be difficult for outsiders to say that it is simply common faith that keeps these groups together, that would be the first thing they would say themselves. And this is a self-definition that has been immensely important to Western civilization.

"You can clearly see early Christianity as having very strong sectarian overtones," says Robert Bellah, a sociologist at the University of California, Berkeley. "Without that at the beginning, there would not have been any Christian church."

"Sect religions," Bellah says, "are most apt to occur in relatively low-status groups, relatively low-educated groups where a combination of intense religious experience and rather high group discipline create a kind of separate world in which the people in the sect live.

"They largely reject the surrounding culture," Bellah explains, "rejecting many of the prevailing cultural forms. In other words, the whole round of life tends to be bound up in the sect itself. Social contact is limited by the sect, and the meaning of most things that one does derives from the sect."

Bellah makes the distinction that churches involve a structure which attempts to encompass the whole of society. They include a range of social classes and do not oppose the dominant social power but view themselves as having influence on how that power is exercised.

"The church accepts the culture while working within it," Bellah says. "A sect is exclusive."

It is the current public concern about total-commitment sects that has many people wondering about what have come to be called cults these days.

"'Cult' is really not a sociological category," Bellah explains. "It is a pejorative, popular term that we use for groups that are unfamiliar to our culture." Generally, the word is used to describe groups that are non-American in origin or are aberrantly individualistic—"in that they have been created by some 'kook,'" Bellah says. "Like Jim Jones."

"We are not a cult," Sizemore says forcefully. "If Jim Jones had been right with Jesus, those people wouldn't have died."

And still, outsiders want to know why anyone handles serpents.

"Because it is written," says Bishop Kelly Williams of the Jesus Only Church of Micco, West Virginia. "The main purpose in our doing this is to obey the Word of God." He points out that it really does not matter that the snakes symbolize Satan.

"Now, we don't think that everyone has to take up serpents," Williams says. "It doesn't say that all signs will follow all believers.

"You," he says to a visitor. "You might only speak in tongues. It depends on how God moves."

"Everyone knows that it is the nature of those serpents to bite you," he nods toward the buzzing pine box of copperheads and rattlers. "But you saw last night that God gave us a victory over those serpents. They were new serpents. They'd never been handled."

In fact, in a quiet evening before, Williams had gone to the box, lifted out a rattler and held it close to his body. He stared down at the serpent in his palms, smiled calmly for about three minutes, and set the snake back down in the box.

"You've seen that there is no one way to handle serpents," he says. "You been around enough to know that. There's no trick."

The styles of serpent handling are as various as the vocabulary of new tongues. On one Sunday afternoon, several men lift up the box, shake it hard and dump a mass of rattlers and copperheads onto the wooden floor without much caution. One man reaches into the tangle of scales and rattles and pulls up a timber rattler about four feet long. The other snakes, as well as a microphone cord, are tangled up with it, so he just shakes the whole mess until the excess snakes drop off. They coil stunned on the floor, as if paralyzed by the bad manners of it all.

The serpents are passed between hands. Loose copperheads wander around the box, uncertain of any particular route of escape. The heavy beat of gospel music picks up and the floor vibrates, the congregation claps and wails and sings.

In the back row of pews a girl about 11 turns back over her shoulder to talk to the visitor. She is playing hairdresser with another girl and holds an unfinished braid in her hands. She smiles halfheartedly and rolls her eyes heavenward as though her patience were limitless and unconditional.

"Boring," she says, "isn't it?"

---

# QUESTIONS 8.1

**1.** Which of the functions of religion discussed in this chapter are supplied by the religious practices of the Holy Ghost People?

**2.** What would Karl Marx say about the Holy Ghost People? How would he explain their behavior?

**3.** Analyze the Holy Ghost People in terms of religious organizations— which category are they and why?

**4.** It is obvious that the Holy Ghost People are different from most, but what *similarities* do you find between them and other religious groups?

## Reading 8.2

# THE LURE OF THE CULT

*Richard Lacayo*

*Religion has been defined by what it does (renews hope, provides answers for the unknown), and by what it is (a set of sacred beliefs and symbols). The power, ritual, and belief system of a religion is often difficult for an outsider to understand. This ability to comprehend another's belief system was sorely tested in March 1997 with the news of the actions of 39 members of Heaven's Gate. In this article, Richard Lacayo, who writes for* Time, *discusses Heaven's Gate and cult behavior in general.*

On Saturday, March 22, around the time that the disciples of Heaven's Gate were just beginning their quiet and meticulous self-extinction, a small cottage in the French Canadian village of St.-Casimir exploded into flames. Inside the burning house were five people, all disciples of the Order of the Solar Temple. Since 1994, 74 members of that group have gone to their death in Canada, Switzerland and France. In St.-Casimir the dead were Didier Quèze, 39, a baker, his wife Chantale Coupillot, 41, her mother and two others of the faithful. At the last minute the Quèze children, teenagers named Tom, Fanie and Julien, opted out. After taking sedatives offered by the adults, they closeted themselves in a garden shed to await their parents' death. Police later found them, stunned but alive.

For two days and nights before the blast, the grownups had pursued a remarkable will to die. Over and over they fiddled with three tanks of propane that were hooked to an electric burner and a timing device. As many as four times, they swallowed sedatives, then arranged themselves in a cross around a queen-size bed, only to rise in bleary frustration when the detonator fizzled. Finally, they blew themselves to kingdom come. For them that would be the star Sirius, in the constellation Canis Major, nine light-years from Quebec. According to the doctrines of the Solar Temple, they will reign there forever, weightless and serene.

Quite a mess. But no longer perhaps a complete surprise. Eighteen years after Jonestown, suicide cults have entered the category of horrors that no longer qualify as shocks. Like plane crashes and terrorist attacks, they course roughly for a while along the nervous system, then settle into that part of the brain reserved for bad but familiar news. As the bodies are tagged and the families contacted, we know what the experts will say before they say it. That in times of upheaval and uncertainty, people seek out leaders with power and charisma. That the established churches are too faint-hearted to satisfy the wilder kinds of spiritual hunger. That the self-denial and regimentation of cult life will soften up anyone for the kill.

The body count at Rancho Santa Fe is a reminder that this conventional wisdom falls short. These are the waning years of the 20th century, and out on the margins of spiritual life there's a strange phosphorescence. As predicted, the approach of the year 2000 is coaxing all the crazies out of the woodwork. They bring with them a twitchy hybrid of spirituality and pop obsession. Part Christian, part Asian mystic, part Gnostic, part *X-Files,* it mixes immemorial longings with the latest in trivial sentiments. When it all dissolves in overheated computer chat and harmless New Age vaporings, who cares? But sometimes it matters, for both the faithful and the people who care about them. Sometimes it makes death a consummation devoutly, all too devoutly, to be desired.

So the worst legacy of Heaven's Gate may yet be this: that 39 people sacrificed themselves to the new millennial kitsch. That's the cultural by-product in which spiritual yearnings are captured in New Age gibberish, then edged with the glamour of sci-fi and the consolations of a toddler's bedtime. In the Heaven's Gate cosmology, where talk about the end of the world alternates with tips for shrugging off your fleshly container, the cosmic and the lethal, the enraptured and the childish come together. Is it any surprise then that it led to an infantile apocalypse, one part applesauce, one part phenobarbital? Look at the Heaven's Gate Website. Even as it warns about the end of the world, you find a drawing of a space creature imagined through insipid pop dust-jacket conventions: acrodynamic cranium, big doe eyes, beatific smile. We have seen the Beast of the Apocalypse. It's Bambi in a tunic.

By now, psychologists have arrived at a wonderfully elastic profile of the people who attach themselves to these intellectual chain gangs: just about anybody. Applicants require only an unsatisfied spiritual longing, a condition apt to strike anyone at some point in life. Social status is no indicator of susceptibility and no defense against it. For instance, while many of the dead at Jonestown were poor, the Solar Temple favors the carriage trade. Its disciples have included the wife and son of the founder of the Vuarnet sunglass company. The Branch Davidians at Waco came from many walks of life. And at Rancho Santa Fe they were paragons of the entrepreneurial class, so well organized they died in shifts.

The U.S. was founded by religious dissenters. It remains to this day a nation where faith of whatever kind is a force to be reckoned with. But a free proliferation of raptures is upon us, with doctrines that mix the sacred and the tacky. The approach of the year 2000 has swelled the ranks of the fearful and credulous. On the Internet, cults multiply in service to Ashtar and Sananda, deities with names you could find at a perfume counter, or to extraterrestrials—the Zeta Reticuli, the Draconian Reptoids—who sound like softball teams at the *Star Wars* cantina. Carl Raschke, a cult specialist at the University of Denver, predicts "an explosion of bizarre and dangerous" cults. "Millennial fever will be on a lot of minds."

As so often in religious thinking, the sky figures importantly in the New Apocalypse. For centuries the stars have been where the meditations of religion, science and the occult all converged. Now enter Comet Hale-Bopp. In an otherwise orderly and predictable cosmos, where the movement of stars was charted confidently by Egyptians and Druids, the appearance of a comet, an astronomical oddity, has long been an opportunity for panic. When Halley's comet returned in 1910, an Oklahoma religious sect, the Select Followers, had to be stopped by the police from sacrificing a virgin. In the case of Hale-Bopp, for months the theory that it might be a shield for an approaching UFO has roiled the excitable on talk radio and in Internet chat rooms like—what else?—*alt.conspiracy.*

Astronomical charts may also have helped determine the timing of the Heaven's Gate suicides. They apparently began on the weekend of March 22–23, around the time that Hale-Bopp got ready to make its closest approach to Earth. That weekend also witnessed a full moon and, in parts of the U.S., a lunar eclipse. For good measure it included Palm Sunday, the beginning of the Christian Holy Week. Shrouds placed on the corpses were purple, the color of Passiontide, or, for New Agers, the color of those who have passed to a higher plane.

The Heaven's Gate philosophy added its astronomical trappings to a core of weirdly adulterated Christianity. Then came a whiff of Gnosticism, the old heresy that regarded the body as a burden from which the fretful soul longs to be freed. From the time of St. Paul, some elements of Christianity have indulged an impulse to subjugate the body. But like Judaism and Islam, it ultimately teaches reverence for life and rejects suicide as a shortcut to heaven.

The modern era of cultism dates to the 1970s, when the free inquiry of the previous decade led quite a few exhausted seekers into intellectual surrender. Out from the rubble of the countercultures came such groups as the Children of God and the Divine Light Mission, est and the Church of Scientology, the robotic political followers of Lyndon LaRouche and the Unification Church of the Rev. Sun Myung Moon. On Nov. 18, 1978, the cultism of the '70s arrived at its dark crescendo in Jonestown, Guyana, where more than 900 members of Jim Jones' Peoples Temple died at his order, most by suicide.

Since then two developments have fostered the spread of cultism. One is the end of communism. Whatever the disasters of Marxism, at least it provided an outlet for utopian longings. Now that universalist impulses have one less way to expend themselves, religious enthusiasms of whatever character take on a fresh appeal. And even Russia, with a rich tradition of fevered spirituality and the new upheavals of capitalism, is dealing with modern cults.

Imported sects like the Unification Church have seen an opening there. Homegrown groups have also sprung up. One surrounds a would-be messiah named Vissarion. With his flowing dark hair, wispy beard and a sing-song voice full of aphorisms, he has managed to attract about 5,000 followers to his City of the Sun. Naturally it's in Siberia, near the isolated town of Minusinsk. According to reports in the Russian press, Vissarion is a former traffic cop who was fired for drinking. In his public appearances, he speaks of "the coming end" and instructs believers that suicide is not a sin. Russian authorities are worried that he may urge his followers on a final binge. In the former Soviet lands, law enforcement has handled cults in the old Russian way, with truncheons and bars. Some have been banned. Last year a court in Kiev gave prison terms to leaders of the White Brotherhood, including its would-be messiah, Marina Tsvigun.

The second recent development in cultism is strictly free market and technological. For the quick recruitment of new congregations, the Internet is a magical opportunity. It's persuasive, far reaching and clandestine. And for better and worse, it frees the imagination from the everyday world. "I think that the online context can remove people from a proper understanding of reality and of the proper tests for truth," says Douglas Groothuis, a theologian and author of *The Soul in Cyberspace*. "How do you verify peoples' identity? How do you connect 'online' with real life?"

"The Internet allows different belief systems to meet and mate," adds Stephen O'Leary, author of *Arguing the Apocalypse,* which examines end-of-the-world religions. "What you get is this millennial stew, a mixture of many different belief systems." Which is the very way that the latest kinds of cultism have flourished. As it happens, that's also the way free thought develops generally. Real ideas sometimes rise from the muck, which is why free societies willingly put up with so much muck.

In Gustave Flaubert's story *A Simple Heart,* an old French woman pines for a beloved nephew, a sailor who has disappeared in Cuba. Later she acquires a parrot. Because it comes from the Americas, it reminds her of him. When the parrot dies, she has it stuffed and set in her room among her items of religious veneration. On her deathbed, she has a vision of heaven. The clouds part to reveal an enormous parrot.

The lessons there for Heaven's Gate? The religious impulse sometimes thrives on false sentiment, emotional need and cultural fluff. In its search for meaning, the mind is apt to go down some wrong paths and to mistake its own reflection for the face of God. Much of the time, those errors are nothing more than episodes of the human comedy. Occasionally they become something worse. This is what happened at Rancho Santa Fe, where foolish notions hardened into fatal certainties. In the arrival of Comet Hale-Bopp, the cult members saw a signal that their lives would end soon. There are many things about which they were badly mistaken. But on that one intuition, they made sure they were tragically correct.

---

# QUESTIONS 8.2

**1.** The author of this article cites two developments that help explain the spread of cults. What are these? Is he right? Analyze and critique.

**2.** Analyze Heaven's Gate and cults in general from the functionalist approach and from the conflict approach.

**3.** What is the relationship (if any) between social class and the emergence of cults?

**4.** From information provided in this chapter, show how Heaven's Gate fits the definition of a cult. Could it be a sect?

---

# NOTES

1. From "An Accident" by G. K. Chesterton, in *Tremendous Trifles* (New York: Sheed and Ward, 1955), pp. 29–33.

2. Andrew Greeley, *Religion: A Secular Theory* (New York: Free Press, 1982), chapter 1.

3. Much of our discussion in this section is drawn from Charles Glock and Rodney

Stark, *Religion and Society in Tension* (Chicago: Rand McNally, 1965). This definition of religion is close to that of Glock and Stark. In their book, Glock and Stark trace the background of their definition, which includes the works of Durkheim, Yinger, Parsons, Nottingham, and Williams.

4. See "Religious Community, Individual Religiosity, and Health: A Tale of Two Kibbutzim," by Ofra Anson, Arieh Levenson, Benyamin Maoz, and Dan Bonneh, *Sociology* 25 (February 1991), pp. 119–132.

5. This project was researched by the National Opinion Research Center in Chicago, directed by Andrew Greeley. See Andrew Greeley, William McCready, and Kathleen McCourt, *Catholic Schools in a Declining Church* (Mission, Kan.: Sheed and Ward, 1976).

6. This comes from Andrew Greeley's book *The Denominational Society* (Glenview, Ill.: Scott, Foresman, 1972), chapter 2.

7. The categories and quotations discussed in this paragraph and in Table 8.3 come from Andrew Greeley's book *The Denominational Society*, pp. 23–24 (see note 6).

8. These paragraphs on types of religious organizations make use of material from Glock and Stark, *Religion and Society in Tension*, chapter 13 (see note 3), and from J. Milton Yinger's book *Religion, Society and the Individual* (New York: Macmillan, 1957), chapter 6, especially pp. 142–155.

9. George Gallup conducts a yearly poll on church attendance.

10. See "Durkheim, Suicide, and Religion: Toward a Network Theory of Suicide," by Bernice Pescosolido and Sharon Georgianna, *American Sociological Review* 54 (February 1989), pp. 33–48.

11. Some of the statistics in this and the following paragraph come from Gallup's study of religion in America in 1983. Also see Will Herberg's interesting book, *Protestant-Catholic-Jew* (Garden City, N.Y.: Doubleday, Anchor, 1960), chapter 1. Also see Glock and Stark, *Religion and Society in Tension*, chapter 4 (see note 3).

12. Herberg, *Protestant-Catholic-Jew*, chapter 11, especially p. 268 (see note 11).

## CHAPTER NINE

AP/Wide World Photos

# Institutions:
# Economic and Political

**Economic Institutions**

*Capitalism • Socialism • Economic Change and the American Economy*

**Political Institutions**

*Distribution of Power • Political Systems and American Politics*

**The Political Economy: Separate Institutions?**

## TWO RESEARCH QUESTIONS

**1.** *Do large U.S. companies (like IBM, Exxon, Ford, and General Motors) make more of their money in the United States or abroad?*

**2.** *From what social class come the people who run the country, who have the highest positions in government? Are they selected from all walks of life, from the middle class predominantly, or from some other group?*

IN CHAPTER SEVEN we defined social institution as the norms, roles, and patterns of organization that develop around societies' central needs or problems. In Chapters Seven and Eight, family and religious institutions were examined. In this chapter we turn to political and economic institutions. *powers do like change*

Institutions are interrelated, with change in one inevitably affecting another. The relationship between political and economic institutions is perhaps closer than most. Looking at their central functions might help us understand this better. Economic institutions arise from societies' efforts to produce and distribute goods and services. Political institutions refer to the social relationships surrounding the assignment and use of power. And the connection between the two? Goods and services are limited and unequally distributed in society. The distribution of goods and services is determined by those with the power to make such decisions. Who has the power to make such decisions? Probably those who benefit most from the existing distribution of goods and services. So we see that the political distribution of power and the economic distribution of goods and services are closely related. Remember Murphy's Golden Rule: Whoever Has the Gold Makes the Rules.

## ECONOMIC INSTITUTIONS

Economic institutions arise from societies' efforts to produce and distribute goods and services. Important issues surface: What is the meaning and value of work? How does a society deal with the situation in which only limited supplies of goods are available and demands for those goods by citizens are unlimited? Is unequal distribution of goods and wealth in society beneficial or detrimental? Several quite contrasting views exist regarding the most appropriate economic system to deal with such issues.

### Capitalism

Capitalist systems cherish the ideas of private property, the profit motive, and free enterprise. *Private property* means that property, especially that used to produce goods and services, is held by private individuals, not by the state. *Profit motive* means that individuals are motivated by the idea of making a profit; if they produce a commodity that people want and that is a better product at a lower price than their competition's, they will capture the market. *Free competition* means that each private entrepreneur is free to compete with others in a marketplace unencumbered by any rules and regulations other than fair play. The market and the economy are self-correcting: Buyers make free choices among goods and services; bad products aren't purchased; workers paid too little go elsewhere or do poor work, whereas those paid too much cause goods to be priced too high to be sold. Good businesses succeed, bad ones fail. Build a better mousetrap at a good price, and the world will beat a path to your door.

However, not many—if any—capitalist systems really work this way. Take the United States, for example. Private property? Some property is publicly held, and many privately held businesses are owned by many scattered stockholders who have little say in ongoing activities. Further, the government has a great, although indirect, influence; many businesses are dependent on the government aid they receive through subsidies and tax breaks. Profit motive? The profit motive is certainly strong. However, many unprofitable but necessary businesses are supported by the government through tax dollars and are not allowed to fail. Free enterprise? Laws regulate prices, minimum wages, and fairness in hiring. Government regulations limit or control entry to occupations, and buyers' "free" choices are limited by what they can afford. Domination of whole segments of the economy by one or two large corporations allows them to "create" markets.

### Socialism

In the 1850s, Karl Marx saw much about capitalism that he didn't like. Workers were exploited, and capitalism seemed to benefit only the capitalists, not the rest of society. The apparent evils of capitalism led Marx to suggest a socialist system in which property is collectively, not privately, owned. Goods and services are produced according to people's needs, regardless of profitability. Centralized planning determines what is produced and how much. Wages and benefits should be equally distributed, and there should be no accumulation of private wealth. As with capitalism, a pure socialist system is difficult to find. In the former Soviet Union, for example, there was some private property and some private free enterprise, and income differences between the managerial level and workers created unequal distribution of wealth. Actually, most economic systems represent a mix of the extremes of capitalism and socialism. In many societies you find private property and the profit motive, government regulation of the economy, and government ownership of key industries such as mass media, transportation, and health care.

### Economic Change and the American Economy

The evolution of economic systems has been remarkable. In early societies, people spent all of their time finding food. There was little specialization or division of labor—everybody did most everything. Specialization gradually increased as societies became larger and more complex. By the time of the Industrial Revolution in the 19th century, the division of labor was extensive; workers would perform one small task continually for the whole of their workday. One of Marx's concerns, however, was the alienation, the meaninglessness, and the powerlessness a worker must feel when he or she is a powerless cog with no meaningful input in the final product. Work is a major and important part of life, but if one has no sense of accomplishment—if there is no meaning or creativity in work—then it can end up being unsatisfying and demoralizing. Some businesses have approached this problem by allowing workers to change jobs or follow the product through from start to finish, by humanizing the workplace, and by giving workers greater voice in management.

The increasing specialization and division of labor caused businesses to become bigger and bigger. Does this also mean better? Specialization and size did mean that tasks earlier unimagined could now be performed. Can you imagine, for example, primitive hunters and gatherers reaching the moon? Or even wanting to? But it has also led to some criticisms and problems (as we discussed in Chapter Four in the section on large organizations). One of the dominant features of the American economy in the mid- and late-20th century was bigger and bigger businesses and corporations that bought other corporations and became conglomerates. In the late 1990s into 2000, these mergers became more extensive than ever before. In March 2000, Deutsche Bank combined with Dresdner Bank to form a company with assets of more than $1.2 trillion. In January 2000, AOL and Time Warner merged and the market value of the new company was estimated at over $290 billion. When Citicorps merged with Travelers Group, the new company (Citigroup) became the seventh largest company in the United States with assets of close to $670 billion (see Table 9.1). One or two of these large corporations can influence or dominate large segments of the economy. The dominance of big organizations such as IBM, AT&T, and General Motors can be illustrated in the following ways: First, they employ many individuals and have great wealth. Table 9.1 profiles 10 of the largest United States companies in 1998. These companies are ranked by their overall sales. The 500 largest industrial corporations employed approximately 20 million people in 1998 and had sales of $5.52 trillion. (Remember when you look at the figures in Tables 9.1 and 9.2 that those numbers are in billions of dollars!) There is clearly a tremendous concentration of wealth at the top.

Second, many of the large corporations are called **multinationals** because they go beyond national boundaries and have connections throughout the world. Many are truly huge in scope. Table 9.2 shows the 10 largest (based on foreign sales) United States–based multinationals in 1998. The total revenues in 1998 from just these 10 companies were almost $830 billion. Oil, car, and computer companies are prominent on the list. Note that the multinationals are huge—Citigroup's total assets, for example, are more than $600 billion. Notice also that many of the multinationals make a substantial part of their profit (especially when compared with

**TABLE 9.1**   The Ten Largest United States Companies, 1998

| Company | Sales (billions) | Profits (billions) | Assets (billions) | Employees |
|---|---|---|---|---|
| General Motors | $161 | $3 | $257 | 594,000 |
| Ford | 144 | 22 | 238 | 345,000 |
| Wal-Mart | 139 | 4 | 49 | 910,000 |
| Exxon | 101 | 6 | 93 | 79,000 |
| General Electric | 100 | 9 | 356 | 293,000 |
| IBM | 82 | 6 | 86 | 291,000 |
| Citigroup | 76 | 6 | 669 | 170,000 |
| Philip Morris | 58 | 5 | 60 | 144,000 |
| Boeing | 56 | 1 | 37 | 227,000 |
| AT&T | 54 | 6 | 60 | 108,000 |

Adapted from *Fortune*, April 26, 1999.

**TABLE 9.2**   The Ten Largest United States–Based Multinationals, 1998

| Company | Foreign Sales (billions) | Foreign Sales as % of Total | Total Assets (billions) | Foreign Assets as % of Total |
|---|---|---|---|---|
| Exxon | $81 | 80% | $65 | 62% |
| IBM | 46 | 57 | 35 | 47 |
| Ford | 44 | 30 | 44 | 42 |
| General Motors | 41 | 31 | 33 | 41 |
| Texaco | 31 | 79 | 15 | 58 |
| General Electric | 31 | 31 | 356 | 36 |
| Mobil | 28 | 59 | 25 | 65 |
| Citigroup | 26 | 34 | 619 | 46 |
| Hewlett-Packard | 26 | 54 | 34 | 59 |
| Philip Morris | 20 | 34 | 22 | 29 |

Adapted from *Forbes*, July 26, 1999.

assets) *outside* the United States. This relates to the first research question asked at the beginning of the chapter. Oil companies like Exxon and Mobil make more money abroad than at home, as does computer giant IBM. Car companies like Ford and General Motors face much greater competition from companies in other countries, and their foreign sales are a smaller percentage of their total sales.

The behavior of the multinational corporations raises interesting issues. Their wealth is often greater than that of the countries in which they do business. Multinationals might encourage development in underdeveloped countries, but whether their efforts are welcome or not is much debated. They have been charged with meddling in local politics to encourage a friendly climate for their investments. They often bring their own employees rather than using native workers, which is of little help to the local employment situation. They often produce products for their own economy that might be of little use to the local society.

## POLITICAL INSTITUTIONS

The social relationships surrounding the assignment and use of power is the central issue in the study of political institutions. **Power** refers to the ability of one party, either an individual or a group, to affect the behavior of another party. (We discussed power in Chapter Five as one of three variables, along with wealth and prestige, that determine one's socioeconomic status or rank in society.) Here we find that power is the main force in political institutions. Power is probably a factor in all social situations: The doctor has power over the patient, the teacher over the student, the parent over the child; and even in dating situations, one party often has superior bargaining power. In this chapter, however, we focus on the power that has a more general impact—power that influences actions and decisions at the societal level.

There are three major types of power: *authority, influence,* and *resource control.*[1] These types of power generally overlap; having one type means you'll probably have another type as well. **Authority** refers to socially approved power. It has legitimacy, meaning it is socially acceptable. People believe it is right and appropriate to drive their cars on the right-hand side of the road and to stop at stop lights, and they would probably do so even if no one were around. Authority is often impersonal in the sense that it is linked to position rather than to the character of the occupant. Students pay attention in class not because of the instructor's fantastic personality, but because the "teacher" position holds a particular authority. We said that authority was viewed as legitimate—people are socialized to obey the rules. At base, however, authority rests on the threat of force and coercion. The relationship is complicated: Does one go along because one wants to, or because of fear of the consequences? Would you stop at the stoplight if no one were looking?

Authority comes in different shapes, and Max Weber has identified three specific types: *traditional, legal-rational,* and *charismatic.* **Traditional authority** means that power is granted according to custom. Leaders are selected by inheritance, and people go along because "it has always been done this way." Hereditary monarchies and tribal chieftainships are examples. The leaders appear to have much power, laws tend to be understood but unwritten, and there is a sacred quality to the whole operation. Historically this has been a common type of authority but is increasingly difficult to find today. **Legal-rational authority** is more typical of modern societies. Leaders are selected to fill positions, the boundaries of which are carefully designated by a formal set of rules and procedures. The *position* has the power rather than the person occupying it. If one goes beyond the boundaries, something bad can result. The example of Watergate and Richard Nixon is typically used to show how the legal-rational system is supposed to work; he went beyond the legitimate boundaries and was forced to leave office. Legal-rational authority is typical not only of governments but of businesses and other types of organizations as well.

**Charismatic authority** is based on the unique personality of a particular person. This person—Mao, Gandhi, Martin Luther King, Jr., or Castro, for example—has authority over others based on his or her personal appeal.

Charismatic authority is very unstable; it tends to emerge out of crisis and is tied to a particular person. **Pseudocharisma** is a recent addition to Weber's idea.[2] The term refers to public figures who appear to have charisma but who actually have a packaged and carefully created image manufactured by public relations techniques. Making politicians into TV celebrities is a clear example of the development of pseudocharisma.

**Influence** refers to subtle, informal, indirect power based on persuasion rather than coercion and force. Lines of influence are very widespread and less easily recognized than those of authority. Parents influence behavior through a look or tone of voice, or because their children should "know how they would feel." Admirers of rock stars and other entertainment personalities can be influenced to ingest particular restricted substances or wear wild and peculiar garments. Friends exercise influence, and so do the mass media. Holding an important position like "member of the board of trustees" or "chair" naturally enhances one's influence. It follows that authority and influence often go together, but occasionally their boundaries differ. A person can have narrow authority but much influence, or vice versa.

Power as **resource control** refers not to things people do, but to their possession or control of valued items. In this case, power often represents a capacity, potential, or threat to act rather than actual decision-making behavior. If you possessed great wealth, control of a major energy source, or ownership of a string of radio and TV stations, you would influence others' behavior whether you intended to or not. Think of the amazing power shift in the direction of the oil-rich Middle East in recent years.

### Distribution of Power

Power is stratified in society; it is distributed unequally. Those few at the top who have it are called elites; those without it are called the masses. But beyond this, how is power distributed? What type of people have it? What is the relationship between elites and the masses? Three theories provide differing answers to these questions, and as with most competing theories, there is evidence to support and refute each of them. These three approaches are called the **pluralistic model,** the **elite model,** and the **class model.**

**The Pluralistic Model**     Pluralists believe that power is diffuse, spread across many diverse interest groups. An individual's interests are protected because everyone is potentially represented by one or more interest groups. Continuous competition and conflict for power among these interest groups lead to a balance—an equilibrium—and thus no radical or rapid social change takes place. The groups resolve differences by bargaining and compromise, and because they are specialized, no single interest group has absolute authority. Some have political power, some have economic power, and some have military or religious power. Political parties, voluntary associations such as the PTA and the League of Women Voters, all levels of government, and finally even the voters—all exercise power.

This traditional pluralist approach, which sees the masses producing interest groups that deal with the state, has been modified in a modern version called elite-pluralism. **Elite-pluralism** recognizes that mass political participation is impossible in modern complex societies and that small elite groups do develop and make major decisions. It is suggested, however, that these elites have competing interests and tend to balance or neutralize each other so that no one elite group gains dominance.

Does the model fit the facts of modern society? Critics have some doubts. They suggest that most people are not represented by interest groups, and those who are don't have much voice unless they are part of the leadership. Interest groups that do exist tend to be dominated by business and monied interests. Finally, the balance of power suggested by the pluralists just doesn't exist, argue the critics; the power differences among various interest groups are vast.

**The Elite Model**　　The elite model, which is sharply different from the pluralist model, sees relatively unrestrained power in the hands of a few who rule the masses, who, in turn, are apathetic and incapable of self-rule—incapable because they are poorly trained, uninformed, and unorganized. Of the early elite theorists, Vilfredo Pareto viewed elites as the highest achievers in society; they govern through physical coercion (the "lions") and through cunning and intellectual persuasion (the "foxes"). Gaetano Mosca felt that elites gain and keep power because they are organized (whereas the masses are not), they have particular personal attributes (strength of personality or intelligence), and they control valued resources or "social forces." Robert Michels viewed rule by a few as an inevitable consequence of large complex organizations. Because a large number of people can't make decisions efficiently, a few are given the power to make decisions on behalf of the whole, and a leadership group develops. Once in power, the elite spends more and more of its time and energy maintaining its power rather than concentrating on other organizational activities. For example, think of how much of a politician's time (from the president on down) is spent on activities designed to keep himself or herself in office.

Modern elite theorists see the masses as manipulated and exploited by an elite motivated by self-interest. Whereas the earlier theorists felt the development of an elite to be inevitable, modern theorists think that such a condition can and should be avoided. Theorists today see the elite as a cohesive group of people bolstered by their control of key resources, including wealth, government authority, and communication facilities. In sociologist C. Wright Mills's view, the top layer of power in society is in the hands of "the warlords, the corporation chieftains, and the political directorate," who work together to form a power elite in America.[3] According to Mills, decisions are made in this country by a few people, and they govern a fragmented mass of people who have no power. Between the two is a group whose power is semiorganized and stagnant. The middle is there, but it neither represents the masses nor has any real effect on the elite. The elite is highly organized and made up of military leaders (admirals and generals), politicians, and

## C. WRIGHT MILLS (1916–1962)

C. Wright Mills earned his Ph.D. in Sociology at the University of Wisconsin. Mills's sociological importance was his almost single-handed effort to keep Marxian ideas alive in sociological theory.

Mills seemed to be constantly at war in both his social and personal life. As a graduate student at Wisconsin, he challenged a number of his professors. In one of his early essays, he engaged in a thinly disguised critique of the ex-chairman of the Wisconsin sociology department and even called the senior theorist at Wisconsin, Howard Becker, "a real fool." Mills's radicalism placed him on the periphery of American sociology, and he received much criticism.

He critiqued sociology, which resulted in a book, *The Sociological Imagination.* The sociological imagination referred to the capacity to shift from one perspective to another and, in the process, to build the most appropriate view of society and its components.

Mills was at odds with people; he was also at odds and seemingly disappointed with American society. This can be seen at the end of his book *The Power Elite,* where he says, "The men of the higher circles are not representative men; their high position is not a result of moral virtue; their fabulous success is not firmly connected with meritorious ability. Commanders of power unequaled in human history, they have succeeded within the American system of organized irresponsibility." Mills was a professor of sociology at Columbia University.

In addition to his ideas on elites mentioned in this chapter, we return to Mills's "sociological imagination" in Chapter Fifteen.

---

business leaders. Mills suggests that a system in which so much power is held by a few who are not responsible to anyone but themselves is both immoral and irresponsible.

Does the model fit the facts? Elites unquestionably exist, and the forces in large groups and organizations work much as Michels predicted. But critics raise specific questions: Must elites inevitably emerge? Are elites cohesive, and do they cooperate to the extent suggested by Mills and others? Are elites only self-serving or do they work in the public interest? And are the masses as totally powerless as the elite model predicts?

**The Class Model**   The class and elite models are similar in that they see power concentrated in the hands of a few. They differ in their description of the elite group and its origins. The class model brings us back again to the ideas of Karl Marx. Marx starts with the idea that economic institutions are the key to society; all else stems from such economic activities as finding food, shelter, and the other necessities of life. Classes emerge based on the means of production, and in industrialized societies, two classes will dominate: the bourgeoisie, or capitalists, who own the means of production, and the proletariat, who own only their own labor. Classes become political groups when class members realize that they need to gain and use power to protect their common interests. Continual domination and exploitation of the proletariat by the bourgeoisie means continuing class conflict. Because economic matters are so important, economic domination is the key to power and control in other noneconomic elements in society.

Contemporary class theorists grant that the workers' revolution hasn't happened and that the classes are not as polarized as Marx predicted. In fact, the classes have become increasingly complex and diverse; in many societies the growing middle class has become a dominant feature. Despite these deviations from Marx's predictions, contemporary class theorists say that the current corporate elite has the power of a ruling class.

Answering the second research question posed at the beginning of this chapter—Who runs the country?—is no problem for William Domhoff. In his book *Who Rules America?*, Domhoff describes a governing class that "owns a disproportionate amount of a country's wealth, receives a disproportionate amount of a country's yearly income, and contributes a disproportionate number of its members to the controlling institutions and key decision-making groups of the country." Excessive wealth and income are the key; having these elements, which other classes do not, allows the upper class to *control* other individuals and institutions. This upper-class elite has well-established ways of training and preparing new members for future service. Bright young people are processed into the upper class by means of "education at private schools, elite universities, and elite law schools; through success as a corporate executive; through membership in exclusive gentlemen's clubs; and through participation in exclusive charities."[4]

Domhoff holds that the leaders of the executive branch of the federal government are either members of the upper class or former employees of institutions controlled by members of the upper class. He cites an impressive array of evidence that describes who has the wealth and how it is used to control power. Domhoff also traces the occupants of various government positions and finds, for example, that members of the upper-class power elite dominate the president's cabinet, especially in the departments of State, Treasury, Defense, Commerce, and even Labor. Of the thirteen men who were Secretary of Defense or Secretary of War between 1932 and 1964, eight have been listed in the *Social Register,* and six of the eight people who were Secretary of State in the same period were members of the upper-class power elite. Through the CIA and other agencies, the upper class has manipulated intellectuals by giving money to individuals, schools, and foundations to exert influence, to start organizations, to hold conferences, and to publish magazines and books. As a consequence, the elite manages ideas, co-opts potential critics, and spreads its influence internationally. Do they act for themselves, or do they act for the good of the country? Domhoff's view on this point is clear, for he states that no matter how much they plead otherwise, "they are primarily self-interested partisans, their horizons severely limited by the ideologies and institutions that sustain and justify their privilege, celebrity, and power."[5]

Does this model fit the facts? As we mentioned previously, Marx missed on several points. The workers' revolution hasn't happened, and classes have developed in ways other than those Marx predicted. Critics also suggest that the class model overemphasizes class and economic factors and ignores the fact that elites continually emerge in modern complex societies regardless of their relationship to the means of production. At the same time, impressive data support the class viewpoint.

The major aspects of the three theories of power are summarized in Table 9.3. Each theory makes important points about the nature of power and the relationship between power and other segments of society. Each model speaks to particular characteristics of society, and each is criticized, usually from the vantage point of another model.

### Political Systems and American Politics

The **state** is the political body organized to govern; it is the center of legitimate power within a society. States can exercise their power in various ways. Look at several contrasting approaches: In a **totalitarian system,** the state has absolute or total control, and power is highly centralized in one party or group; dissent or conflicting viewpoints are not allowed; and all parts of society—education, mass media, and so on—are controlled to maintain purity of viewpoint. In a **democratic system,** the power of the state is limited or modified by the wishes of the people. The people are seen as having direct input into the decision-making process. They can themselves be involved, or they can participate indirectly through their elected officials. Ideally, there are restraints on power—"checks and balances"—that keep the state in line. If the government goes too far, the people have an obligation to oppose it and make their feelings heard. Democratic systems seem to be more inclined to succeed in economically developed societies that have a well-educated and well-informed citizenry and in societies in which there is diffused or decentralized power and an absence of major political conflicts or differences of opinion over issues. As is usually the case with ideal or pure types, the complete totalitarian or democratic state is seldom, if ever, seen; systems generally fall somewhere between the extremes.

The political system in the United States has several curious characteristics. It is surely a democracy, yet where is the participation? In 1996, 49 percent of the voting-age population voted in the presidential race. In the off-year elections of 1998, the voter turnout was 33 percent. This low turnout is not unusual in the United States, where perhaps individuals feel that a single vote in a mass society is of little consequence. However, voter participation is much higher in western European countries. The **political party,** a number of people organized to gain legitimate governmental power, represents another path into participation. It was assumed for a time that American political parties were divided along class, racial, and ethnic lines. The Republican Party was the party of the middle class and of big business, and the Democratic Party represented the working class and the common people. In recent elections, however, this seems to have broken down, with much crossing of party lines. The importance or influence of parties is muddled because both parties have liberal and

**TABLE 9.3**  Societal Power as Seen by Three Models

| Model | Chief Source of Power | Key Power Group(s) | Role of Masses | Function of State |
|---|---|---|---|---|
| Class | Control of society's productive resources (wealth) | Ruling class (owners and controllers of the corporate system) | Manipulated and exploited by the ruling class | Protect capitalist class interests; reproduce class system |
| Elitist | Control of key institutions, primarily of the corporation and of the executive branch of the federal government | Relatively cohesive power elite, made up of top corporate and government leaders | Manipulated and exploited by the power elite | Protect interests of dominant elites and their institutions |
| Pluralist | Various political resources, including wealth, authority, and votes | Elected political officials; interest groups and their leaders | Indirectly controls elites through competitive elections and interest-group pressures | Referee the arena of interest groups; create political consensus |

Martin Marger, *Elites and Masses* (New York: Van Nostrand, 1981), p. 112.

conservative wings and because of the tendency for people to vote for candidates rather than for issues in many situations. The **interest group,** a number of people organized to influence decisions on issues important to them, is another type of participation in American politics. A regular alphabet soup of interest groups flourishes on the political scene—AMA, NRA, NAACP, Right to Life, and so on. Once a significant bill is proposed, the interest groups possibly affected by that bill immediately become geared up. Interest groups hire lobbyists to aid their cause. **Lobbyists** are people in the business of knowing, influencing, and persuading the right people in government. The final product of any governmental decision-making process is inevitably greatly influenced by the pressures of interest groups and lobbyists. The final product is a compromise, manipulated by the concerns of politicians who wanted to be faithful to their party and get reelected by the citizens and by the pressures exerted by interest groups and lobbyists on politicians—an interesting example of the democratic process in action.

## THE POLITICAL ECONOMY: SEPARATE INSTITUTIONS?

The connection between economics and politics is clear: One of the three major sources of power is resource control, and the economic theme runs through both the class and elite theories of power. The huge corporations, with the wealth and workforce they control, have tremendous power to influence governmental decisions. And the concentrated power of the corporation is further strengthened by other factors. Corporations are often linked to each other through interlocking directorates. An individual sitting on the board of

## DWAYNE'S WORLD

Dwayne Andreas is a grandfather in his late seventies who owns and runs the biggest food and agriculture company in the world. The name of the company is Archer Daniels Midland (ADM), and in its ads ADM calls itself the "supermarket to the world." ADM products are present in thousands of items found in supermarkets, liquor stores, and gas stations. Texturized vegetable protein in burritos and meatless burgers, corn sweetener in Coca-Cola, ethanol if your car runs on gasohol, and grain alcohol used to make gin and vodka are all ADM products. ADM ranked 92nd on the 1995 Fortune 500 list of largest U.S. companies on sales of $12 billion and profits of close to $800 million. ADM gets substantial help from the government in terms of price supports for corn, limiting of production and price setting on sugar, and a 54-cent-per-gallon tax credit on ethanol. ADM buys corn at subsidized prices, uses it to make corn sweeteners which are subsidized by the sugar program, and uses the remainder for the big subsidy, tax credits on ethanol. And it all adds up to billions for ADM.

How does this happen? Some suggest that it is related to the contributions made to politicians by Andreas, his family, and ADM. Andreas has been a friend to politicians of all parties for decades, including Hubert Humphrey, George Bush, Richard Nixon, Bill Clinton, and Boris Yeltsin. During the 1992 election Andreas gave more than $1.4 million to party organizations and another $345,000 to candidates. In the 1996 campaign, ADM gave heavy support to Bob Dole, who was one of Ethanol's biggest champions.

All is not completely happy in Dwayne's world however. In October 1996, ADM was fined $100 million by the Justice Department for price fixing—this was the largest fine ever in a criminal anti-trust case.

*—See "Dwayne's World," by Dan Carney,* Mother Jones *(July/August 1995), pp. 44–47.*

one organization is likely a director of several others as well. This encourages cooperation, reduces competition among groups, and smooths the path toward greater profits. In addition, although large companies appear to be publicly owned by stockholders, ownership is in fact highly concentrated. Most stock is owned by a select few, which are often banks and other corporations. Because corporate leaders sit on the governing boards of universities, higher education naturally responds to the needs of business and industry. Corporations influence the mass media either by direct ownership or because they purchase advertising. In summary, the power of corporations is great and concentrated; their influence extends to all segments of society.[6]

One might expect that government—the political sector—would act as a restraint, but the contrary seems to be the case: The two institutions (economic and political) seem to work together. Corporations influence political decisions through lobbying and campaign financing and by having their leaders move freely back and forth between the two job worlds.

For example, in 1971 a professor was appointed head of an insurance regulatory agency in Pennsylvania. His task, with the support of the powers of the state government, was to reform insurance company practices. He was ambitious, interested in reform, and had power, but he ran into difficulty. Nearly all the government staff and managerial people he worked with had a background in the insurance industry. Furthermore, the insurance industry had a large lobbying operation; lobbyists attended legislative committee meetings and helped draft legislation favorable to the industry. Finally, good jobs in insurance companies were promised to state employees who were loyal to insurance industry needs. During the professor's three years in office (he resigned seven months early to be replaced by a candidate acceptable to the insurance industry), no legislation put forth by his department passed if it was opposed by the insurance industry.[7]

Corporate leaders serve as presidential advisors, cabinet officials, and heads of regulatory agencies and later move back to the corporate world. Government, in return, supports a tax structure and gives tax breaks that are favorable to individuals and corporations. Louis B. Mayer, founder of MGM, saved $2 million in taxes when he retired by hiring a Washington lawyer who obtained a special tax provision that applied only to him. Government is also the corporation's best customer as a major consumer of goods and services; the Department of Defense, for example, is the largest single consumer in American society. And government pursues a foreign policy favorable to corporate interests: Home markets are protected, foreign markets encouraged, and continued supply of necessary foreign raw materials assured. Government and economic elites clearly have mutual interests; as one prospers, so does the other.

Members of the government and economic elite are generally wealthy, male, college-educated with advanced degrees, and "well-connected." Some from lower social classes occasionally sneak in, but not often. "The outstanding fact of elite recruitment in the United States and other Western industrial societies is that leaders are chosen overwhelmingly from socially dominant groups, and have been for many generations."[8] These elites differ on some issues but agree on the important ones, which should be expected given their common interests, backgrounds, and needs. The masses have little authority or influence in selecting or controlling the elites. There is potential control through the political process by voting or working for candidates, but few participate significantly, and those who do tend to be from the higher social classes. Social movements (see Chapter Thirteen) might be effective in controlling elites, but they are seldom a direct challenge.

## THEORY AND RESEARCH: A REVIEW

The study of social institutions starts with *functional analysis:* Economic institutions function to organize production and distribution of goods and services in society. Governmental institutions function to maintain some semblance of order in a complex society, to enforce norms that people feel are important, to settle disputes, and to deal with other states. All are important and necessary functions. Again we see the emphasis on the basic elements of functional analysis—order, stability, structure, and function. *Conflict theorists* see other aspects of these institutions: Economic and governmental institutions operate to benefit powerful groups. Economic institutions clearly illustrate the basic battle between capitalists and workers that Marx saw long ago. Governmental institutions aid tremendously by helping the privileged to insulate and protect their position. The two institutions work together for the benefit of the established order. Both theories seem to "fit the facts"; both provide important insights toward understanding these institutions.

As for the data cited in this chapter, Domhoff used records in a *nonreactive approach*. He used information available in *Who's Who* and the *Social Register*, and he obtained lists of the directors, trustees, and partners from records of the largest companies, banks, foundations, and law firms. The economic data on corporations and multinationals were collected by business magazines from information collected and reported by the businesses themselves. Should we be suspicious? What would a conflict theorist say?

## SUMMARY

In this third chapter on social institutions we have focused on economic and political institutions. We have discussed the economic systems of capitalism and socialism, and we have looked at some of the distinctive features of the American economy—large corporations and multinationals. After turning to political institutions, we have defined power, discussed types of power, and explored different models explaining the distribution of power. In addition, we have described totalitarian and democratic political systems and have discussed some of the characteristics of politics in America. We then followed Martin Marger's argument from his book *Elites and Masses,* in which he suggests that the key factor in understanding modern industrialized societies is their domination by elites from the political and economic realms. These elites have common interests, come from common backgrounds, and work together to their mutual benefit. The masses are relatively powerless and only marginally involved in the political process. There is considerable debate about the power and influence of elites, and one has to decide which of the various models and explanations best fits the facts of modern society.

To understand the American political economy, we need to look at the power and influence of large corporations. Class and elite theorists believe that a powerful group of people, well known to each other, run this country. The reading in this chapter profiles Bill Gates, the world's richest man, and his company Microsoft which in June 2000 was the country's second most valued company with a market value of more than $360 billion.

# TERMS FOR STUDY

authority (296)

capitalism (292)

charismatic authority (296)

class model (297)

democratic system (301)

elite model (297)

elite-pluralism (298)

influence (297)

interest group (302)

legal-rational authority (296)

lobbyist (302)

multinationals (294)

pluralistic model (297)

political party (301)

power (296)

pseudocharisma (297)

resource control (297)

socialism (293)

state (301)

totalitarian system (301)

traditional authority (296)

For a discussion of Research Question 1, see page 295.
For a discussion of Research Question 2, see page 300.

 **INFOTRAC COLLEGE EDITION**
**Search Word Summary**

To learn more about the topics from this chapter, you can use the following words to conduct an electronic search on InfoTrac College Edition, an online library of journals. Here you will find a multitude of articles from various sources and perspectives:

**www.infotrac-college.com/wadsworth/access.html**

Socialism

Exxon Corporation

Elite

Totalitarianism

Government subsidies

Department of defense

Multinational corporations

## Reading 9.1

# IN SEARCH OF THE REAL BILL GATES

*Walter Isaacson*

*Bill Gates is the world's richest person. How rich? In late 1996 (see below) he was worth $23.9 billion. By June 2000, his net worth was $60 billion. His company, Microsoft, had a market value of more than $360 billion in mid 2000. Walter Isaacson profiled Gates for* Time *in January 1997. This reading is an excerpt from Isaacson's article.*

When Bill Gates was in the sixth grade, his parents decided he needed counseling. He was at war with his mother, Mary, an outgoing woman who harbored the belief that he should do what she told him. She would call him to dinner from his basement bedroom, which she had given up trying to make him clean, and he wouldn't respond. "What are you doing?" she once demanded over the intercom.

"I'm thinking," he shouted back.

"You're thinking?"

"Yes, Mom, I'm thinking," he said fiercely. "Have you ever tried thinking?"

The psychologist they sent him to "was a really cool guy," Gates recalls. "He gave me books to read after each session, Freud stuff, and I really got into psychology theory." After a year of sessions and a battery of tests, the counselor reached his conclusion. "You're going to lose," he told Mary. "You had better just adjust to it because there's no use trying to beat him." Mary was strong-willed and intelligent herself, her husband recalls, "but she came around to accepting that it was futile trying to compete with him."

A lot of computer companies have concluded the same. In the 21 years since he dropped out of Harvard to start Microsoft, William Henry Gates III, 41, has thrashed competitors in the world of desktop operating systems and application software. Now he is attempting the audacious feat of expanding Microsoft from a software company into a media and content company.

In the process he has amassed a fortune worth (as of last Friday) $23.9 billion. The 88 percent rise in Microsoft stock in 1996 meant he made on paper more than $10.9 billion, or about $30 million a day. That makes him the world's richest person, by far. But he's more than that. He has become the Edison and Ford of our age. A technologist turned entrepreneur, he embodies the digital era.

His success stems from his personality: An awesome and at times frightening blend of brilliance, drive, competitiveness and personal intensity. So too does Microsoft's. "The personality of Bill Gates determines the culture of Microsoft," says his intellectual sidekick, Nathan Myhrvold. But though he has become the most famous business celebrity in the world, Gates remains personally elusive to all but a close circle of friends.

Part of what makes him so enigmatic is the nature of his intellect. Wander the Microsoft grounds, press the Bill button in conversation and hear it described in computer terms: He has "Incredible processing power" and "unlimited bandwidth," and agility at "parallel processing" and "multitasking." Watch him at his desk, and you see what they mean. He works on two computers, one with four frames that sequence data streaming in from the Internet, the other handling the hundreds of E-mail messages and memos that extend his mind into a network. He can be so rigorous as he processes data that one can imagine his mind may indeed be digital, no sloppy emotions or analog fuzziness, just trillions of binary impulses coolly converting input into correct answers.

In 1986, after Microsoft became successful, Gates built a four-house vacation compound dubbed Gateaway for his family. There his parents would help him replicate his summer activities on a grander scale for dozens of friends and co-workers in what became known as Microgames. "There were always a couple of mental games as well as performances and regular games," says Bill Sr. as he flips through a scrapbook. These were no ordinary picnics: One digital version of charades, for example, had teams competing to send numerical messages using smoke signal machines, in which the winners devised their own 4-bit binary code.

"We became concerned about him when he was ready for junior high," says his father. "He was so small and shy, in need of protection, and his interests were so very different from the typical sixth grader's." His intellectual drive and curiosity would not be satisfied in a big public school. So they decided to send him to an elite private school across town.

Learning BASIC language from a manual with his pal Paul Allen, Trey [Gates' nickname] produced two programs in the eighth grade: One that converted a number in mathematical base to a different base, and another (easier to explain) that played tic-tac-toe. Later, having read about Napoleon's military strategies, he devised a computer version of Risk, a board game he liked in which the goal is world domination.

Trey and Paul were soon spending their evenings at a local company that had bought a big computer and didn't have to pay for it until it was debugged. In exchange for computer time, the boys' job was to try (quite successfully) to find bugs that would crash it. "Trey got so into it," his father recalls, "that he would sneak out the basement door after we went to bed and spend most of the night there."

The combination of counseling and the computer helped transform him into a self-assured young businessman. By high school he and his friends had started a profitable company to analyze and graph traffic data for the city. "His confidence increased, and his sense of humor increased," his father says. "He became a great storyteller, who could mimic the voices of each person. And he made peace with his mother."

"In ninth grade," Gates recalls over dinner one night, "I came up with a new form of rebellion. I hadn't been getting good grades, but I decided to get all A's without taking a book home. I didn't go to math class, because I knew enough and had read ahead, and I placed within the top 10 people in the nation on an aptitude exam. That established my independence and taught me I didn't need to rebel anymore." By 10th grade he was teaching computers and writing a program that handled class scheduling, which had a secret function that placed him in classes with the right girls.

His best friend was Kent Evans, son of a Unitarian minister. "We read *Fortune* together; we were going to conquer the world," says Gates. "I still remember his phone number." Together with Paul Allen, they formed the official-sounding Lakeside Programmers Group and got a job writing a payroll system for a local firm. A furious argument, the first of many, ensued when Allen tried to take over the work himself. But he soon learned he needed the tireless Gates back to do the coding. "O.K., but I'm in charge," Gates told him, "and I'll get used to being in charge, and it'll be hard to deal with me from now on unless I'm in charge." He was right.

Steve Ballmer, big and balding, is bouncing around a Microsoft conference room with the spirit of the Harvard football-team manager he once was. "Bill lived down the hall from me at Harvard sophomore year," he says. "He'd play poker until 6 in the morning, then I'd run into him at breakfast and discuss applied mathematics." They took graduate-level math and economics courses together, but Gates had an odd approach toward his classes: he would skip the lectures of those he was taking and audit the courses of those he wasn't, then spend the period before each exam cramming. "He's the smartest guy I've ever met," says Ballmer, 40, continuing the unbroken sequence of people who make that point early in an interview.

"Bill brings to the company the idea that conflict can be a good thing," says Ballmer. "The difference from P&G is striking. Politeness was at a premium there. Bill knows it's important to avoid that gentle civility that keeps you from getting to the heart of an issue quickly. He likes it when anyone, even a junior employee, challenges him, and you know he respects you when he starts shouting back." Around Microsoft, it's known as the "math camp" mentality: a lot of cocky geeks willing to wave their fingers and yell with the cute conviction that all problems have a right answer. Among Gates' favorite phrases is "That's the stupidest thing I've ever heard," and victims wear it as a badge of honor, bragging about it the way they do about getting a late-night E-mail from him.

The contentious atmosphere can promote flexibility. The Microsoft Network began as a proprietary online system like CompuServe or America Online. When the open standards of the Internet changed the game, Microsoft was initially caught flat-footed. Arguments ensued. Soon it became clear it was time to try a new strategy and raise the stakes. Gates turned his company around in just one year to disprove the maxim that a leader of one revolution will be left behind by the next.

If Ballmer is Gates' social goad, his intellectual one is Nathan Myhrvold (pronounced Meer-voll), 37, who likes to joke that he's got more degrees than a thermometer, including a doctorate in physics from Princeton. With a fast and exuberant laugh, he has a passion for subjects ranging from technology (he heads Microsoft's advanced-research group) to dinosaurs (he's about to publish a paper on the aerodynamics of the apatosaurus tail) to cooking. He sometimes moonlights as a chef at Rover's, a French restaurant in Seattle.

"There are two types of tech companies," Myhrvold says in between pauses to inhale the aroma of the food. "Those where the guy in charge knows how to surf, and those where he depends on experts on the beach to guide him." The key point about Gates is that he knows—indeed loves—the intricacies of creating software.

"Every decision he makes is based on his knowledge of the merits. He doesn't need to rely on personal politics. It sets the tone."

Myhrvold describes a typical private session with Gates. Pacing around a room, they will talk for hours about future technologies such as voice recognition (they call their team working on it the "wreck a nice beach" group, because that's what invariably appears on the screen when someone speaks the phrase "recognize speech" into the system), then wander onto topics ranging from quantum physics to genetic engineering. "Bill is not threatened by smart people," he says, "only stupid ones."

Microsoft has long hired based on I.Q. and "intellectual bandwidth." Gates is the undisputed ideal: talking to most people is like sipping from a fountain, goes the saying at the company, but with Gates it's like drinking from a fire hose. Gates, Ballmer and Myhrvold believe it's better to get a brilliant but untrained young brain—they're called "Bill clones"—than someone with too much experience. The interview process tests not what the applicants know but how well they can process tricky questions: If you wanted to figure out how many times on average you would have to flip the pages of the Manhattan phone book to find a specific name, how would you approach the problem?

Gates is ambivalent about his celebrity. Although he believes that fame tends to be "very corrupting," he is comfortable as a public figure and as the personification of the company he built. Like Buffett, he remains unaffected, wandering Manhattan and Seattle without an entourage or driver. Nestled into a banquette one Sunday night at 44, a fashionable Manhattan restaurant, he is talking volubly when another diner approaches. Gates pulls inward, used to people who want his autograph or to share some notion about computers. But the diner doesn't recognize him and instead asks him to keep his voice down. Gates apologizes sheepishly. He seems pleased to be regarded as a boyish cutup rather than a celebrity.

The phone in Gates' office almost never rings. Nor do phones seem to ring much anywhere on the suburban Microsoft "campus," a cluster of 35 low-rise buildings, lawns, white pines and courtyards that resemble those of a state polytechnic college. Gates runs his company mainly through three methods: he bats out a hundred or more E-mail messages a day (and night), often chuckling as he dispatches them; he meets every month or so with a top management group that is still informally known as the boop (Bill and the Office of the President); and most important, taking up 70% of his schedule by his own calculation, he holds two or three small review meetings a day with a procession of teams working on the company's various products.

Gates does not hide his cutthroat instincts. "The competitive landscape here is strange, ranging from Navio to even Web TV," he says. He is particularly focused on Navio, a consumer-software consortium recently launched by Netscape and others designed to make sure that Windows and Windows CE (its consumer-electronics cousin) do not become the standard for interactive television and game machines. "I want to put something in our products that's hard for Navio to do. What are their plans?" The group admits that their intelligence on Navio is poor. Gates rocks harder. "You have to pick someone in your group," he tells Mundy, "whose task it is to track Navio full time. They're the ones I worry about. Sega is an investor. They may be willing to feed us info." Then he moves on to other competitors. "What about the Planet TV guys?" Mundy explains that they are focusing on video games, "a platform we haven't prioritized." Gates counters: "We can work with them now, but they have other ambitions. So we'll be competitive with them down the line."

His mother may have come to terms with this competitive intensity, but much of the computer world has not. There are many websites dedicated to reviling him, law firms focused on foiling him and former friends who sputter at the mention of his name. Companies such as Netscape, Oracle, and Sun Microsystems publicly make thwarting his "plan for world domination" into a holy crusade.

The criticism is not just that he is successful but that he has tried to leverage, unfairly and perhaps illegally, Microsoft's near monopoly in desktop operating systems in ways that would let him dominate everything from word processing and spreadsheets to Web browsers and content. The company is integrating its Internet Explorer browser and Mircosoft Network content into its Windows operating system, a process that will culminate with the "Active Desktop" planned for Windows 97, due out in a few months. Critics see a pattern of Microsoft's playing hardball to make life difficult for competing operating systems and applications: Microsoft Word has been buggy on Macintosh operating systems, users have found it tricky to make Netscape their default browser when going back and forth from Windows to the Microsoft Network, and application developers have complained that they don't get the full specs for the new releases of Windows as quickly as Micosoft's own developers do.

Esther Dyson, whose newsletter and conferences make her one of the industry's fabled gurus, is another longtime friend and admirer who shares such qualms. "He never really grew up in terms of social responsibility and relationships with other people," she says. "He's brilliant but still childlike. He can be a fun companion, but he can lack human empathy." "If we weren't so ruthless, we'd be making more creative software? We'd rather kill a competitor than grow the market?!?!" Gates is pacing around his office, sarcastically repeating the charges against him. "Those are clear lies," he says coldly. "Who grew this market? We did. Who survived companies like IBM, 10 times our size, taking us on?" He ticks off the names of his rivals at Oracle, Sun, Lotus, Netscape in an impersonal way. "They're every bit as competitive as I am."

"We win because we hire the smartest people. We improve our products based on feedback, until they're the best. We have retreats each year where we think about where the world is heading." He won't even cop a plea to the charge that Mircosoft tends to react to competitor's ideas—the graphical interface of Apple, the Web browser of Netscape—more than it blazes new trails of its own. "Graphical interfaces were done first at Xerox not Apple. We bet on them early on, which is why Microsoft Office applications became the best."

He hopes to be running Microsoft for another 10 years, he says, then promises to focus as intensely on giving his money away. He says he plans to leave his children about $10 million each. "He will spend time, at some point, thinking about the impact his philanthropy can have," Buffett says. "He is too imaginative to just do conventional gifts." Already he's given $34 million to the University of Washington, partly to fund a chair for human genome-project researcher Leroy Hood; $15 million (along with $10 million from Ballmer) for a new computer center at Harvard; and $6 million to Stanford. An additional $200 million is in a foundation run by his father, and he has talked about taking over personally the funding of Microsoft's program to provide computers to inner-city libraries, to which he's donated $3 million in book royalties. "I've been pushing him gently to think more about philanthropy," his father says. "I think his charitable interests will run, as they do now, to schools and libraries."

# QUESTIONS 9.1

**1.** "Microsoft represents an important change in the American economy." Analyze and critique this statement by comparing Microsoft with the top 10 companies in Table 9.1.

**2.** Some other competing companies fear Microsoft. Why? What type of power does Microsoft have?

# NOTES

1. The major source for this section is *Elites and Masses* by Martin Marger (New York: Van Nostrand, 1981). I have followed Marger's organization and viewpoint on most topics. His discussion of power and the three theoretical models comes from his chapters 2–5.

2. The idea of pseudocharisma comes from Joseph Bensman and Bernard Rosenberg in *Mass, Class, and Bureaucracy: An Introduction to Sociology* (New York: Praeger, 1976), pp. 431–433, and is discussed by Marger (see note 1) on pp. 20–21.

3. See C. Wright Mills, *The Power Elite* (New York: Oxford University Press, 1959).

4. Both quotations in this paragraph come from *Who Rules America?* by G. William Domhoff (Englewood Cliffs, N.J.: Prentice-Hall, 1967), p. 5.

5. This is from G. William Domhoff's later book *The Higher Circles* (New York: Random House, 1970), p. 275.

6. The source for these paragraphs is Marger, *Elites and Masses* (see note 1), especially chapters 6–8 and 11.

7. This case is described in *The Powers That Be* by G. William Domhoff (New York: Vintage, 1978), pp. 32–36. The original source is David L. Serber, "Regulating Reform: The Social Organization of Insurance Regulation," *The Insurgent Sociologist,* Spring 1975. The Mayer example in the next paragraph is also from Domhoff, p. 29.

8. Marger, *Elites and Masses* (see note 1), p. 207.

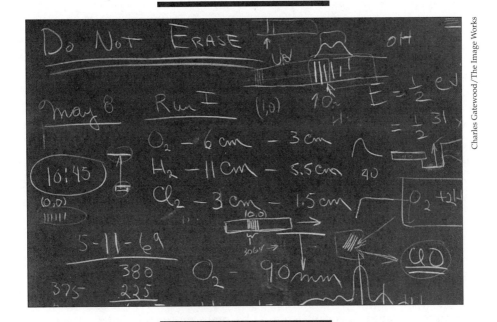

# Institutions: Education

Education and Schooling Defined

The Functional Approach to Education

The Conflict Approach to Education

Education in America

Is Change Coming in Education?

## TWO RESEARCH QUESTIONS

**1.** *Is the high-school dropout rate in a state related to the amount of money spent on education in that state? In other words, if a state has high expenditures on education, will it have a lower high-school dropout rate?*

**2.** *Do American elementary and secondary students spend more, less, or about the same number of days in school each year as students from other countries?*

IN CHAPTER SEVEN we defined social institutions as the norms, roles, and patterns of organization that develop around central needs or problems that societies experience. In previous chapters we discussed the family, religion, and economic and political institutions. In this chapter we look at education.

One basic problem that all societies face is socialization and training of the young. The young must *learn* how to become adults, to count to 100, to drive cars, and to become rocket scientists. Who is going to teach them these things? Early parts of the socialization process are carried out primarily in the family. In primitive societies, parents and those close to the family probably carried most of the load. But as societies become more complex and the path to becoming a rocket scientist becomes more convoluted, other elements are necessary to continue the process. All these elements, then—from small groups like a family, a kindergarten class, or a graduate seminar, to larger organizations like the University of California or the Boston school system—are part of the institution of education.

As we have learned, there are overlaps among institutions. Socialization and training of the young take place in the institutions of the family and religion, and perhaps in others as well. However, by focusing on education in this chapter, we may be able to identify some of the problems and issues that are unique to this institution.

## EDUCATION AND SCHOOLING DEFINED

A distinction should be made between education and schooling. **Education** is the more general term and refers to the process of socialization—learning the knowledge, skills, and appropriate ways of behaving to exist in a larger social world. Education (socialization) is thus most intense for the very young, but it continues to occur throughout the life span. **Schooling** is the narrower term and refers to formal instruction in a classroom setting. Our major focus in this chapter will be on schooling.

The United States pioneered the creation of the "common school," a publicly supported elementary school for the education of lower- and middle-class children. The comprehensive high school grew out of the elementary-school program. We started what has been called a "schooling revolution" in the modern world, a revolution that represented a fundamental change in how young people were prepared for life in the adult world.[1] No longer did the child receive informal teaching in the community;

the school became the setting for most of the child's preparation for the adult world. Following the lead of the rest of society, the small, folk-society, gemeinschaft, isolated-village approach to education was left behind; thus the "schooling revolution" was but one aspect of our move to a more complex, industrial and scientific, secondary, gesellschaft type of society.

## THE FUNCTIONAL APPROACH TO EDUCATION

The functional approach makes several assumptions about schooling that require some attention.[2] Functionalists see modern society as a system in which success is *achieved;* ability and hard work are more important than are being given something or inheriting it. Children of the poor should have the same opportunities to get high-status jobs as do the children of the rich. Functionalists see society as an *expert society*, one that needs people who are highly trained and who have developed specialized skills. Modern society is becoming more and more complex, and this implies an ever-increasing need for expertise and technical know-how. Functionalists also see modern society as a *democratic society* in which an educated citizenry is an informed citizenry composed of people who are more likely to make responsible and informed decisions, are likely to be less prejudiced and more tolerant, and will be more concerned with the quality of life and social justice.

In this context, functionalists view schooling as providing a number of obvious and important functions:

- Passing on the culture of a society

- Facilitating the mixing and merging of different classes and ethnic groups

- Creating the informed citizenry necessary in a complex society

- Producing cognitive skills

- Screening, sorting, and selecting of talent and ability

- Developing new knowledge and providing scientific discoveries

Obvious functions like these could be called *manifest functions*. Perhaps equally important but less obvious consequences (*latent functions*) of schooling might include the following:

- Delaying young people from entering the job market

- "Baby-sitting" young children, thereby freeing parents to work outside the home

- Teaching habits of discipline and obedience to authority that are necessary for survival in a bureaucratic society

How good is the fit between what the functionalist approach proposes and the current facts? Christopher Hurn suggests that it is not as good as it might be. According to the functionalist approach, modern society needs an ever-increasing number of highly skilled and trained individuals. Those who perform best will reap the highest rewards in modern society. And yet in studies of college graduates, grade point average does not consistently predict either occupational status or future earnings! Actually, it might be that the presence of a *degree* or *credential* is the crucial factor for being hired, not the knowledge or skill that is assumed to go with it. This suggests that rather than teaching skills, the education system more likely rewards *persistence*—the student who sticks it out to the B.A., M.A., or Ph.D. gathers the benefits. As job candidates become more educated, entrance standards for the job seem to get raised. Does this mean that the job has become more complex, or that entrance standards are used as ways of controlling admission to the occupation?

Many schools use a procedure called **tracking** (or curriculum tracking) by which students are placed in certain programs based on curricular needs or students' abilities. They are assigned to these tracks based on test scores and on input from teachers, parents, and school administrators. Some students are placed in a college-preparatory track, others in a vocational-career track. Some schools have three or four curricular tracks; other schools may have more depending on such factors as school size, budget, type of community being served, and diversity of faculty. According to the functionalist view, people should *achieve* success, not be given it. People should get ahead according to their ability, not their birthright. Tracking makes sense to functionalists because it provides equal opportunity for all and rewards ability and hard work. Functionalists hope for a "sorting mechanism" that is truly class- and color-blind and in which the sharper and harder-working students are rewarded by being placed in the higher (college-prep) classes. The question, of course, is whether tracking is truly color-blind and in fact treats everybody equally, separating only on the basis of ability and hard work. Research shows a relationship between I.Q. (ability) and later earnings, but it also shows that one's social class is the overpowering factor in the equation. In other words, the evidence seems to indicate that one's social class is a better predictor of future economic success than is one's measured I.Q. With these criticisms of the functionalist approach in mind, let's turn to a discussion of the conflict approach.

## THE CONFLICT APPROACH TO EDUCATION

Conflict theorists see education and schooling quite differently than the functionalists do. First, conflict theorists see a struggle among competing groups for the control of schooling. The struggle for control represents each group's attempts to advance its own set of interests and instill its own values in the education system. Second, the struggle among groups is an unequal one. The elite—the "haves," the privileged class with its superior power and resources—always has the upper hand. Third, conflict theorists are convinced that even though employers recognize the need for a certain minimum level of skill, they are far more interested in having workers who are loyal, obedient, and submissive. What the employers really want from schooling, then, is workers who have "appropriate" attitudes and values.

Conflict theorists emphasize the role schools play in maintaining the existing social-class system. Tracking is one way of doing this because, from a conflict perspective, there is clearly no equality of opportunity. Higher-class children are tracked into colleges and universities and thus into positions of economic and political dominance like those their parents hold. Children with lower-class or working-class origins tend to be placed in tracks that will move them toward blue-collar work in adulthood.

Another way in which schools reinforce the existing class system is by teaching the values and personality characteristics necessary in certain positions in society. A **hidden curriculum** exists in schools whereby these important cultural values and attitudes are taught. These values could be either compliance, punctuality, and conformity, or independence, self-reliance, and nonconformity. A hidden curriculum is not part of the regular or intended curriculum, and teachers may not even be aware it exists, but in subtle and effective ways students come to understand their social place in school and in society as a whole.

Conflict theorists Samuel Bowles and Herbert Gintis suggest specifically that schooling serves the interests of capitalist society.[3] They propose that for capitalist societies to work efficiently, production-line workers must have a different approach to their jobs than scientists, heads of corporations, or presidential advisors need to have to theirs. Schools recognize this, and schools whose graduates enter lower-level occupations stress following rules, being punctual, having respect for authority, having fewer choices, and showing less creativity. Schools whose graduates enter higher-level occupations are likely to encourage independent thinking, creativity, less obedience to authority, and more freedom and flexibility. If you were to compare lower-track high-school and junior-college courses with college-prep high-school courses and courses at elite universities, you would probably find that the former had more assignments, less flexibility, more supervision, and less encouragement of initiative or creativity. According to Bowles and Gintis, then, schools are conferring on students the values and attitudes they will need for their place in society, a place not much different from the place their parents hold.

In much the same vein, Randall Collins suggests in his book *The Credential Society* that educational credentials that are largely unnecessary are used to control access to desirable jobs. Collins believes that most jobs are not particularly complex, that they are not becoming more complex, and that the skills most jobs require can easily be learned on the job. He contends that schools are not particularly good at teaching intellectual skills and that much of what is taught is quickly forgotten. Collins concludes that various groups use the credentialing that education provides as a way of elevating their own status, *not* because the complexity of the job requires it. For example, many police departments now want their candidates to have a college education, when not so long ago a high-school

## RANDALL COLLINS (1941–    )

Randall Collins earned his Ph.D. in Sociology at the University of California at Berkeley. Of the three major sociological theories, two of them—functional analysis and conflict theory—could be called macro in that they focus on broad social factors that involve society as a whole. The third theory—symbolic interaction—is micro in that it holds that society is best understood by focusing on the everyday interactions of individuals. Collins was a major contributor in developing a more integrative conflict theory. His book, *Conflict Sociology,* was highly integrative because it moved in a much more micro-oriented direction than the macro conflict theories of Dahrendorf and others. Since his theoretical roots lie in areas such as ethnomethodology and phenomenology,

Collins approached conflict from the individual point of view. Collins himself says of his early work, "My own main contribution to conflict theory . . . was to add a micro level to these macro-level theories. I especially tried to show that stratification and organization are grounded in the interactions of everyday life."

Collins's road to becoming a sociologist was well traveled. At Harvard he studied subjects such as literature and mathematics before majoring in social relations. He later moved on to Stanford where he received an M.A. in psychology before finally receiving his Ph.D. in sociology at U.C. Berkeley. Randall Collins is a professor of sociology at the University of Pennsylvania.

---

education (or even less) was seen as adequate. Has the job gotten that much more complex? Is a college-educated police officer a significantly better officer than a high-school-educated officer? Or are the new requirements a way of elevating the profession—a way of giving it higher status and controlling admission to the job? Collins agrees with other conflict theorists who say that much of what schools teach are good manners and how to behave, not cognitive skills.

You might think of the discussion in the previous several paragraphs as a debate between the relative strengths of two models: the social reproduction model and the democratic model. The **social reproduction model** refers to the tendency of the institution of education to reproduce the social class of the student. The **democratic model** refers to the struggle to provide equal opportunity for all and to reward ability and hard work regardless of one's origin. The constant tension between these two models is nicely illustrated in the two competing theories, functionalist and conflict.

The conflict perspective sees tension and conflict as normal in social situations, so it is no surprise that they exist in schools. Teachers organize into unions and educational associations for negotiating salaries and work conditions. Teachers use punishment (grades, detention) to ensure conformity. Students may resort to informal subversive activities to counter discipline and classwork demands made by teachers. Accordingly, a recent high-school student recalls seeing graffiti on walls, bringing a Walkman or books to class to help pass the time, figuring out clever ways to get out of class or to cut class, vandalizing equipment (clogging the toilets, stuffing paper or pencils into gas outlets, or breaking lockers), and carving or spray-painting slogans or initials on various surfaces. It is estimated that 130,000 children bring guns to school, and many schools now have security guards and metal detectors.

Another type of behavior that can be a form of resistance or rebellion is to drop out of school entirely. It is estimated that about 600,000 young-sters drop out of school each year—in many urban schools, as much as half of the enrollment drops out.[4] The next stop for many is the correc-tional system. Table 10.1 shows percentages of high-school dropouts for two different years and for different groups. Why do students drop out? When one study asked students about it, the most frequently given rea-sons for dropping out were: "School was not for me"; had poor grades; was offered a job and chose to work; got married or planned to get married; couldn't get along with teachers; had to support family; and was pregnant.

High-school dropout rates vary across the country, as can be seen in Table 10.2. North Dakota had the lowest dropout rate in 1990 and Nevada had the highest. The first research question at the beginning of this chapter asked whether the amount of money spent on education might be related to the dropout rate: Do states that spend less per student have higher dropout rates? Expenditures on education might be part of the explana-tion, but as the data in Table 10.2 indicate, expenditures don't appear to be the whole answer. Although the average spent by the low dropout states is more than the average spent by the high dropout states, neither group does particularly well: All five of the high dropout states and three of the five low dropout states are below the national average on money spent. The lowest dropout state (North Dakota) is one of the lowest of all states in expenditures. Clearly, certain demographic factors are also at work. More urbanized states will probably have higher dropout rates reflecting prob-lems of inner-city schools. Hispanic Americans tend to have a higher school dropout rate and this might explain some of what is happening in Arizona, Florida, and California. It should be noted that the data in Table 10.2 are not as good a fit as we'd like—the dropout information is from 1990 and the expenditures data are from 1995, but these rankings are con-sistent with those in earlier years.

Conflict theorists suggest that one reason for students' resistance or rebel-lion, especially dropping out of school, is that some students, particularly working-class students, detect the "hidden curriculum" and decide that they won't go along with it. They refuse to accept the school's view that conformity to the school's rules and regulations will lead to economic opportunity. They see that the system is stacked against them, that there "ain't no making it," and they rebel.

Read the box entitled "The Typical Dropout" and decide whether this is an example of the functionalist or conflict interpretation of why students drop out.

**TABLE 10.1**   Percentages of High-School Dropouts Aged 14 to 24 Years

| | 1980 | 1997 |
|---|---|---|
| U.S. (all groups) | 12 | 9.1 |
| White | 11.3 | 8.8 |
| Black | 16 | 11.2 |
| Hispanic | 29.5 | 21.0 |

United States Bureau of the Census, *Statistical Abstracts of the United States, 1999* (119th ed.), Washington, D.C.

**TABLE 10.2**   High-School Dropout Rates (1990) and Educational Expenditures, 1995 (U.S. Average = $5,907)

| Fewest Dropouts | Expenditures (per student) | Most Dropouts | Expenditures (per student) |
|---|---|---|---|
| North Dakota | $4,606 | Nevada | $5,126 |
| Minnesota | $6,033 | Arizona | $4,252 |
| Iowa | $5,560 | Florida | $5,717 |
| Wyoming | $6,070 | California | $4,731 |
| Nebraska | $5,384 | Georgia | $5,396 |

United States Bureau of the Census, *Statistical Abstracts of the United States, 1996* (116th ed.), Washington, D.C.

# EDUCATION IN AMERICA

Are we schooling our young people as well as we should be? We hear a lot about Scholastic Aptitude Test (SAT) scores. Both math and verbal SAT scores were lower in 1998 than scores of 30 years earlier. At the same time, curiously, the number of students doing very well on the test increased. This led one expert to comment that we have an "educational elite" and an academic "underclass" that is "ill-prepared for the demands of college or the workplace." The International Association for the Evaluation of Educational Achievement (IEA) has been trying to gauge the quality of education in various countries. In their comparison of eighth-grade math students in 20 countries in 1981 through 1982, American students ranked 10th in arithmetic, 12th in algebra, and 16th in geometry.[5] Japan ranked first in all three of these categories. The IEA examined the science knowledge of 10-year-olds in 15 countries from 1983 to 1986; American students ranked 8th. Other studies that suggested that American students fall farther behind each year were confirmed when the IEA examined 14-year-olds and found them tied for 14th (with students from Singapore and Thailand) out of 17 countries. Most countries don't try to educate as many of their young people as we do, so perhaps these comparisons aren't completely valid. However, in another study, IEA compared 12th-grade students who were engaged in the serious study of math (were in their fourth year of high-school math), clearly a more

## THE TYPICAL DROPOUT

Compared with those who will graduate, the future dropout shows a clear indication of academic problems by the third grade. Achievement test scores are below the scores of his or her classmates and also below the level one would expect given the student's ability. The poor attendance and underachievement increases as the student goes into middle school, and by the seventh grade failing grades are present. By the ninth grade a pattern of high absences, failing grades, and a low overall GPA is well established, and it continues until the student drops out of high school.

Regarding family background, the dropout's father tends to have a job of lower social status, although not significantly lower, than that of the graduate's father (about 1.5 points on the 9-point Hollingshead scale). Although a lower percentage of the dropout group lived in families with two parents present than did the graduates, more than two thirds (69 percent) were living at home with both parents at the beginning of the ninth grade. This finding evidently questions the assumption that the typical dropout is from a "broken home." ...

The typical dropout was no more likely to be either male or female than the student who graduated, and similarly being in a more flexible modular scheduled high school or following a more traditional schedule had no significant effect on the likelihood of being a dropout.

The high accuracy with which the elementary school data identified potential dropouts probably reflects family attitudes toward education. The elementary school student who is not in school is absent with the parent's knowledge and at least tacit consent. The parents who are uninterested in their children's attendance in elementary school, and probably their achievement, likely not only convey their values to the child, but also are willing to agree when the child later decides to leave high school.

*—From "Differentiating Characteristics of High School Graduates, Dropouts, and Nongraduates," by B. L. Barrington and B. Hendricks,* Journal of Educational Research *82 (July/August, 1989), pp. 309–319.*

---

elite group of students. A portion of the results of this international comparison is shown in Table 10.3.

Maybe a crucial factor has to do with how education is structured in the United States. The Third International Mathematics and Science Study released a report in late 1996 that examined the math and science scores of thousands of eighth-grade students from many countries. Singapore, Korea, and Japan ranked highest in math, and Singapore, the Czech Republic, and Japan ranked the highest in science. The United States was slightly below the international average in math and slightly above average in science. Surprisingly, teachers in Japan were found to assign *less* math and science homework than do U.S. instructors. Further, Japanese eighth graders watch as much television and videos after school as do their American counterparts! The experts believe the differences in scores have to do with how the subjects are taught in each country. For example, U.S. students are taught procedures

**TABLE 10.3**    Rankings of Student Achievement in Two Math Subjects
in Various Parts of the World

| Advanced Algebra | Geometry |
|---|---|
| 1. Hong Kong | 1. Hong Kong |
| 2. Japan | 2. Japan |
| 3. Finland | 3. England/Wales |
| 4. England/Wales | 4. Sweden |
| 5. Flemish Belgium | 5. Finland |
| 6. Israel | 6. New Zealand |
| 7. Sweden | 7. Flemish Belgium |
| 8. Ontario | 8. Scotland |
| 9. New Zealand | 9. Ontario |
| 10. French Belgium | 10. French Belgium |
| 11. Scotland | 11. Israel |
| 12. British Columbia | 12. United States |
| 13. Hungary | 13. Hungary |
| 14. United States | 14. British Columbia |
| 15. Thailand | 15. Thailand |

"The Case for More School Days," by Michael J. Barrett, *The Atlantic Monthly* 266 (November 1990), pp. 78–106.

for solving practice math problems whereas in Japan, the goal is conceptual understanding. An additional structural difference in education is that in some countries (for example, England, France, Germany, and China), students must pass a national examination to graduate from high school.

Another structural question has to do with amount of time children spend in school. The second research question asked at the beginning of this chapter concerned the number of days children spend in school in different countries. It might surprise you to learn that our children spend *fewer* days in school than do the children in many other countries—especially in those countries with which we increasingly find ourselves competing (see Table 10.4). This fact would seem to explain the tendency of American students to get farther behind students from other countries with each passing year. For more than 40 years, Gallup has polled Americans on their attitudes about lengthening the school year, and the response has always been negative. The gap has been closing recently as more people realize that some nations' students spend as much as 25 percent more time in school than do students in this country, but people are still not enthusiastic about the idea.

Does the problem lie with teachers? We hear that they burn out, that they are bored, that they aren't happy in their work. It is probably true that the best college-bound high-school seniors aren't headed for careers in education. But why should they be? Education is not widely seen as a challenging, interesting, or valuable career. It's difficult to get the best people to go into a profession in which the pay isn't great, there isn't much prestige, and the working conditions are often far from ideal. Teachers' starting salaries are typically far lower than those of other college majors. You would need to be dedicated to take less money, and you may not get the best and the brightest to go into teaching under those circumstances. Education is always

**TABLE 10.4**   Number of Days in a Standard School Year in Various Countries

| | | | |
|---|---|---|---|
| Japan | 243 | Finland | 190 |
| West Germany | 226+ | New Zealand | 190 |
| South Korea | 220 | Nigeria | 190 |
| Israel | 216 | France | 185 |
| Soviet Union | 211 | Ireland | 184 |
| Netherlands | 200 | Spain | 180 |
| Scotland | 200 | Sweden | 180 |
| Thailand | 200 | United States | 180 |
| England/Wales | 192 | French Belgium | 175 |

"The Case for More School Days," by Michael J. Barrett, *The Atlantic Monthly* 266 (November 1990), pp. 78–106.

squeezed for money, and the failure of yet another bond measure to raise funds for schools seems to be an annual occurrence. Teachers are very well paid in Japan, earning salaries generally higher than those of pharmacists or engineers. In a typical year, there are five applicants for every teaching job.

Maybe our values need reviewing. It is not unusual to find states cutting funding on education to pay for prisons and corrections programs. Maybe we are complacent and take education for granted. A few years ago a study compared the performance and attitudes of students and parents from Japan, Taiwan, and the United States. The American fifth graders performed poorly (the worst of the three groups) on mathematics tests. When the mothers were asked how their children were doing in school, the American mothers were more likely to report that they were "very satisfied," whereas the Chinese and Japanese mothers were more likely to say that they were "not satisfied." Ninety-one percent of the American mothers said the schools were doing an "excellent" or a "good" job, whereas only 42 percent of the Chinese mothers and 39 percent of the Japanese mothers were that positive.[6] The author of the study despaired at the practicality of pushing educational reform if the parents and citizens who provide the money and the force for change aren't interested enough to urge change.

Much public discussion has focused on the declining quality of American education, and many ideas have been offered about how to change it. Leon Botstein, president of Bard College, suggests some adjustments in the box entitled "How to Change the System" (see page 324).

## IS CHANGE COMING IN EDUCATION?

So far this chapter has sounded quite negative. Is there another side to the story? There continues to be tremendous interest in educational reform. We continually hear about individual teachers, particular schools, and cities in which innovative programs are working or being developed. Traditional policies and procedures are being reviewed. For example, there is increasing interest in getting away from the memorization of facts and focusing instead on developing critical thinking skills, even in the lower grades. California is getting away from A, B, C grading in elementary schools; instead, teachers

## HOW TO CHANGE THE SYSTEM

How do we go about changing the system? First of all, I would simplify the curriculum and focus on a few very basic things: a command of language, reading and writing, mathematical reasoning. Science and mathematics, as opposed to the humanities, should be at the center of the curriculum from the beginning. I take this radical position because what is taught in the humanities—in reading and writing and history—has become so politicized that science is the last common ground we have. It's also the easiest curiosity to sustain; every child wants to know how the world works. College is too late to develop a curiosity in science and mathematics; it can only be done early in life.

I would eliminate the American high school, because it is not fixable in its current design. High school is a nineteenth-century structure that has run its course: it is no longer congruent with the rapid development and autonomy of adolescents, their physical maturation, and the independence that society now permits them. I would divide schooling into five

or six years of elementary school and then middle school. At the end of what is now the sophomore year of high school, students would feed into the community college system, which works very well by comparison, doing what high schools should be doing.

I think we also have to face the fact that we must make teaching a desirable profession. In order to do this, we have to raise teachers' salaries. But more importantly, we have to change the working conditions in our public schools. Why do good people go to private schools to teach, even though salaries are considerably lower than at public schools? Because in public schools the teaching profession is bureaucratized, teachers are not treated as professionals, there is enormous regulation, the conditions of work are horrendous, there is no professional community. The administration and organization of schools need a radical overhaul.

—From "Educating in a Pessimistic Age," Harper's, *August 1993.*

---

are using written reports. This approach eliminates some of the competition present in the early years and probably gives a more complete evaluation of the child. This reminds me of the time several years ago when the parents of a beginning student proudly told me that their child had made the prekindergarten honor roll.

The types of tests being given have come under question. What do the SAT, ACT, or I.Q. tests actually measure? Their ability to predict future academic or job success is being increasingly questioned. It has been recognized for some time that they are culture-bound and class-linked. The tests are being revamped and revised to accommodate cultural diversity.

Tracking is currently in disrepute and is widely criticized, especially by those with the conflict perspective. There is a growing interest in looking at educational stages rather than tracks and in emphasizing developmental age rather than chronological age. We have tended to place children in school based on **chronological age**—kindergarten at 5, first grade at 6, and graduation from high school at 18. The Gesell Institute has long held that

## VOUCHERS—THE ANSWER?

School voucher programs became a controversial educational (and political) issue in the late 1990s, likened in terms of social and educational impact to the 1954 U.S. Supreme Court *Brown* decision. Voucher programs would provide taxpayer funds for parents who, because they feel that public schools are failing, wish to send their children to private and parochial schools. A 1999 Gallup poll of parents with children in public schools found that the parents supported school choice 60 percent to 38 percent, and vouchers found strongest support among nonwhites and urban parents. Legislators in many states are pushing through voucher programs, and the major political parties are lining up on both sides of the issue in election year 2000. There have been legal setbacks—court decisions in Florida and Ohio—and there are cases pending in other states. Experts believe that the issue will ultimately go before the U.S. Supreme Court.

Supporters argue that parents should have free choice to select the best school for their children—after all, it's their tax money. They believe that voucher programs may act as motivators for public school teachers and administrators to get their act together and improve their schools. Opponents of voucher programs say that public schools, in general, are not failing miserably, and that voucher programs are defeatist, an acknowledgment by the government that it has given up on public schools. Opponents also say that programs are abused (some voucher schools are not open to all as they are supposed to be), and that voucher programs to religious schools violate the constitutionally mandated separation of church and state. Currently, there is an absence of evidence that voucher programs actually improve schooling.

children should enter school and be promoted based on their actual behavior, or **developmental age,** rather than on their chronological age. What parents should ask themselves is, "Is my child (regardless of his or her actual age) *ready* for kindergarten?" Starting a child too early may be a disservice; many contend that overplacement is a major factor in a child's subsequent decision to drop out of school. Or the overplaced child may develop perfectly well in one area (intellectually, for example) but be left hopelessly behind in other areas (socially, for example). Emphasis on developmental age recognizes that people don't all mature at the same rate.

Finally, we might look back at some of the ideas that functionalist and conflict theorists debate. Is a mass of college-educated people needed for our society to work well? The current job market contains mostly service (nontechnical) types of jobs that don't need a high level of proficiency in math and science. Perhaps we need to focus our resources on workforce needs rather than on educating everyone to the greatest extent possible. Or, if we want to continue with our goal of educating everyone, a task few societies attempt, we must consider the possibility that, given that goal, perhaps we are doing the best we can do.

## THEORY AND RESEARCH: A REVIEW

This chapter has been structured around a comparison of the two dominant theoretical approaches, *functional analysis* and *conflict theory*. Functional theorists see society as a meritocracy in which all have an equal chance to achieve success. Ability and hard work will be rewarded, and the educational system will produce the number of highly skilled and trained people that an expert society needs. Functionalists see schooling as providing a number of important functions: selecting people with talent and ability, creating an informed citizenry, and passing on the culture of a society to the next generation. They also recognize the presence of other (latent) functions: baby-sitting young children so parents can work, and teaching habits of discipline and obedience.

Conflict theorists ask, as they usually do, who benefits and who is deprived. They see a social-class bias in our educational system, which means that privileged children benefit and working- and lower-class children are the losers. They believe that the social reproduction model operates in that schooling reproduces the social class of the student. Children of working-class parents go to schools that teach them to be obedient and follow rules, and students of the elite go to schools that encourage creativity and flexibility. Conflict theorists believe that working-class students drop out at least in part because they see that the system does not benefit them, and they give up. Conflict theorists would be skeptical about educational reform; they fear that given the type of changes suggested and the continuing lack of funding, the beneficiaries of reform will be those who are already privileged.

Much of the data discussed in this chapter comes from records: *reactive* and *nonreactive surveys* of numbers of days in school, salary by occupation, and high-school dropout rates by state and ethnicity. Some data came from *reactive questionnaires* and *interviews*—for example, asking students why they drop out of school. The cross-cultural studies on student performance were clearly *reactive* and were probably *experiments*.

## SUMMARY

The chapter focused on education, the fifth of five institutional areas explored in this book. Previous chapters examined the family, religion, and economic and political institutions. All societies must socialize and train their young—the central issue for the institution of education. We distinguished between education and schooling. The major focus of the chapter was on the unique and contrasting ways in which functional analysis and conflict theory view education and schooling. Finally, we discussed some current issues in education in the United States.

The reading that follows is a study of the school environment. Donald Ratcliff uses the school hallways as a laboratory and describes and organizes the behavior that occurs there.

# TERMS FOR STUDY

chronological age (324)                hidden curriculum (317)

democratic model (318)                 schooling (314)

developmental age (325)                social reproduction model (318)

education (314)                        tracking (316)

For a discussion of Research Question 1, see page 319.
For a discussion of Research Question 2, see page 322.

 ## INFOTRAC COLLEGE EDITION
### Search Word Summary

To learn more about the topics from this chapter, you can use the following words to conduct an electronic search on InfoTrac College Edition, an online library of journals. Here you will find a multitude of articles from various sources and perspectives:

**www.infotrac-college.com/wadsworth/access.html**

High school dropouts                   Bureaucracy

Educational vouchers                   Home schooling

Functional analysis                    Social interaction

## Reading 10.1

# MY EXOTIC TRIBE: CHILDREN IN A SCHOOL HALLWAY

*Donald Ratcliff*

*Sociologists have long maintained that the most everyday sort of situation can provide interesting insights if you just look carefully enough. Donald Ratcliff shows that in this reading about what goes on in elementary-school hallways. In so doing, he illustrates a number of the concepts and theories we have talked about throughout this book. Professor Ratcliff teaches sociology at Toccoa Falls College in Georgia.*

Sociologists sometimes study the strange and unusual situations that most people never see. On the other hand, sociologists may also study groups of people in commonplace situations and discover new understandings and meanings that those studied have in everyday life.

I took the second approach, concentrating my attention on a public school hallway. The school was a fairly typical elementary school in the Southeastern United States, with kindergarten through third grade in one wing and another third grade through sixth grade in another wing. I spent most of my time in the second wing of the school, concentrating my attention on third through fifth graders. For about four months I came in at least twice a week, spent several hours each visit, and watched what happened in the hallway. In the process I logged well over one hundred hours of observation. Later in the study I interviewed 52 children in groups of two to six, seeing each group at least four times. This is termed *field research* because I was in a real-life setting with people who spent their everyday lives at this school.

I approached the hallway as if I were an anthropologist and as if this hallway were a strange, exotic tribe that had never been studied previously. Actually, there *are* very few research studies of school hallways, so I did not know what all to expect. I tried to think of myself as an alien from another planet, seeing this situation for the first time. I described in detail everything I saw, and pushed myself to see the unexpected as well as the things that most people might take for granted (I used paper, pen, and a camcorder to record my observations). I left behind my role of college teacher, and instead became a learner. The hallway and the children were my teachers. This is called *participant observation* because I was also in the hallway and occasionally walked in the hallway as did the children I observed. The youngsters considered me to be a participant as well, stopping to chat occasionally—especially when teachers were not looking.

As I watched in the hallway and viewed the videotapes, I asked myself, "Of what is this an example?," "Does this always happen?, and "What does this mean?" Before long I began to see some things that seemed to stand out more than others. Eventually I showed some of my videotapes from the hallway to some fellow graduate students as well as some of my own students at a nearly college, and asked them what they thought was important. But perhaps most crucial, I met with my groups of youngsters from the school I studied and asked them to comment on what we saw on the tapes. These are ways of double-checking the importance placed on certain events that are observed.

While I tried to develop my own theories about what occurred in the hallway, using children's comments and my own description, I also reflected on the three major theories of sociology. Did my notes and children's comments fit the three theories? What aspects of those theories fit the activities in the school hallway?

*Functionalist* theory encouraged me to discover the obvious purposes of the hallway, as well as the more latent or hidden aspects of this part of the school. The obvious purpose of a hallway is to provide a way for children to move from one part of the school to another. But it was very obvious from my observations and what children told me that the latent functions were at least as important as the obvious purpose. Other purposes included carrying messages, avoiding classes, using restrooms or drinking fountain, studying, sitting for punishment, and seeing friends. When teachers were not present, youngsters often formed groups and talked with one another. Sometimes they would dramatize a basketball game or even perform as a stand-up comedian in front of the other kids. Clearly the hallway contained many social events and sometimes going from one place to another was a relatively minor aspect of what occurred.

Functionalism also suggests that some things that occur are *dysfunctional* or somehow counteract the main functions of the school. The school I observed had eliminated recess for everyone except the third graders in this wing of the school. This action, I concluded from what I saw and heard, probably contributed to the high energy activities I often observed. I concluded that at least some of the fighting, dramatizing, and social activity in the hallway were more likely to occur because it was one of the very few places in which children could have relatively unsupervised activities. I suspect there would have been fewer problems in the hallway if youngsters could release some of their energy and socialize on a playground. In England, for example, children have a "break time" much like the American recess even into their teen years. Increased concentration and learning in the classroom is associated with a recess break. It seemed to be a terrible mistake to remove recess, and the consequences appeared to be the high energy activity I saw in the hallway. The high activity is particularly dysfunctional when teachers are drawn away from educating and spend time disciplining children. This, in turn, may also be dysfunctional for children because they may develop unhealthy attitudes toward school as a result.

A second approach to looking at my research data is reflected in *conflict theory.* In school there is a dominant culture, sometimes described as "school culture," marked by formality, routine, and submission to adults and adult values. Children are expected to control their actions by sitting quietly at their desks, answering questions when the teacher asks them, and avoiding interaction with other children except when requested by the teacher. Youngsters react by creating their own "peer culture," marked by play, multiple options, peer norms and values, and flexibility. Specific actions reflecting peer culture include "play fighting"—sometimes called "rough and tumble play"—and the more or less spontaneous games and rhymes of childhood, such as the classic "ring around a rosie" jump-rope activity. Friendly teasing, mocking adults, and "fooling around" are also expressions of a culture that is a reaction to the sometimes oppressive actions of the teacher and other adults in the school.

The tension and conflict between school culture and peer culture is often below the surface; teachers tend to interpret that conflict as a problem between the teacher and a specific student, rather than a conflict of cultures. Children socialize their peers into the "tribe" of children by teaching their friends aspects of peer culture, while teachers and principals attempt to teach the "hidden curriculum" of adult supremacy and control. When teachers observe the rough and tumble play of peer culture, it is usually misinterpreted to be fighting and usually punished. Boys are characteristically more likely to engage in rough and tumble play, although I found many girls—most often minority girls—also participated in "play fighting." The more common expression of peer culture among girls—clapping rhymes—is less likely to be punished, perhaps because girls omit verses with sexual innuendo when adults are nearby, and possibly because elementary teachers are disproportionately women and thus more likely to ignore the aspects of peer culture in which they participated as children.

Youngsters are not often allowed to express "peer culture" in the classroom, and with the removal of recess, "peer culture" tends to be expressed in the hallway. I found in my interviews that children resent the many rules teachers imposed upon them in the classroom and hallway. They freely admitted to me that they regularly broke the rules when they thought they could get away with it, and that they enjoyed "play fighting," informal games, and talking. While a few youngsters—mostly girls—told me they would never engage in rough-and-tumble play and were sometimes afraid of those that did, nearly everyone wanted more time for their own social activities. The conflict between school culture and peer culture regularly emerged in the hallway, such as when children lined up at the teacher's command, yet a few youngsters near the end of the line might dance and playfully jostle one another when the teacher was not looking. Their view of school teachers as being too controlling made sense to me, and fit many of my observations. Yet I also realized that teachers were fearful of what children might do if not controlled, especially when other teachers and the principal could easily hear and see those activities in the hallway.

*Symbolic interaction theory* suggests that the symbols that arise in interactions are important, and there are underlying meanings held by those who are observed. Many, many times I observed one kind of symbol, the ritual, occur in the hallway. For example, I noticed children regularly brushed their hands against the wall as they walked down the hallway. Erik Erikson has described in detail the importance of everyday rituals and ceremonies to children. I found several variations on this activity, including rubbing arms or legs as they walked, touching finger tips against the wall, and even sliding their bottoms against the wall! What could this mean to children?

Usually the youngsters told me that all the rubbing was just the result of boredom. But, considering all they told me about the oppression of teachers, I wondered if there was more to it than just boredom. Perhaps the walls symbolized the division between the classroom and outside world. The ritual touching may represent their affirmation of that separation, and that at least occasionally the hallway could be a place where they could be more themselves and relatively free from adult rules.

I saw three general ways children gathered together in groups in the hallway. These might be considered ceremonies in which children regularly participated. The first ceremony was initiated and carefully monitored by teachers, the school line. Phases of school lining include: 1. positioning, which could be done in several possible ways, 2. waiting in the classroom until everyone is in line, 3. moving through the doorway, 4. waiting outside the room until cued by the teacher, 5. moving down the hallway until the destination is reached. School lines, I discovered, were very different from the lines—technically termed "queues"—that occur at fast food restaurants and many other places in everyday life. Queues determine priority of service and only move when someone has been served. School lines, in contrast, do not always indicate priority of service, as the primary purpose is to get children to move from one location to another in a quiet and controlled manner. On several occasions, the solemnity of children in line reminded me of a funeral procession or a ceremony in church. I found in my reading that lines that move people from one place to another are fairly rare in Western society, generally found in prisons, mental hospitals, and the military!

A second ceremony is when children group together in circles or semicircles, usually so they can talk together informally. I was impressed with how the physical shape encouraged interaction, as they watched one another's faces and body movements, as well as being able to hear what was being said. These interaction clusters would form fairly often when teachers were not present, and sometimes teachers would overlook these group activities if the talking was very quiet. Girls were more likely to get away with this kind of grouping.

A third ceremony was by far the most frequent variety, and it exists in everyday life for nearly everyone in Western society. This involved walking side-by-side, often talking in the process, as they walk down hallways. I was impressed with the physical coordination required to do this, as children made tiny adjustments to speed and direction to align themselves with one another in the process. At the same time, conversation is maintained or at least some social interaction occurs, such as smiling and looking at one another occasionally. It is a taken-for-granted aspect of real life that nearly everyone experiences, but has never been studied in detail.

What meanings did children attribute to these three ceremonies? Most of the children saw the imposed ceremony of school lines as unpleasant, a constant reminder that adults could force them to obey. It was an oppressive ritual, most concluded, although a few children—mostly girls—felt lines kept them safe from the misbehavior of other children.

Curiously, even though the other two ceremonies looked quite different to me— the stationary circle or semicircle and the side-by-side walking—children emphasized a common aspect of both: friendship. You walk with people you like, and you stop and talk with those who are friends. These two ceremonies both point to the importance of peers; they symbolize friendship.

Children in a school hallway is a small part of human life, yet to children it is an important and meaningful place where hundreds of things occur during the day. I talked with several teachers at the school after the research was concluded, and they all affirmed the importance of the hallway for youngsters. One even told me "What happens in the hallway is more important than what we do in the classroom." Perhaps that is an exaggeration, but I am sure the children I studied would agree.

# QUESTIONS 10.1

1. What is meant by "hidden curriculum"? Give examples from the reading and from your own experience.

2. Identify and list at least five concepts from this book that you can find in this reading.

3. What type of research techniques were used in this study? Would other techniques have worked? Why or why not?

4. All three sociological theories are discussed—which do you find the most interesting/useful?

5. Is this a macro or a micro study? Explain.

# NOTES

1. This discussion comes from chapter 3 of Christopher Hurn's book, *The Limits and Possibilities of Schooling: An Introduction to the Sociology of Education* (Boston: Allyn & Bacon, 1985).

2. This section on functionalism and conflict theory follows from Hurn's discussion in his chapter 2 (see note 1).

3. See *Schooling in Capitalist America* by Samuel Bowles and Herbert Gintis (New York: Basic Books, 1976).

4. Estimates on guns and dropouts come from "America Skips School," by Benjamin Barber, *Harper's Magazine* 288 (November

1993), p. 39. Other information on dropouts in this paragraph comes from "High School Dropouts; Descriptive Information from High School and Beyond," U.S. Department of Education, National Center for Educational Statistics, Bulletin NCES 83 221b, November 1983.

5. "The Case for More School Days," by Michael J. Barrett, *The Atlantic Monthly* 266 (November 1990), pp. 78–106.

6. This work by Harold Stevenson is discussed in Michael Barrett's article (see note 5), p. 94.

## CHAPTER ELEVEN

Stock • Boston/Rick Smolan

# *Population and Ecology*

**Population**

*Fertility • Mortality • Migration*
*• Other Population Measures*

**Human Ecology and the City**

*Patterns in the United States • World Patterns*

**Population and Ecology: Problem Aspects**

*A Population Bomb? • Other Ecological Problems*

## TWO RESEARCH QUESTIONS

**1.** *Some states are gaining population rapidly; others are losing population. How is population being redistributed in the United States? Are people moving to the West and South or to the Midwest and East, or is there any pattern?*

**2.** *Is there really a world population explosion? If so, what is the cause? Is it fertility (increases in birth rate), mortality (declines in death rate), or migration (population relocation)?*

PREVIOUS CHAPTERS IN this section on social organization have examined concepts dealing with groups, categories, aggregations, and institutions. In this chapter, we will show how social organization can be analyzed through the study of population and human ecology. Population refers to the number of people in a given unit, as in a state, society, world, or universe. **Human ecology** refers to the adaptation of people to their physical environment, their location in space. **Demography** is the study of human population—its distribution, composition, and change.

As an introduction to some of the important techniques and variables, let's try a demographic analysis of a college sociology class. What does the professor see as he or she looks out at the mass of eager young faces? The total population of the class selected is 130, of which 80 are females and 50 are males. There are 4 African Americans, 11 Asian Americans, and 115 Caucasian Americans. The age distribution of this "society" is overwhelmingly in the 19 to 23 year category, with trace amounts of ages up to 45. Both birth and death rates for this "society" are exceedingly low. Apparently this represents a very healthy but nonfertile tribe. The life expectancy for the majority will be one academic term, although for some it may be at least double that. Migration variables are peculiar for this tribe. Temporary immigration ("in-migration") occurs on those occasions when the class is planning to discuss sex or deviant behavior. Permanent emigration ("out-migration") occurs just before tests are scheduled or papers are due.

Ecologically, this class is also interesting. After several meetings class members arrange themselves in space consistently and predictably. There are occasional shifts as dating alliances change, but generally they sit in the same seats throughout the term. Part of the patterning can be explained by classical ecological processes. Cooperation: *X* takes great notes, so she is constantly surrounded by *A, B,* and *C* (at least I think that's the reason). Competition: When asked why they sit in front, several have said they pay more attention to what is being said and feel the instructor is more likely to remember them; consequently, they will get better grades. Films of migrating apes have shown ecological processes at work: Older males proceed in front, younger males at the rear, and females are protected in the middle. Do these principles apply in your classroom?

## THE WHOLE WORLD AS 100 PEOPLE

If we shrunk the world to a village of 100 people and kept the existing proportions the same, it would look something like this:

There would be . . .

—57 Asians, 21 Europeans, 14 from the Western Hemisphere, 8 Africans

—52 females, 48 males

—70 nonwhites, 30 whites

—70 non-Christians, 30 Christians

—6 people would possess 59% of the entire world's wealth; all 6 would be from the United States

—80 would live in substandard housing

—70 would be unable to read

—50 would suffer from malnutrition

—1 would be near death

—1 would be near birth

—1 would have a college education

—1 would own a computer

## POPULATION

Demographers study populations from several viewpoints. They might want to look at population growth or its general characteristics—a statistical portrait, so to speak. A number of different variables can be used. For example, if the issue is population growth, the important variables are births, deaths, and migration. (Population growth = births - deaths + migration.) If the concern is a general description of the population, then age, gender, race, ethnicity profiles, and perhaps patterns of internal migration may be examined. Growth potential and general characteristics are related; the number of young and old people in a population as well as the number of women in childbearing ages will have an effect on birth and death rates. In the following paragraphs we will look at some of the major population variables and show how they are used in the United States and in some other societies.

### Fertility

**Fertility,** which refers to the number of children born, may be measured in several ways. The **birth rate** (or crude birth rate) is the number of live births per 1,000 people in the population. These data are often used because they are easily collected, but they ignore the age structure of the population. The **general fertility rate,** which is a better measure, is defined as the number of births per 1,000 women aged 15 through 44. The **total fertility rate** is an estimate of the average number of children born to each woman based on current rates.

Fertility has both a biological and a social aspect. The physical or biological ability to reproduce, also known as **fecundity,** is present in most of us, and yet actual fertility rates differ greatly. A 19th-century Russian woman gave birth to 69 children; the Hutterite women (of North and South Dakota and Canada) were averaging more than 12 children each in the 1930s; currently, American women are averaging about 2 children each. The social influences on fertility are factors that vary from one society to another. For example, fertility would be affected by age at marriage, the proportion of people who never marry, frequency of sexual intercourse, and whether birth-control techniques are used and how effective they are. These factors are influenced by tradition, lifestyle, standard of living, and countless other social values.[1]

Trends in birth, general fertility, and death rates in the United States are shown in Figure 11.1. Our birth rate has been around 15 per 1,000 since the mid-1970s (14.4 in 1998). Remember that although birth and general fertility rates are similar, the general fertility rate is the better measure because it focuses on the number of women in the childbearing ages (15 to 44). The rates follow the same pattern, but the fluctuations are seen more clearly in the general fertility rate. The rates were high in the early 1900s, dropped in the 1920s and 1930s, went up after World War II, dropped sharply in the 1960s and 1970s, and were fairly flat in the 1980s. The general fertility rate dropped to 65.4 in 1986, the lowest it had ever been. In 1990 it increased to 71 but started down again and hit a new record low of 65.3 in 1996 and 1997.

The total fertility rate in 1983 was 1,799 children per 1,000 women and in 1997 increased to 2,040, which means that women averaged about 2 children each. This is slightly below the **replacement level,** which is defined as the average number of births necessary per woman over her lifetime (say 70 years) for the population eventually to reach zero growth. Although the United States has been below the replacement level, population continues to increase because of large increases in the number of women of reproductive age. The Census Bureau reports that this means that even if the current low fertility rates were to persist for some time, the population of the United States would still continue to grow by natural increase until well into the 21st century.

Understanding why fertility rates fluctuate helps us predict social conditions of the future, but the factors are many and complicated. This is what seems to be happening: More young women are remaining single, and divorce rates remain high (see the discussion in Chapter Seven). Women, especially those in the prime childbearing ages of 20 through 34, are either not having or are putting off having children. The increased freedom of women to work outside the home, the increased use of contraceptives, and changing viewpoints about family roles have contributed to the changing birth rate. Unemployment rates and a sluggish economy are probably also contributing factors.

World birth rates in 1996 showed an average of about 3 children per woman, compared with about 2 in the United States. Some countries averaged more than 7 children per woman (Ethiopia, Mali, Niger, Somalia, and

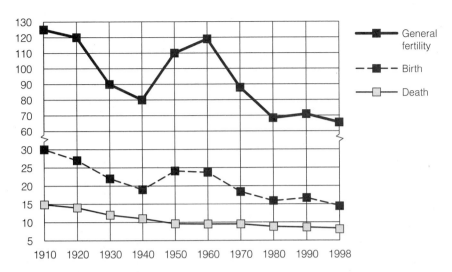

**FIGURE 11.1**   *General fertility rates (per 1,000 women aged 15 to 44) and birth and death rates (per 1,000 people) in the United States.*

United States Bureau of the Census, *Statistical Abstracts of the United States, 1999* (119th ed.), Washington, D.C.

Yemen) and some countries averaged 1.3 children or less per woman (Bulgaria, Germany, Hong Kong, Italy, Rumania, and Spain). (Some of these comparisons can be seen later in this chapter in Table 11.3, page 351.) High fertility rates can, if mortality rates are low, lead to rapid population growth, and later in this chapter we will examine some of the implications and consequences.

## Mortality

**Mortality** is the number of deaths occurring in a population. It is typically represented by the death rate (or crude death rate), which is the total deaths per 1,000 people in the population. Because death rates vary by age and gender, a more precise measure is the age-gender-specific death rate, which categorizes deaths of males and females in five-year groups—for example, males 15 through 19 and females 15 through 19. A third useful measure is **life expectancy,** which predicts one's expectation at birth of length of life based on risks of death for people born in that year.

Mortality has two major components: life span and longevity. *Life span* refers to the oldest age to which human beings can live. There is some uncertainty about just what the limits are; reports of people living to 110 and even to 120 years of age are not unheard of, and medical science continues to increase these outer limits. *Longevity* refers to the ability to remain alive from one year to the next—that is, to resist death. Both biological and social factors, of course, are integral to one's longevity. Biological determinants include inherited genetic characteristics, strength of organs, and resistance to disease. Social factors include lifestyle and the amount of stress and conflict it produces, the way society takes care of its members—especially the elderly—and so on. The Abkhasians in Russia and the Vilcabambas in Ecuador seem to live very long lives; it has been suggested that this is due to their simple, unchanging lifestyle, their avoidance of stress, and their tendency to remain physically and mentally active throughout their lives.[2]

The importance of social factors as explanations of longevity and mortality can be seen in other ways. Studies have shown that as occupational prestige goes up, death rates go down. This might be caused by greater work hazards in lower-class occupations. However, because the mortality levels of "nonworking" wives seem to parallel those of their husbands, lifestyle is apparently a major factor as well. Studies have also found that deaths from communicable diseases such as tuberculosis and pneumonia are higher among lower-class people, whereas deaths from coronaries, diabetes, and stroke increase with upward social mobility. Yet another study has found that for virtually every major cause of death, white males with at least one year of college had lower risks of death than those with less education. The existence of mortality-rate differences among various racial and ethnic groups is also probably best explained by income, educational, and lifestyle differences.

Death rates in the United States are also shown in Figure 11.1. We see that they have declined, but during the last 50 years they haven't changed much, and over the last decade, hardly at all. Death rates have leveled at between 8.5 and 8.8 from 1975 to the present. Sharp drops in the death rate occurred in the late-19th and early-20th centuries as medical science gained control of many infectious diseases. Life expectancy has risen to a current level of about 73 years for males and 79 for females. Death rates in European countries are very similar to ours, but rates in some other countries (see Table 11.3) are twice ours. This is changing rapidly, however, as health care and disease-control measures spread and countries share their discoveries. So we witness a growth in population caused not by rising fertility, but by declining mortality.

## Migration

**Migration** is defined as a permanent change of residence, with the consequent relocation of one's interests and activities. The number of dimensions—who moves, how often, where, and why—makes migration the most complex of the population processes. Internal migration generally describes migration within a country, whereas international migration refers to migration from one country to another. People migrate for many reasons, and these have been summarized into what is called the *push-pull theory*. This theory suggests that one reason people move is that they are pushed out of their original home—by a bad climate, by unpleasant treatment by others, by legal harassment, and so on. The other reason people move is that they are pulled by the attractions of another place. Those who advocate this theory suggest that motivation to improve one's lot is stronger than the desire to escape existing conditions; we could conclude that the *pull* motivation is more influential than is the *push*. It does seem, as Weeks points out, that migration associated with career advancement is a common theme in society today.[3]

Americans are a mobile group: About 42 million people, 16 percent of all people one year old and over, moved during the period between 1995 and 1996. Sociologists have found that some types are more prone to migration than others. Young adults aged 20 to 24 and 25 to 29 have the highest rates. Often these are young married couples with no children or with very young children. College graduates are more likely to move than high-school graduates, who in turn are more mobile than those with less education. All this tends to support the idea that people are being pulled to areas that are more economically attractive for them.

Patterns of internal migration continually change. People used to migrate from farm to city and later from city to suburb. More recently a new trend developed: the rapid growth of small communities near metropolitan areas and increasing migration to rural areas. The first question at the beginning of this chapter dealt with population redistribution. A major trend of the 1980s and 1990s was a migration flow from "frostbelt" and "rustbelt" to "sunbelt." People left northeastern and midwestern states and moved to the West and South. The five biggest gainers in numbers between 1990 and 1998 were, in order, California, Texas, Florida, Georgia, and Arizona. North Dakota, Rhode Island, and Connecticut actually lost population during that period. Frostbelt states are declining in population, not only because of out-migration, but also because they typically have an older population and lower birth rates. The states gaining the most and least population on a *percentage* basis are shown in Table 11.1. Looking at those states, patterns of population redistribution become very clear. This probably reflects differing economic opportunities as well as changing recreation and leisure patterns.

International migration to the United States has gone through several stages. Before World War I, immigrants were admitted freely. Then, as people became concerned about "racial purity," job competition, and political "undesirables," immigration laws became very restrictive. Policies were relaxed slightly in the mid-1960s, but the overall number of immigrants allowed into this country was relatively small and constant throughout the 1970s and 1980s. Then another change took place in the early 1990s. Immigration, especially illegal immigration, became a potent political issue and the number of immigrants declined from more than 1.8 million in 1991 to 798,000 in 1997. A major part of this reduction was immigrants from Mexico, which went from 900,000 in 1991 to 147,000 in 1997. Figure 11.2 shows the 10 countries supplying the most immigrants in 1997. It is interesting that the countries comprising the former Soviet Union supplied more than 63,000 immigrants in 1996—if taken as a group, the second most after Mexico. Immigration patterns are clearly affected by economic conditions and standard of living as well as by political turmoil and instability.

Another side of migration that further underscores the point that Americans are a mobile group is the following: Americans are the second largest immigrant group to Italy and the third largest to England. More than a half million Americans live permanently in Mexico.

**TABLE 11.1**   Population Change in the United States 1990–1998
(National Population Increased 8.7 Percent)

| States Gaining Most Population | Percentage Change | States Gaining Least Population | Percentage Change |
|---|---|---|---|
| Nevada | 45 | Rhode Island | −1.5 |
| Arizona | 27 | Connecticut | −0.4 |
| Utah | 22 | North Dakota | −0.1 |
| Idaho | 22 | Pennsylvania | 1 |
| Colorado | 21 | West Virginia | 1 |

United States Bureau of the Census, *Statistical Abstracts of the United States, 1999* (119th ed.), Washington, D.C.

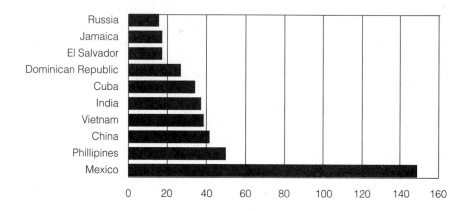

**FIGURE 11.2**   *Immigrants by country of birth, 1997.*

United States Bureau of the Census, *Statistical Abstracts of the United States, 1999* (119th ed.), Washington, D.C.

The consequences of migration are many. For the individual, uprooting oneself and trying to adapt to a new environment involves stress, and the burden of coping is greater in the absence of familiar support groups. The community left behind might lose a valuable segment of its population. The community to which people are migrating has a growing population to assimilate, with the accompanying problems of housing, mass transit, health care, job opportunities, municipal services, and ways to pay for it all. This has led some communities and even states to devise policies to limit migration to their areas.

### Other Population Measures

Fertility, mortality, and migration are the major demographic variables, but other measures help describe certain population characteristics. One of the simplest and most direct measures is *population size*. The world population passed the 6 billion mark on October 12, 1999. Population clock watchers at the United Nations (U.N.) wanted to note the exact historic moment when the 6 billionth person arrived, but the number went by so quickly that one U.N. official said, "Somebody must have had triplets." By October 24th, 12 days later, the world's population had increased by another 1.5 million. The population of the United States in 1998 was 271 million. This means that we are smaller than China (1.2 billion) and India (1 billion) but larger than Canada (31 million) and France (59 million).

More detailed profiles based on sex, age, race, religion, and ethnicity are also of interest. **Sex ratio** is defined as the number of males per 100 females. The sex ratio at birth in the United States is 105, or 105 males for every 100 females. Because males have a higher death rate than women at all ages, however, the sex ratio for the population as a whole is 95. Because there are 6 million more women than men in the United States, nearly all states show that same sex ratio, with Massachusetts, New York, Rhode Island, and Pennsylvania having the greatest number of women compared with men. Alaska, Hawaii, Wyoming, Nevada, and California were the only states in 1990 to have more men than women. The *median age* of a population, which designates the midpoint of the age range, is another population measure. The United States median age in 1970 was 28, in 1980 was 30, in 1990 was 32.8, and in 1998 it was 35.2. Our population is getting older; we have fewer teenagers and more elderly people and more people in their thirties and forties. The fastest-growing segments of the population are those aged 35 to 44 and those aged 85 years and older. The number of young people (15 to 24 years old) actually declined between 1980 and 1998.

Several variables, such as age and sex, can be profiled together graphically in a figure called a **population pyramid.** Figures 11.3 and 11.4 illustrate the distinctive age and sex profiles presented by two countries, Mexico and Sweden. Mexico has a high birth rate and a young population, whereas Sweden has a low birth rate and a balanced population image. These types of profiles also can be done on cities—imagine the pyramid you might get in St. Petersburg, Florida (older, retired people and more females), Norfolk, Virginia (influenced by the number of young males at the large military base), or Davis, California (a university community).

We have examined the major population variables—fertility, mortality, and migration—and other measures that allow demographers to predict future population changes and to provide a statistical portrait of societies. Population data, carefully collected and analyzed, are useful in social planning.

Population profiles by age and sex enable planners to determine the number of schools needed, the number and proportion of apartments and homes to build, the types of recreation and childcare facilities required, and so forth. For example, when we hear that a group of city planners is developing a community with no schools (because there are no children), no tennis courts (because tennis is too strenuous for the existing population), and many recreation centers and shuffleboard courts, it is safe to assume that the population data evidenced a need for a planned community for the elderly.

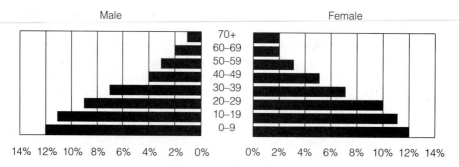

**FIGURE 11.3**   *Population pyramid for Mexico, 1995.*
*Demographic Yearbook,* 1997.

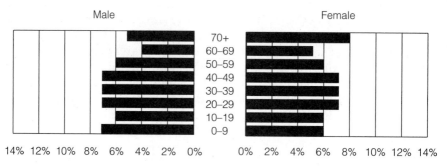

**FIGURE 11.4**   *Population pyramid for Sweden, 1996.*
*Demographic Yearbook,* 1997.

## HUMAN ECOLOGY AND THE CITY

Biologists have for some time studied the interrelationships and interdependence of plants and animals—the balance of nature. Living organisms are dependent on other organisms and distribute themselves in space accordingly. Why don't porcupines live in the desert? It may be because they would get sand in their quills, but most certainly because their food supply—the bark of specific types of trees—does not grow there. Likewise, all organisms live in particular places and patterns but not in others. Social scientists noticed that people, too, locate themselves in space in specific patterns, and this gave rise to the study of human ecology.

Flying over the United States, we see that people are not equally distributed but are grouped in clusters. There are fewer clusters in the West and more in the East. Most of the clusters are near rivers, lakes, or oceans. Highways run through the clusters, and various natural resources are nearby. There are fewer clusters where the climate is severe or where the terrain is rugged. People also distribute themselves in space in predictable ways. They group together in cities, where mutual cooperation and a complex division of labor allow people to accomplish much more than they could as

## ERNEST W. BURGESS (1886–1966)

Ernest Burgess earned his Ph.D. in Sociology at the University of Chicago. Burgess's *concentric zone hypothesis* was one of the most notable and well-known attempts to develop a comprehensive description of the forces that act upon urban development. His model was founded on the view that various elements of a heterogeneous and economically complex urban society actively competed for favorable locations within the city. The model assumed steady urban growth and the desirability of central location for commercial and financial interests. Burgess's goal was to anticipate and predict development within the city.

Burgess also conducted a great deal of research on the nature of the family and the quality of marital stability. His books in this area include *Predicting Success or Failure in Marriage* and *The Family: From Institution to Companionship*. His *Introduction to the Science of Sociology* became a classic and forged new directions in sociology.

Burgess taught at the University of Chicago from 1916 to 1952 where he was a major influence on the development of the Chicago School, focusing on the value of empirical research, delinquency, family life, and pioneering the study of the city.

isolated individuals. These cities are located to take maximum advantage of natural factors—waterways for transportation, resources for technological needs, and so on.

If we look more closely at the cities, we see that they also have consistent patterns. One city patterning first suggested by Ernest Burgess has five concentric circles or zones (see Figure 11.5). The small inner circle is called the *central business district* and contains the major stores, banks, offices, and government buildings—the downtown area of the city. The next circle is called the *zone in transition* and contains rundown buildings, slums, a high proportion of minority groups, some industry, and higher rates of unemployment. The third ring is called the *workingman's zone* and contains small apartment buildings and older single-dwelling units. The *middle-class residential district* is the fourth zone and contains better private residences and high-class apartments. Finally, the *suburban* or *commuters' zone* contains larger estates and golf courses and makes up the upper-class residential district. Generally speaking, as one moves from the inner zones outward toward the suburbs, population density decreases, home ownership increases, the proportion of foreign-born decreases, social class rises, crime rate decreases, land cost decreases, and family size decreases. The pattern of the city can be affected by how fast the city develops, by the influence of cars, and by natural obstacles—hills, oceans, rivers. So a given city might not fit this design exactly, but many cities are close to this general pattern.

Other consistent city patternings have been suggested and are also depicted in Figure 11.5. Homer Hoyt contends that cities are more accurately

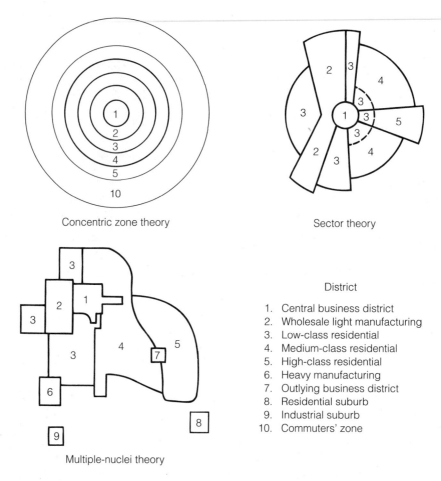

Concentric zone theory

Sector theory

District

1. Central business district
2. Wholesale light manufacturing
3. Low-class residential
4. Medium-class residential
5. High-class residential
6. Heavy manufacturing
7. Outlying business district
8. Residential suburb
9. Industrial suburb
10. Commuters' zone

Multiple-nuclei theory

**FIGURE 11.5** *Three generalizations of the internal structure of cities.*

From C.D. Harris and Edward L. Ullman, *Annals* 242 (November 1945), pp. 7–17.

described as *sectors* with a specific type of land use than as concentric zones. He believes that expensive high-rent areas follow patterns of city growth outward from the center like wedges of pie or spokes of a wheel. High-rent areas develop along the paths of major streets, highways, or other major transportation lines. This type of outward growth could be along rivers or rapid-transit systems, for example, and would often cut across the concentric circles suggested by Burgess.

*Multiple nuclei* describes a city in which a number of specialized centers or downtowns develop, each of which exerts dominance over its particular area. These centers might exist from the beginning of a city's development, as in London, or might develop later as the city grows and expands. Industry and residential areas grow up around each nucleus. Los Angeles, which has been called "a bunch of suburbs in search of a city," is an example of multiple-nuclei development.

More recent theories of city patterning include social area analysis and factorial ecology. These techniques focus on a city's census tracts and obtain information across many variables, such as median rent, median education, percentage of the population married and employed, and the racial and ethnic mix. Tracts are then compared in an attempt to isolate the factors that distinguish among tracts and that explain city patterning. These techniques are usually computerized and are very precise.[4]

How are these patterns explained? People arrange themselves in space according to social and cultural values that they believe to be important. It is apparent from previous paragraphs that economic competition has much to do with spatial arrangement in cities. Land is in greatest demand and, therefore, is most expensive in the center of the city. Banks, large stores, businesses, office buildings, and government buildings can afford the cost of locating there because of their importance, business volume, or financial base. Structures whose economic effects are less dramatic—homes, apartment buildings, schools, country clubs—are usually located away from the center of the city where land is cheaper. Other social values are also reflected. Many cities have set aside large green areas in those parts of the city where land is most expensive. Parks near the centers of London and Paris, the Boston Common, and Central Park in New York are but a few of numerous examples illustrating that economic competition is not the sole determining factor in spatial arrangement. Racial and religious minority groups generally reside in separate areas of the city. This separation or segregation sometimes occurs by choice of the minority group, but more often it represents the wishes and values of the majority group. Social-class groupings also separate themselves into different parts of the city, and inevitably areas develop reputations that attract certain types of people and discourage others. As Otis and Beverly Duncan have pointed out, "spatial distances between occupation groups are closely related to their social distance."[5] In a variety of ways, then, we see that spatial arrangement reflects social and cultural values.

It is both important and interesting to note that spatial arrangement affects social interaction. People interact with those who live next door or across the street, unless it's a wide and busy street. People living toward the middle of the block have more social contacts than those living on corners. Friendships develop more on the side of the house where the driveway is located, especially if there are adjacent driveways; driveways are natural areas for gathering and talking. A family that doesn't mix or fit in or that is unfriendly often acts as a boundary; neighbors have difficulty interacting past this boundary as they would across a busy street. Such habits become so ingrained that when a new family moves into a house, it might inherit the previous family's reputation as a social boundary.[6]

William Michelson, in his book *Man and His Urban Environment*, outlines other factors that correlate with spatial arrangements. He finds, for example, that physical and social pathologies—crime and social and physical illness—are related to certain spatial factors. The health of women and their children living in self-contained houses was compared with the health of those living in three- and four-story apartments. The sickness rate of the apartment dwellers was 57 percent greater than the sickness rate of those living in houses, as measured by first-consultation rates for any ailment. The differences were attributed to the cramped space and greater isolation of women in apartments removed from the ground. This conclusion is underlined by the fact that the women in apartments who did not have young children (and were therefore able to come and go from their homes more freely) had excellent health. High noise levels are related to the incidence of diseases that involve tension. Studies of overcrowding (many people per room) and high density (many people per acre) have found generally that *density* is more related to pathology than is overcrowding. For example, in one study the researchers found that overcrowding at home was not related to students' achievement in school, but the number of families on the block (density) was. Michelson reminds us that although spatial arrangement can be important, its effects are modified by cultural values.

Different types of people have different spatial preferences. Michelson reports that people who highly value convenience are likely to prefer more mixed land uses and small lot sizes; people who highly value individualism prefer larger lot sizes. Research indicates that children to about the age of seven living in high-rise apartments cling much more closely to their parents than do children in single-family homes, who become independent at an earlier age. But past that age the patterns reverse: When they become more mobile, children living in high-rise apartment buildings spend much more time away from home than do children living in single-family houses. Finally, one study reports that although apartment dwellers suffer more from noise disturbance than do people living in homes, their main complaint is the restriction they feel on making noise themselves; this affects their leisure patterns and leads them to such sedentary practices as watching television.[7]

Ecologists use the terms *centralization, decentralization, segregation, invasion,* and *succession* to describe specific types of spatial arrangements. **Centralization** describes the tendency of people to gather around some central or pivotal point in a city. Centralization allows citizens to better fulfill social and economic needs and functions and is represented in most American cities by the central business district. **Decentralization,** on the other hand, refers to the tendency to move away from the central focus. Decentralization probably occurs at least partly because of dissatisfaction with some of the consequences of centralization—traffic, crowds, noise, concrete, and so on—and is seen in the United States in the rush to the suburbs. **Segregation** refers to the clustering together of similar people. These similarities can be along the lines of occupation, race, religion, nationality, ethnicity, or education. **Invasion** refers to the penetration of one group or function into an area dominated by another group or function. **Succession** refers to the complete displacement or removal of the established group and represents the end product of invasion. The invasion-succession process often produces tension, hostility, and sometimes overt conflict.

A brief look at the ongoing drama of the local faculty dining room illustrates some of these ecological concepts. The room was originally segregated: Faculty ate there, students ate elsewhere. Segregation also existed within the room: Academic faculty generally sat on the south side of the room, and nonteaching staff—secretaries and administrators—sat on the north side of the room. As the college population grew, students were crowded out of other eating areas and began moving into the faculty room. Invasion had begun. Students found that the room made a convenient study area, and they brought stacks of books and papers, which further crowded the faculty. Finally, the faculty gave up and moved elsewhere as the students took over the room. Alas, succession had occurred.

## Patterns in the United States

Ecological arrangement in the United States has changed. People who live in large population clusters (50,000 people or more) and in other smaller cities of 2,500 people or more are defined as *urban* by the Census Bureau. It's probably no surprise to learn that, like the rest of the world, we are becoming more and more an urban nation. In 1860 our population was 20 percent urban; in 1990 we were more than 75 percent urban. That 75 percent of the population occupies just 2.5 percent of the nation's land. The number of people living on farms, about 4.6 million in 1991, has been steadily declining during the past 30 years. As nations undergo technological and industrial development, it is probably inevitable that they become more and more urbanized. The transformation from rural to urban and the increased problems of the urbanized society have led to comparisons between the two ways of life. Urban life is seen as secondary, tense, complicated, and anonymous, and rural life is seen as primary, simple, peaceful, and benign. This tendency to idealize rural life might partially explain the current rush to the suburbs. At times in our past, differences between people living in rural areas and those living in cities were pronounced; those differences in religious observance (rural: higher), education (urban: higher), birth rate (rural: higher), and divorce (urban: higher) have been particularly noted. It is likely today, however, that with the effects of migration, technological change, and the mass media, the differences between rural and urban people have become slight.

A related ecological phenomenon is the exodus from the cities to the suburbs, especially by whites. The area near the center of the city (zone in transition or workingman's zone) often offers the least expensive housing in the city. As a result, lower-class and minority-group members migrate into these areas, and as they move in, others who can afford to do so leave for the suburbs. Suburbanites outnumbered city dwellers for the first time in 1970, and this trend has continued. Many people who can afford to move are escaping the central city for the suburbs. The heavy migration of minorities *to* cities, which started in the 1950s, and whites *away* from cities more recently has meant that a number of cities now have a population that is more than 60 percent black and Hispanic; Atlanta, Chicago, Detroit, Miami, New Orleans, Newark, Oakland, San Antonio, and Washington, D.C., are examples.

## EARTH TRENDS—HOW ARE WE DOING?

**What's Good?**

• Life expectancy on the planet has gone from 46 years in 1950 to 65 years in 1993

• Production of chlorofluorocarbons has been reduced

• Worldwide bicycle sales are triple those of cars

• The number of nuclear weapons has been reduced

**What's Not So Good?**

• Birds are disappearing—two thirds of all species are on the decline, and 1,000 species are threatened with extinction

• Destructive insects are developing resistance to poisons—now at least 17 species are unaffected by any insecticide on the market

• The sea is yielding about all the food it can—the average yield per person is dropping

• Grain stocks are at their lowest level since the 1970s

• Loss of tropical rain forests is now exceeded by the loss of temperate forests

• There are 19 million refugees worldwide, a new high

• Population continues to grow at alarming rates

• The AIDS virus has now infected more than 40 million people with no cure in sight

• Americans are 5 percent of the world population but consume 25 percent of the energy

• The average American uses as much energy as 6 Mexicans, 153 Bangladeshis, or 499 Ethiopians

• The U.S. population is growing faster than that of any other industrialized nation

*—From Worldwatch Institute and the Population Reference Bureau.*

---

In the late 1990s, much of the population growth in metropolitan areas actually took place in suburbs and surrounding communities rather than in central cities. Some urban experts, once critical of life in the suburbs, have begun to change their tune. Suburbs and surrounding communities are experiencing job growth, economic expansion, and even a sense of community. Many of the functions and pleasures of the city are available in the suburbs without the population density. The two fastest-growing metropolitan areas between 1990 and 1998 were Las Vegas, Nevada, which increased in population by 55 percent, and Laredo, Texas, which increased by 41 percent.

Another ecological trend that occurred in the United States in the latter half of the 20th century was the development of *strip cities* or *megalopolises*. One city and its suburbs merge and grow into an adjacent city and its suburbs. The eventual result is an unbroken series of cities for tens, and even hundreds, of miles. Travelers flying at night over the East Coast report an unbroken chain of lights from north of Boston to south of Richmond, a 600 mile-long city. If urban or near-urban populations continue to grow, strip cities can be anticipated along the West Coast from San Francisco to San Diego, in the Midwest around the Great Lakes, in the Pacific Northwest, along both Florida coasts, and in many other places throughout the country.

**World Patterns**

Like the United States, the world is also becoming more urbanized. In many areas, people are moving from the countryside to the city in great numbers. In 1999, 47 percent of the world's population lived in cities; this will be 50 percent by 2005 and more than 60 percent by 2015. Sometimes the motivation is war, but more often it is poverty, the lure of better job opportunities, and improved living conditions. Often the reality is quite the opposite—menial jobs (if any), increased poverty, and squalid living conditions. This migration, combined with high population growth (high birth rates), is leading to the development of "mega-cities," as is shown in Table 11.2. Notice several interesting factors in Table 11.2. First, the *huge size* of cities of the future. It's hard to imagine metropolitan areas nearly as large as some of our most populous states (in 1998, California had 33 million, Texas had 20 million, and New York had 18 million). How can such cities work? Can you imagine the traffic? Also, notice the population shifts—the dominant urban centers switched from the developed nations of the United States and Europe to the less-developed but high-growth countries of Asia. These patterns of growth and shift bring us to a discussion of some of the problem aspects associated with population and ecology.

## POPULATION AND ECOLOGY: PROBLEM ASPECTS

*Somewhere on this globe every 10 seconds, there is a woman giving birth to a child. She must be found and stopped. —Sam Levinson*

As well as providing important information about the characteristics of societies, the study of population and human ecology calls attention to several important problems. One is what many say will become our most serious social problem: the population explosion. The second question at the beginning of this chapter asked if we have a world population problem, and if we do, what caused it: rising birth rates, dropping death rates, or migration patterns? The answer to the first part of the question is easy: *yes!* During the next 24 hours the world will add more than 210,000 people—it's getting very crowded in many parts of the world. The answer to the second part of the question—what caused it—is more complicated.

For ages the human population remained fairly stable. Such things as disease, wars, and high rates of infant mortality kept births and deaths pretty much in balance. Then some 200 years ago, a demographic transition occurred as death rates began to fall and population growth suddenly accelerated. A general overview of the growth of the human population over the past 10,000 years is shown in Figure 11.6.

**TABLE 11.2**    The World's Ten Largest Urban Areas—Population in 1960 and 1998

| | 1960 | | 1998 | |
|---|---|---|---|---|
| Rank | Urban Area | Population (millions) | Urban Area | Population (millions) |
| 1 | New York | 14.2 | Tokyo | 28.8 |
| 2 | London | 10.7 | Mexico City | 17.8 |
| 3 | Tokyo | 10.7 | Sao Paulo | 17.5 |
| 4 | Shanghai | 10.7 | Bombay | 17.4 |
| 5 | Beijing | 7.3 | New York | 16.5 |
| 6 | Paris | 7.2 | Shanghai | 14 |
| 7 | Buenos Aires | 6.9 | Los Angeles | 13 |
| 8 | Los Angeles | 6.6 | Lagos | 12.8 |
| 9 | Moscow | 6.3 | Calcutta | 12.7 |
| 10 | Chicago | 6.0 | Buenos Aires | 12.3 |

U.N. Population Fund 1999 Report and Statistical Abstracts

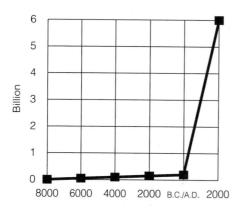

**FIGURE 11.6**    *World population growth.*

**Demographic transition** refers to the movement of a population through three stages: (1) high birth rates and high death rates, (2) high birth rates and low death rates, and (3) low birth rates and low death rates. Western industrialized nations are in the third stage, but nations in the underdeveloped regions of the world are generally in the second stage, which means rapid population growth. Table 11.3 shows birth rates, death rates, life expectancy, and population density for several underdeveloped, or Third World, nations and for several industrialized, or developed, nations. Of the underdeveloped countries, Angola, Ethiopia, Mozambique, and Nigeria remain on the fringes of the first stage. Their death rates are higher and life expectancy is lower than in other countries. The remaining underdeveloped nations are clearly in stage two, just as the industrialized countries shown are clearly in stage three—low birth and death rates. It is interesting to note that population density is not as closely related to industrialization as one

**TABLE 11.3**  Birth Rates and Death Rates (per 1,000 People), Life Expectancy, and Population Density in Selected Countries, 1999

| | Birth Rate | Death Rate | Life Expectancy at Birth | Density: People per Square Mile |
|---|---|---|---|---|
| **Underdeveloped Nations** | | | | |
| Angola | 43 | 16 | 48 | 23 |
| Egypt | 27 | 8 | 62 | 175 |
| Ethiopia | 44 | 21 | 41 | 138 |
| Guatemala | 36 | 7 | 67 | 295 |
| India | 25 | 9 | 63 | 872 |
| Iran | 21 | 5 | 70 | 103 |
| Mexico | 25 | 5 | 72 | 135 |
| Mozambique | 43 | 17 | 46 | 63 |
| Nigeria | 42 | 13 | 53 | 324 |
| **Developed Nations** | | | | |
| Australia | 13 | 7 | 80 | 6 |
| Canada | 12 | 7 | 79 | 9 |
| France | 11 | 9 | 79 | 280 |
| Japan | 10 | 8 | 80 | 828 |
| Russia | 10 | 15 | 65 | 22 |
| Spain | 10 | 10 | 78 | 203 |
| United Kingdom | 12 | 11 | 77 | 634 |
| United States | 14 | 9 | 76 | 77 |

United States Bureau of the Census, *Statistical Abstracts of the United States, 1999* (119th ed.), Washington, D.C.

might think. Some of the industrialized countries have high densities and some do not, and likewise with the nonindustrialized countries. If the non-industrialized nations stay in stage two, however, their population densities will probably increase substantially. In 1996, 4.6 billion people—80 percent of the world's population—lived in the underdeveloped countries.

With increasing control of disease and death, population is exploding in many countries of the world. For example, the birth rate in Mexico is nearly twice that of the United States; Mexico's death rate, however, has dropped to even less than ours. The result of this lowered death rate but higher birth rate means population growth. World population is growing at a rate of almost 1.5 million people per week! Every year the world adds about 80 million people—the number of people in California, Texas, New York, and Florida—to the population. The world population is twice what it was 20 years ago. At current growth rates, it is reported that in 600 years or so we will have about one square foot of land per person. (*Then* maybe further reproduction will be impossible.) The world has witnessed headlines of famine and starvation in Bangladesh, in Ethiopia, and in many other parts of Africa. Food production is not keeping up with population growth in many areas of the world. High oil prices have hurt production of fertilizers, and there is worldwide soil erosion. Population pressure means poverty, disease, starvation, and war.

## A Population Bomb?

An influential essay by Thomas Malthus published in 1798 forecast trouble. Malthus stated that people have a natural urge to reproduce and that population would increase faster than would people's ability to produce food. The situation could be helped by preventing births, but Malthus felt that postponing marriage and other types of "moral restraint" were the only acceptable means. Other methods of birth control—contraception, abortion, infanticide—were considered improper means. To Malthus, rational planning was the best hope for controlling the population cycle that would ultimately lead to starvation and poverty. Marx and Engels didn't go along with Malthus, primarily because Malthus felt that poverty was the fault of the poor. Marx and Engels saw poverty as the fault of a poorly organized (capitalist) society. They felt it unnecessary to limit births because advances in science and technology would enable food supplies to keep up with population growth. Modern Malthusians have suggested the idea of moral restraint as the only acceptable means of birth control and see far more disastrous consequences of population growth than Malthus did. Modern Marxists are more likely to recognize the presence of population-growth problems.[8]

Many population experts, although not wholly agreeing with Malthus's analysis, are concerned about population growth. Many organizations—United Nations, Worldwatch, Population Institute, Zero Population Growth—have been attempting to inform and enlighten people about the problem and the means to deal with it. Educational programs throughout the world have focused on family planning and birth control. Apparently these programs are having some success. An optimistic Populations Institute report, issued in late 1996, reported faster declines in fertility rates than had been anticipated and raised the hope that it might be possible to stabilize the world's population at 8 billion by 2025 if we keep on the current track. The United Nations was also positive—it reported that poor countries were increasingly developing plans to curb population growth, and rich countries were spending more money to help them. China's birth rate has dropped significantly following their strict one-child family planning policy, and Mexico has also reported a slowing in population growth, indicating that their family-planning program might be working.

Most demographers are cautious, however. Local customs often make new programs difficult. Distrust of doctors, ignorance about biological facts or about the consequences of overpopulation, desire for sons to farm the land, and numerous other cultural values have frustrated population-control efforts. To further complicate matters, political motivations lead some countries to believe that it is in their best interests to *increase* population. Concerns about having enough workers to run the factories, or soldiers to fight the wars, or voters to provide political power have led countries to try to increase their birth rates. They have offered financial incentives and have introduced a variety of "family-friendly" measures and pro-birth information.

Some suggest that population pressures can be mitigated by technological advances in agriculture and food production. We are making great strides in these areas, but it must be noted that even with a current

## THOMAS ROBERT MALTHUS (1766–1834)

Thomas Malthus earned his B.A. in Economics at Cambridge University. Malthus's main contribution was his theory of population, published in *An Essay on the Principle of Population*. Malthus stated that whenever a relative gain occurs in food production over population growth, a higher rate of population increase is stimulated; on the other hand, if population grows too much faster than food production, famine, disease, and war check the growth.

Many who supported this theory used it against efforts to better the conditions of the poor. Malthus himself was opposed to English Poor Laws. He said, "A man who is born into a world already possessed, if he cannot get subsistence from his parents on who he has a just demand, and if the society does not want his labour, has no claim of *right* to the smallest portion of food, and in fact, has no business to be where he is. At nature's mighty feast there is no vacant cover for him." Malthus's work was more influential than that of any other single individual in helping secure the passing of the new Poor Law of 1834, a measure that reflected the interests of the industrial bourgeoisie. His work received much criticism from Karl Marx who said, "The hatred of the English working class against Malthus is therefore entirely justified. The people were right here in sensing instinctively that they were confronted not with a *man of science* but with a *bought advocate*, a pleader on behalf of their enemies, a shameless sycophant of the ruling classes."

Malthus's work encouraged the first systematic demographic studies. From 1805 until his death, he was a professor of political economy and modern history at the college of the East Indian Company at Haileybury.

worldwide surplus of food and fiber, 10 million babies die of malnutrition every year. Appropriate distribution of available food supplies remains a vexing problem. In early 1997, political differences between Japan and North Korea meant that Japan was spending hundreds of millions of dollars storing tons of unneeded surplus rice while in nearby North Korea 20 million people were hungry and suffering from malnutrition.

### Other Ecological Problems

Measurements around the globe indicate that atmospheric dustiness is increasing by perhaps 30 percent per decade, and this poses a serious threat to the climatological balance. Fuels are being burned at rates that will exhaust estimated world reserves in less than a century. Pesticide residues are found everywhere, even in the fat of Antarctic penguins and in the Greenland ice cap. Chemicals developed to control pests have turned out to be more effective than anticipated and are now destroying other forms of wildlife.

Migration is placing tremendous strains on urban areas. Lima, Peru, is adding 1,000 people a day, most of them from the countryside. The U.N.

## JAPAN—NOT ENOUGH PEOPLE?

Birth rates in Japan are down to an all-time low of 1.39 per woman. The number of children aged 14 or under is at its lowest since 1920. (In an amazing prediction, a recent Health Ministry report estimated that if the current birth trend continued, the Japanese population could disappear sometime around the year 3500!) Within a decade, the population will begin to fall, meaning fewer workers and taxpayers. The Japanese government is worried that this will mean a less prosperous and lonelier country and will squeeze funds needed to provide for the rapidly increasing number of elderly. Japan is slowly trying to change to make it easier to raise children. Men are helping out more at home as women increasingly go to work. More baby-sitting and day care agencies are appearing and more companies are allowing workers time off to care for children. But change is slow—traditional roles indicating that married women are supposed to stay home and take care of the family are conflicting with the aspirations of the growing number of well-educated career women.

reports that in 1992, a former garbage dump in Lima was changed in less than six months into a straw-hut shantytown with a population of 10,000 people and no running water, sanitation, or police. The U.N. estimates that 100 million people worldwide are homeless and 1 billion live in inadequate housing. In the near future, 730 million new workers in the developing world will need jobs.

In the United States, the automobile seems to have taken over the city. Most of the land area in cities must be devoted to car-related functions: streets, highways, parking lots. Drive-in movies, restaurants, banks, stores, church services, marriage ceremonies, and funeral parlors make it almost possible to go from birth to death without getting out of the car. (On a positive note, mass transit seems to be making a comeback in some cities.) Cities are having increasing difficulty disposing of the huge amounts of garbage and refuse that people produce. Cities have problems supplying and paying for necessary services, city centers are ugly, smog and pollution are continuing problems, and people flee to the suburbs if they can afford to.

Perhaps we came too quickly into an urban world. Living in cities, especially large cities, is a relatively new experience. In 1800, only about 3 percent of the world's population lived in cities of 5,000 people or more. By 1900, the proportion living in cities had grown to 14 percent, and by 1950 it had increased to 30 percent. It is projected that by 2005, half of all the people in the world will live in urban areas. The United States has changed from a sparsely settled rural country to an extremely urbanized nation. Perhaps our problems are to be expected, given our lack of experience with the phenomenon of urbanization.

In his book *The Human Zoo,* zoologist Desmond Morris suggests that the human being is basically a tribal or small-group animal. People are used to interacting on a localized, interpersonal basis. Relatively recently in our history have we, because of population growth and economic advantage, been forced into a supertribal or urban existence. Morris describes this new existence as "a human zoo." We have adapted well to some aspects of the supertribe; the urban environment has stimulated "man's insatiable curiosity, . . . inventiveness, . . . intellectual athleticism." We have adapted less well to other aspects, however, and we therefore see suicide, war, crime, mental illness, and destruction of the environment. The simple tribesman became a citizen, but according to Morris, we have not had time to evolve into a new, genetically civilized species. We have adapted somewhat to our new environment through learning and conditioning, but biologically we are still simple tribal animals. Morris concludes, "If [man] is given the chance he may yet contrive to turn his human zoo into a magnificent human game-park. If he is not, it may proliferate into a gigantic lunatic asylum, like one of the hideously cramped menageries of the last century."[9]

## THEORY AND RESEARCH: A REVIEW

This chapter was mainly descriptive, and no strong theoretical cast was present. Still, in some instances specific viewpoints could be detected. In the discussion on city patterning, we learned that cities are units and that they develop certain patterns or structures because of functions provided, such as transportation, low rent, and industry; this sounds consistent with *functional analysis.* Analysis of the effects of spatial arrangement on patterns of interaction is moving in the direction of *symbolic interaction. Conflict theory* would have much input in the matter of overpopulation—overpopulation consequences, the tactics of governments, and the types of people who are victims. The data are census data, mainly *nonreactive surveys* of national and international birth, death, and migration records. The following questions might arise from this type of data: Do countries in different parts of the world use the same standards? Do they use equal care in collecting their data, and are the data comparable?

## SUMMARY

In the previous chapters in this section on social organization we dealt with groups, categories, aggregations, and institutions. These are abstract concepts, but they are valuable for their capacity to describe and analyze the social organization of society. In this chapter we continue to study social organization, but in a somewhat different manner. The focus changes to population and human ecology, and we move from the abstract to the concrete.

Population and human ecology involve the study of numbers of people and their locations in space. The major population variables are *fertility, mortality,* and *migration.* Ecological variables that we examined include study of *city patterns* (slums, suburbs, strip cities), *ecological processes,* and the reasons for and consequences of *spatial patterning.* The study of population and human ecology supplies descriptive data about a society. These data, usually of a statistical nature, develop a picture of a society that often provides an interesting contrast to the picture supplied by the more abstract studies of groups and institutions. Finally, the study of population and ecology has today called attention to several serious problems facing us—*overpopulation, urbanization,* and the *destruction of the environment*—and has encouraged development of the social planning needed to solve these problems.

The first reading that follows examines some of the problems that some of our oldest and finest cities are having and makes the novel suggestion that maybe we should try making cities smaller rather than larger. The second reading examines China's one-child policy of 1979—how the policy works and some of its consequences.

## TERMS FOR STUDY

birth rate (335)

centralization (346)

decentralization (346)

demographic transition (350)

demography (334)

fecundity (336)

fertility (335)

general fertility rate (335)

human ecology (334)

invasion (346)

life expectancy (337)

migration (338)

mortality (337)

population pyramid (341)

replacement level (336)

segregation (346)

sex ratio (341)

succession (346)

total fertility rate (335)

For a discussion of Research Question 1, see page 339.
For a discussion of Research Question 2, see page 349.

## INFOTRAC COLLEGE EDITION
### Search Word Summary

To learn more about the topics from this chapter, you can use the following words to conduct an electronic search on InfoTrac College Edition, an online library of journals. Here you will find a multitude of articles from various sources and perspectives:

**www.infotrac-college.com/wadsworth/access.html**

Longevity

Decentralization

Grain

Rural-urban migration

Fertility

Population

Immigrants

Pesticides

## Reading 11.1

# DOWNSIZING CITIES

*Witold Rybczynski*

*Different cities face different problems. Some cities are vital, expanding, and healthy whereas others, some of our finest and oldest cities, are losing population and experiencing inner-city blight. One way to revitalize cities is to annex nearby suburbs, but this isn't always possible and doesn't always work when it is possible. The author of this article suggests another answer—make the city smaller. Treat it like an unsuccessful shopping mall—clear out vacant space, consolidate successful stores, run a smaller but more lucrative operation. He thinks downsizing and redesigning have real possibilities for making cities work better.*

Urban America is changing, and as has so often happened, the change is largely unplanned. Fields turn into housing developments; housing developments turn into suburban towns; suburban towns grow into cities; commercial strips and malls grow into full-fledged urban centers. Moreno Valley and Santa Clarita, both in California, not even incorporated in 1980, are today among the fastest-growing cities in the country. Forty years ago Mesa, Arizona, was a town outside Phoenix; today, with 325,000 inhabitants, Mesa is bigger than Louisville or Tampa. As for Phoenix, it is now the eighth largest city in the United States, with more than a million people, and makes up only part of a metropolitan area that covers more than 400 square miles.

A metro area—or a metropolitan statistical area (MSA), as the Census Bureau calls it—is defined as a central city of at least 50,000 inhabitants together with adjacent communities with which it has a high degree of social and economic integration. A metro area without a central city must contain at least 100,000 people in all (or, in New England, 75,000, as the term is defined there); major metro areas of more than a million, which in practice always contain central cities, may be designated consolidated metropolitan statistical areas. Although the concept of a metro area was introduced in 1949 as a demographic convenience, the city proper remains a distinct legal entity.

In *Cities Without Suburbs* (1993), David Rusk, a former mayor of Albuquerque, convincingly demonstrates that those central cities that have expanded their limits to annex suburbs, or that have enough vacant land to accommodate suburban growth, do better than static cities in a number of significant ways. The two Ohio metropolitan areas of Columbus and Cleveland serve as an example. Since 1950 the city of Columbus has been aggressively expanding, and now covers about 200 square miles; Cleveland's area—seventy-seven square miles—is almost unchanged. Although metro Cleveland is comparable in income level and racial composition to metro Columbus, and both metro areas have grown significantly in the past four decades, the present situation of the two cities is very different. Cleveland has fallen behind Columbus in economic growth and job creation; it is more racially segregated; and it has a significantly lower per capita income, more poverty, and a lower municipal-bond rating. The city of Cleveland, despite being the center of a consolidated metropolitan statistical area, is anything but integrated with its surrounding communities. By every measure (income, education, employment, poverty) the city is less well off than its suburbs. The same pattern is repeated across the country in cities like Detroit, St. Louis, and Chicago, which are increasingly isolated—economically and racially—within their metro areas.

Can old manufacturing cities and their suburbs be unified? Rusk concedes that true metropolitan government is unlikely (the only American metro area within a directly elected regional government is Portland, Oregon) and suggests revenue sharing between rich and poor jurisdictions, metro-wide housing-assistance programs, and economic-development plans. But helping ailing center cities by transferring funds from the suburbs is unlikely to garner much political support.

What is more likely is a federal program, such as the recently created Empowerment Zones, aimed at revitalizing inner-city slums. This is another version of what used to be called urban renewal, or enterprise zones, and it is as unlikely to succeed as its predecessors. Nicholas Lemann made this point forcefully last year in a *New York Times Magazine* article titled "The Myth of Community Development." "Nearly every attempt to revitalize the ghettos has been billed as a dramatic departure from the wrong-headed Government programs of the past," he wrote, "even though many of the wrong-headed programs of the past tried to do exactly the same thing."

Even if it were possible to expand the tax base of isolated central cities to capture at least some of the wealth of surrounding suburbs, doing so would not solve another problem. A city like Cleveland is not just less dynamic, poorer, and less racially heterogeneous than it used to be; it is also considerably smaller. Since 1970 Cleveland's population has decreased by 33 percent—one of the most precipitous declines of any large city. (During the same period Columbus grew by 17 percent.)

Cleveland is an example of an urban phenomenon that is increasingly common: The shrinking city. From 1970 to 1990 the total population of the 200 U.S. cities with more than 100,000 inhabitants apiece increased from 59 million to 64 million—a nine percent growth rate. (The nation's population grew by 22 percent in the same period.) But growth was not experienced equally by all cities. In fact, cities fell into two distinct categories: about two thirds grew vigorously, and the other third actually lost population. Over the two decades the average population decline was 12 percent and the average growth was an astonishing 81 percent. The cities that lost population tend to be the older manufacturing cities; the gainers are newer suburban-style cities, chiefly in the West and the Southwest. The trend has continued; today nine of the twenty largest cities in the country have smaller populations than they did even in 1950, and some cities, such as Boston and Buffalo, are smaller today than they were in 1900.

These and many smaller cities now have what is referred to in the real-estate business as a low occupancy rate. In a shopping mall with a low occupancy rate the owner can decide to refurbish the mall or to offer special leases. If these strategies don't work, in the short run the owner will absorb the loss; in the long run either rents must be raised or the owner will go bankrupt. But if rents are raised too high, more tenants will leave and bankruptcy will only be accelerated.

Like a mall owner, the administration of a city with a low occupancy rate can try to increase its tax base by refurbishing the downtown area to make it more attractive to business. It can organize riverboat gambling and build aquariums and world trade centers. It can stimulate employment by enlarging the public sector. (It cannot, however, create the sort of manufacturing jobs that were the basis for the earlier prosperity and growth of great cities like New York, Chicago, and Philadelphia.) It can also try to balance its budget by raising revenues through higher property taxes, business taxes, and income taxes, and by curtailing services—although these tactics, like raising rent, will eventually only hasten population decline.

The mall owner who has tried everything and finds that there is simply no demand for space has a final option: make the mall smaller. Consolidate the successful stores, close up an empty wing, pull down some of the vacant space, and run a smaller but still lucrative operation. Many cities, such as New York, Detroit, and Philadelphia, don't stand a chance of annexing surrounding counties. Downsizing has affected private institutions, public agencies, and the military, as well as businesses. Why not cities?

Two things happen when a city loses population. The reason for the first is that although a city is often said to be shrinking, its physical area remains the same. The same number of streets must be policed and repaired, sewers and water lines maintained, and transit systems operated. With fewer taxpayers, revenues are lower, often leading to higher taxes per capita, an overall deterioration of services, or both. More people depart, and the downward spiral continues.

The reason for the second is that urban vitality has always depended on an adequate concentration of people. In 1950 the average density in cities like Detroit, Cleveland, and Pittsburgh was more than 12,000 people per square mile; by 1990 it was around 6,000 or 7,000—a dramatic decline. The reality is even worse than it sounds, because the decline was not distributed equally across the city, and certain areas experienced much more dramatic reductions.

Without sufficient concentrations of people, not only is the provision of normal municipal services extremely expensive but urban life itself begins to break down. There are not enough customers to support neighborhood stores and services, or even to provide a sense of community. Empty streets become unsafe, and abandoned buildings become haunts for drug dealers and other criminals. A national study of housing abandonment found that the "tipping point" in a neighborhood occurred when just three to six percent of the structures were abandoned. Vacant lots and empty buildings are more than just symptoms of blight—they are also causes of it. Central cities of metro areas that have aggressively expanded their borders face these problems too, even if the cities have a broader and richer tax base.

The first need of a city whose population has declined radically is to consolidate those neighborhoods that are viable. Rather than mounting an ineffectual rearguard action and trying to preserve all neighborhoods, as is done now, the de facto abandonment that is already in progress should be offered in other parts of the city, partly occupied public housing vacated and demolished, and private landowners offered land swaps. Finally, zoning for depopulated neighborhoods should be changed to a new category—zero occupancy—and all municipal services cut off. Efforts should be made to concentrate in selected areas resources such as housing assistance and social programs.

Inevitably consolidation would involve the movement of individuals and families from one part of the city to another. It is true that private freedoms would be sacrificed for the common good in the process, just as they are when land is expropriated to build a highway or a transit system. Does this sound heartless? Surely it is less so than the current Pollyannaish pretense of providing services to many poor and depopulated neighborhoods, which are occasionally half revived with community-development projects and then left on their own to decay even further.

The comprehensive downsizing of cities faces formidable obstacles—not only the lack of legal and bureaucratic tools (town planning has traditionally dealt with growth, not decline) but also political opposition. Historic preservationists will likely oppose the demolition of many old buildings. Local politicians whose electoral base would be eroded can also be expected to resist attempts at consolidation.

In 1976 Roger Starr, a past administrator of New York City's Housing and Development Administration, published an article suggesting "planned shrinkage," a reduction in New York's municipal services. "I profoundly wish I'd never coined that phrase," he says today. "It was a most unfortunate term that was misinterpreted by many people as a plan to drive poor blacks out of New York, and I still receive calls that accuse me of being a racist." It is important to underline that the act of consolidating neighborhoods would not move people out of the city, and no loss of benefits and services would ensue. Nevertheless, since the most depopulated parts of the city tend to be the poorest, and since the poorest city inhabitants are predominantly black and Hispanic, relocation would undoubtedly affect members of these groups more than others. Much would depend on the ability of minority leaders to see that given the lack of real alternatives, abandoning half-empty neighborhoods is not necessarily a political defeat.

Much of the housing in older industrial cities is as dense as it is because of high land costs, a history of property speculation, and the greed of the original tenement builders. Consolidating neighborhoods would provide an opportunity to relieve congestion and encourage variety in kinds of housing. Two-story attached and semidetached houses can be built at net densities of sixteen to twenty-four houses per acre, instead of the higher densities that characterize nineteenth-century neighborhoods of three- to five-story buildings. There may even be an opportunity, in some cases, for cities to compete directly with the suburbs by providing detached single-family housing. In cities such as New York and Philadelphia detached single-family housing, built with a mixture of public and private support, is appearing in previously high-density areas.

Oscar Newman wrote in *Community of Interest* (1980) that "architects and architectural historians have been damning the suburban tract development since the 1930s, but social scientists and realtors will tell you that tract houses continue to be the most sought after and the most successful form of moderate- and middle-income housing ever built." The success of the tract house is one reason that cities have been losing people to suburbs. If cities are to attract—or at least keep—moderate- and middle-income citizens, they will have to provide the sort of housing those citizens want. If cities cannot annex suburbs, they can do the next best thing: rezone areas to permit the construction of suburban-style housing.

The consolidation of residential neighborhoods would produce stretches of empty blocks where buildings would have to be knocked down, services stopped, streets closed, and bus lines rerouted. It would be comforting to think that such vacant land would find new use as recreational parks and wilderness areas. On a small scale, empty lots could be converted into allotment gardens or playing fields. But there is little likelihood of creating large urban parks on the scale of Frederick Law Olmsted's nineteenth-century creations. The costs associated with the removal of basements and old foundations, soil amelioration, drainage, and planting would probably be out of the reach of cities already in financial difficulties. Vacated urban land would more likely be left empty, streets and buried infrastructure in place, available for use at some future date.

The objection may be raised that zero-occupancy zoning would create large tracts of empty land whose presence would disrupt the proper functioning of a city. In fact many cities have grown up around cemeteries, or have enveloped earlier industrial areas such as quarries and railroad yards, and most cities already have large areas of land such as tank farms and container depots that are cut off from everyday use. The only difference between these areas and zero-occupancy zones would be that the latter would be unused, and would not create noise or air pollution.

Cities have another option: divestiture. Contiguous parcels of, say, at least fifty acres could be put up for sale, with one of the conditions of sale being that the land would cease to be part of the city. According to Peter Linneman, the director of the Wharton School's Real Estate Center, large tracts in proximity to cities but without the burden of city taxes and bureaucracy would very likely attract developers who would otherwise shun them. The new developments, whether residential or commercial, would be responsible for their own municipal services, as new developments already are in suburban locations. Assuming that questions of ownership of rights-of-way and underground infrastructure could be worked out, the prospect is attractive. Cities would increase their income (although they would not gain taxpayers), and they would divest themselves of unproductive land. At the same time, people and economic activities would be attracted back into the urban vicinity.

The idea of downsizing goes against the progressive, optimistic American grain. Surely, conventional wisdom says, one should attack the root of the urban problem, not merely its symptoms. The solution to urban decline has always been assumed to be a combination of economic development (leading to renewed growth, more jobs, a larger population, and a larger tax base) and financial aid from federal and state governments. The problem is that population loss has gone on for too long: generally cities that have stopped growing and are losing population have been doing so for more than forty years. If metro-wide tax sharing were to be implemented, and if a more equitable allocation of financial resources finally gave cities a chance to deal more effectively with problems of poverty and immigration, the rate of population decline might be reduced or, at best, the population might be stabilized. Great manufacturing cities like St. Louis, Detroit, Philadelphia, and Cleveland will not return to their earlier predominance or their earlier size. Downsizing alone would not solve the problems of unemployment and urban poverty, but it would permit shrinking cities to marshal their resources more effectively as they make their way to a stable—albeit smaller—future.

---

## QUESTIONS 11.1

**1.** This article makes the comparison between a city and a shopping mall. In what ways does this argument make sense and in what ways doesn't it make sense?

**2.** Define and illustrate or discuss the following: "tipping point," "zero-occupancy zoning," "shrinking city," and "MSA."

**3.** What is happening to the metropolitan area with which you are most familiar? Is it getting bigger or smaller? Are the ideas suggested in this article needed there? Would they work?

**4.** List and discuss the sort of problems that you might face as a city manager or city planner if you decided to downsize your city.

---

## Reading 11.2

# THE LITTLE EMPERORS, PART 2

### *Daniela Deane*

---

*China's population in 1993 was about 1.2 billion, about 300 million larger than that of India and about five times larger than that of the United States. China has recognized its emerging population problem and in 1979 passed a one child per family policy. Enforcing the policy has been a struggle. It worked better in the crowded cities than it did in the countryside. A growing group of wealthy farmers resisted the policy because they wanted sons to carry on their businesses, and because of their wealth, they were less bothered by the fines levied for violation of the one-child rule. Daniela Deane, a writer now based in Italy, examines some of the other problems and consequences of the policy in this article she wrote for the* Los Angeles Times Magazine. *(Also see another part of this article—Reading 7.2—in which Deane describes how the "one child" has become the center of the Chinese family.)*

China's Communist government adopted the one-child policy in 1979 in response to the staggering doubling of the country's population during Mao Tse-tung's rule. Mao, who died in 1976, was convinced that the country's masses were a strategic asset and vigorously encouraged the Chinese to produce even larger families.

---

But large families are now out for the Chinese—20 percent of the world's population living on just 7 percent of the arable land. "China has to have a population policy," says Huang Baoshan, deputy director of the State Family Planning Commission. With the numbers ever growing, "how can we feed these people, clothe them, house them?"

Thirteen years of strict family planning have created one of the great mysteries of the vast and remote Chinese countryside. Where have all the little girls gone? A Swedish study of sex ratios in China, published in 1990, and based on China's own census data, concluded that several million little girls are "missing"—up to half a million a year in the years 1985 to 1987—since the policy was introduced in late 1979.

In the study, and in demographic research worldwide, sex ratio at birth in humans is shown to be very stable, between 105 and 106 boys for every 100 girls. The imbalance is thought to be nature's way of compensating for the higher rates of miscarriage, stillbirth and infant mortality among boys.

In China, the ratio climbed consistently during the 1980s, and it now rests at more than 110 boys to 100 girls. "The imbalance is evident in some areas of the country," says Stirling Scruggs, director of the United Nations Population Fund in China. "I don't think the reason is widespread infanticide. They're adopting out girls to try for a boy, they're hiding their girls, they're not registering them. Throughout Chinese history, in times of famine, and now as well, people have been forced to make choices between boys and girls, and for many reasons, boys always win out."

With the dismantling of collectives, families must, once again, farm their own small plots and sons are considered necessary to do the work. Additionally, girls traditionally "marry out" of their families, transferring their filial responsibilities to their in-laws. Boys carry on the family name and are entrusted with the care of their parents as they age. In the absence of a social security system, having a son is the difference between starving and eating when one is old. To combat the problem, some innovative villages have begun issuing so-called "girl insurance," an old-age insurance policy for couples who have given birth to a daughter and are prepared to stop at that.

"People are scared to death to be childless and penniless in their old age," says William Hinton, an American author of seven books chronicling modern China. "So if they don't have a son, they immediately try for another. When the woman is pregnant, they'll have a sex test to see if it's a boy or a girl. They'll abort a girl, or go in hiding with the girl, or pay the fine, or bribe the official or leave home. Anything. It's a game of wits."

Shen Shufen, a sturdy round-faced peasant woman of 33, has two children—an 8-year-old girl and a 3-year-old boy—and lives in Sihe, a dusty, one-road, mud-brick village in the countryside outside Beijing. Her husband is a truck driver. "When we had our girl, we knew we had to have another child somehow. We saved for years to pay the fine. It was hard giving them that money, 3,000 yuan ($550 in U.S. dollars), in one night. That's what my husband makes in three years. I was so happy when our second child was a boy."

The government seems aware of the pressure its policies put on expectant parents, and the painful results, but has not shown any flexibility. For instance, Beijing in 1990 passed a law forbidding doctors to tell a couple the results of ultrasound tests that disclose the sex of their unborn child. The reason: Too many female embryos were being aborted.

And meanwhile, several hundred thousand women—called "guerrilla moms"—go into hiding every year to have their babies. They become part of China's 40-million-strong floating population that wanders the country, mostly in search of work, sleeping under bridges and in front of railway stations. Tens of thousands of female children are simply abandoned in rural hospitals.

And although most experts say female infanticide is not widespread, it does exist. "I found a dead baby girl," says Hinton. "We stopped for lunch at this mountain ravine in Shaanxi province. We saw her lying there, at the bottom of the creek bed."

Death comes in another form, too; neglect. "It's female neglect, more than female infanticide, neglect to the point of death for little girls," says Scruggs of the U.N. Population Fund. "If you have a sick child, and it's a girl," he says, "you might buy only half the dose of medicine she needs to get better."

Hundreds of thousands of unregistered little girls—called "black children"—live on the edge of the law, unable to get food rations, immunizations or places in school. Many reports are grim. The government-run China News Service reported last year that the drowning of baby girls had revived to such an extent in Guangxi province that at least 1 million boys will be unable to find wives in 20 years. And partly because of the gender imbalance, the feudalistic practice of selling women has been revived.

The alarming growth of the flesh trade prompted authorities to enact a law in January that imposes jail sentences of up to 10 years and heavy fines for people caught trafficking. The government also recently began broadcasting a television dramatization to warn women against the practice. The public-service message shows two women, told that they would be given high-paying jobs, being lured to a suburban home. Instead, they are locked in a small, dark room, and soon realize that they have been sold.

Li Wangping is nervous. She keeps looking at the air vents at the bottom of the office door, to see if anyone is walking by or, worse still, standing there listening. She rubs her hands together over and over. She speaks in a whisper. "I'm afraid to get into trouble talking to you," Li confides. She says nothing for a few minutes.

"After my son was born, I desperately wanted another baby," the 42-year-old woman finally begins. "I just wanted to have more children, you understand? Anyway, I got pregnant three times, because I wasn't using any birth control. I didn't want to use any. So, I had to have three abortions, one right after the other. I didn't want to at all. It was terrible killing the babies I wanted so much. But I had to."

By Chinese standards, Li (not her real name) has a lot to lose if she chooses to follow her maternal yearnings. As an office worker at government-owned CITIC, a successful and dynamic conglomerate, she has one of the best jobs in Beijing.

"I had to keep everything secret from the family-planning official at CITIC, from everyone at the office," continues Li, speaking in a meeting room at CITIC head-quarters. "Of course, I'm supposed to be using birth control. I had to lie. It was hard lying, because I felt so bad about everything."

She rubs her hands furiously and moves toward the door, staring continuously at the air slats. "I have to go now. There's more to say, but I'm afraid to tell you. They could find me."

China's family-planning officials wield awesome powers, enforcing the policy through a combination of incentives and deterrents. For those who comply, there are job promotions and small cash awards. For those who resist, they suffer stiff fines and loss of job and status within the country's tightly knit and heavily regulated communities. The State Family Planning Commission is the government ministry entrusted with the tough task of curbing the growth of the world's most populous country, where 28 children are born every minute. It employs about 200,000 full-time officials and uses more than a million volunteers to check the fertility of hundreds of millions of Chinese women.

"Every village or enterprise has at least one family-planning official," says Zhang Xizhi, a birth-control official in Changping county outside Beijing. "Our main job is propaganda work to raise people's consciousness. We educate people and tell them their options for birth control. We go down to every household to talk to people. We encourage them to have only one child, to marry late, to have their child later."

China's population police frequently keep records of the menstrual cycles of women of childbearing age, on the type of birth control they use and the pending applications to have children. If they slip up, street committees—half-governmental, half-civilian organizations that have sprung up since the 1949 Communist takeover—take up the slack. The street committees, made up mostly of retired volunteers, act as the central government's ear to the ground, snooping, spying and reporting on citizens to the authorities.

When a couple wants to have a child—even their first, allotted one—they must apply to the family-planning office in their township or workplace, literally lining up to procreate. "If a woman gets pregnant without permission, she and her husband will get fined, even if it's their first," Zhang says. "It is fair to fine her, because she creates a burden on the whole society by jumping her place in line."

The official Shanghai Legal Daily last year reported on a family-planning committee in central Sichuan province that ordered the flogging of the husbands of 10 pregnant women who refused to have abortions. According to the newspaper, the family-planning workers marched the husbands one by one into an empty room, ordered them to strip and lie on the floor and then beat them with a stick, once for every day their wives were pregnant.

"In some places, yes, things do happen," concedes Huang of the State Family Planning Commission. "Sometimes, family-planning officials do carry it too far."

Large billboards bombard the population with images of happy families with only one child. The government is desperately trying to convince the masses that producing only one child leads to a wealthier, healthier and happier life. But foreigners in China tell a different story, that the people aren't convinced. They tell of being routinely approached—on the markets, on the streets, on the railways—and asked about the contraceptive policies of their countries. Expatriate women in Beijing all tell stories of Chinese women enviously asking them how many sons they have and how many children they plan to have.

## QUESTIONS 11.2

**1.** From the viewpoint of this article, what is the main stumbling block to population control in China?

**2.** Why is such high value placed on sons in China? Is it similar in the United States? Why or why not?

**3.** The one-child policy and its implementation might seem harsh.

Construct arguments (both sides) on the necessity of a population-control policy.

**4.** Construct and describe a policy (different from the one-child approach) that achieves the same end—population control—that China is trying to achieve.

## NOTES

1. The source for this paragraph is *Population* by John Weeks (Belmont, Calif.: Wadsworth, 1978), chapter 4.

2. The major source for these paragraphs on mortality is *Population* by John Weeks (see note 1), chapter 6, especially pp. 106–108 and 114–119.

3. See *Population* by John Weeks (see note 1), chapter 7.

4. Original sources referred to in this section include: Ernest Burgess, "The Growth of the City," in Robert Park and Ernest Burgess, *The City* (Chicago: University of Chicago Press, 1925), p. 51; Homer Hoyt, *The Structure and Growth of Residential Neighborhoods in American Cities*, United States Federal Housing Administration (Washington, D.C.: United States Government Printing Office, 1939), chapter 6; and C. D. Harris and Edward L. Ullman, "The Nature of Cities," *The Annals* 242 (November 1945), pp. 7–17.

5. The article by Otis Dudley Duncan and Beverly Duncan, "Residential Distribution and Occupational Stratification," is quoted here from Ralph Thomlinson's book *Urban Structure* (New York: Random House, 1969), pp. 12–13.

6. For a detailed description of these interaction patterns, see William H. Whyte, Jr., *The Organization Man* (New York: Simon & Schuster, 1956), chapter 25.

7. See William Michelson, *Man and His Urban Environment* (Reading, Mass.: Addison-Wesley, 1970). The comments in these two paragraphs are drawn especially from pp. 98–99, 158, 161, and 193–195.

8. The major source for this discussion of Malthus and Marx is *Population* by John Weeks (see note 1), pp. 16–21.

9. Desmond Morris, *The Human Zoo* (New York: McGraw-Hill, 1969), especially chapter 1. The quotation is from p. 248.

# Social Change and Social Deviance

In Parts One and Two we dealt
with the individual as a member of society and with the social
organization of society. We saw that social organization could be
studied using the following concepts: group, category and aggregation, institution, population, and human ecology. In Part
Three we will focus on change and deviance in society. The concepts discussed in this section suggest that all is not as predictable and organized as previous sections may have implied.
Uncertainty and instability exist in all societies and are the
result of a variety of factors. Social change occurs and affects the
ways individuals and groups relate to each other. If social change
is rapid and extreme, the organization of society might break
down, and social disorganization can result. In some situations, behavior occurs that is spontaneous and unstructured but
not necessarily disorganized. This type of behavior is called collective behavior. We learned in Part One that norms and roles
describe what people are expected to do in specific situations and
positions. But inevitably, many individuals do not behave in the
ways expected of them. Societies, in turn, devise ways to encourage or even force conformity. This leads us to the final topics of
deviant behavior and social control.

# Social Change and Social Disorganization

**Factors Related to Social Change**

*Rates of Change and Planned Change*

**Theories of Social Change**

*Cyclic Theories* • *Evolutionary Theories* • *Functional Theories*
• *Conflict Theories* • *Neoevolutionary Theories*

**Social Change and the Future: Two Viewpoints**

**Social Disorganization**

## A RESEARCH QUESTION

*Patterns of violence (for example, lynching of blacks in the deep South)
change over time. Are these social changes random, or are they related to other
factors? What might such factors be?*

*[News from Beijing, April 1984] Nomadic herdsmen in northwestern China
are trading in their yak-hair tents for a new polyvinyl model, which features
optional folding furniture and a wind-powered generator. The new tent,
which has more than 70 options, was offered to 500,000 herdsmen in remote
Qinghai province as part of a government home-improvement program.
Herdsmen interested in the wind-powered generator hope it will make it possible for them to watch television.*

WE HAVE LOOKED at the social organization of society now for the last
eight chapters. At this point, society might appear to be a well-oiled
machine: groups interacting, highly polished institutions dealing efficiently
with society's central issues and conducting business as usual, populations
reproducing, and people continuing to distribute themselves in space in
very predictable ways. Focusing on the social organization of society is an
extremely important perspective, but it unfortunately implies that conditions are constant, that all behave as they are expected to, and that there are
never unpredictable occurrences or failures—either in people or in social
conditions.

The fact is that all is *not* constant—life is not the same today as it was yesterday. Right now I'm looking at storm patterns over France on a satellite
weather map that is posted on the Internet. Next, I click over to the NBC
home page to check the latest news. Then perhaps to airline schedules to see
when the next flight to Los Angeles departs. And I haven't left my desk,
used the phone, or turned on the television. This sort of behavior was
unheard of four or five years ago. Today, Internet use has become so common that it's getting crowded, and legislation is being considered to govern
access to certain types of material. Today, most people understand when I
say my address is jlandis@csus.edu. The Concorde travels from the East
Coast to Europe in three hours; I can remember when it took more than
twelve hours to cross the Atlantic by propeller plane. I used to *dial* my telephone and talk to *people*; now I punch buttons and talk to answering
machines. Recently they cloned a sheep in Scotland, and people are wondering if they will clone a Scot next. Scientists and environmentalists are concerned about the expansion of holes in the ozone layer over the earth's
poles. Space stations are being developed, and manned trips to Mars are
being discussed. The Berlin Wall and the Soviet Union don't exist anymore,
but Myanmar, cellular phones, and fax machines do. I used to write lectures
by hand; now my desktop computer produces transparencies almost without my help. Our students used to stand in long lines to register for classes
at the beginning of each semester; now they register by phone through our
mainframe computer while sitting at home watching one of 200 channels on

cable television. AIDS, a disease that didn't seem to exist a short time ago, looms as the major health issue of today. In 1998, there were 48,000 AIDS cases reported in the United States and 20,000 deaths from the disease. Worldwide, it is estimated that 40 million people have HIV/AIDS and that by the end of 1999, 16 million people had died of it. Life expectancy in Africa will drop from 59 years to 45 by 2010 because of AIDS.

The point, of course, is that all societies are dynamic and constantly changing. Some societies change more rapidly than others, and within a society some parts change more rapidly than other parts. Some changes lead to improvement, to a better life for some or most of the members of a society. Change can also lead to problems because, although some individuals and institutions adjust rapidly, others have difficulty adapting to change. We can anticipate that social change will occasionally lead to periods of social disorganization as old ways erode and collapse and new ways of behaving are developed.

**Social change** refers to significant alterations in social relationships and cultural ideas.[1] Social and cultural factors are combined in this definition because they are tightly interrelated. Changes in ideas and values quickly lead to changes in social relationships, and vice versa. Scientific and technological developments have led to new roles (astronauts, spacecraft experts) and changed social relationships. People's concerns about pollution have led to new ideas and material inventions, such as smaller cars with fewer harmful emissions. The concern about AIDS mentioned earlier is quickly leading to major changes in other areas; we see changes in laws (is it a crime to knowingly put someone else at risk of infection?), in the nature of health care and how to finance it, in medical research and in tracking the course of a disease, and of course in the area of social and sexual patterns of interaction.

Social change takes place continually and, in modern society, at ever-increasing rates. Change is such a constant factor in our existence that we frequently lose sight of its extent. One way to put change in perspective is to imagine that your own life started 10 years earlier than it did or to compare your life experiences with those of your parents. Technological changes probably come to mind first: the growth of jet aviation, voyages to outer space, atomic power, organ transplants, a computerized society. But other changes have occurred as well. Look at the matters in which so much has transpired after only 10 years: the role of women; the average family size; educational philosophies and techniques; the control of particular diseases and life expectancy; the likelihood of getting married and the average age at which one marries; demonstrating for or against a cause (apartheid, legalized abortion, the death penalty); deciding whether to go to college, and if so, choosing a course of study; the amount of leisure time available and what to do with it; and so on. And if we can find these considerable differences across 10 years, imagine the contrast that would develop over 20 or 30 years.

Change factors are related to each other in complex patterns; often the path of influence is difficult to detect clearly. The development of the contraceptive pill (new idea, invention) led to changes (more freedom) in sexual behavior. Or did it? Perhaps concerns with increased freedom for women and with family size led to the development of contraceptive devices. Keeping in mind the complex nature of social change, perhaps we can outline some of the major factors related to change.

## FACTORS RELATED TO SOCIAL CHANGE

Various attempts have been made to relate social change to other factors in the physical and social environment.[2] Some of these hypothesized relationships seem farfetched; others seem more reasonable. For example, social change has been related to the biological characteristics of a nation. Adherents of master-race theories believe that societies fortunate enough to be graced by the presence of a superior race will have greater progress (and more rapid social change) than will "less fortunate" nations. Sociologist Robert Bierstedt points out that such theories—present at all times in all societies—are only a primitive form of ethnocentrism. A more realistic approach might hold that people who differ by race or ethnicity have different social and cultural patterns and that particular mixtures of these types of people in a given society could lead to greater or lesser conflict and, correspondingly, greater or lesser social change. Geographical factors can also be related to change. Some countries are more favored by climate and natural resources than are others, and this can affect their rates of technological advancement and social change. As we discussed in the previous chapter, demographic factors can lead to change as well. When a country faces population growth to the point of overcrowding, social relationships change, both within that society and with surrounding societies that may be less crowded.

Social change can be a result of the influence of individuals. Recall how societies, and even the world, were affected and changed by Adolf Hitler, Napoleon Bonaparte, Albert Einstein, Abraham Lincoln, Karl Marx, and Julius Caesar. The course of events in the United States in the mid-20th century was altered by men such as Franklin D. Roosevelt and Martin Luther King, Jr. It is sometimes argued that significant social changes are initiated only by great and unique men and women; but others maintain that such people are merely products of their society and that if they had not appeared, someone else of the same caliber would have. The truth probably lies somewhere between these views. The cultural and societal conditions might be seen as a stage on which people perform, a few in unique ways having lasting effects, many others in more commonplace and predictable ways. And, certainly, the historical aspect cannot be overlooked as a factor in change. Many great men and women were made by historical conditions—accidents of history. Had they appeared at another time, they probably would have passed unnoticed. Also, historical accidents or unpredictable events—the assassination of a leader or a population-decimating catastrophe such as an earthquake or epidemic—can lead to change.

The research question at the beginning of this chapter inquired into changes in patterns of violence over time. A recent paper looked at the pattern of lynchings of blacks in the deep South between 1882 and 1930. The researchers found that the number of lynchings was related to the price of cotton! In years when the price of cotton was down, the number of lynchings increased. Economic downturns affected all cotton producers, especially the small farmers. In hard times they were probably more frustrated by competition from black laborers, and they favored replacing blacks with unemployed whites. Given the racial caste structure existing in the deep South, white workers would react to other supposed "infractions" and vent their hostility on blacks. Lynchings were one result. Undoubtedly there was anger at the white elite as well, but the small farmers couldn't do anything to them for fear of retaliation.[3]

Ideological and technological factors are related to social change. The appearance of and commitment to ideas of socialism, democracy, Christianity, science, or progress in general have led to major social changes. The Industrial Revolution of the 18th and 19th centuries was a major force for change in the United States and elsewhere; its dramatic consequences are still with us today. Technological innovations and inventions that have produced tremendous social change include the wheel, the car, the birth-control pill, the atom bomb, the gun, the telephone, the airplane, and the computer. It is easy to point to the dramatic inventions that affect societies, probably because their emergence seems sudden rather than slow and evolutionary, as are many changes. Sometimes the effects of inventions are less obvious. Sociologist William Ogburn recorded numerous examples of the widespread effects of technological innovations.[4] The self-starter on cars aided in the emancipation of women: It allowed women to use cars as easily as do men. The invention of the elevator made possible the construction of tall apartment buildings. Living in these buildings, in turn, changed the family: The rearing of large families was more difficult, and consequently the urban birth rate declined. The invention of the cotton gin was a major factor leading to the Civil War. How? Ogburn suggests that the invention of the cotton gin led to the need for more slaves to cultivate more cotton, and the greater production led to increased trade with England. This in turn led the South, which supported free trade, into conflict with the economic interests of the North, which wanted protective tariffs and restricted foreign imports, and this conflict led to war. Invention of the six-shooter, barbed wire, and the windmill made possible the settling of the Great Plains. To imply that inventions have such far-reaching effects no doubt seems as much of an oversimplification as does the "great men and women" theory. These examples suggest that we probably cannot single out any one element or invention in a culture and attribute change to it alone.

Social change is introduced into a culture through two processes: invention and diffusion. **Invention** refers to the creation of a new object or idea. Although the end product of the invention (car, airplane, telescope) is usually new, the parts from which it is made are not. The elements have been around, but they are now arranged or put together in a new and significant way. **Diffusion** refers to the spread of objects or ideas from one society to another (Egypt to England) or from one group to another within the same society (upper class to lower class or black to white). Although innovation

## AIBO AND POKÉMON

Pikachu, Magikarp, Snorlax, Rhyhorn? Forget Beanie Babies. In late 1999, Pokémon ruled. A movie, a TV show, Nintendo games, and then trading cards and assorted other products for worldwide sales of $6 billion and climbing. What Satoshi Tajiri started in Japan in 1991 when he combined his two passions—monster movies and the study of insects—became the rage of the 4- to 12-year-old set. There are 155 different Pokémon characters and more every day. Why so popular? A game developer at Nintendo says, "Pikachu and the other Pokémon characters certainly look very cute, but if anything happens, they are ready to fight. That may be why they appeal to Americans."

If you're tired of Pokémon, perhaps what you need is an electronic artificial intelligence robotic dog. SONY has just the product for you, and it's called AIBO. At $2,500, this digital pet will play ball with you, roll over, and learn

commands. It even has its own synthetic motivations and behavior patterns that will continue to develop over time. AIBO gathers information from the world in a number of ways: there are microphones in its ears, infrared sensors in its eyes, touch sensors in its paws, an internal gyroscope for balance, and a temperature sensor that prevents it from overheating in hot weather.

As the AIBO dog gathers more information, it becomes more advanced. If it's kept active, the dog reaches maturity in three to four months. For an extra $450, you can get a performer kit that will permit you to teach your dog new tricks. These tricks can even be shared with other electronic dogs via e-mail and the Web. True, AIBO can't carry a newspaper and is basically useless, but it's cute, and SONY hopes this electronic companion will make you chuckle. The first 5,000 units were snapped up immediately.

---

or invention attracts the most attention, most change is introduced through the process of diffusion. Recall Ralph Linton's "One Hundred Percent American" in Chapter Three.

In sum, social change represents the coming together of a number of events: the conditions in a particular culture at a given time in history and the change agent or catalyst (a particular person or the invention or diffusion of a new technique, object, or idea).

### Rates of Change and Planned Change

The various segments of a given society frequently change at different rates. Ogburn called this condition **cultural lag.** A new element can be introduced that requires change in other areas of society, and when these other areas are slow to change, problems result. Faster cars were built before highways could really handle them. Jumbo jets arrived before airports could accommodate the increased passenger load. People's attitudes usually change far more rapidly than laws change, especially in the area of morality laws, and a cultural lag exists.

## WILLIAM GRAHAM SUMNER (1840–1910)

William G. Sumner earned his B.A. from Yale in 1863. He continued his education and was ordained an Episcopal priest in 1869. He taught the first course in the United States called sociology. Sumner was a major exponent of Social Darwinism, adopting a survival-of-the-fittest approach to the social world. He saw people struggling against their environment, and the fittest were those who were successful. Success was rewarded to those who deserved it and failure to those who did not. Sumner was opposed to government efforts to help those in need because they operated against natural selection in which the fit survive, the unfit perish. As Sumner put it, "If we do not like the survival of the fittest, we have only one possible alternative, and that is the survival of the unfittest." This view modeled well with capitalism and provided legitimacy for the great differences in wealth and power during the time.

Sumner's most important book was *Folkways*. He felt that the force of customs and mores (norms) made social reform efforts useless. He also originated the concept of ethnocentrism—the belief in the superiority of one's own group over others—that we discussed in Chapter Three. Sumner was a professor of sociology at Yale University from 1872 to 1909.

A society's receptiveness to change is an important factor affecting the rate of social change. If a society has a strictly defined system of stratification or a rigid institutional structure, change will be slower. If a society is isolated from others, change will be slower. Change will be more rapid in a society that emphasizes the values of individualism and self-determination than in a society that emphasizes conformity and reverence for the past and custom. William Graham Sumner believed that there is a fundamental resistance to change in all societies in that people are basically conservative and seek stability or they "**strain for consistency**." Resistance to change is balanced to a certain extent by people's curiosity. They continue to seek new knowledge and better ways of doing things, so change must occur.

Perhaps social change is an evolutionary process beyond our control; we are at the mercy of forces we can do nothing about. Many people believe, however, that we *can* do something more than just describe and accept our social environment. Rather, the environment can actually be manipulated for our benefit. One view of the process of change, as expressed by Sumner, is that social change must be slow, and change in people's attitudes must precede change in legislation. Sumner believed that laws should not move ahead of the customs of the people, or, in his words, "stateways cannot change folkways." An opposite view suggests that new laws can lead the way—they can change people's attitudes and behavior in necessary and beneficial ways. From 1954 on, court decisions and legislative acts have attempted to change the pattern of interaction between the races in the United States. Civil rights groups, feeling that change had been too slow, tried sit-ins, nonviolent demonstrations, and passive resistance to speed up the process. Both sides—those who felt change was beyond control and those who felt that change could be planned and controlled—had fuel for their arguments. The latter group could state that the conditions of minorities in America had improved. The former group, however, could call attention to unanticipated changes that also occurred: increasing violence between the races, and probably a hardening of racial attitudes by many people. At any rate, social scientists today feel that change within limits can be a planned phenomenon. As a result, we see the development of commissions and agencies focusing on a number of matters in which it is felt that planned social change is needed.

## THEORIES OF SOCIAL CHANGE

Is there any pattern to social change? Is its direction or speed predictable? These questions have fascinated people for ages, and scientists and philosophers have applied themselves energetically to the task of unraveling the mysteries of social change. Numerous theories have emerged. We will briefly examine five types of theories: cyclic, evolutionary, functional, conflict, and neoevolutionary.[5]

### Cyclic Theories

One way of understanding social change is by looking at cycles. Chinese historians gave world history neither a beginning nor an end, but instead saw periods of order and disorder, of prosperity and decline. A 14th-century Arab scholar, Ibn Khaldun, based his theory of change on the clash between nomadic and sedentary peoples. Desert nomads seek the luxuries of the city and therefore continually attack cities and towns. The sedentary city-dwellers are no match for the fierce nomads and are quickly conquered. The conquerors form a new empire, but as they settle in the cities they become comfortable and sedentary like those before them, and they are eventually overrun by a new horde of nomads.

Arnold Toynbee saw civilizations arising from primitive societies through a process that involved challenge and response. Not all primitive societies become civilizations; those that do, do so because of their response to the challenge of adverse conditions, such as a difficult physical environment, land that has not been settled and tilled, sudden military defeats, and continuing external threats. Two factors are crucial: the severity of the challenge and the development of an elite to manage the response. Most of the society is tied to the past but led by the elite; the civilization can grow by responding successfully to continuing challenges. Eventually civilizations come apart, dividing into several groups, which later end up battling with each other. For Toynbee, civilization's cycle included birth, growth, stagnation, and disintegration. When he looked at the future of Western civilization, Toynbee saw several problems or challenges he felt we had to solve: war, class conflict, population growth, and abundance of leisure time.

Pitirim Sorokin's theory of change had societies fluctuating back and forth like a pendulum through three orientations: ideational, sensate, and idealistic. The **ideational culture** emphasizes feelings and emotions; it is subjective, expressive, and religious. The **sensate culture** emphasizes the senses; it is objective, scientific, materialistic, profane, and instrumental. The **idealistic culture** is an intermediate stage emphasizing logic and rationality. Change from one type to another happens because change is normal and continual in any active system. Sorokin believed that our society was in a stage of decline—that we were an "overripe" sensate culture. He suggested that the future looks bad for us, what with loss of freedom, growth of tyranny, deterioration of the family, and loss of creativity. But good things are ahead: "the magnificent peaks of the new Ideational or Idealistic culture."

## Evolutionary Theories

Many scholars have seen social change as linear—a line going in one direction, and usually that direction was progress. Auguste Comte viewed societies as moving through three stages: the theological, or fictitious; the metaphysical, or abstract; and the scientific, or positive. Each stage represents a higher level of existence with the ultimate—the positive stage—resulting in the emergence of priest-sociologists who lead society to a harmonious existence! Comte felt that population increase and density were especially important in influencing change. Herbert Spencer, another evolutionist, likened society to an organism, always growing and increasing in complexity, with its parts dependent on each other. Émile Durkheim saw societies moving from **mechanical solidarity** (small, primitive, homogeneous) to **organic solidarity** (large, complex, specialized).

As all these theories differed, so did the various views of the future: Comte saw increasing progress, with sociologists leading the way, and Spencer saw chances for progress *or* regress.

## Functional Theories

Functional analysis sees society as a system of interrelated parts, and it tends to emphasize social equilibrium, stability, and integration of the various pieces. Social change is difficult to accept given this basic perspective. Functional theorists like Talcott Parsons and Neil Smelser suggested that there are tremendous forces resisting change and that these forces might be overcome slowly and adaptively because of changes outside the system, growth by differentiation, and internal innovations. Generally, however, stability and inertia triumph, and change is viewed as deviant and traumatic.

## Conflict Theories

Conflict is universal in the world, and as a force leading to change, it figures in several of the theories we have already mentioned. Current **conflict theories** spring from the ideas of Karl Marx, who viewed the economic structure to be the foundation of society; changes or contradictions in the economic realm led to changes in other social relations. Capitalist systems were especially liable to produce conflict. The conflict would come from within as the society polarizes into two antagonistic groups: the rulers (bourgeoisie) and the ruled (proletariat). Ultimately the outcome of this struggle between classes would be a revolution and the emergence of a classless society. One attraction of Marxism and conflict theory as an explanation for change is that it promises people a role in determining their future. Far from being pawns manipulated by unseen forces, Marx suggests that people can recognize the conditions of their life, react to them, and change them; they can create their own destiny.

Ralf Dahrendorf, like Marx, believed that in every society some members are subject to coercion by others. This means that class conflict is inevitable, and that social change, which emerges out of this conflict, is also inevitable. Dahrendorf believed that human development, creativity, and innovation emerge mainly from conflicts between groups and individuals. The connection between conflict and social change is well documented. Accepting the conflict approach is tempting for those who are frustrated with the slowness of change but lack resources to battle the "system" effectively. Over the last decade, the use of violence and terrorism to produce change has become increasingly common. As Chairman Mao suggested, "Anything can grow out of the barrel of a gun. . . ."

## Neoevolutionary Theories

Some theorists have begun to drift back to and rework the evolutionary ideas of yesteryear. Early evolutionary theories explained or analyzed change by focusing on one central factor in a unilinear approach. This broke down when the theory didn't fit the facts of a particular culture and when various theorists came up with different central factors. Modern, or neoevolutionary, theorists still suggest that societies move through a series of evolutionary stages, but now the process is seen as multilinear. By this they mean that change is influenced by varied factors and that although there are similarities or parallels, not all cultures change in the same direction or at the same speed. The newer evolutionary theories also place more importance on cultural borrowing and diffusion than did the earlier ones.

Gerhard Lenski is a contemporary evolutionary theorist. He believes that societies move through a series of forms (hunting and gathering, horticultural, agrarian, industrial) based on their mode of subsistence. Continuity, innovation, and extinction are key elements for Lenski. *Continuity* refers to the persistence of particular elements in society and emphasizes that, even in a society that appears to be changing rapidly, most of its elements are not changing. In our society, for example, think of the alphabet, the concept of God, the calendar, or driving on the right-hand side of the road. Elements remain because they are useful, because they answer society's needs, or because the cost involved in changing them would be too high. *Innovation* results from inventions and discoveries from within as well as from diffusion from other cultures. Innovation occurs at different rates in different societies because of the amount of information available, the number of people disseminating it, whether it is fundamental (the steam engine is more fundamental than is the can opener), and whether society is receptive to it. *Extinction* refers to the disappearance of cultural elements or of whole societies. The consequence of these three processes is diversity and progress. Societies become more and more diverse as the processes work at different rates in different societies. Progress occurs, not necessarily in the direction of happiness or some higher state of being, but in the sense of continuing technological developments.

Lenski believes the process of change goes something like this: Diversity among societies causes them to change at different rates, they develop increasing needs for resources, and this leads to increasing competition among societies. This in turn leads to the survival of some societies and the extinction of others. Because societies with the more advanced technologies are more likely to survive, and because technological changes are basically responsible for other changes, technology becomes the crucial factor in social change. Although the technological advances of the last 150 years have generally raised the levels of freedom, justice, and happiness for people in industrial societies, Lenski is not wildly optimistic about the future. His concern with the continuing destruction of the environment, with rapid population growth (especially in Third World nations), and with increasing international tensions and the growing potential for nuclear war led him to conclude that societies *must* take a more active role in planning their futures. They must learn to control population and technology rather than continuing to be the victims of uncontrolled growth.[6]

Although the social change theories presented here might appear distinct unto themselves, there is actually much overlap among them. Some evolutionary theories seem to incorporate cycle theory, the cycles occasionally look evolutionary, the functionalists borrow from the evolutionists, and conflict seems to appear everywhere. So what is the best explanation for social change? Does any one theory fit the facts best? That, of course, depends on which set of facts you select; explanations for change in nomadic societies must be different from those attempting to explain change in postindustrial society. On the whole, I find the ideas of the conflict theorists and Lenski's neoevolutionary approach most interesting. Conflict seems to be a major element in social change, increasingly so in recent years. Antagonism among classes is commonplace in many societies and acts as a catalyst for change. Lenski's focus on the importance of technology also makes sense as I write these words on a computer, look up information and order products on the Internet, and read of new breakthroughs in genetic engineering and of "Mafiaboy," a 15-year-old computer hacker from Canada who brought down the CNN Internet site. We are even releasing people from prison early on the condition that they be "tethered" to their homes by electronic "handcuffs."

# SOCIAL CHANGE AND THE FUTURE: TWO VIEWPOINTS

Every year or so, experts review the state of the world and predict the future. Alvin Toffler, who analyzed social change in his popular books *Future Shock* and *The Third Wave*, reports that we are about to be engulfed by a third wave of change (the first two waves were the agricultural revolution and the industrial revolution). In this third wave, we will see the increasing importance of electronics and computers: Toffler envisions the home as an electronic cottage in which people no longer commute to offices but instead work at home, with results sent to the office by electronics. He predicts space manufacturing—the development of industries that can take advantage of zero-gravity and vacuum environments. He sees more development of ocean resources; algae and fish farming will supply oil and protein, and valuable minerals will be mined from the ocean floor. Toffler also predicts developments in the gene industry that would produce high-yielding crops that could be grown anywhere; perhaps even human beings could be improved somewhat. More and more people will raise their own food, weave their own rugs, and make their own clothing with the new home technologies, according to Toffler.

In their book *Megatrends 2000,* John Naisbitt and Patricia Aburdene analyze major trends in American society and the world, and they try to predict where these trends might take us in the future. Their predictions include the following:

**1.** *A global economic boom.* The world is becoming a single economy, and factors like increasing worldwide free trade, changing taxes, an Asian consumer boom, and few wars have fueled growth.

**2.** *A renaissance in the arts.* As evidence Naisbitt and Aburdene cite the growth in the number of symphony orchestras and attendance at performances, the increasing number of new book titles, and the booming interest in art and in museums both here and in Europe. New careers and business opportunities seem to be developing in the arts.

**3.** *A global lifestyle.* The volume of air travel will continue to increase, foods are transported easily around the world—you can get a Big Mac anywhere (São Paulo, Brazil, has 16 McDonald's), and fashion, film, and entertainment are more international than ever (Disneylands in Japan and France).

**4.** *Rise of the Pacific Rim.* Today, Asia has nearly two thirds of the world's population. Demographically, economically, and culturally, it will continue to replace the Atlantic-Europe connection as a powerful global force. Even now, think of the economic and political influence of Japan, China, Taiwan, Korea, Thailand, and Singapore.

**5.** *Women in leadership.* The authors predict that as we continue to pass from an industrial age (male-dominated) to an informational/service age, women will become an increasing economic force.

**6.** *Age of biology.* The view of science and technology as having a role in making life "better" is more prevalent. Plants (a super tomato!), animals (more milk from fewer cows), even humans (genetic engineering) can be biologically manipulated in perhaps positive ways (if we can sort out the ethical issues).

**7.** *A religious revival.* Naisbitt and Aburdene believe that although there will be a decline in mainline religion, this decline will be countered by increasing interest in "fringe" religions. Catholic, Protestant, and Jewish groups will shrink, but fundamentalist, evangelical, and spiritualist types of religions will increase in what the authors see as a swing from organized religion to spirituality.

**8.** *Triumph of the individual.* In a theme carried through from Naisbitt's 1982 book *Megatrends,* the authors express faith that the power of the individual and individual initiative will continue to be recognized. As one example, they believe that recent inventions (computers, cellular phones, fax machines) have enhanced the power of individuals rather than oppressing them, as it was originally feared they might.

If Naisbitt and Aburdene are correct, these trends are among those that will shape our future.

## SOCIAL DISORGANIZATION

**Social disorganization** refers to the breakdown of norms and roles with the result that customary ways of behaving no longer operate. Suppose the star halfback in the weekend football game is given the ball to run through the opposing team's line, but instead he turns and runs 50 yards the wrong way over his own goal line. The spectators are aghast. Several plays later, a player is running with the ball (in the right direction) only to be tackled by the referee. Again, consternation, because "that's not the way the game is played." The norms are so well known that when they are flagrantly violated and seem to be breaking down, effects on participants are marked. So unusual are these events that sometimes the "norm violators" become famous. In a college game some years ago, a football player had broken loose on a touchdown run. No one had a chance to get him. Suddenly an opposing player, who was out of the game and sitting on the bench, raced onto the field and tackled the ball carrier. The incident made the sports pages throughout the country the next day, and the two players (ball carrier and tackler from the bench) appeared on a nationwide television program a few days later to describe the great event. The event was remarkable because the game just isn't played that way.

But let's get back to the spectators at our peculiar game in which something strange happens on every play: A player runs the wrong way, the wrong person is tackled, eight players come into the game as substitutes but only three leave, and the quarterback runs up into the stands to lead cheers. How long would the spectators tolerate this behavior? It is my guess that very soon they would start leaving the stands: "This is ridiculous—the rules seem to change on every play—you don't know what to expect. . . ." In like manner, extensive breakdown of a society's norms and roles can produce a condition of social disorganization. Extreme social disorganization can lead to personal disorganization, in which the individual becomes upset, disaffected, demoralized, and apathetic; he might want to "leave the game."

Social disorganization is often used to describe living conditions in ghetto and slum areas, which are damaged by the effects of poverty, anonymity, overcrowding, and absence of roots or ties to the community. The person living in these conditions might, like the person watching the peculiar football game, be disgusted at the way the game is being played and decide to leave or fight back. Socially disorganized inner cities have higher rates of crime, mental illness, infant mortality, disease, and family instability. This is not meant to imply that social disorganization causes these to occur, but rather that social disorganization describes a general set of conditions under which these social problems flourish.

**Anomie,** a particular type of social and personal disorganization, has also been used to describe a generalized condition of "normlessness." French sociologist Durkheim used the concept of anomie to describe a condition in societies in which norms and rules governing people's aspirations and moral conduct had eroded.[7] Solidarity is lost as old group ties break down, and consensus is lost as agreement on values and norms disappears. It is difficult for individuals to know what is expected of them or for them to feel a sense of close identity with the group. Anomie describes a demoralized society in which norms are rapidly changing, uncertain, or conflicting. Durkheim felt that a particular type of suicide occurs in such societies: *anomic suicide,* in which the individual feels lost and disaffected because of an absence of clear-cut rules and standards for behavior. Durkheim believed that abrupt economic changes—sudden inflation or depression, poverty, or wealth—might lead to anomie. This concept has received much attention recently, especially from sociologists in the areas of criminology and delinquency.

As social disorganization occurs, perhaps from rapid social change, new norms replace older ones, role behavior is modified, conflict in values occurs, and institutions assume different forms and functions. Individuals, groups, institutions, and societies must adapt to change, and for some of these it can be difficult, especially if change is rapid or if change is defined as unacceptable and is therefore resisted.

Some sociologists feel that the concept of social disorganization is not very useful. They maintain that the organization of society is constantly in flux and is therefore always changing, and that the result is not disorganization but *reorganization.* They believe that we are really talking about another stage in the social-change process rather than about social disorganization. Further, the critics of the concept argue that there is a tendency to apply the label of social disorganization to those things we dislike, to things that might contradict our own particular values or what we are used to. For example, suppose that in the United States we have high divorce and illegitimacy rates, birth-control pills are freely available, and teenagers commit more crime than any other age group. Does this mean

that the institution of the family in America is a victim of social disorganization, or that it is changing to a different form of organization? Is the current situation regarding relations between blacks and whites in America
symptomatic of a disorganized society, or of a society undergoing change
and reorganization?

## THEORY AND RESEARCH: A REVIEW

Functional analysts and conflict theorists have differing viewpoints on
social change, as can be seen in the section on theories of social change.
Typically, *conflict theory* is much more applicable to the topic of change.
Conflict among groups with different interests, to be expected in any social
situation, is an important cause of change. *Functional analysis,* on the other
hand, emphasizes stability and order; social change becomes of peripheral
interest. The data used in this chapter differed from the survey type of data
we have seen in most of the preceding chapters. The "data" of the theorists
were analyses of historical records and archives, which led them to their
impressions about stages (cyclical, evolutionary, and so on) through which
societies pass. Toffler and Naisbitt and Aburdene also analyze this type of
data, thereby emphasizing current conditions. For his earlier book,
*Megatrends,* Naisbitt used local newspapers, analyzing more than two million articles about events in American cities and towns over a 12-year period. This is another type of *nonreactive survey.*

## SUMMARY

In the second section of this book we focused on social organization—the
*social fabric* of society. There is a tendency in studying social organization to
emphasize the constant, recurring, normal, stable nature of society. In fact,
however, all is not as organized, predictable, and fixed as the two previous
sections might have implied. Change and deviance are present in all societies, and this section addresses itself to those topics.

This chapter dealt with the concepts of social change and social disorganization. Social change occurs at varying rates in all societies; on this at
least there is general agreement. On some other issues there is more
debate: Which, if any, factors are related to change? Is change automatic
and inexorable, or subject to intervention and control? We have examined
cyclic, evolutionary, functional, conflict, and neoevolutionary theories
explaining social change. One conclusion is that social disorganization is
the condition that sometimes results when norms have broken down and
behavior becomes unpredictable. Although there is some disagreement
over the usefulness of the concept of social disorganization, the term is
often helpful in describing what might happen in periods of rapid and
traumatic social change.

In the first of the readings that follow, anthropologist John Yellen
describes the !Kung Bushmen of the Kalahari Desert and shows how recent
social change is destroying a lifestyle that has existed for thousands of

years. The second reading is a study of the social disorganization that occurred in the aftermath of a flood in the mining country of West Virginia.

## TERMS FOR STUDY

anomie (381)

conflict theories (376)

cultural lag (373)

cyclic theories (375)

diffusion (372)

evolutionary theories (376)

functional theories (376)

idealistic culture (376)

ideational culture (376)

invention (372)

mechanical solidarity (376)

neoevolutionary theories (377)

organic solidarity (376)

sensate culture (376)

social change (370)

social disorganization (380)

strain for consistency (374)

For a discussion of the Research Question for this chapter, see page 371.

 **INFOTRAC COLLEGE EDITION**
**Search Word Summary**

To learn more about the topics from this chapter, you can use the following words to conduct an electronic search on InfoTrac College Edition, an online library of journals. Here you will find a multitude of articles from various sources and perspectives:

**www.infotrac-college.com/wadsworth/access.html**

Julius Caesar

Genetic engineering

Internet

Spiritualism

Social fabric

## Reading 12.1

# BUSHMEN

*John Yellen*

*Anthropologists led by Richard Lee went to the Kalahari Desert in South Africa in the 1950s and 1960s to study the !Kung Bushmen, one of the last remaining groups in the world still living by hunting and gathering. To their surprise, the anthropologists found the tribe's life to be pleasant and food gathering to be easy, and one of the scientists dubbed the Bushmen "the original affluent society."*

*Social change is inevitable, however, and anthropologists visiting in the 1970s and 1980s found change among the Bushmen. Drinking vessels made from ostrich eggshells were replaced by pop-top aluminum cans. Before they stalked giraffe and wildebeest; now they gathered at the mission station for their daily rations. Before village huts were close together, facing each other and intimate; now they were separated and enclosed by fences demonstrating the newly acquired ethic of privacy. Views about owning property changed. They used to have few possessions and carried them on their backs; now property was collected and stashed in locked trunks. In several decades, a way of life thousands of years old had started coming apart.*

*John Yellen is an anthropologist at the National Science Foundation. He has visited the Kalahari four times since 1968.*

By the mid-1970s, things were different at Dobe. Diane Gelburd, another of the anthropologists out there then, only needed to look around her to see how the Bushman lifestyle had changed from the way Richard recorded it, from how Marshall Sahlins described it. But what had changed the people at DBC 12 who believed that property should be commonly held and shared? What had altered their system of values? That same winter Diane decided to find out.

She devised a simple measure of acculturation that used pictures cut from magazines: an airplane, a sewing machine, a gold mine in South Africa. (Almost no one got the gold mine right.) That was the most enjoyable part of the study. They all liked to look at pictures, to guess.

Then she turned from what people knew to what they believed. She wanted to rank them along a scale, from traditional to acculturated. So again she asked questions:

"Will your children be tattooed?"

To women: "If you were having a difficult childbirth and a white doctor were there, would you ask for assistance?"

To men: "If someone asked you for permission to marry your daughter, would you demand (the traditional) bride service?"

Another question so stereotyped that in our own society one would be too embarrassed to ask it: "Would you let your child marry someone from another tribe—a Tswana or a Herero—a white person?"

First knowledge, then belief, and finally material culture. She did the less sensitive questions first. "Do you have a field? What do you grow? What kind of animals do you have? How many of what?" Then came the hard part: She needed to see what people actually owned. I tagged along with her one day and remember the whispers inside one dark mud hut. Trunks were unlocked and hurriedly unpacked away from the entrance to shield them from sight. A blanket spread out on a trunk revealed the secret wealth that belied their statements: "Me? I have nothing." In the semidarkness she made her inventory. Then the trunks were hastily repacked and relocked with relief.

She went through the data, looked at those lists of belongings, itemized them in computer printouts. Here's a man who still hunts. The printout shows it. He has a bow and quiver and arrows on which the poison is kept fresh. He has a spear and snares for birds. He has a small steenbok skin bag, a traditional carryall that rests neatly under his arm.

He also has 19 goats and two donkeys, bought from the Herero or Tswana, who now get Dobe Bushmen to help plant their fields and herd their cows. They pay in livestock, hand-me-down clothing, blankets, and sometimes cash. He has three large metal trunks crammed full: One is packed to the top with shoes, shirts, and pants, most well-worn. He has two large linen mosquito nets, 10 tin cups, and a metal file. He has ropes of beads: strand upon strand—over 200 in all, pounds of small colored glass beads made in Czechoslovakia that I had bought in Johannesburg years earlier. He has four large iron pits and a five-gallon plastic jerry can. He has a plow, a gift from the anthropologists. He has a bridle and a bit, light blankets, a large tin basin. He has six pieces of silverware, a mirror and hairbrush, two billycans. His wife and his children together couldn't carry all that. The trunks are too heavy and too large for one person to carry so you would have to have two people for each. What about the plow, those heavy iron pots? Quite a job to carry those through bush, through the thick thorns.

But here is the surprising part. Talk to that man. Read the printout. See what he knows, what he believes. It isn't surprising that he speaks the Herero language and Setswana fluently or that he has worked for the Herero, the anthropologists. Nothing startling there. A budding Dobe capitalist. But then comes the shock. He espouses the traditional values.

"Bushmen share things, John. We share things and depend on each other, help each other out. That's what makes us different from the black people."

But the same person, his back to the door, opens his trunks, unlocks them one by one, lays out the blankets, the beads, then quickly closes each before he opens the next.

Multiply that. Make a whole village of people like that, and you can see the cumulative effect: You can actually measure it. As time goes on, as people come to own more possessions, the huts move farther and farther apart.

In the old days a camp was cosy, intimate and close. You could sit there by one fire and look into the other grass huts, see what the other people were doing, what they were making or eating. You heard their conversations, the arguments and banter.

We ask them why the new pattern?

Says Dau: "It's because of the livestock that we put our huts this way. They can eat the grass from the roofs and the sides of our houses. So we have to build fences to keep them away and to do that, you must have room between the huts."

I look up from the fire, glance around the camp, say nothing. No fences there. Not a single one around any of the huts, although I concede that one day they probably will build them. But why construct a lot of separate small fences, one around each hut? Why not clump the huts together the way they did in the old days and make a single large fence around the lot? Certainly a more efficient approach. Why worry about fences now in any case? The only exposed grass is on the roofs, protected by straight mud walls and nothing short of an elephant or giraffe could eat it.

Xashe's answer is different. Another brief reply. An attempt to dispose of the subject politely but quickly. "It's fire, John. That's what we're worried about. If we put our houses too close together, if one catches fire, the others will burn as well. We don't want one fire to burn all our houses down. That's why we build them so far apart."

But why worry about fire now? What about in the old days when the huts were so close, cheek by jowl? Why is it that when the huts were really vulnerable, when they were built entirely of dried grass, you didn't worry about fires then?

You read Diane's interviews and look at those lists of how much people own. You see those shielded mud huts with doors spaced, so far apart. You also listen to the people you like and trust. People who always have been honest with you. You hear their explanations and realize the evasions are not for you but for themselves. You see things they can't. But nothing can be done. It would be ludicrous to tell these brothers: "Don't you see, my friends, the lack of concordance between your values and the changing reality of your world?"

Now, years after the DBC study, I sit with data spread out before me and it is so clear. Richard's camp in 1963: just grass huts, a hearth in front of each. Huts and hearths in a circle, nothing more. 1968: more of the same. The following year though the first *kraal* appears, just a small thorn enclosure, some acacia bushes cut and dragged haphazardly together for their first few goats. It's set apart way out behind the circle of huts. On one goes, from plot to plot, following the pattern from year to year. The huts change from grass to mud. They become larger, more solidly built. Goats, a few at first, then more of them. So you build a fence around your house to keep them away from the grass roofs. The *kraals* grow larger, move in closer to be incorporated finally into the circle of huts itself. The huts become spaced farther and farther apart, seemingly repelled over time, one from the next, people, families move farther apart.

The bones tell the same story. 1947: All the bones from wild animals, game caught in snares or shot with poisoned arrows—game taken from the bush. By 1964 a few goat bones, a cow bone or two, but not many. Less than 20 percent of the total. Look then at the early 1970s and watch the line on the graph climb slowly upwards—by 1976 over 80 percent from domesticated stock.

But what explains the shattering of this society? Why is this hunting and gathering way of life, so resilient in the face of uncertainty, falling apart? It hasn't been a direct force—a war, the ravages of disease. It is the internal conflicts, the tensions, the inconsistencies, the impossibility of reconciling such different views of the world.

At Dobe it is happening to them all together. All of the huts have moved farther apart in lockstep, which makes it harder for them to see how incompatible the old system is with the new. But Rakudu, a Bushman who lived at the Mahopa waterhole eight miles down the valley from Dobe, was a step ahead of the rest. He experienced, before the rest of them, their collective fate.

When I was at the Cobra Camp in 1969, Rakudu lived down near Mahopa, off on his own, a mile or so away from the pastoral Herero villages. He had two hats and a very deep bass voice, both so strange, so out of place in a Bushman. He was a comical sort of man with the hats and that voice and a large Adam's apple that bobbed up and down.

The one hat must have been a leftover from the German-Herero wars because no one in Botswana wore a hat like that—a real pith helmet with a solid top and a rounded brim. It had been cared for over the years because, although soiled and faded, it still retained the original strap that tucks beneath the chin. The second hat was also unique—a World War I aviator's hat, one of those leather sacks that fits tightly over the head and buckles under the chin. Only the goggles were missing.

I should have seen then how out of place the ownership of two hats was in that hunter-gatherer world. Give two hats like that to any of the others and one would have been given away on the spot. A month or two later, the other would become a gift as well. Moving goods as gifts and favors along that chain of human ties. That was the way to maintain those links, to keep them strong.

When I went to Rakudu's village and realized what he was up to, I could see that he was one of a kind. The mud-walled huts in his village made it look like a Herero village—not a grass hut in sight. And when I came, Rakudu pulled out a hand-carved wood and leather chair and set it in the shade. This village was different from any of the Bushman camps I had seen. Mud huts set out in a circle, real clay storage bins to hold the corn—not platforms in a tree—and *kraals* for lots of goats and donkeys. He had a large field, too, several years before the first one appeared at Dobe.

Why shouldn't Bushmen do it—build their own villages, model their subsistence after the Herero? To plant a field, to tend goats, to build mud-walled houses like that was not hard to do. Work for the Herero a while and get an ax, accumulate the nucleus of a herd, buy or borrow the seeds. That year the rains were long and heavy. The sand held the water and the crickets and the birds didn't come. So the harvest was good, and I could sit there in the carved chair and look at Rakudu's herd of goats and their young ones and admire him for his industry, for what he had done.

Only a year later I saw him and his eldest son just outside the Cobra Camp. I went over and sat in the sand and listened to the negotiations for the marriage Rakudu was trying to arrange. His son's most recent wife had run away, and Rakudu was discussing a union between his son and Dau the Elder's oldest daughter who was just approaching marriageable age. They talked about names and Dau the Elder explained why the marriage couldn't take place. It was clear that the objection was trivial, that he was making an excuse. Even I could see that his explanation was a face-saving gesture to make the refusal easier for all of them.

Later I asked Dau the Elder why he did it. It seemed like a good deal to me. "Rakudu has all that wealth, those goats and field. I'd think that you would be anxious to be linked with a family like that. Look at all you have to gain. Is the son difficult? Did he beat his last wife?"

"She left because she was embarrassed. The wife before her ran away for the same reason and so did the younger brother's wife," he said. "Both brothers treated their wives well. The problem wasn't that. It was when the wives' relatives came. That's when it became so hard for the women because Rakudu and his sons are such stingy men. They wouldn't give anything away, wouldn't share anything with them. Rakudu has a big herd just like the Herero, and he wouldn't kill goats for them to eat."

Not the way Bushmen should act toward relatives, not by the traditional value system at least. Sharing, the most deeply held Bushman belief, and that man with the two hats wouldn't go along. Herero are different. You can't expect them to act properly, to show what is only common decency; you must take them as they are. But someone like Rakudu, a Bushman, should know better than that. So the wives walked out and left for good.

But Rakudu understood what was happening, how he was trapped—and he tried to respond. If you can't kill too many goats from the herd that has become essential to you, perhaps you can find something else of value to give away. Rakudu thought he had an answer.

He raised tobacco in one section of his field. Tobacco, a plant not really adapted to a place like the northern Kalahari, has to be weeded, watered by hand, and paid special care. Rakudu did that and for one year at least harvested a tobacco crop.

Bushmen crave tobacco and Rakudu hoped he had found a solution—that they would accept tobacco in place of goats, in place of mealy meal. A good try. Perhaps the only one open to him. But, as it turned out, not good enough. Rakudu's son could not find a wife.

Ironic that a culture can die yet not a single person perish. A sense of identity, of a shared set of rules, of participation in a single destiny binds individuals together into a tribe or cultural group. Let that survive long enough, let the participants pass this sense through enough generations, one to the next, create enough debris, and they will find their way into the archeological record, into the study of cultures remembered only by their traces left on the land.

Rakudu bought out. He, his wife, and his two sons sold their goats for cash, took the money and walked west, across the border scar that the South Africans had cut, through the smooth fence wire and down the hard calcrete road beyond. They became wards of the Afrikaaners, were lost to their own culture, let their fate pass into hands other than their own. At Chum kwe, the mission station across the border 34 miles to the west, they were given numbers and the right to stand in line with the others and have mealy meal and other of life's physical essentials handed out to them. As wards of the state, that became their right. When the problems, the contradictions of your life are insoluble, a paternalistic hand provides one easy out.

Dau stayed at Dobe. Drive there today and you can find this mud-walled hut just by the waterhole. But he understands: He has married off his daughter, his first-born girl to a wealthy Chum kwe man who drives a tractor—an old man, more than twice her age, and by traditional Bushmen standards not an appropriate match. Given the chance, one by one, the others will all do the same.

## QUESTIONS 12.1

**1.** List and illustrate the concepts (invention, diffusion, and so on) mentioned in this chapter with examples from this reading.

**2.** Which theory of social change best fits the Bushmen's situation? Why?

How would conflict theorists and functional theorists describe what happened?

**3.** How do the anthropologists seem to have done their research? List the different techniques they used.

---

## Reading 12.2

# EVERYTHING IN ITS PATH

*Kai Erikson*

---

*On February 26, 1972, a makeshift mining-company dam in the hills of West Virginia gave way, and a flood of water swept down Buffalo Creek. Communities along Buffalo Creek were destroyed and 125 people were killed. For the survivors, the disaster turned what had been a tightly knit community into a collection of disorganized people with no roots, no neighborhood, and little of the feeling they formerly held for each other. Sociologist Kai Erikson described the effects of the disaster in* Everything in Its Path. *In this excerpt from his book, Erikson describes the trauma of loss of communality. By communality, Erikson means a state of mind shared among a particular gathering of people.*

In most of the urban areas of America, each individual is seen as a separate being, with careful boundaries drawn around the space he or she occupies as a discrete personage. Everyone is presumed to have an individual name, an individual mind, an individual voice, and, above all, an individual sense of self—so much so that persons found deficient in any of those qualities are urged to take some kind of remedial action such as undergoing psychotherapy, participating in a consciousness-raising group, or reading one of a hundred different manuals on self-actualization. This way of looking at things, however, has hardly any meaning at all in most of Appalachia. There, boundaries are drawn around whole groups of people, not around separate individuals with egos to protect and potentialities to realize; and a person's mental health is measured less by his capacity to express his inner self than by his capacity to submerge that self into a larger communal whole.

I am going to propose, then, that most of the traumatic symptoms experienced by the Buffalo Creek survivors are a reaction to the loss of communality as well as a reaction to the disaster itself, that the fear and apathy and demoralization one encounters along the entire length of the hollow are derived from the shock of being ripped out of a meaningful community setting as well as the shock of meeting that cruel black water.

- It is almost like a ghost town now.

- It has changed from the community of paradise to Death Valley.

- Some reason or other, it's not the same. Seems like it's frozen.

- I have found that most of the people are depressed, unhappy, mournful, sick. When you go up Buffalo Creek the only remains you see is an occasional house here and there. The people who are living in the trailers have a depressed and worried look on their faces. You don't see children out playing and running as before. Buffalo Creek looks like a deserted, forsaken place.

- What I miss most is the friendliness and closeness of the people of Buffalo Creek. The people are changed from what they were before the disaster. Practically everyone seems despondent and undecided, as if they were wait-ing for something and did not know what. They can't reconcile themselves to the fact that things will never be the same again.

- It's kind of sad around there now. There's not much happiness. You don't have any friends around, people around, like we had before. Some of them are in the trailer camps. Some of them bought homes and moved away. Some of them just left and didn't come back. It's like teeth in an old folk's mouth down there now.

- People don't know what they want or where they want to go. It is almost as though they don't care what happens anymore.

- My husband and myself used to enjoy working and improving on our home, but we don't have the heart to do anything anymore. It's just a dark cloud hanging over our head. I just can't explain how we feel.

- I don't know. I just got to the point where I just more or less don't care. I don't have no ambition to do the things I used to do. I used to try to keep things up, but anymore I just don't. It seems I just do enough to get by, to make it last one more day. It seems like I just lost everything at once, like the bottom just dropped out of everything.

The clinical name for this state of mind, of course, is depression, and one can hardly escape the conclusion that it is, at least in part, a reaction to the ambiguities of post-disaster life in the hollow. Most of the survivors never realized the extent to which they relied on the rest of the community to reflect back a sense of meaning to them, never understood the extent to which they depended on others to supply them with a point of reference. When survivors say they feel "adrift," "displaced," "uprooted," "lost," they mean that they do not seem to belong to anything and that there are no longer any familiar social landmarks to help them fix their position in time and space. They are depressed, yes, but it is a depression born of the feeling that they are suspended pointlessly in the middle of nowhere. "It is like being all alone in the middle of a desert," said one elderly woman who lives with her retired husband in a cluster of homes. As she talked, the voices of the new neighbors could be heard in the background; but they were not *her* neighbors, not *her* people, and the rhythms of their lives did not provide her with any kind of orientation.

This failure of personal morale is accompanied by a deep suspicion that moral standards are beginning to collapse all over the hollow, and in some ways, at least, it would appear that they are. As so frequently happens in human life, the forms of misbehavior people find cropping up in their midst are exactly those about which they are most sensitive. The use of alcohol, always problematic in mountain society, has evidently increased, and there are rumors spreading throughout the trailer camps that drugs have found their way into the area. The theft rate has gone up too, and this has always been viewed in Appalachia as a sure index of social disorganization. The cruelest cut of all, however, is that once close and devoted families are having trouble staying within the pale they once observed so carefully. Adolescent boys and girls appear to be slipping away from parental control and are becoming involved in nameless delinquencies, while there are reports from several of the trailer camps that younger wives and husbands are meeting one another in circumstances that violate all the local codes. A home is a moral sphere as well as a physical dwelling, of course, and it would seem that the boundaries of moral space began to collapse as the walls of physical space were washed down the creek. The problem is a complex one. People simply do not have enough to do, especially teenagers, and "fooling around" becomes one of the few available forms of recreation. People have old memories and old guilts to cope with, especially the seasoned adults, and drinking becomes a way to accomplish that end. And, for everyone, skirting the edges of once-forbidden territory is a way to bring new excitement and a perverse but lively kind of meaning into lives that are otherwise without it.

A retired miner in his sixties speaking of himself:

- I did acquire a very bad drinking problem after the flood which I'm doing my level best now to get away from. I was trying to drink, I guess, to forget a lot of things and get them moved out of my mind, and I just had to stop because I was leading the wrong way. I don't know what the answer is, but I know that's not it. I don't want to drink. I never was taught that. I've drunk a right smart in my life, but that's not the answer.

And a woman in her late twenties who had recently moved out of the largest of the trailer camps:

- There was all kinds of mean stuff going on up there. I guess it still does, to hear the talk. I haven't been back up there since we left. Men is going with other men's wives. And drinking parties. They'd play horseshoes right out by my trailer, and they'd play by streetlight until four or five in the morning. I'd get up in the morning and I'd pick up beer cans until I got sick. The flood done something to people, that's what it is. It's changed people. Good people has got bad. They don't care anymore. "We're going to live it up now because we might be gone tomorrow," that's the way they look at it. They call that camp "Peyton Place," did you know that? Peyton Place. I was scared to death up there. I don't even like to go by it.

Many families have been affected:

- Many marriages have broken up that seemed secure before the flood. My husband and I can agree on only one thing: we won't go back to Lorado. When the time comes to buy us a house, we both agree that we will face a major problem in our marriage. I hope we can agree on where to live. If not, then we may have to come to a parting of the ways after twenty-six years of marriage.

- My husband and I, we was happy before the flood. We got along real good, other than just a few quarrels that never amounted to nothing. But after the flood we had fights, and it was constantly we were quarreling about something or other. We had fights. He would hit me and he would choke me and he would slap me around.

- My children are changed. I sit and try to talk to them, tell them they are a family and should love each other and treat each other like brothers and sisters. But most of the time they treat each other like enemies. They're always on the firing line at each other. It's always screaming and yelling.

- My grandchildren. It used to be we was the loveliest people you ever seen. We was, together. Now my grandchildren won't hardly give me a look. I don't know what's wrong. They seem like they are moody or something. My grandson there, used to be he loved me better than anything, and now he won't even look at me. He don't want to be around me. One of the granddaughters, too, is about the same. She has spells that way. I don't know. I can't understand it.

- I have good new neighbors, but it's not the same. The neighbors I had before the flood shared our happiness when our babies were born, they shared our troubles and our sorrows. Here is the change. My husband has been sick going on three weeks. My old neighbors would ask about him or go see him or send him a get-well card. But he only got one card, and it was from someone away from here. The day the flood came, the people of Buffalo Creek started running, and they are still running inside their minds. They don't have time to stand and talk.

One result of all this is that the community, what remains of it, seems to have lost its most significant quality—the power it generated in people to care for one another in times of need, to console one another in times of distress, to protect one another in times of danger. Looking back, it does seem that the general community was stronger than the sum of its parts. When the people of the hollow were sheltered together in the embrace of a secure community, they were capable of extraordinary acts of generosity; but when they tried to relate to one another as individuals, separate entities, they found that they could no longer mobilize whatever resources are required for caring and nurturing.

- It used to be that everyone knew everyone. When you were hitchhiking, you just put out your thumb and the first car along would pick you up. But it's not like that now. They just don't care about you now. They got problems of their own, I guess.

- The changes I see are in the people. They seem to be so indifferent toward their fellow man. I guess it's because they had to watch a whole lifetime go down the drain.

- I'm getting old, too, and I can't get no help. Nobody'll help you do nothing. You have to pay somebody, and they'll come and start a project for you, but then they'll walk off and leave you. It's just too much.

In general, then, the loss of communality on Buffalo Creek has meant that people are alone and without very much in the way of emotional shelter. In the first place, the community no longer surrounds people with a layer of insulation to protect them from a world of danger. There is no one to warn you if disaster strikes, no one to rescue you if you get caught up in it, no one to care for you if you are hurt, no one to mourn you if the worst comes to pass. In the second place—and this may be more important in the long run—the community can no longer enlist its members in a conspiracy to make a perilous world seem safe. Among the benefits of human communality is the fact that it allows people to camouflage what might otherwise be an overwhelming set of realities, and the question one should ask about Buffalo Creek is whether the people who live there are paralyzed by imaginary fears or paralyzed by the prospects of looking reality in the eye without the help of a communally shared filter.

---

## QUESTIONS 12.2

**1.** Apply the terms *social disorganization* and *anomie* to this selection.

**2.** Analyze and comment on the following statement: "It was not social disorganization but reorganization

that the people of Buffalo Creek were experiencing."

**3.** Forecast the future of the Buffalo Creek area using the social-change theories discussed in this chapter.

---

## NOTES

1. See the chapter on social change by Alvin Boskoff in Howard Becker and Alvin Boskoff, *Modern Sociological Theory* (New York: Dryden, 1957).

2. This discussion is drawn in part from Robert Bierstedt, *The Social Order*, 4th ed. (New York: McGraw-Hill, 1974), chapter 20.

3. "The Killing Fields of the Deep South: The Market for Cotton and the Lynching of Blacks. 1882–1930," by E. M. Beck and Stewart Tolnay, *American Sociological Review* 55 (August 1990), pp. 526–539.

4. Ogburn's works include *Social Change* (New York: Viking, 1922, 1950); *The Social Effects of Aviation* (Boston: Houghton Mifflin, 1946); *Machines and Tomorrow's World*, rev. ed., Public Affairs Pamphlets, no.

25, 1946; and Ogburn and Nimkoff, *Sociology*, 2d ed. (Boston: Houghton Mifflin, 1950).

5. The major source for this section is *Perspectives on Social Change*, 2d ed., by Robert Lauer (Boston: Allyn & Bacon, 1977), especially chapters 2, 3, 6, 9, and 12, and pp. 47 and 89.

6. See *Human Societies*, 4th ed., by Gerhard Lenski and Jean Lenski (New York: McGraw-Hill, 1982), especially chapters 2–4 and 14; also see Lauer (see note 5), chapters 2 and 12.

7. Émile Durkheim, *Suicide*, translated by John A. Spaulding and George Simpson (New York: Free Press, 1951).

AP/Wide World Photos

# Collective Behavior

## TWO RESEARCH QUESTIONS

**1.** *Does the information spread in a rumor become more and more distorted as the rumor is passed on, or does the information become more accurate?*

**2.** *Why do people wear T-shirts with messages and pictures on them? Are there any meaningful patterns among the types of T-shirt messages?*

COLLECTIVE BEHAVIOR is group behavior that is spontaneous, unstructured, and unstable. It can be either sporadic and short-term or more continuous and long-lasting.[1] Collective behavior is often difficult to predict because it is not rooted in the usual cultural or social norms. Spontaneous and unstructured behavior is difficult to observe or record objectively and is, therefore, difficult to study. Ethically and practically, the researcher cannot yell "Fire!" in a crowded theater, start a downtown riot, or produce a natural disaster and then observe how people behave. Although some artificial or laboratory-created studies of rumor and panic have been conducted, most studies of collective behavior by social scientists are after-the-fact analyses and discussions with people who happened to be involved.

These studies have revealed that collective behavior may follow reasonably consistent patterns. For example, Neil Smelser[2] has suggested that specific conditions must be present for collective behavior to occur:

**1.** *Structural conduciveness.* Given the setting or structure of a specific group or society, a particular type of collective behavior, such as panic, riot, craze, or lynching, *could* happen; although the behavior would not necessarily be encouraged, it would still be possible because of historical or other structural reasons.

**2.** *Structural strain.* Conditions in society place strain on people; a general feeling of deprivation or conflict is produced by such things as economic failure, hostility among races, social classes, or religions, or sudden changes in the existing order.

**3.** *Generalized belief.* A set of feelings, beliefs, or rationalizations must be present to explain the cause of the strain, to create a common culture prepared for action.

**4.** *Precipitating factors.* A specific event provides a concrete reason for taking action.

**5.** *Mobilization for action.* The participants become organized and act.

**6.** *Social control.* Factors that occur after the event has started can prevent or inhibit the effect of the previous conditions from continuing and can change its course.

Look at Reading 13.1, which describes an episode of collective behavior, and use Smelser's conditions to analyze it: (1) *Conduciveness:* American society has a tradition of violence, of taking the law into one's own hands to

## NEIL SMELSER (1930– )

Neil Smelser earned his Ph.D. in Social Relations at Harvard University. Smelser has had a very active professional career. At the age of 26 and still a graduate student, he coauthored *Economy and Society* with Talcott Parsons. After receiving a Harvard doctorate in 1958 while studying under Parsons, Smelser went on to Oxford where he earned a master's degree in 1959. By the time he was 28, he had authored several books and was an associate professor of sociology at Berkeley. Smelser is the author, coauthor, or editor of numerous books and articles on a wide range of topics. He has written on social movements, economic sociology, and British social history. Smelser also has formal psychoanalytic training and an ongoing counseling practice. These aspects of his interests are reflected in his most recent book, *The Social Edges of Psychoanalysis* (1999). Psychoanalytic theory and sociology have always had an uneasy relationship, and in this book Smelser provides arguments about how and why psychoanalytic approaches can deepen the sociological perspective.

Neil Smelser is the director of the Center for Advanced Study in the Behavior Sciences at Stanford University, and he is Professor Emeritus at the University of California, Berkeley.

---

deal with situations in which it is felt that the law is moving too slowly; this is not legally encouraged, but it happens nevertheless; (2) *Strain:* Perhaps strain developed because of poor crops, bad economic conditions, or racial enmity between blacks and whites; (3) *Beliefs:* Perhaps minority groups were seen as the cause of the situation—"they are stepping out of their place and citizens had to respond to put the situation right"; (4) *Precipitating factors:* Over some drink and discussion, the idea emerged that something had to be done, and a lynching seemed the appropriate response; (5) *Mobilization:* A group distracted the sheriff to another part of the county and, with collars up and hats pulled down, gathered in front of the jail; (6) *Social control:* The lynching was stopped by an unanticipated type of informal social control; what was it?

We will look at several types of collective behavior in the following pages. Keep in mind Smelser's model. It might also be helpful to consider collective behavior along a continuum as suggested in Table 13.1. Those types of collective behavior toward the top are more structured and stable; those toward the bottom are more spontaneous, unstructured, and unstable.

## CROWDS

An outline of the major collective behavior concepts would start with the crowd. A **crowd** is a temporary collection of people in close physical contact reacting together to a common stimulus. For example, the passengers on the flight from New York to San Francisco whose pilot suddenly decides he

**TABLE 13.1** A Continuum of Collective Behavior

| Behavior Is More Structured and Stable | |
| --- | --- |
| ↑<br><br>↓ | Audiences<br><br>Social Movements<br><br>Fads<br>Crazes<br>Fashions<br><br>Mobs<br><br>Riots<br><br>Disasters |
| Behavior Is More Spontaneous,<br>Unstructured, and Unstable | Panic<br>Mass Hysteria |

Suggested by Constance Verdi.

would like to go to the North Pole might be transformed from an aggregation into a crowd (or even a mob). Crowds have certain characteristics in
common. **Milling** usually occurs as a crowd is being formed. In one sense
milling refers to the excited, restless physical movement of the individuals
involved. In a more important sense, milling refers to a process of communication that leads to a definition of the situation and possible collective
action. Not long ago, a classroom I was in suddenly started shaking with the
first tremors of an earthquake. Almost at once the people began turning,
shifting, and looking at each other, at the ceiling, and at the instructor. They
were seeking some explanation for the highly unusual experience, and
whether spoken aloud or not, the questions on their faces were clear: "What
is it?" "Did you feel it?" "What should we do?" Buzzing became louder
talking, and someone shouted "Earthquake!" The students began to get up
and move toward the doors. Many continued to watch the ceiling. . . .
Milling may involve the long buildup of a lynch mob or the sudden reaction
in a dark and crowded theater when someone shouts "Fire!" Milling helps
ensure the development of a common mood for crowd members.

When they are part of a crowd, people tend to be *suggestible*. They are less
critical and will readily do things that they would not ordinarily do alone.
This is true in part because, as members of a crowd, they are *anonymous*. The
prevailing feeling is that the crowd is responsible, not the individual. Once
one becomes a member of an active crowd, it is extremely difficult to step
back, to get perspective, and to evaluate objectively what one is doing.
Crowd members have a narrowed focus, a kind of tunnel vision. The physical presence of a crowd is a powerful force; people almost must separate
themselves from the crowd physically before they can critically examine
their own behavior. There is also a *sense of urgency* about crowds. Crowds
are oriented toward a specific focus or task: "We've got to do *this*, and we've
got to do it *now!*" Some form of leadership usually appears in the crowd,

but as the mood of the crowd changes, the leadership can shift quickly from one individual or group to another.

There are many different types of crowds. Some are passive, such as those watching a building burn or those at the scene of an accident. Some are active, such as a race riot or a lynch mob. Some crowds have a number of loosely defined goals. Other crowds are focused on a specific goal. Turner and Killian distinguish between crowds that direct their action toward some external object—harassing a speaker until he leaves the platform or lynching a person—and expressive crowds that direct their focus on the crowd itself—cheering at a football game or speaking in tongues at a church service.

Controlling the behavior of a crowd is difficult because of the mass of people involved and the spontaneous nature of their behavior. Some methods of dealing with a potentially riotous crowd have been suggested, however: Remove or isolate the individuals involved in the precipitating incident. Reduce the feelings of anonymity and invincibility of the individuals; force them to focus on themselves and on the consequences of their actions. Interrupt patterns of communication during the milling process by breaking the crowd into small units. Remove the crowd's leaders if it can be done without use of force. Finally, attempt to distract the attention of the crowd by creating a diversion or a new point of interest, especially if this can be accomplished by someone who is considered to be in sympathy with the crowd.

## Audiences, Mobs, and Riots

Audiences and mobs are specific types of crowds. An **audience** at a concert, football game, lecture, religious service, or burning building can usually be likened to a passive crowd. Emotional contagion is possible in such situations, and individuals in a group are responsive in ways that they would not be as individuals. Audiences at performances of rock-and-roll stars and in Pentecostal church services can become very expressive in a variety of possibly unpredictable ways. Comedians expect audiences to laugh at their jokes. But most comedians have "bits" that they deliver when audiences are unpredictable and don't laugh. Much of audience behavior is predictable, or at least predictably unpredictable. The football fan knows he is going to cheer at the game, the comedian's rejection bits are well prepared, and rock concerts are adequately staffed with police to protect the musicians and with paramedics to minister to injured fans. At the same time, audiences demonstrate collective behavior characteristics in that their behavior is frequently spontaneous, and members are suggestible and anonymous.

A **mob** is a focused, acting crowd. It is emotionally aroused, intent on taking aggressive action. A lynch mob is an example of such a crowd.

Mob behavior has also occurred when unusual events, such as a severe winter storm in New Jersey and power outages in New York City, have been followed by looting and vandalism. The lights went out in New York City at about 9:35 in the evening on July 13, 1977. The looting and property damage that followed cost businesses $135 to $150 million. Robert Curvin and Bruce Porter studied the mob behavior that occurred during the blackout, and they discovered some interesting patterns. Different types of people looted at different stages of the blackout. Stage-one looters started shortly after the lights went out and were generally criminal types; more than 80 percent of them had previous arrest records. They moved in quickly and selected valuable merchandise that they could sell quickly and profitably. Here's a typical example: A man gathered several friends and a 38-foot moving van and filled the truck twice with such goods as Pampers, baby food, color TVs, leather coats, and sneakers. Stage-two looters started a little later and were primarily teenagers out looking for fun and excitement. Stage two also included unemployed, poorly educated ghetto poor younger than 35. Their looting tended to be random and unorganized. Stage-three looting started several hours later and lasted into the next day. Looters at this stage were likely to be working-class people who got caught up in the street activity and felt social pressure to get involved. Better-off employed people with little or no experience in crime were common. Unfortunately for them, police were now becoming more effective, and more arrests were taking place. As one participant (a salesman with a family income of $375 a week) said, "I wasn't thinking about a color TV or anything that the professional dudes were after, like cars and things. I just wanted to snatch something." A store was broken into, items were thrown out onto the street, he gathered up a pile of women's clothes, and then a police car pulled up and he was arrested. Curvin and Porter suggest that because the looters arrested during the blackout had very high unemployment rates, the main cause of the theft and destruction was the blackout opportunity combined with the national economic decline, high unemployment, high prices for essentials, and the worsened living conditions of the poor.[3]

A **riot** describes the situation in which mob behavior has become increasingly widespread and destructive. Riots can involve a number of mobs acting independently. Throughout our history, the United States has had riots over the issues of race, poverty, and social class—examples include New York, 1863; Chicago, 1919; and Detroit, 1943; and Los Angeles, 1965 and 1992. There have been riots in prisons, and, increasingly, riots related to sporting events. Teams win a championship, their fans celebrate, and small-scale riots occur. Or, as happened in an overcrowded stadium at a World Cup soccer match in Guatemala City in October 1996, attempts to escape a drunken brawl through a jammed exit led people to panic and pile against a fence. Seventy-eight people were killed and more than 100 were injured. Recent sports riots appear to be fueled by a hot evening, an overcrowded stadium, and widespread consumption of alcohol.

England had its worst riots in a century during July 1981. Gangs of white, black, and Asian youths burned cars, stoned police, and smashed shops. In the first week, 300 policemen were injured, millions of dollars' worth of property was destroyed, and more than 1,000 people were arrested in disturbances in London, Liverpool, Manchester, and other cities. In the analysis that followed, there was much disagreement over causes, but the following conditions were mentioned most often: high rates of unemployment (especially among black youth), racism, and young people's resentment of police and authority.

On April 29, 1992, four Los Angeles police officers who had been charged with assault in the beating of Rodney King were acquitted by a jury in Simi Valley, California. The verdict, which surprised, angered, and saddened many, was announced at 3:10 P.M. West Coast time. Around 4 P.M., five young men stole some bottles from a Korean-owned liquor store near the intersection of Florence and Normandie avenues in South Central Los Angeles. A little after 5 P.M., motorists in the area reported being attacked and having their car windows broken out. Police responded in numbers, felt overwhelmed by the crowd, and left the area. Around 6 P.M., looting, especially of liquor stores, began. At 6:46 P.M. and watched by hundreds of thousands of Americans on television, truck driver Reginald Denny was pulled from his truck, assaulted, and nearly killed. Thus began the 1992 Los Angeles riot, which, by the time it ended on May 4, resulted in 53 deaths, 2,325 reported injuries, more than 5,000 arrests, more than 600 buildings completely destroyed by fire, and approximately $735 million in total damages. The causes are still being studied—clearly the Simi Valley verdict was a flash point, but it played against a backdrop of the 1990–1992 recession that worsened already bad economic conditions. There was also continuing anger toward a criminal justice system that appeared to be racially biased; there was resentment toward Korean merchants who had been buying out black store owners in South Central; perhaps even the presence and use of alcohol in the area was a factor—the health department in Los Angeles reported that four times as many blacks died in 1990 from alcohol-related causes as from cocaine use and that two of three accidental drug deaths in the county were alcohol-related. South Central has an amazing number of liquor outlets, and looters and arsonists made liquor stores a prime target. It is interesting that the apparent causes of the L.A. riot in 1992 are similar to those suggested for the riots in England a decade earlier (see previous paragraph).

Social scientists have studied riots extensively since the activities of the 1960s. Ralph Conant has examined a number of urban conflicts, and he suggests four phases that describe the possible life history of a riot. Conant believes that not all civil disturbances go through all four stages; in fact, most do not even reach Phase Three but die out earlier. Typically, the temper or mood of the participants is ambivalent and unstable throughout the riot:

> *Phase One—The precipitating incident:* Some gesture, act, or event is taken by the community as concrete evidence of injustice. The incident—use of force by police or eviction of a tenant, for example—is inflammatory because it is seen as typical of the antagonist. It is seen as ample excuse for striking back, and it tends to draw together a large number of people. In a community in which there is much unrest, the precipitating incident might be a minor event to others, but it is very important to the members of the community.

*Phase Two—Confrontation:* More and more people gather. Individuals suggest targets for violence and attempt to focus crowd reaction, a process called **keynoting.** Other individuals attempt to calm the crowd and encourage moderation. Police officials might appear and attempt to disrupt the keynoting by dispersing the crowd. This might have the opposite effect and raise a hostile keynoter to an even more prominent position. Conant states that the outcome of Phase Two is of crucial importance. The temper of the crowd can calm, or it can escalate explosively. The responses of civil authorities, of local leaders, and of news media (whether they overreport and sensationalize or show restraint) are all critical.

*Phase Three—Roman holiday:* Success of hostile keynoting and an explosion of activity mark the beginning of Phase Three. The crowd, dominated now by younger people, begins moving. Conant describes their mood as an angry, gleeful intoxication. Buildings and cars are attacked. Targets are not necessarily random but more likely are selected and patterned, representing objects that rioters resent or fear. Police are taunted and attacked. Later in Phase Three, excitement subsides and systematic looting begins. According to Conant, behavior in the carnival-like atmosphere of the Roman holiday is explained at least in part by the amazing contradictions and ambivalences present in the urban ghetto: stores owned by hated white people and the beloved contents of the stores; despised police and their admired weapons. This ambivalence extends to the riot participants, who show great changes in behavior. They can be violent one day and calm the next day. These ambivalences toward violence and in mood make it unlikely that most riots will reach Phase Four.

*Phase Four—Siege:* Polarization of the two sides signals the end of effective communication. Curfew is declared. Warlike acts occur. Snipers attack, buildings are firebombed, and a general state of siege exists.[4]

## PANIC AND MASS HYSTERIA

Panic represents a particular type of reaction in a crowd situation. Sociologically, **panic** is defined as nonadaptive or nonrational flight resulting from extreme fear and loss of self-control. Usually there is a severe threat, a limited number of ways out, and a feeling of being trapped. Flight in itself does not necessarily mean panic. Some flight is rational and sensible, as when people leave in an orderly manner an area that is threatened by hurricane or flood. Panic refers specifically to nonadaptive flight—people stampeding through a burning building and attempting to fight their way out a door that is already hopelessly blocked. Panic differs somewhat from other forms of collective behavior in that although it is frequently a result of the crowd situation (spontaneous and contagious), it is essentially an individualistic and competitive reaction. During a panic, each person is desperately trying to obtain an objective on his or her own, and because many others are doing the same thing, there is a good chance that some will not make it.[5]

## A SNAKE WITH BAD BREATH

A venomous snake has been haunting the people of the Pokharan desert area of Rajasthan in India. The snake, known as *peevana* or *piana*, is about 20 inches long. It steals into huts at dawn and delicately crawls on the chest of someone asleep. It than opens it mouth wide and holds it in front of the sleeping person's nose, and the victim inhales the snake's poisonous breath. The snake appears to be sadistic as well; it lets its victims know of their fate by whipping them in the face with its tail as it slips away. Newspapers in Rajasthan recently reported that five people fell victim to *peevana* in one village in a single day.

The *peevana* myth has endured for centuries. According to the local maharaja's gazette, the ruler once offered 1,000 rupees to anyone who could produce the reptile, dead or alive. No one has yet been able to do so.

---

The Iroquois Theater fire in Chicago provides an extreme example of panic and its possible consequences. During a performance on a December afternoon in 1903, draperies on the stage caught fire. Somebody yelled "Fire," and most of the audience panicked. Actors and musicians attempted to calm the crowd, but did not succeed. There were many exit doors, but some were poorly marked and some were difficult to open. People were crushed against doors and on stairways; others jumped to their deaths from fire escapes. The fire was not very serious, and the fire department arrived quickly—the panic lasted only eight minutes. The death toll, however, was 602.

**Mass hysteria** is in some ways similar to panic and describes the situation in which a particular behavior, fear, or belief sweeps through a large number of people—a crowd, a city, or a nation. Examples of mass hysteria might include fainting at a rock concert, the fear that flying saucers are after us, or the belief that particular women are practicing witchcraft.

## RUMORS

Rumor is a type of collective behavior that might or might not be crowd-oriented. Much **rumor** is merely a form of person-to-person communication. Rumor is defined as unconfirmed, although not necessarily false, communication. It is often related to some issue of public concern. Rumors change constantly as they spread. They tend to grow shorter, more concise, and more easily told. Some attractive details of the rumor become magnified. New details are manufactured to complete the story or to make it internally consistent. For example, a rumor that Beatle Paul McCartney had been killed in an automobile accident swept the country in the fall of 1969. Fans of the rock-music group frantically sought further proof of Paul's demise by playing Beatle records backwards and by finding clues on album covers and in photographs of the group. Even several personal appearances by the supposedly dead Beatle did little to quash the rumor. In fact, some are still suspicious today of that fellow passing himself off as Paul McCartney.

People pass rumors for many reasons. The rumor might fit with what we want to believe, with what we know to be true: "Teachers are absentminded" or "Sex education is Communist-inspired." We might pass rumors on to increase our status in the eyes of others. If we were the first to get the news, we want others to know it. Or it's such a tremendous story that others are bound to think more of us when we tell it to them. Did you hear about the flying saucers over Minneapolis? About those giant birds perching on trees in the Southwest? About Bigfoot wandering through downtown Yakima? You *didn't*? Well, let me tell you. . . .

Rumor can have a variety of functions. It may be gossip and serve to enliven the Thursday-afternoon bridge club, or it can provide a core of communication during a crowd's milling period and thus be the stimulus for crowd action.

As is implied in the first question asked at the beginning of this chapter, there is a common assumption that rumors become less accurate as they are passed on. Taylor Buckner makes the point that in some situations rumors become *more* accurate. The important variables to consider are rumor set and interaction setting. Buckner suggests three rumor sets: critical, uncritical, and transmission. *Critical set* occurs when people have knowledge about the rumor's subject matter, have personal experience with it, or are habitually suspicious of rumors anyway. These people tend to eliminate inaccurate and irrelevant parts of rumors they hear and pass on only the most important parts. An *uncritical set* can exist when people have no knowledge about a situation, when passing the rumor fills a need, or during crisis situations when little information is available. People with an uncritical set can modify the rumor so it sounds better, change it to fit their own needs, devise totally new rumors to fit the situation, or in some other way contribute to the inaccuracy of the rumor. *Transmission set* means that the content of the rumor is irrelevant to the person. The person's only interest is to pass the rumor on. People who merely transmit a rumor forget or reword parts but usually do not purposely distort or correct the rumor.

Group characteristics—the interaction setting—also affect a rumor's accuracy. Buckner suggests that a rumor is heard more often and evaluated more carefully in a close, more primary group (college dorm, sorority, army unit, social club, clique) than in a more diffuse, or secondary, group. It also follows that the more interested and involved a group is with a rumor, the more the rumor is told and evaluated. The rumor that then-Beatle Paul McCartney was dead was undoubtedly more actively discussed among rock-music and Beatles fans than it was in the boardroom at General Motors or among people passing on rumors about flying saucers. Buckner concludes that the multiple interaction likely to occur in a close, highly involved group combined with a critical set leads to increased accuracy in the rumor. However, if the critical set is missing, greater inaccuracy results.[6]

## FADS, CRAZES, AND FASHIONS

Fads, crazes, and fashions are forms of collective behavior that are usually more widespread and long-lasting than is crowd behavior. **Fads** and **crazes** refer to the relatively short-term obsessions that members of society or members of specific groups have toward specific mannerisms, objects, clothes, or ways of speaking. Among the more recent fads (in California at least) are in-line skates, tattoos and body piercing, baggy low-slung pants, and gangsta rap music. Past examples include pet rocks, the hula hoop, swallowing goldfish, and stuffing phone booths. Fads and crazes can interest many and can burn brightly for a while, but they usually die out quickly. **Fashions** are similar to fads and crazes but are more widespread and last longer. Examples of fashions might include miniskirts, two-piece bathing suits, long hair on men, wide neckties, Mustang style in cars, ranch style in homes, and rock and folk rock in popular music.

The second research question at the beginning of this chapter asked a fashion question: Why do people wear distinctive T-shirts? In a mass society, one way of establishing your identity—who you are, what you've done, where you've been—is through the clothes you wear. T-shirts fit these conditions well. They are generally acceptable, easily available, and inexpensive. Collecting and wearing T-shirts has become something of an international clothing fad. In an attempt to learn more about T-shirts and the search for identity, the authors of a recent paper asked a number of college students to describe their favorite T-shirts and why they liked them. Four categories of T-shirts emerged: idiosyncratic, commercialized, elitist, and utilitarian. *Idiosyncratic* T-shirts announce uniqueness and individualism. Often these shirts are designed and made by the wearer. These shirts might show off-beat humor or might be souvenirs from places others haven't heard of. These shirts make the individual "stand out"; they demonstrate his or her uniqueness. *Commercialized* T-shirts were the majority. "Corona Beer," "Hard Rock Cafe," "Gold's Gym," and "Bart Simpson" are easily recognized, are available to everyone, and are often good conversation starters with strangers. *Elitist* shirts tell others that the wearer is a member of a group that most others can't get into. They might announce an exotic and expensive place (Club Med—Phuket, Thailand), an athletic event (Pikes Peak Invitational), an accomplishment (I Survived Kilimanjaro), or a school or organization (Dartmouth '92 or Phi Beta Kappa). *Utilitarian* shirts are the plain type and just another article of clothing. T-shirts, then, are both fad and fashion. They have become a typical and popular style of clothing, and they are used in part to establish one's identity.[7]

Turner and Killian report an interesting peculiarity of fads and fashions: Fashions follow the social-class structure, but fads *do not* necessarily follow the social-class structure; that is, fashions usually start at upper-class levels and seep downward. Fads, on the other hand, can appear anywhere in the class structure and can be adopted more quickly by members of lower social classes than by members of the upper classes. Because following fads and fashions brings prestige to people, fashion supports the *status quo* (the current prestige system), whereas fads might upset the *status quo* by granting prestige to people who otherwise have low status.[8]

## DISASTERS

Studies of disasters provide us with data about collective behavior. A **disaster** is defined as a situation in which there is a basic disruption of the social context within which individuals and groups function—a radical departure from the pattern of normal expectations. Such a situation almost by definition results in collective behavior.[9]

Disaster research reveals three general time periods: immediately before, during, and after the disaster. *Predisaster* studies have noted that making people aware of impending disaster is extremely difficult unless the group concerned has already experienced a disaster—a flood, a bombing, or an earthquake. Otherwise, people do not take the warning seriously. They look around, note that others don't seem bothered, and define the situation as normal or at least not serious. During Hurricane Camille in Mississippi, some 20 people were killed in an apartment house where they had decided to ride out the storm and have a "hurricane party." People tend to interpret disaster cues as normal or familiar events.

Studies of behavior *during* a disaster indicate that, contrary to popular belief, panic and loss of control are rare. Flight is a frequent response, but it usually represents an adaptive rational reaction to the situation rather than panic. Studies of behavior *after* a disaster indicate that people tend to underestimate the scope and destructiveness of the disaster. Again, people define events in terms of the familiar and normal. Often those who have clearly defined tasks in case of disaster—civil-defense workers, nurses—experience severe role conflict. The conflict is between helping people in general, as they are trained to do, or finding and helping family and loved ones. Frequently the latter choice is made. People tend to flock *to*, rather than away from, a disaster area. Within 24 hours after the atom bomb was dropped on Hiroshima, thousands of refugees streamed *into* the city. Also, airliners in trouble at major airports attract crowds of people to the scene. One such incident at New York's Kennedy Airport brought so much automobile traffic that emergency vehicles had difficulty getting through. Communication lines often become clogged with calls coming *into* disaster areas. Material convergence occurs as all sorts and varieties of material are shipped into the disaster area in an effort to help the victims. Within 48 hours after a tornado in Arkansas, truckloads full of material began arriving. Among the mass of material that people sent were button shoes, derby hats, a tuxedo, and a carton of falsies. It took 500 workers two weeks to sort it all out. After Hurricane Camille, when supplies began pouring into Mississippi, authorities had to make television appeals to stop the shipments. The same thing happened after an earthquake in Guatemala. People in an area heavily hit by the quake who usually eat corn and beans were sent canned goods they could not open and would not eat. They were also sent jars of peanut butter, boxes of cornflakes, and cans of anchovy paste.

Other useless items that piled up in warehouses after the quake included canned cherries, high-heeled shoes, see-through shirts, and wigs. A Red Cross official reported that the same thing happened after an earthquake in Nicaragua and after a hurricane in Honduras; they concluded that the public simply responds, partly out of a desire to be helpful in a bad situation, partly because they just want to empty their closets. Finally, the studies report that in many cases, disaster as an immediate threat produces solidarity among those experiencing it. People tend to grow closer together as a result of their attempts to fight and survive the situation.

## PUBLICS, PUBLIC OPINION, AND PROPAGANDA

Publics and public opinion represent another aspect of collective behavior. A **public** is a number of people who have an interest in, and difference of opinion about, a common issue. A public engages in communication and discussion (more often indirectly than face-to-face), and, contrary to the crowd, a public is dispersed or scattered rather than in close physical contact. **Public opinion** refers to opinions held by a public on a given issue. **Propaganda** refers to attempts to influence and change the public's viewpoint on an issue.

In a complicated mass society, a great number of issues appear, and each issue has its concerned interest group or public. Membership in these publics is transitory and constantly changing. Some members of a public are vitally interested, others only marginally so. Members of a public might communicate by special magazines, newspapers, television, and letter writing. Professional football and baseball have their publics; those people interested in gun control or their opposites interested in the free right to bear arms are each a public. The issues of socialized medicine, legalization of prostitution, and elimination of the death penalty all have publics. Some publics are widespread; if the issue is the type of job the president is doing, the whole nation can become the public. Other publics might be much smaller.

Public opinion is registered in a variety of ways. It is registered when an incumbent running for office is defeated. It is recorded in the letters-to-the-editor column of a newspaper or magazine that prints a story with which the general public agrees or disagrees. More recently, public opinion polls have become a popular way of determining what the general public is thinking. Polls are avidly followed by politicians as well as by people on the street. Polls have even been criticized because it is feared that some people look at the polls before deciding how to vote or what to think on a given issue.

Public opinion polling has many pitfalls, and occasionally even the best pollsters go wrong. Probably the most famous miss was the *Literary Digest* poll of 1936, which predicted that Landon (Republican) would defeat Roosevelt (Democrat) for president. The *Literary Digest* obtained a biased sample by mailing ballots to people whose names, for the most part, were selected from telephone directories or from lists of automobile owners. Only 40 percent of all homes had phones in 1936, and only 55 percent of all families had cars, so the magazine was sampling people who were economically well-off. Another error was introduced by *mailing* ballots; people of higher income and education are more likely to return mailed questionnaires than are people with lower incomes and education. Ballots were mailed early (September) so that last-minute changes were missed. The upshot was that *Literary Digest* asked a group of reasonably well-to-do people (who generally tend to vote Republican) for whom they intended to vote. Their prediction was so grossly wrong that it is given as the reason that the magazine went out of business.[10]

Because it is usually impossible in a public opinion poll to talk to all members of the public, a sample is taken. Frequently the success of the poll hinges on the accuracy of the sampling technique. The timing of the poll is also important: Opinions change rapidly, and if the latest political poll is three or four weeks before the election, pollsters can expect to have missed many last-minute changes. A more basic fault in polling is that some people are very interested and know a lot about the issue, and others are only slightly interested and know little or nothing. Yet in most polls, the responses of these two types are given the same weight. Other difficulties in sampling public opinion occur when people give pollsters answers they think the pollsters want, or when a person is a member of several conflicting publics at once, as most of us are. What if a member of the Catholic Church who is a single mother with more kids than she can comfortably support is asked her opinion of government-supported artificial birth control? She might find herself a member of conflicting publics. Finally, as far as action is concerned, it is often much more important to know how a few powerful people or opinion leaders feel about an issue than it is to know what many relatively powerless people think.

In conclusion, let's examine how the public, public opinion, and propaganda might relate to a given issue, say capital punishment. Members of the public interested in this issue would include police officers, prison officials, sociologists, psychologists, social workers, and future criminals. To obtain a measure of public opinion, the pollsters might randomly select 1,000 people and ask them what they think about keeping the death penalty. The sample includes interested members of the general public and disinterested bystanders, and all opinions are counted equally. Suppose we find that about 75 percent of those polled want to keep the death penalty? The other side decides to hire a public-relations firm to persuade the public a little. Propaganda begins to appear. For example, a public-interest story in a popular magazine tells how the family of an executed man is getting along. A newspaper story appears describing a famous case in which a condemned man was cleared of the crime he was supposed to have committed—but just too late. An ex-warden of San Quentin is interviewed on a nationwide television news broadcast, and he relates in detail his opposition to the death penalty. A reporter, witness to an execution in the electric chair, describes it in vivid detail in a Sunday newspaper supplement. The effectiveness of the propaganda will depend on its ability to reach the audience, on the sophistication of the audience, and on the receptiveness of the audience to the new viewpoint. Let's assume that the propaganda has been cleverly conceived, and a new sampling indicates that public opinion has changed; now only 40 percent want to keep the death penalty. Still, nothing happens until power is applied. A lobbyist in Washington, D.C., working for the Anti-Death Penalty League, puts pressure on legislators. Thirty thousand anti-capital-punishment letters (all remarkably similar) flood in. Finally a new law is passed, more because of the League, the lobbyist, and the letters than because of public opinion. We can see from this imaginary and oversimplified example that boundaries of what constitutes a public are obscure, polls can be misleading, public opinion can change rapidly, propaganda is frequently very useful at building or changing public opinion, and finally, public opinion must be backed by political power if meaningful action is to result.

## SOCIAL MOVEMENTS

In the chapter on social change (Chapter Twelve), we mentioned that many people believe that through conscious effort the environment can be manipulated and changed. One way this can happen is through a form of collective behavior called a social movement. A **social movement** is defined as a group of people acting with some continuity to promote or to resist a change in their society or group. Social movements are lasting rather than temporary, they have a distinct perspective or viewpoint, they are oriented toward a specific goal or goals, and members have a sense of solidarity or *esprit de corps*.[11] Social movements seem to be especially popular in modern mass society. In mass society there is mass confusion; that is, there are a variety of viewpoints on every issue. The United States, like other mass societies, has a multiplicity of groups from numerous backgrounds, and each group has its own values, its own way of looking at things. Individual discontent, anxiety, and frustration about the condition of the world and about one's own opportunities are all conducive to social movements. Mass communication exposes us to all the confusing aspects of the mass society and puts us immediately in touch with other people who might also be frustrated, anxious, and looking for a way to change things. The combination of these elements—complex societies with a variety of peoples and viewpoints, numbers of people who feel discontented or shortchanged, and a system of mass communication to tie them together—provides fertile soil for social movements.

Current examples of social movements in the United States include pro- and antiabortion groups, the Moral Majority, antinuclear power groups, people interested in gay rights, and groups on both sides of the gun-control issue. Our recent history also includes social movements centered on civil rights, women's rights, peace, anticommunism, religion, and many other issues. Some of these are **revolutionary movements,** in that they demand a complete change in the social order. Others are **reform movements,** in that they seek modification of certain aspects of society. Social movements of either type, however, provide an important impetus for social change.

Social movements frequently make use of public demonstrations to bring attention to an issue. The pro-life and pro-choice movements and the gay-rights movement are examples. Farmers have descended on Washington,

D.C., in several massive "tractorcades" to bring attention to their economic problems. Sometimes, certain pressures lead groups that otherwise would not do so to use public protest as a means for change.

Often one social movement will develop to counter the effects of another. A current example focuses on the legalization of abortion. Until 1966, abortion was illegal throughout the United States, although a few states would allow it if the mother's life was in danger. A social movement developed among people who, for a variety of reasons—rights of women, population control, objection to laws that discriminated against the poor—felt that abortion should be legalized. Pamphlets to citizens, letters to senators, and articles in newspapers and magazines made people aware of the issue. In 1967, Colorado passed a liberalized abortion law, and soon California, New York, Hawaii, and other states followed. By early 1972, 16 states and the District of Columbia had changed their laws. Then, in January 1973, the U.S. Supreme Court overruled all state laws that prohibited or restricted women from obtaining abortions during the first three months of pregnancy. Seldom is social change, especially legal social change, so rapid, and seldom are the effects of a social movement so dramatic. There was an immediate outcry from those who objected to the Supreme Court's action. Groups such as Right to Life began forming, and a social movement aiming toward restrictive abortion legislation was born. Assaults on doctors, firebombings of abortion-providing clinics, and "abortion planks" in political platforms provide current evidence of the deep feeling on this issue.

Collective-behavior experts feel that social movements have a life history. One description of the life span of movements sees them going through four stages.[12] The *preliminary stage* involves individual excitement, discontent, and unorganized and unformulated restlessness. Mechanisms of suggestion, imitation, and propaganda are important, and an "agitator" type of leader appears. The *popular stage* involves crowd or collective excitement or unrest. The unrest is now open and widespread, and *esprit de corps* and an ideology or viewpoint become important mechanisms. The leader at this stage is a "prophet" who puts forth a general ideology, or a "reformer" who focuses on specific evils and develops a clearly defined program. Eric Hoffer believes that the early stages of social movements are dominated by "true believers." These are people of fanatical faith for whom mass movements have an almost irresistible appeal. The true believer is ready to sacrifice his or her life for a cause, any cause.[13] The *formal stage* of the social movement involves the formulation of issues and the formation of publics. The movement becomes "respectable" to gain wider appeal. Issues are discussed and debated. The leaders are "statesmen." The *institutional stage* involves the legalization of the movement; the movement becomes part of the society. The leaders are of the "administrator-executive" type.

Remember that this is a generalized description; some movements follow this pattern, some do not. The time involved in any one stage is variable. The unsuccessful social movement disappears before it can progress through all stages. The successful social movement that attains its objectives must adapt and broaden its goals, or it too will die out.

Finally, the *charismatic leader* figures prominently in both sociological and popular literature. As we discussed in Chapter Four, **charisma** refers to a particular quality an individual has that sets that person apart from other people. Charismatic leaders are treated as though they are superhuman and capable of exceptional acts. The charismatic movement follows the charismatic leader, who is a symbol for the movement and is above criticism. Some movements are so closely identified with the leader that they have no goal beyond that of following the leader. Other movements, although they follow such a leader, have well-defined goals that allow the movement to continue even if the charismatic leader is lost. Examples of charismatic leaders might include Gandhi, Joan of Arc, Martin Luther King, Jr., Hitler, and Charles de Gaulle.

## THEORY AND RESEARCH: A REVIEW

Conflict-theory and symbolic-interactionist perspectives are useful in this chapter. *Symbolic interactionists* will be interested in looking at patterns of communication and the person-to-person aspects of collective behavior. There is a contagion in collective behavior; waves of feeling sweep through a group of people and the group becomes a crowd or a mob. How does this happen? Are there consistent patterns? Is there a leader? If so, what are his or her special traits? How are feelings transmitted? What are the communication processes in rumor? Are recurrent patterns present? *Conflict theorists* might have some ideas about the underlying causes of riot behavior. They might suggest, noting that participants in urban and sports riots often seem to be of the working class or the lower class, that perhaps the class structure is a good place to look for the sources of the unrest and alienation that lead to such events. Revolutionary and reform social movements attempting to bring about social change would be of central interest to conflict theorists.

The study of the New York blackout was a *reactive survey* using interviews. Data obtained from polls are also a product of reactive surveys. As examples in this chapter have illustrated, careful *sampling technique* is essential. Disaster research can also involve surveys, but some research teams collect information by arriving at disaster scenes as quickly as possible and then using *nonreactive observations* of people's behavior: reviewing the scene, observing disaster agencies at work, watching the victims' behavior, listening to the media reports, and so on.

## SUMMARY

In the previous chapter in this section on change and deviance in society we dealt with social change and social disorganization. In this chapter we examined another form of activity that is outside the organized and the ordinary: collective behavior. *Collective behavior* refers to spontaneous and somewhat unstructured actions by groups of people. A number of concepts and terms central to the study of collective behavior were discussed. We examined types of collectivities: crowds, audiences, and mobs. We described types of behaviors: panic and mass hysteria, rumors, fads, crazes, and fashions. Special situations involving collective behavior were noted: riots and disasters. Our discussion of collective behavior concluded with an analysis of publics, public opinion, propaganda, and social movements.

A mob is a focused, acting crowd. A mob is difficult to deal with once set in motion, but the first reading that follows relates how one of the characters in Harper Lee's novel *To Kill a Mockingbird* distracted a lynch mob. The second reading describes the beginning of the 1992 Los Angeles riot.

## TERMS FOR STUDY

audience (398)

charisma (410)

collective behavior (395)

craze (404)

crowd (396)

disaster (405)

fad (404)

fashion (404)

keynoting (401)

mass hysteria (402)

milling (397)

mob (398)

panic (401)

propaganda (406)

public (406)

public opinion (406)

reform movement (408)

revolutionary movement (408)

riot (399)

rumor (402)

social movement (408)

For a discussion of Research Question 1, see page 403.
For a discussion of Research Question 2, see page 404.

## INFOTRAC COLLEGE EDITION
### Search Word Summary

To learn more about the topics from this chapter, you can use the following words to conduct an electronic search on InfoTrac College Edition, an online library of journals. Here you will find a multitude of articles from various sources and perspectives:

**www.infotrac-college.com/wadsworth/access.html**

Riots

Rumor

Pro-choice Movements

Rodney King

Fads

Race

Hurricanes

Social control

# Reading 13.1

# MAYCOMB JAIL

*Harper Lee*

*A mob is a focused, acting crowd. It is emotionally aroused and intent on tak-ing action. A mob is a powerful force, and it is difficult to change its direction. When the direction is changed (as it is in this fictional account), the change can be understood in collective-behavior terms. Here, Scout and Jem Finch help their father, Atticus, protect a black man from a lynch mob in Harper Lee's novel* To Kill a Mockingbird.

The Maycomb Jail was the most venerable and hideous of the county's build-ings. Atticus said it was like something Cousin Joshua St. Clair might have designed. It was certainly someone's dream. Starkly out of place in a town of square-faced stores and steep-roofed houses, the Maycomb jail was a miniature Gothic joke one cell wide and two cells high, complete with tiny battlements and fly-ing buttresses. Its fantasy was heightened by its red brick facade and the thick steel bars at its ecclesiastical windows. It stood on no lonely hill, but was wedged between Tyndal's Hardware Store and *The Maycomb Tribune* office. The jail was Maycomb's only conversation piece: its detractors said it looked like a Victorian privy; its supporters said it gave the town a good solid respectable look, and no stranger would ever suspect that it was full of niggers.

As we walked up the sidewalk, we saw a solitary light burning in the distance. "That's funny," said Jem, "jail doesn't have an outside light."

"Looks like it's over the door," said Dill.

A long extension cord ran between the bars of a second-floor window and down the side of the building. In the light from its bare bulb, Atticus was sitting propped against the front door. He was sitting in one of his office chairs, and he was reading, oblivious of the nightbugs dancing over his head.

I made to run, but Jem caught me. "Don't go to him," he said, "he might not like it. He's all right, let's go home. I just wanted to see where he was."

We were taking a short cut across the square when four dusty cars came in from the Meridian highway, moving slowly in a line. They went around the square, passed the bank building, and stopped in front of the jail.

Nobody got out. We saw Atticus look up from his newspaper. He closed it, folded it deliberately, dropped it in his lap, and pushed his hat to the back of his head. He seemed to be expecting them.

"Come on," whispered Jem. We streaked across the square, across the street, until we were in the shelter of the Jitney Jungle door. Jem peeked up the sidewalk. "We can get closer," he said. We ran to Tyndal's Hardware door—near enough, at the same time discreet.

In ones and twos, men got out of the cars. Shadows became substance as lights revealed solid shapes moving toward the jail door. Atticus remained where he was. The men hid him from view.

"He in there, Mr. Finch?" a man said.

"He is," we heard Atticus answer, "and he's asleep. Don't wake him up."

In obedience to my father, there followed what I later realized was a sickeningly comic aspect of an unfunny situation: the men talked in near-whispers.

"You know what we want," another man said. "Get aside from the door, Mr. Finch."

"You can turn around and go home again, Walter," Atticus said pleasantly. "Heck Tate's around somewhere."

"The hell he is," said another man. "Heck's bunch's so deep in the woods they won't get out till mornin'."

"Indeed? Why so?"

"Called 'em off on a snipe hunt," was the succinct answer. "Didn't you think a' that, Mr. Finch?"

"Thought about it, but didn't believe it. Well then," my father's voice was still the same, "that changes things, doesn't it?"

"It do," another deep voice said. Its owner was a shadow.

"Do you really think so?"

This was the second time I heard Atticus ask that question in two days, and it meant somebody's man would get jumped. This was too good to miss. I broke away from Jem and ran as fast as I could to Atticus.

Jem shrieked and tried to catch me, but I had a lead on him and Dill. I pushed my way through dark smelly bodies and burst into the circle of light.

"Hey, Atticus!"

I thought he would have a fine surprise, but his face killed my joy. A flash of plain fear was going out of his eyes, but returned when Dill and Jem wriggled into the light.

There was a smell of stale whiskey and pigpen about, and when I glanced around I discovered that these men were strangers. They were not the people I saw last night. Hot embarrassment shot through me: I had leaped triumphantly into a ring of people I had never seen before.

Atticus got up from his chair, but he was moving slowly, like an old man. He put the newspaper down very carefully, adjusting its creases with lingering fingers. They were trembling a little.

"Go home, Jem," he said. "Take Scout and Dill home."

We were accustomed to prompt, if not always cheerful acquiescence to Atticus's instructions, but from the way he stood Jem was not thinking of budging.

"Go home, I said."

Jem shook his head. As Atticus's fists went to his hips, so did Jem's and as they faced each other I could see little resemblance between them: Jem's soft brown hair and eyes, his oval face and snug-fitting ears were our mother's, contrasting oddly with Atticus's graying black hair and square-cut features, but they were some-how alike. Mutual defiance made them alike.

"Son, I said go home."

Jem shook his head.

"I'll send him home," a burly man said, and grabbed Jem roughly by the collar. He yanked Jem nearly off his feet.

"Don't you touch him!" I kicked the man swiftly. Barefooted, I was surprised to see him fall back in real pain. I intended to kick his shin, but aimed too high.

"That'll do, Scout," Atticus put his hand on my shoulder. "Don't kick folks. No—" he said, as I was pleading justification.

"Ain't nobody gonna do Jem that way," I said.

"All right, Mr. Finch, get 'em outa here," someone growled. "You got fifteen seconds to get 'em outa here."

In the midst of this strange assembly, Atticus stood trying to make Jem mind him. "I ain't going," was his steady answer to Atticus's threats, requests, and finally, "Please Jem, take them home."

I was getting a bit tired of that, but felt Jem had his own reasons for doing as he did, in view of his prospects once Atticus did get him home. I looked around the crowd. It was a summer's night, but the men were dressed, most of them, in overalls and denim shirts buttoned up to the collars. I thought they must be cold-natured, as their sleeves were unrolled and buttoned at the cuffs. Some wore hats pulled firmly down over their ears. They were sullen-looking, sleepy-eyed men who seemed unused to late hours. I sought once more for a familiar face, and at the center of the semi-circle I found one.

"Hey, Mr. Cunningham."

The man did not hear me, it seemed.

"Hey, Mr. Cunningham. How's your entailment gettin' along?"

Mr. Walter Cunningham's legal affairs were well known to me; Atticus had once described them at length. The big man blinked and hooked his thumbs in his overall straps. He seemed uncomfortable; he cleared his throat and looked away. My friendly overture had fallen flat.

Mr. Cunningham wore no hat, and the top half of his forehead was white in contrast to his sunscorched face, which led me to believe that he wore one most days. He shifted his feet, clad in heavy work shoes.

"Don't you remember me, Mr. Cunningham? I'm Jean Louise Finch. You brought us some hickory nuts one time, remember?" I began to sense the futility one feels when unacknowledged by a chance acquaintance.

"I go to school with Walter," I began again. "He's your boy, ain't he? Ain't he, sir?"

Mr. Cunningham was moved to a faint nod. He did know me, after all.

"He's in my grade," I said, "and he does right well. He's a good boy," I added, "a real nice boy. We brought him home for dinner one time. Maybe he told you about me, I beat him up one time but he was real nice about it. Tell him hey for me, won't you?"

Atticus had said it was the polite thing to talk to people about what they were interested in, not about what you were interested in. Mr. Cunningham displayed no interest in his son, so I tackled his entailment once more in a last-ditch effort to make him feel at home.

"Entailments are bad," I was advising him, when I slowly awoke to the fact that I was addressing the entire aggregation. The men were all looking at me, some had their mouths half-open. Atticus had stopped poking at Jem: they were standing together beside Dill. Their attention amounted to fascination. Atticus's mouth, even, was half-open, an attitude he had once described as uncouth. Our eyes met and he shut it.

"Well, Atticus, I was just sayin' to Mr. Cunningham that entailments are bad an' all that, but you said not to worry, it takes a long time sometimes . . . that you all'd ride it out together. . . ." I was slowly drying up, wondering what idiocy I had committed. Entailments seemed all right enough for living-room talk.

I began to feel sweat gathering at the edges of my hair; I could stand anything but a bunch of people looking at me. They were quite still.

"What's the matter?" I asked.

Atticus said nothing. I looked around and up at Mr. Cunningham, whose face was equally impassive. Then he did a peculiar thing. He squatted down and took me by both shoulders.

"I'll tell him you said hey, little lady," he said.

Then he straightened up and waved a big paw. "Let's clear out," he called. "Let's get going, boys."

As they had come, in ones and twos the men shuffled back to their ramshackle cars. Doors slammed, engines coughed, and they were gone.

I turned to Atticus, but Atticus had gone to the jail and was leaning against it with his face to the wall. I went to him and pulled his sleeve. "Can we go home now?" He nodded, produced his handkerchief, gave his face a going-over and blew his nose violently.

"Mr. Finch?"

A soft husky voice came from the darkness above: "They gone?"

Atticus stepped back and looked up. "They've gone," he said. "Get some sleep, Tom. They won't bother you any more."

From a different direction, another voice cut crisply through the night: "You're damn tootin' they won't. Had you covered all the time, Atticus."

Mr. Underwood and a double-barreled shotgun were leaning out his window above *The Maycomb Tribune* office.

It was long past my bedtime and I was growing quite tired; it seemed that Atticus and Mr. Underwood would talk for the rest of the night, Mr. Underwood out the window and Atticus up at him. Finally Atticus returned, switched off the light above the jail door, and picked up his chair.

"Can I carry it for you, Mr. Finch?" asked Dill. He had not said a word the whole time.

"Why, thank you, son."

Walking toward the office, Dill and I fell into step behind Atticus and Jem. Dill was encumbered by the chair, and his pace was slower. Atticus and Jem were well ahead of us, and I assumed that Atticus was giving him hell for not going home, but I was wrong. As they passed under a streetlight, Atticus reached out and massaged Jem's hair, his one gesture of affection.

---

# QUESTIONS 13.1

**1.** Outline how mob and crowd characteristics are illustrated in this selection.

**2.** Describe in collective-behavior terminology why and how the lynch mob was broken up.

### Reading 13.2

# LOOKING FOR JUSTICE IN L.A.

*Peter J. Boyer*

*The riot in South Central Los Angeles that started on April 29, 1992, left 53 dead, more than 600 buildings destroyed by fire, and $735 million in damage. Five thousand people were arrested, one of whom was Damian Williams. Peter Boyer wrote "Looking for Justice in L.A." for* The New Yorker *in which he examines the aftermath of the L.A. riots and the impending trial of the four defendants in the beating of Reginald Denny. This excerpt from the beginning of his article describes how the riot got started.*

Last April 29th, while Georgiana Williams was sitting in a hair-dresser's chair at Clinton Young's Beauty Shop, she heard on the radio that there was a disturbance at the intersection of Florence Avenue and Normandie Avenue, a block from her house, on Seventy-first Street, in the section of Los Angeles known as South Central. When her hair was done, she hurried home, but the streets were so clogged that it was hard for her to make her way through. She was stunned by what she saw: "People running and hollering and crying and screaming, truckloads of people, vans of people," she recalls. "So I said, 'What in the world's going on?' And they told me something had happened up on Florence." Her youngest son, Damian, was at home, with a group of his friends. They told her that her oldest son, Mark Jackson, had been arrested that afternoon along with two other young men from the neighborhood, for fighting with a police officer. She recalls that she went into her room to pray, then went to bed. Although she was at ground zero, Georgiana Williams, who watches only the Christian Trinity Broadcasting Network, did not see any of the innumerable replays of the violence of that day, nor did she know that her son Mark's arrest had ignited the worst riot of the century and that her son Damian would become its most notorious figure.

Georgiana Williams is a soulful woman with an easy smile and a strong will—strong enough to have carried her from Mississippi, where as a girl she worked the fields of a plantation with twelve brothers and sisters, to California, where she got a job as a nurse. She bought the house on Seventy-first Street twenty-two years ago, for eight hundred and fifteen dollars. In other circumstances, Miss Williams, as she is called in the neighborhood, would be considered one of those quiet heroines of the American inner city, a hardworking, churchgoing mother who brought up her four children mostly alone, and brought up some others who weren't hers as if they had been. Now, unexpectedly—and although there is nothing in her manner to suggest it—Georgiana Williams has become the mother of the L.A. riot.

In fact, the narrative of the early hours of the riot sounds like an entry in the Williams family chronicle. While Mark was in his mother's back yard that afternoon, installing brakes in a neighbor's car, and Damian was hanging out with friends at a vacant house next door, a neighbor told them that the jury had just acquitted the four white policemen charged in the beating of the black motorist Rodney King. "We was all saying that was jacked up, it was fucked up, we was all mad about what happened," Mark Jackson says. "We just sat down on the porch and said, 'Aw, man, that was crazy. That was messed up, what they did.' Everybody just kept on talkin' about it." All around the intersection of Florence and Normandie, the mood was an eerie combination of holiday and natural disaster: people pouring outdoors, jamming the sidewalks, shouting "Rodney King!"

"If the police hadn't started harassing us, we'd have all walked back to the house and the riot would never have started," Jackson says. According to police reports, rocks were thrown at two squad cars driving up Normandie; one officer chased down a teen-age boy from the neighborhood, Shondell Daniels. When Daniels was apprehended, the crowd got close and vocal. "Everybody was just whooping and hollering and going off," Jackson says. "They say I had the biggest mouth in the crowd. They say I was trying to provoke people to start a fight." When officers tried to arrest Jackson, he resisted, and after a scuffle he was hand-cuffed and brought to a squad car, where his head slammed into one of the car's windows. By now, several more squad cars had been called to the scene. The offi-cers formed a "skirmish line"—standing their ground, nightsticks in hand. Shots of the scene on neighborhood home videos show that the police were fairly restrained, but the crowd pressed in, shouting insults at the officers. Damian Williams was there, too, and at one point, in a gesture of disapproval of the police presence, he pulled down his pants and underwear and mooned the police. By the time Jackson and two others were taken away, the crowd had grown to more than a hundred and fifty, and the police numbered about twenty-five. At that point, the officer in charge ordered his people to leave the scene. Bart Bartholomew, a photographer working for the *New York Times,* was among the first to feel their absence. He headed back to his car, which was parked on Seventy-first, near the Williams house, but as he walked he acquired a crowd. He reached his Volvo and somehow got inside; someone pulled open the passenger door and stole his cam-era equipment, while others jumped up and down on the hood. As Bartholomew frantically started his car and drove away, a rock smashed through the window on the driver's side, shattering glass over his face.

With the police gone, there was suddenly a power vacuum in South Central. The crowd, Damian included, moved a block south, to Florence and Normandie, and soon Tom's Liquor Store, an Asian-owned business, was hit with rocks and overrun by looters. Liquor fueled the anger and the urge to act on it, and passing cars were soon being pelted with rocks and bottles. One motorist was dragged out of his car and beaten, and then another. The crowd, at first indiscriminate in its attacks, began targeting whites. "Leave the Mexicans alone," one voice yelled from the street. "We're only getting the Buddha-heads and white boys!" The white driver of a medical-supply delivery truck was stopped, pulled out, beaten, and struck with an oxygenator stolen from his load. A mile or so away, Reginald Denny, hoping to avoid the afternoon rush hour, exited from the Harbor Freeway onto

Florence Avenue, en route to Inglewood, where he was to deliver a load of sand and gravel. Before he had gone far on Florence, he noticed the medical-supply truck stopped in the right lane, with its rear loading door open, and people helping themselves to its cargo. Denny thought of turning around and heading back to the freeway, but he was driving an eighteen-wheeler, a forty-ton lummox of a truck without power steering, and he was afraid he couldn't make a U-turn, so he eased into the left lane and crept toward Normandie. At the intersection, people walked in front of his truck, and he stopped. Denny, who had been listening to music on the radio, didn't know about the Rodney King verdict, or that it had anything to do with the object that shattered the passenger window of his truck, which is the last thing he remembers about that day.

---

## QUESTIONS 13.2

**1.** Which of Conant's riot phases are illustrated in this reading?

**2.** Using collective-behavior concepts, come up with some ways the tense situation in L.A. might have been defused and the following riot and destruction avoided.

**3.** Construct an argument that supports the idea that the incident described in this reading is best described as (1) a mob, or (2) a riot, or (3) a social movement.

---

## NOTES

1. Treatment of concepts in this chapter in general follows the description given by Ralph Turner and Lewis Killian in their book, *Collective Behavior*, 2d ed. (Englewood Cliffs, N.J.: Prentice-Hall, 1972).

2. Neil Smelser, *Theory of Collective Behavior* (New York: Free Press, 1962).

3. See *Blackout Looting: New York City, July 13, 1977,* by Robert Curvin and Bruce Porter (New York: Gardner Press, 1979), and their article "Blackout Looting," *Society,* (May/June 1979), pp. 68–76.

4. Ralph Conant, "Rioting, Insurrection, and Civil Disobedience," *The American Scholar* 37 (Summer 1968).

5. This discussion of panic is drawn from Duane Schultz, *Panic Behavior* (New York: Random House, 1964), and from Turner and Killian, *Collective Behavior* (see note 1), chapters 5 and 6.

6. H. Taylor Buckner, "A Theory of Rumor Transmission," in *Readings in Collective Behavior,* 2d ed., edited by Robert Evans (Chicago: Rand McNally, 1975), pp. 86–102. Also see Warren A. Peterson and Noel P. Gist's article, "Rumor and Public Opinion," *American Journal of Sociology* 57 (September 1951), pp. 159–167, for a summary and analysis of some theories on rumor.

7. "Identity Announcement in Mass Society: The T-Shirt," by Donna Darden and Steven Worden, *Sociological Spectrum* 11 (January–March 1991), pp. 67–79.

8. Turner and Killian, *Collective Behavior* (see note 1), pp. 152–153.

9. This section on disaster is taken from Charles Fritz's chapter on disaster, which appears in Robert Merton and Robert Nisbet's book *Contemporary Social Problems* (New York: Harcourt Brace Jovanovich, 1961), chapter 14.

10. This discussion is summarized from Julian Simon's very complete discussion in his book *Basic Research Methods in Social Science* (New York: Random House, 1969), chapter 8.

11. Turner and Killian, *Collective Behavior* (see note 1), p. 246.

12. This description of the life history of a social movement is from Rex Hopper, "The Revolutionary Process: A Frame of Reference for the Study of Revolutionary Movements," *Social Forces* 28 (March 1950), pp. 270–279.

13. See Eric Hoffer, *The True Believer* (New York: New American Library, 1951).

## CHAPTER FOURTEEN

Stock • Boston/Charles Kennard

# Deviance and Social Control

*THREE RESEARCH QUESTIONS*

*1. Suppose a person's favored method of committing suicide is not available. Will the person choose another method, or not commit suicide?*

*2. When capital punishment (an execution) takes place and is publicized, does the homicide rate drop (deterrent effect), rise (brutalization effect), or do nothing?*

*3. Suppose that a group of inmates who are on death row and are about to be executed are instead released! Will they go on to commit more violent crime on the outside or become law-abiding citizens?*

EXPERTS TELL US that no two people have the same fingerprints. People differ from each other in their appearance and behavior as well. Variations in behavior occur even though we have norms and roles that specify what *should* happen and what people *should* do in almost any situation. Social differentiation is not only tolerated but expected in social interaction. We expect people to be different from ourselves in behavior and appearance, and we would be very surprised if they were not. At a particular point, however, when differences from a group norm become great enough, social differentiation becomes social deviance.

# DEFINING DEVIANCE

There are a number of ways to define deviance. Sometimes deviance and normality are defined statistically. Those around the average are considered normal, and those at the extremes are seen as deviant. If the average height of American males is 5'10", then males from 5'7" to 6'2" are probably viewed as normal, and a fellow 6'8"or 5'2" is considered deviant. Because most people have an I.Q. of between 95 and 115, that I.Q. level is "normal," and a person with an I.Q. of either 145 or 55 is deviant. Because most Americans are heterosexual and fewer are homosexual, heterosexuality is considered normal and homosexuality deviant, if viewed statistically. Sociologists, however, generally do *not* look at deviance as a statistical phenomenon.

Another way to define deviance is to liken it to a disease. The disease, or medical, model sees deviance as something that a person has that we wish he or she could get over, like a cold or the measles. People wonder, "What is *wrong* with Susan? Why does she continue to screw up? She must be crazy...."

The person behaves differently because something is wrong with the *person*. The logical next step: If we study carefully, perhaps we can discover what is wrong with the person and then correct it. The medical analogy fits beautifully with this view of deviance: Find the nature of the illness and prescribe a cure.

This view of deviance is a popular one; it's easily understood and it makes sense. This has meant that for years many scholars, scientists, and graduate students alike have scurried about looking for *the* cause, the explanation for the disease. As you can imagine, a number of answers have been found. Several categories of explanations come to mind.

## Biological Explanations

In 1875, Cesare Lombroso stated that deviant behavior was inherited and that criminals were throwbacks (closer in the evolutionary chain) to apes. Lombroso came to this conclusion by taking a series of body measurements of institutionalized criminals and noncriminals and comparing them with those of primitive humans. (The criminals' measurements were closer to apeman's than those of the noncriminals.) Other scientists had difficulty obtaining the same results when they repeated his work, but Lombroso's research was influential. It explained a complicated activity—crime—in a reasonably simple and straightforward way. Other studies examined criminal families and identical and fraternal twins to try to prove that crime was inherited. Attempts were made to relate physique and body type (fat, skinny, or muscular) to being criminal and even to the type of crime committed. Peculiar chromosome patterns, especially the XYY pattern, were thought to be more prevalent among prison populations and therefore a potential "disease" factor. More recently, researchers have examined brain and body chemistry and diet to see how they are related to behavior.

## Psychological Explanations

The ideas of Sigmund Freud (see Chapter Two) had a great influence on those seeking the causes of deviant behavior. Researchers stressed how early childhood experiences and guilt can affect the personality. The "unconscious" became an important concept. It was seen as that part of the mind in which unpleasant, perhaps antisocial, memories are repressed or stored. Psychologists believed that these memories, even though repressed, would seek expression and would often make their presence known in other ways, such as dreams, slips of the tongue, and other difficult-to-explain behaviors. Or it was surmised that an individual becomes delinquent or criminal because his or her superego never developed adequate strength to deal with the antisocial forces of the powerful id. A whole new vocabulary became available that focused not on biology or heredity, but on the *mind*, and it proved to be a convenient means of explaining antisocial behavior. One outgrowth of this research was the development of personality tests, which could be given to troublemakers, potential delinquents, or people suspected of committing crimes in an attempt to prove guilt or detect the problem ahead of time.

## Sociological Explanations

Sociologists define **deviant behavior** as behavior contrary to generally accepted norms. These norms can be family, group, organizational, or social norms, often established by custom and sometimes supported by law.

Whereas the medical approach to deviant behavior infers weakness in the individual, be it biological, psychological, or whatever, sociologists believe that deviant behavior is *normal* behavior—normal in the sense that, because of individual differences, diversity is to be expected in all societies. Sociologists recognize that biological and psychological factors play a part in the development of the individual (as we outlined in Chapter Two), but they emphasize that deviant behavior, like nondeviant behavior, is learned—developed through the socialization process.

One problem with defining deviance this way, however, is that it is relative; what is deviant to one audience might be normal to another. For example, take the member of an urban delinquent gang; he is conforming to gang norms when he defends his turf, carries a weapon, and saves face by fighting rather than backing down to a threat. His "deviant" behavior is learned and is as conforming in its own way as is the behavior of the middle-class college student who conforms to the fashions, behaviors, and attitudes of those around her. There will always be differences of opinion about norms; likewise, there will be differences over what is deviant and what is normal behavior.

Again, sociologists believe that deviant behavior and crime are learned; they are consequences of social conditions. What the sociologist does, then, is try to discover the situations in society that are most likely to lead to crime and deviant behavior. Poverty, lack of legitimate opportunity to move up the social-class ladder, racism, sexism, and other social factors place tremendous pressures on people and may very well push some of them into delinquency and crime. Conflict between social classes, which results in one class overpowering another, can produce pressure leading to crime. These are *structural* conditions—they are a part of the social structure.

**Three Theories**

Explanations for crime and deviant behavior reflect the three theories that we introduced in Chapter One and have returned to throughout this book. *Conflict* theorists, like Marx, Bonger, and Quinney, believe that crime is one of the consequences of poverty and the battle between "haves" and "have-nots." Elites make laws to reinforce their position in society and enforce these laws to protect themselves from "the dangerous classes."

In an approach more consistent with *functional analysis*, Robert Merton visualized behavior as means and goals. He suggested five patterns, or modes, of adaptation based on goals (what you want) and means (how you get them). The *conformist* follows the rules. For a college student the goal is a degree, and the accepted means are to register, go to class, read books, and pass tests; and that is what the conformist does. But suppose you don't have the time or resources to use the accepted means. This can produce pressure to become an *innovator*. The innovator is one who wants the same goal but uses unacceptable means to get it; the student innovator cheats on tests, buys term papers written by others, copies other people's notes, and so on. The *ritualist*, on the other hand, sticks to the accepted means and does what he or she is supposed to (studies, goes to class), but he or she has lost sight of the goal. The student ritualist wanders around taking course after course, changing majors frequently, becoming a professional student, and never graduating. The *retreatist* abandons both means and goals, gives up, and drops out of the fight totally. The student retreatist has given up the goal of a degree and the means for obtaining it and has dropped out of school. Merton's final mode of adapting, the *rebel*, abandons accepted means and goals like the retreatist; but rather than giving up, he or she substitutes new goals and means. The student rebel might try to set up an alternative college in which various goals are sought using different means (open classes without tests, grades, or assignments). To theorists following Merton, the innovator response was especially interesting as a way of explaining property crime committed by people feeling blocked from achieving desired goals. Merton's ideas influenced many, including several of the gang theorists we discussed earlier (Chapter Five), who also based their explanations of delinquent behavior on structural conditions.

Other sociological approaches address the specific ways in which crime is learned. Edwin Sutherland proposed a *symbolic-interactionist* theory which he called **differential association.** This theory says that behavior, normal and deviant alike, is learned during interaction with others. Sutherland felt that the prostitute, the professional thief, and the white-collar criminal learn to behave through association and communication with people who portray deviant behavior in a favorable light. They are taught the positive rewards of that behavior—that crime does pay.

Before leaving the sociological theories of deviant behavior, we should briefly look at another approach that has attracted wide interest. This approach focuses not so much on the deviant act or deviant person, but rather on the process of defining the act or person as deviant. Who makes the rules? Who enforces them? On whom and why? These questions emerge from what is generally called **labeling theory.**

### Labeling Theory

*A 6'8" basketball player for the University of Arkansas, sporting collar-length hair and a long drooping mustache, fouled out of three straight games. Before his next one, against Baylor, he decided to trim his hair and shave his mustache. In the Baylor game he was called for only two fouls. Two nights later, against Texas Tech, the neat, trim player played forty minutes and was still around at the end of a 93–91 double-overtime victory that knocked Tech out of the Southwest Conference lead.*

*The haircut? The shave?*

*"I think they helped," the player stated. "I hate to say it, but I guess the hairstyle did affect the refs."*

Labeling theory, which is also a symbolic-interactionist theory, sees deviant behavior as a product of group definitions. Becoming a deviant involves a labeling process; one is a deviant because a particular group

# EDWIN SUTHERLAND (1883–1950)

Edwin Sutherland earned his Ph.D. in Sociology from the University of Chicago. Like most Chicago criminologists, Sutherland rejected individualistic explanations of crime. Instead, he was convinced that the social organization, the context in which individuals are embedded, regulates criminal involvement. Sutherland felt this applied to criminals at all levels—the street criminal as well as the corporate criminal. In one aspect of his research, Sutherland revealed that lawlessness is widespread in the world of business, politics, and the professions. He pointed out that these people were involved in a variety of illegal practices which he called "white-collar crimes." Sutherland felt that this observation presented special problems for most theories of his day, which had assumed that "criminal behavior in general is due either to poverty or to the psychopathic and sociopathic conditions associated with poverty." On the contrary, these criminals were typically not psychologically disturbed and usually not poor.

Sutherland is probably best known for his learning theory of crime called "differential association." His book on professional thieves is also widely cited. After teaching at the University of Chicago and several other institutions from 1913 to 1929, Sutherland was appointed chair of the Department of Sociology at the University of Indiana in 1935 and held that post until his death in 1950.

---

labels him or her as such.[1] As with the basketball player, the focus changes from an examination of why the individual does peculiar things to an examination of why these things are called deviant. In 1919, lawmakers succeeded in passing the Eighteenth Amendment to the United States Constitution, which made the sale and manufacture of intoxicating beverages a crime. In 1933, the Twenty-First Amendment repealed the Eighteenth and decreed that drinkers were no longer criminal. Which was the majority view? Probably the latter, but the important point is that both amendments were results of the actions of rule makers who defined the same act as deviant and criminal at one time and permissible at another time. In other words, labeling groups determines who and what is deviant; *deviation is socially defined.*

Why are some acts labeled as deviant and others not? Sometimes the rule makers are motivated by moral or religious beliefs. They are strongly convinced that acts like drinking alcohol, smoking marijuana, or legalizing abortion are morally wrong and then make tremendous efforts to see their beliefs written into law. You can probably think of examples in American history, some quite recent, in which people have successfully crusaded for a cause that ultimately became law.

Perhaps rule makers are motivated by profit. If new clean-air standards are going to mean retooling a company's whole assembly line and spending $5 million on research and development, then you can be fairly sure that the company might spend about $3 million fighting to delay that law. Or turn it

## "IT'S A BOY! . . . I THINK . . ."

For about one in every three thousand births (about one thousand babies a year in the United States) the determination of the newborn baby's sex may be difficult. These are babies who for a variety of reasons have what doctors call ambiguous genitalia. When the external sex organs do not clearly identify the infant's sex, what the doctor does next sets the course for the infant's future.

For the doctor to say, "I don't know yet," or "I'm not sure," will be very traumatic to the parents and ultimately to the baby. The father and mother should have no doubts that they have a boy or a girl. A clear-cut sex announcement is needed, because of the great importance that "the sex of rearing" plays in forming the child's own sense of sexual identity.

What does the doctor do? One obstetrician who has dealt with many such cases tells us that the size of the visible genitals is the key. "If it's small, say it's female. If in doubt, say it's female. If there is any uncertainty about the baby's eventual ability to perform sexually, it would be a serious mistake to call it a male." The doctor believes that a "male" who is incapable of functioning biologically as a male might be better off functioning as a female, even if by other standards (XY chromosomes) he might be judged male.

—*From* The Sacramento Bee, *February 15, 1978.*

---

around. Imagine that I just invented an ingenious little catalytic converter that makes exhaust air *cleaner* than normal air, and I am lobbying for a law to get my invention placed on all cars. Would you believe me if I told you that I am motivated solely by concern for the environment?

So, all sorts of motivations and justifications are used by rule makers to make rules. The result is a set of personal judgments about what is "right" and "wrong," which become justification for attempts to control people's behavior. Now rule enforcers take over. They seek out and punish those who violate the rules. Howard Becker refers to rule creators and rule enforcers—people whose profession centers on defining and regulating deviant behavior—as moral entrepreneurs. Labeling theory looks at it this way. Some people *make* rules and others *enforce* them, and from this process particular individuals are *labeled* deviant. It is a *social* process.

To whom does the deviant label get applied? Well, not to everyone equally, of course. It depends in part on the *seriousness* of the act. On very serious behaviors (homicide, armed robbery) about which there is general agreement, the deviant label is quickly and formally applied. If the act is less serious (belching in public, dressing very strangely), the deviant label, if applied at all, will be applied less stringently. It also depends on the *vulnerability* of the person being labeled. Lower-class people and minorities are more likely to be labeled as deviant than are others. It depends as well on *visibility*. If the act is obvious and occurs in full view of others, it is more likely to get a reaction.

These and other factors interact in complicated ways. Thus, in the view of the labeling theorists, calling people deviant can be an arbitrary and capricious act—an act that tells us more about the person doing the labeling than about the person being labeled. Take the long-haired basketball player once more. As you recall, he cut his hair and no more fouls were called after the Baylor game. Tell me, which of the following would apply? Was he: (1) a mild-mannered, peaceful, and orderly player victimized by referees who have a bias against grubby, long-haired-hippie types, or (2) a brutal "hatchet man" whose behavior became less visible to the rule enforcers (referees) when he cut his hair? We'll never know.

An example of the labeling process in action has been provided by Stanford psychologist David Rosenhan. As you may recall from our discussion of **Question 6** in Chapter One, Rosenhan and seven colleagues each went to the admissions office of a mental hospital and told the same story: They were hearing unclear voices that seemed to be saying words like *empty, hollow,* and *thud.* That was enough for the admitting psychiatrists at 12 of the hospitals; Rosenhan and his friends were diagnosed as schizophrenic and admitted. The label was applied, and it stuck. The diagnoses were not questioned by other staff members, and the pseudopatients were incarcerated for between 7 and 52 days before being released. Although they behaved calmly and normally in the hospital, they were continually given pills (a total of nearly 2,100) to "help" them. None was considered cured upon release—"in remission" was the final diagnosis. When a label is firmly affixed, it is very difficult to escape it. While in the hospitals, the pseudopatients took careful notes about what was happening; the hospital staff either ignored this or saw it as evidence of insane compulsiveness. The pseudopatients' normal behavior was either overlooked entirely or profoundly misinterpreted. Often they would approach hospital staff with reasonable questions. In response, nurses and attendants would move hurriedly away, eyes averted. Psychiatrists also ignored the questioners. This sort of depersonalization is a predominant characteristic in labeling situations. Rosenhan is discouraged about the inability to correctly diagnose mental illness; but more to the point of our study of labeling is his comment that once a label is applied to people it sticks, a mark of inadequacy forever.[2]

## PRIMARY AND SECONDARY DEVIANCE

In an approach to deviance that emphasizes society's *reaction* to behavior as the labeling approach does, the distinction between primary and secondary deviance is of central importance. **Primary deviance** is not too serious and does not affect the individual's self-concept. People can rationalize primary deviant acts as being within the bounds of reasonable, acceptable, and normal behavior for people like them. It fits in with or does not seriously detract from the person's typical status and role. The doctor who splits fees, the person who drinks too much, and the eccentric college professor are all deviant to be sure; but these types of deviance are considered primary because they do not go far beyond the boundaries of the conventional statuses and roles these people occupy.

A process also occurs whereby society reacts to continued primary deviance by expressing moral indignation, rejection, and penalties. **Secondary deviance** refers to deviance that represents a defense, attack, or adjustment to the problems created by the societal reaction to the primary deviance. The individual becomes labeled and stigmatized. The person begins to see himself or herself as others do, and the self-concept begins to change; in short, he or she wears the label and the self-fulfilling prophecy has done its work. It is not long before the clever lawyer who drinks a little too much becomes the drunk who occasionally practices law. As one's self-concept changes in secondary deviance, the person might become involved in a deviant subculture and gain further knowledge and skill regarding the behavior, including techniques for avoiding detection. The important factors in secondary deviance are the redefinition of self that follows society's reaction and the consequences of that redefinition for future behavior.[3]

A paradoxical and important outcome develops from this sequence: The societal reaction has the intent of stopping primary deviance, but instead it might force someone into secondary deviance. This in turn often leads to an *increase* in deviant involvement and behavior. Here are some examples: The basketball player fouls out of a couple of games. The local paper labels him a rough and aggressive player, he begins to see himself differently, and suddenly he fouls out of *every* game. The juvenile is picked up on a minor charge, such as a runaway or curfew violation. She is locked up in a detention home and treated like the other delinquents. She learns from them, starts to see herself differently, and when she is released she is ready for a full-blown career in delinquency.

## DEVIANT BEHAVIOR IN PERSPECTIVE

Deviant behavior is often seen as a malignant element in society to be eradicated at all costs. To say that something is deviant is to imply that something is wrong and needs to be fixed. This is an incorrect view for several reasons. First, it is important to remember that deviant behavior is basically the product of people's definitions rather than something natural or inherent. Knowing this allows us to shift our focus from the person's behavior to the question, "Why is his behavior being called deviant?" Second, Émile Durkheim, Kai Erikson, and others have pointed out that deviant behavior is not necessarily harmful to group life. Actually, deviant behavior often plays an important part in keeping the social order intact. Deviant behavior enables groups to define boundaries; it preserves stability within the group by pointing out the contrast between what is inside the group and what is outside. Erikson suggests that without the battle between "normal" and "deviant," the community would lack a sense of identity and cohesion, a sense of what makes it a special place in the larger world.[4]

Finally, deviant behavior appears to be one of the processes involved in social change. Often, what is deviant today is accepted tomorrow and expected the next day. Maybe the people who seem to be listening to a different drummer just heard the beat before the rest of us. *Some* forms of deviant behavior represent an early adaptation to changing conditions. There are many examples of this: Deviant clothing styles of the past are the fashion of today. Yesterday's pornography is accepted literature today. Obviously there are limits; some behaviors should be considered deviant no matter what. The point is, however, that deviant behavior does not necessarily represent the evil its title implies; often it serves necessary and important functions in society, including those of defining the group's boundaries and helping to bring about social change.

## CATEGORIES OF DEVIANCE

It is often difficult for people to reach agreement about what is deviant and what is normal. At the same time, there are some behaviors and characteristics about which there is more general agreement. In the following paragraphs we will briefly examine some of these categories of deviance.

Individuals who lack intelligence to the extent that they are unable to perform the normal tasks expected of people their age are considered *mentally deficient*. Current practice defines five levels of mental deficiency or retardation: *borderline* (I.Q. range of 83–68), *mild* (67–52), *moderate* (51–36), *severe* (35–20), and *profound* (below 20). The extent of mental retardation in the United States is estimated at approximately 3 percent of the population. More than half of the retarded receive special educational services, and a great majority of those in the 50 to 80 I.Q. range make a satisfactory adjustment in the community as adults. Those with I.Q. levels between 25 and 50 are viewed as trainable in that they can learn certain skills necessary for living in the home, neighborhood, or sheltered workshop. They will, however, need some care and supervision throughout life. About 5 percent of all retarded fall in the profoundly retarded category and require complete care and supervision throughout life. Mental deficiency, especially severe retardation, can be related to genetic factors, birth trauma, or diseases of the mother before birth. Many instances of retardation, particularly at the higher I.Q. levels (50–80), seem to be related to factors in the social environment.[5]

*Mental illnesses* are commonly classified in two categories: neuroses and psychoses. **Neuroses** are the mildest and most common; **psychoses** are more severe and less common. The neurotic is anxious, nervous, and compulsive, but generally able to function adequately in society. The psychotic loses touch with reality. The psychotic might have hallucinations, incoherent speech and thought, or delusions of persecution; or he or she might withdraw into a dream world. Psychoses fall into two categories: organic and functional. **Organic disorders** are those caused by brain injury, hereditary factors, or physiological deterioration (as a result of aging or the effects of alcoholism or syphilis). **Functional disorders** are those without organic cause; they are based on environmental factors. Functional disorders represent a reaction to such things as stress or rejection, or they might be a chosen alternative to an unlivable and intolerable world. Examples of functional disorders include schizophrenia and manic-depressive psychoses. Social scientists are generally more interested in the functional psychoses because of their relationship to social factors. It is important to note that these categories (organic and functional) are neither clear-cut nor mutually exclusive. Some research has indicated, for example, that hereditary factors affect schizophrenia and that psychoses resulting from aging might be as much related to social isolation (functional) as to physiological deterioration (organic).[6]

It is difficult to estimate the exact number of neurotics in the United States. Some say we are all a little neurotic; other estimates of the number of neurotics in the population run from a high of 40 percent to a low of 5 percent. It has been estimated that there are more than two million psychotics in the United States. Half of all hospital beds in the country are used for the care of mental patients, and at current rates one person in ten will be hospitalized for mental illness sometime during his or her life. It is easy to see why mental illness has been called America's *major* health problem.

Drinking alcoholic beverages is relatively common in the United States, but excessive drinking, or *alcoholism,* is viewed as deviant behavior. Surveys tell us that about 60 percent of people older than 18 currently (1997) drink alcoholic beverages at least once a month. Drinking and alcoholism are not identical. Some groups (for instance, those for whom drinking has mainly a ritualistic significance) might have relatively high alcohol-consumption rates but little alcoholism. The major factors here are the group associations and cultural factors associated with drinking.

There are several types of drinkers. *Social drinkers* drink when the occasion suggests or demands it, and they are relatively indifferent to alcohol—they can take it or leave it. Some social drinkers drink regularly, others infrequently. *Heavy drinkers* drink more frequently and consume greater quantities when they drink. They occasionally become intoxicated. *Acute alcoholics* have much trouble controlling their use of alcohol. They might go on weekend binges or drunks. They have sober periods, but they rely more and more on alcohol. *Chronic alcoholics* drink constantly. They "live to drink, drink to live." It is difficult if not impossible for them to hold jobs, and their health will inevitably be affected. It is estimated that 15 to 20 million Americans have serious drinking problems, and the social costs are enormous in lost wages, medical bills, drunk driving injuries and deaths, homicides and suicides, spousal and child abuse, and other damage to family relationships. Of an estimated 14.5 million arrests in 1998, almost 3.5 million were for offenses directly related to alcohol (drunkenness, drunk driving, and disorderly conduct).[7]

A 1999 Gallup poll found that 36 percent of respondents said that drinking had been a cause of trouble in their family. There has been a tremendous increase in the availability and use of treatment centers throughout the country. The efforts of a number of people and organizations, including MADD and SADD (Mothers and Students Against Drunk Driving), are having some success in changing harmful behaviors. Sociologists believe that alcoholism is primarily a learned behavior because drinking patterns tend to be associated with occupational groups, social classes, religious categories, and nationalities. Genetic links to alcoholism continue to be hypothesized by some experts, and current research is attempting to clarify the roles played by environmental and genetic factors.

The use of *narcotics* is a central issue in American society. Drug use is viewed as deviant in a society dominated by the values of the Protestant ethic: hard work, self-control, and self-discipline. But use of drugs is widespread. Many youths are using marijuana, amphetamines, and LSD and other hallucinogens. Their elders are using tranquilizers, pep pills, and sleeping pills. Use of antianxiety drugs like Valium, Xanax, and Prozac is widespread. On the illegal side, cocaine use has become a focus of attention. Cocaine comes in several forms, including a white powder that is sniffed and a solid form called crack that is smoked. Crack is popular because it can be cut into small portions that make it easier to carry, because it's cheaper, and because it can get into the system faster (by smoking). Cocaine generally produces a euphoric state, followed by a variety of different feelings, including depression. Increased amounts may need to be taken to obtain the desired effect, and it is often used in combination with alcohol. Some of the current violent urban youth-gang behavior has been connected with the selling and distribution of crack cocaine.

In the mid and late 1960s, there was great concern about drugs. This led to new laws, increased enforcement, numerous arrests, and harsh punishment. In the mid and late 1970s, public opinion moderated, things changed, and there were fewer arrests, although it's unlikely that people's drug-taking behavior changed. Then in the 1980s the picture changed again and arrests increased. The usual enforcement approach has been to work on supply (trying to stop the production and importation of drugs) rather than on demand (working on the reasons that people want drugs in the first place). This approach hasn't worked well in the United States. In the early 1990s, law-enforcement officials saw an increasing and alarming connection between narcotics and violent crime, especially in big cities like Los Angeles, Miami, and New York. The response was harsher drug laws, longer sentences ("three strikes," mandatory minimum sentences), and more prisons. In California, in the last 15 years violent offenders have gone from 57 percent of the prison population to 42 percent, while drug offenders have gone from 8 percent to 28 percent. There is no doubt that the drug trade, both internationally and within the United States, involves a huge amount of money, and drug use seems widespread. How to effectively deal with it is a complicated question— perhaps we should work more on demand rather than supply, perhaps we should decriminalize certain activities.

In the late 1980s and again in the 1990s use of drugs became a scandal in professional sports. Players in baseball, basketball, and football were suspended and sent to drug rehabilitation programs or kicked out of the sport completely. Several athletes in the recent Olympics, including a gold medalist, were disqualified for steroid use.

*Suicide* is viewed as deviant behavior in the United States, as it is in most societies dominated by Christian or Jewish religions. Reactions vary, however. Some countries have called suicide a crime and have buried suicides in special cemeteries. In some Asian societies, suicide is looked upon with less disfavor. Ceremonial self-destruction, known as *hara-kiri,* has long been a custom in Japan. Several states in the United States have classified attempted suicide as a crime, but it is unlikely that this has any deterrent effect, except possibly to ensure the success of the attempt. It is estimated that 30,000 Americans committed suicide in 1997, a rate of about 11 per 100,000 people. Probably at least five times as many people attempt suicide as actually commit it.[8]

Patterns of deviant behavior often vary substantially from one country to another, and this is certainly true of suicide, as can be seen in Table 14.1. However, suicide statistics might be somewhat inaccurate. Countries vary in how they report suicides, and in most places it is probably underreported. In the United States, the coroner usually determines whether or not a death is reported as suicide. A coroner who is uncertain about the cause is more likely to label the death "due to natural causes." In addition, relatives, motivated by personal or religious beliefs, might press for a "death by natural causes" decision from the coroner.

Suicides in America are more likely to be male; white; over 45; single, divorced, or widowed; and Protestant. Although men commit suicide more often than women do, more women than men attempt suicide. Men more often use guns to commit suicide; women tend to use poison. Some occupational groups—military officers, police officers, psychiatrists—have high suicide rates. And, interestingly, suicide is a leading cause of death for young adults.

In the late 19th century Durkheim studied suicide in Europe and described in detail three types of suicide: egoistic, altruistic, and anomic.[9] **Egoistic suicide** occurs when interpersonal relationships are secondary, distant, and not group-oriented. In such situations, the individual lacks group attachments, and when personal problems appear, the individual, in the absence of emotional support from others, resorts to suicide. The high suicide rate of single people might be an example of egoistic suicide. **Altruistic suicide,** on the other hand, is the result of strong group attachments. The individual commits suicide to benefit the group and to follow group norms. Suicides of Japanese soldiers and airmen during World War II, the self-immolation of Buddhist monks in Vietnam, and more recently, deaths of the members of the Heaven's Gate cult in California are described as altruistic suicides. **Anomic suicide** occurs in situations in which norms are confused or are breaking down. Economic depressions or rapid social changes that lead to disequilibrium in society and a state of "normlessness" can result in anomic suicides.

The first research question asked at the beginning of this chapter concerned suicide. A recent paper discussed some interesting data on how people commit suicide. There clearly are cultural or national variations. For example, suicide by gunshot is favored in Australia and the United States,

**TABLE 14.1**   Suicide Rates (per 100,000 People) in Selected Countries, 1993–1997

| | | | | | |
|---|---|---|---|---|---|
| Hungary | 32 | Australia | 13 | England | 8 |
| Finland | 26 | Canada | 13 | Italy | 8 |
| France | 21 | United States | 11 | Spain | 8 |
| Switzerland | 20 | Ireland | 11 | Israel | 5 |
| Austria | 20 | | | Greece | 3 |
| Japan | 18 | | | Mexico | 3 |
| Denmark | 17 | | | Kuwait | 2 |
| | | | | Phillipines | 1 |

*Demographic Yearbook*, 1997.

the Golden Gate Bridge is used in San Francisco, the most common technique in Holland for a long time was drowning, and the typical method in India involved drinking a cheap and easily available insecticide. It has been claimed that jumping out of tall buildings and suicide by gunshot depend directly on how many tall buildings and guns are available. But what would happen if the favored technique of committing suicide were not available? Would people choose another technique (displacement), or would they not commit suicide?

Between 1963 and 1975 the number of suicides in England dropped suddenly (35 percent) at a time when suicides were on the increase in nearby European countries. It was difficult to figure out why. It turns out that one of the favored techniques for suicide in England was gas poisoning. The remarkable decline in suicide occurred because of the removal of carbon monoxide from the domestic gas supply. Improved production methods and natural gas from the North Sea meant that the gas was less toxic and no longer lethal. The method some people wanted to use didn't work, and they did not switch to another technique. Displacement didn't happen, and the authors of the paper suggest that this might be a lesson in suicide prevention—an argument for a suicide barrier on the Golden Gate Bridge, for restricting the number of guns in circulation, and so on.[10]

## DEVIANT BEHAVIOR AND CRIMINAL BEHAVIOR

A distinction should be made between deviant behavior and criminal behavior. As we have mentioned, deviant behavior, like beauty, is in the eye of the beholder. Behaviors are not naturally deviant; they are defined as deviant by groups. Similarly, reactions to deviance vary greatly. Some deviant behaviors are ignored, some deviant behaviors are tolerated, and some deviant behaviors elicit severely critical reactions from others. As behaviors fall further outside the range of what is defined as normal, societies feel that they must formally proscribe the behaviors. Rule makers decide that particular acts are a threat to the organization and structure of society and must therefore be prohibited. Laws are passed, and these forms of deviance become illegal or criminal if performed.

## BLACK ON BLACK CRIME

Blacks represent 12 percent of our population, but they make up 56 percent of all arrests for homicide, 57 percent of the arrests for robbery, and close to 40 percent of the arrests for aggravated assault (1997). Blacks also run a much greater risk of being victims of these crimes. Much violent crime, especially homicide, is intraracial (within the same race). In 1997, 94 percent of blacks killed were killed by blacks; 85 percent of whites were killed by whites. Why the high incidence of black on black crimes? There are several factors. Homicide in the United States usually happens between friends, relatives, and acquaintances. Homicide rates are higher in big cities and among economically disadvantaged people. According to William J. Wilson, there is a growing split between the black middle and lower classes. As one group thrives and moves upward, economic conditions for the "underclass" worsen. For many blacks, absence of equal educational and economic opportunity, along with the absence in the inner city of a stabilizing black professional middle class, has led to higher unemployment, increased teenage pregnancy and single parenthood, and an increase in crime and violent behavior. Further, organized crime has traditionally supplied a path of upward mobility for those who were blocked from more respectable pursuits. Increased involvement of blacks in organized criminal activity likely increases the incidence of black on black crime. Finally, today in the inner city violence is much more a part of American culture—"murder is in style."

—See "Black on Black Crime," Howard Palley and Dana Robinson, Society 25 (July/August 1988), pp. 59–62.

---

Conflict theorists tell us to be alert here and ask crucial questions. Who are the rule makers? Who passes the laws? The answer is middle- and upper-class representatives of the power structure. Who enforces the laws? Police, who, if they are not members of the middle class, are agents of and are working for the middle class. The theme continues for courts and prison systems. Conflict theorists suggest that we should not be surprised when arrest figures show that those without power and wealth commit most crimes; the system is designed to work that way. According to this view, the whole criminal justice system from law to prison is an instrument the ruling class uses to perpetuate existing patterns of power and privilege.[11]

A law violation is considered **criminal behavior,** and there are several categories of crimes. More serious acts are called **felonies,** punishable by a year or more in a state prison or by death. Less serious crimes are called **misdemeanors.** The FBI has developed a list of eight major crimes: homicide, rape, robbery, aggravated assault, burglary, larceny-theft, motor-vehicle theft, and arson. Trends in crime are studied by noting the number of crimes reported to police throughout the country. Table 14.2 shows the number of major crimes reported in the United States in 1998.

**TABLE 14.2**  Reported Crime in the United States, 1998

| Crime | Number | Type |
|---|---|---|
| Homicide | 15,000 | |
| Forcible rape | 90,000 | Violent = 12% |
| Robbery | 450,000 | |
| Aggravated assault | 975,000 | |
| Burglary | 2,335,000 | |
| Larceny-theft | 7,375,000 | Property = 88% |
| Motor-vehicle theft | 1,235,000 | |
| Arson* | —— | |
| **Total** | 12,475,000 | |

*Total not available.
*Uniform Crime Reports*, 1998.

Aside from the eight major crimes, a number of other acts are illegal: prostitution, gambling, fraud, embezzlement, vandalism, traffic offenses, and so on. Some of the forms of deviant behavior previously discussed in this chapter can fall into criminal categories. If alcoholism means drunk driving, it becomes a crime. The sale, possession, or use of particular drugs becomes a crime.

You might assume that if an act is defined as criminal, there must be general agreement about it. Such a serious move by a society must mean that people are in accord with each other and that group differences and vested interests have been set aside. This, of course, is not necessarily so. One category of behaviors is called crime because it threatens the public order. There is general agreement that acts such as homicide, arson, robbery, and larceny threaten the public order. About other categories of crime, however, there can be less agreement. For example, one such category grew out of the concern, religious beliefs, and moral outrage of our Puritan ancestors, who felt that in some areas of behavior, individuals must be protected from themselves. It is not so much that their behavior might be a threat to society as that their behavior might be a threat to themselves. Consequently, a number of acts, some of which we today call vices, were defined as crimes. These include gambling, prostitution, narcotics use, abortion, and a variety of sexual activities, including homosexuality. Opinions as to the "rightness" or "wrongness" of these acts vary from individual to individual and from group to group. However, there is a substantial demand to engage in many of these acts.

Gambling is legal in some states and is widely sought and available throughout the country even where it is not legal. The popularity and proliferation of casinos on lands owned by American Indians illustrate this. Prostitution is legal in some areas (in all Nevada counties but two, for example). Many young people as well as some of their elders wonder at the current laws punishing marijuana use. Studies show that marijuana is widely used and is relatively easy to obtain. Laws in many of our states outlaw sodomy, the "infamous crimes against nature," and yet Kinsey's classic studies of adult sexual behavior reported that many of these "crimes," which involve various types of sexual behavior, are relatively common practices. People wonder why homosexual acts between consenting adults in private are not legal if heterosexual acts under the same conditions are.

Edwin Schur has commented that because all parties involved in these actions (gambling, prostitution, abortion, and so on) are seeking them or at least consenting to them, these are "crimes without victims."[12] If there is no victim, Schur wonders whether anyone should be punished. He points out that when these activities are called crimes, they are forced underground to a certain extent and become more expensive to obtain. This leads to secondary crime—theft, for example, to obtain funds to support the primary crime, such as narcotics or abortion. So, although some acts are defined as criminal by society, this does not mean that there is universal agreement as to their "wrongness."

**Juvenile delinquency** refers to young people (younger than 21 in some states, younger than 18 in others) who commit criminal acts or who are wayward, disobedient, uncontrollable, truant, or runaways. Delinquency is a confusing concept. Young people *are* heavily involved in certain types of crimes, as can be seen in Table 14.3. At the same time, many young people are defined as juvenile delinquents for committing acts like disobedience or running away that are ignored if committed by adults. American law is such that we are very strict with juveniles: "Get ahold of the bad kids quick and change them so they won't grow up to engage in worse activities." However, as we have seen in the section on primary and secondary deviance, if the juvenile is thrown into a reform school with other hardened delinquents, we almost guarantee the result we were trying to avoid. Most countries are more lenient than the United States is in their treatment of juveniles.

## SOCIAL CONTROL

**Social control** refers to the processes, planned or unplanned, by which people are made to conform to collective norms. Some amount of conformity seems to be essential in all societies. Predictability and order are necessary to the social organization of group behavior. If people could act in complete isolation, possibly they could ignore the existence of any norms. However, it is almost impossible to imagine such a situation because people do behave in groups, and although individuality and nonconformity are very popular, we should remember that they are acceptable only when most people, most of the time, conform.

Social control can be provided in a variety of ways. We could, for example, hope that people will naturally conform, that conformity is inherent in their nature. It seems, however, that people must be *trained* to conform. Basic social control is taught through the socialization process. Shortly after a child is born, the family begins telling him or her what to do and what not to do, what is right and what is wrong. Sanctions or punishments are applied if the child misbehaves. Children who fail to conform are told, in

**TABLE 14.3** Arrests of Young People, 1997

| | Percentage of Arrests for People: | |
| --- | --- | --- |
| | Under Age 15 | Under Age 18 |
| Robbery | 8 | 30 |
| Burglary | 14 | 37 |
| Larceny-theft | 14 | 34 |
| Motor-vehicle theft | 10 | 40 |
| Arson | 33 | 50 |
| Vandalism | 19 | 43 |

*Uniform Crime Reports*, 1997.

effect, that mother and father won't love them or will beat them, depending on the family's view of proper child-rearing practices. Later, influences outside the family continue the process. Peers, teachers in the school system, the church, the mass media all have a profound effect on inculcating group and societal norms and encouraging conformity to group expectations. Finally, if all else fails, the society can pass laws to ensure conformity. Laws force people to conform and are a response to breakdowns in other forms of social control. Again, sanctions are applied for failure. These sanctions can range from a fine to imprisonment to death.

Social-control mechanisms vary depending on the type of group or society. In a primary group or small primitive society, rules are not written and social control is informal. Violators might be subjected to gossip or ridicule or possibly even ostracism; although these techniques do not have the dramatic effect of a raid by the FBI, they are surprisingly effective, even in modern mass society. Concern over what family and friends will think, potential disgrace and loss of reputation, and other informal sanctions are probably the first line of defense in producing conformity. Large societies add more formal types of social control—written laws, police, courts, prisons—making an exceedingly complicated legal system to guarantee an acceptable degree of conformity.

Social-control mechanisms also vary depending on the type of norm violated. If the violated norm is a folkway regarded as insignificant or harmless to the group, informal and minor sanctions are elicited: whispering, giggling, or a little mild ridicule. If one of the mores of the group is being abused, however, more formal sanctions might be used.

The most extreme of the formal sanctions is the death penalty. Both the second and third research questions asked at the beginning of this chapter concern the death penalty. Question 2 concerns what happens when executions take place and are publicized: Do homicide rates go down or up? One of the arguments in favor of the death penalty is that its use will deter other people from committing serious crimes, which should result in a subsequent decline in the homicide rate. One of the arguments in opposition to the death penalty is that executions tend to brutalize society in that they suggest a continuance of violence and that the taking of revenge is appropriate. If this second argument were true, the homicide rate could be expected to go up after executions. A paper examined this issue by looking at executions, the amount of publicity about the execution (days of coverage, amount of TV air time), and the homicide rate in the United States. The idea was to see if more executions and more publicity had an effect on the homicide rate. What happened? In somewhat of an anticlimax, the researchers found that homicide rates were *unaffected* by executions or publicity. They neither went up nor down significantly. Maybe the two arguments (deterrence and brutalization) canceled each other out, or maybe neither is valid, or perhaps the number of executions performed is too small to be meaningful.[13]

The third research question at the beginning of this chapter wondered what might happen if death-row inmates—a group scheduled to be executed—were released, a frightening thought perhaps. This actually happened in 1972, when the U.S. Supreme Court ruled in *Furman v. Georgia* that the death penalty was unconstitutional. This meant that 600 death-row inmates nationwide went back into the general prison population. In 1972 there were 47 "Furman" inmates on death row in Texas; a study followed their behavior from 1973 to 1986. First they were transferred to the general prison population. In spite of what we might expect, they committed few serious rule violations, and most were model inmates during their time in prison. Thirty-one of the "Furman" inmates were eventually released on parole. Most of those released did not commit other crimes. Four of those released did commit new felonies: one murder, one rape, and two burglaries. Twenty-three inmates were successfully completing their parole at the time the article was written (1988). The success rate, in and out of prison, in a group that was seen as enough of a danger to be isolated from other prison inmates and enough of a threat to society to be executed, is surprising.[14]

Individual social control probably emerges from both internal and external sources. Internal constraints are those aspects of the normative system that one internalizes through the socialization process. Internalized norms become the individual's *conscience.* External social control refers to the external mechanisms—*rules and laws*—applied by society. It would be fortunate if the two were in balance, but as we saw in the previous section on deviant behavior, this is frequently not the case. Take, for example, the boy whose peers happen to be little hoods; his most important primary group is a delinquent gang. He internalizes a set of values and norms from them. He obeys the norms of the gang or he is ridiculed and ostracized. Yet in following gang social control he runs afoul of external societal social control, which in many cases makes opposing demands. To complete the dilemma, it is obvious that the social-control system of the primary group, the gang, is much more important for one's behavior than is any external social-control system. This type of gap occurs when the values of one segment of society are vastly different from the values of the rule makers of society.

Another type of gap or difference between internal and external social control occurs when people's attitudes, values, and norms change more rapidly than laws change. A cultural lag results. For example, there are

## ANOTHER TYPE OF PUNISHMENT

The Reformed Mennonite Church imposed its "doctrine of avoidance" on Robert Bear of Carlisle, Pennsylvania. Commonly called shunning, this meant no one in the church including his family could speak or have anything to do with him. His wife and six children left him when the sanction was imposed. His parents died shortly after the church action and, according to Bear, "To the moment they died they looked at me with disgust. I was already burning in hell." Five years later the punishment remained in effect.

sharp differences of opinion regarding whether to and how to enforce existing gambling and prostitution laws. Laws traditionally have been slow to change, and perhaps they should be, so as not to respond to every whim or alteration in public opinion. We might predict a pattern through which change in laws occurs: (1) Attitudes change; (2) laws relating to the changed attitudes are not enforced; and (3) laws change. It is likely, however, that there is a large lapse of time between step 2 and step 3. Consequently, when attitudes change, laws tend to be kept on the books, but they are just not enforced as vigorously as they might be. Law-enforcement agencies and individual citizens often find themselves in a difficult position because of this situation. The marijuana laws in California—first strictly enforced with heavy penalties, then less strictly enforced, now changing—are a good example of this sequence.

White-collar crime and organized crime provide examples of another type of conflict between society's laws and people's attitudes. The public favors and encourages prosecution of offenses such as burglary, assault, larceny, homicide, and sex offenses. They are indignant with the small-time thieves and quick to lock them up for their sins. On the other hand, white-collar crime is seldom punished. White-collar crime refers to the illegal acts committed by middle- and upper-class people during the course of their regular business activities and includes such examples as the bank vice-president who embezzles funds, the physician who splits fees or performs unnecessary operations, the executive who cheats on his or her income tax, and the disk jockey who accepts payola.

Individual white-collar crime is only part of the picture. We know that some corporations are engaged in unethical business practices, such as price fixing by computer, steel, and electrical companies; false advertising of drugs and food; collusion between government regulatory agencies and the companies they are supposed to be regulating; campaign payoffs in return for favors; bribery; cost overruns on defense contracts; land and computer fraud; and on and on. Take one case: General Electric, Westinghouse, and some 20 other electrical companies cooperated to fix prices of electrical equipment. Bids, supposed to be secret and competitive, were neither. The companies worked out ahead of time which company would bid lowest on

---

## A TOMBSTONE EPITAPH—GIRARD, PENNSYLVANIA, 1870

In memory of
Ellen Shannon
Aged 26 years
Who was fatally burned
March 21st 1870
by the explosion of a lamp
filled with "R. E. Danforth's
Nonexplosive
Burning Fluid"

—*From* Folklore on the American Land,
    *by Duncan Emrich (Boston: Little, Brown, 1972).*

---

a contract and what the other bids should be to make it look good. The resulting lowest bid was far higher than it should have been, and the companies made good money—in the neighborhood of $2 billion more than they should have in one seven-year period. But (surprise) they were caught. An outraged judge levied $2 million in fines (a drop in the bucket compared to the take) and sentenced seven men to 30 days in jail. See whether the average thief who steals a car or burglarizes a house gets off with 30 days. However, people were amazed at the "harshness" of the sentences in the great electrical conspiracy, for corporate criminals almost never go to jail, even for 30 days.

Criminologist Edwin Sutherland has maintained that more money is lost yearly through white-collar crime than through all other forms of crime combined and that the damage to social relationships that results (no one can be trusted, everyone is on the take) is even more serious. Yet, as we said, the larceny of the white-collar criminal is seldom punished, and even when it is, the penalties are slight compared with those for other crimes. Why? Well, the activity might have involved substantial skill and ingenuity and it wasn't violent, so the public tends to identify with it or at least not object to it. Many times the victim, often a large organization or the public in general, is invisible or not easy to identify with. Often the culprit is a large corporation that has spent much money over the years to cultivate its image, and the public finds it difficult to believe that such an upstanding important company could be guilty of such dastardly deeds. And finally, the white-collar criminal is supposedly a "good" person—middle-class and respectable. So, the public prefers to look the other way.

Likewise, the activities of organized crime go largely unpunished, but the reasons for breakdown in social control in this case are more basic and easier to understand. Many people *want* what organized crime supplies: gambling, narcotics, tax-free cigarettes and alcohol, prostitution, and loan rackets. The conclusion is clear: The attitudes and values of society are reflected even more in punishment practices than they are in the written laws of society.

## THEORY AND RESEARCH: A REVIEW

All three theoretical viewpoints—functional analysis theory, symbolic interactionist theory, and conflict theory—can be seen in this chapter. The comment that deviant behavior is not necessarily harmful to group life but is actually helpful in defining group boundaries and thus helps preserve group stability represents the *functional analysis* view. Labeling theory is *symbolic-interactionist* in that it focuses on the definitions of people's behavior and sees the definitions as the product of a type of interaction between two or more parties. Conflict theorists would be interested in the question, Who gets labeled? From the *conflict theory* perspective, the criminal justice system—lawmakers, police, courts, prisons—is in the hands of the ruling class and is used to protect existing patterns of power and privilege. Conflict theorists see conflict among classes as producing unemployment, poor working conditions, and family breakdown, which are the important pressures that often lead to crime.

The major source of the data in this chapter is records: *reactive* and *nonreactive surveys* of information collected by police departments, the United Nations, census agencies, and organizations dealing with drugs, alcohol, and mental illness. Self-reports of people's activities are also used. The problem with this sort of information is that people, agencies, and governments don't always report accurately for a series of reasons ranging from embarrassment to self-incrimination to a desire to "look good" (encourage tourism) or "look bad" (increase the budget). Statistics on crime and deviant behavior might therefore get manipulated.

## SUMMARY

The previous chapters in this section have dealt with social change, social disorganization, and collective behavior. In this chapter we returned to the topic of norms, which was first discussed in Chapter Three. A *norm* is defined as the accepted or required behavior in a specific situation. Most behavior is predictable because it is in accordance with generally accepted norms. However, some behavior is contrary, and this is called *deviant behavior*. Societies attempt to restrict the amount of deviant behavior through *social-control* measures. This chapter addressed itself to these concepts: deviance and social control.

Individual variation and individual differences are a basic part of the human condition. When differences from group norms reach a particular point, however, social differentiation becomes social deviance. Many sociologists believe that deviance is best understood as a *labeling* process. Behaviors or characteristics of individuals are deviant when they are defined as such by members of society. Deviance is neither a natural nor a statistical phenomenon but is instead a product of group viewpoints and definitions.

Criminal and delinquent behaviors are also seen as products of a defining or labeling process. If society is especially concerned about specific types of behavior, it might define these behaviors as not only deviant but criminal as well. Frequently differences of opinion develop about the "rightness" or "wrongness" of particular acts. Such differences are more likely to occur in large, complex societies in which a variety of viewpoints exist about what is deviant and what is normal. These variations in definitions make it possible within a society for a given act to be viewed as criminal and deviant, as criminal but not deviant, as deviant but not criminal, or as "normal."

We briefly examined several categories of deviance—mental illness, mental deficiency, alcoholism, drug use, and suicide—as well as some descriptions of and explanations for criminal and delinquent behavior. Finally, we discussed social control, which refers to the processes by which people are made to conform to collective norms. Social-control mechanisms are practiced by all societies and include norms internalized through the socialization process as well as laws enforced by the police and courts of a society.

Two readings follow, both of which deal with social control. Who goes to prison and who stays in prison? You might be surprised. The first reading looks at some of the problems confronting the California prison system and some of the possible consequences of the "three strikes" legislation. The second reading describes a different type of social control. It sounds like vigilante justice out of the Old West, but this article is about modern-day Northern Ireland.

---

# TERMS FOR STUDY

altruistic suicide (432)

anomic suicide (432)

criminal behavior (434)

deviant behavior (422)

differential association (424)

egoistic suicide (432)

felony (434)

functional disorders (429)

juvenile delinquency (436)

labeling theory (424)

misdemeanor (434)

neurosis (429)

organic disorder (429)

primary deviance (427)

psychosis (429)

secondary deviance (428)

social control (436)

For a discussion of Research Question 1, see page 432.
For a discussion of Research Question 2, see page 437.
For a discussion of Research Question 3, see page 438.

## INFOTRAC COLLEGE EDITION
### Search Word Summary

To learn more about the topics from this chapter, you can use the following words to conduct an electronic search on InfoTrac College Edition, an online library of journals. Here you will find a multitude of articles from various sources and perspectives:

**www.infotrac-college.com/wadsworth/access.html**

| | |
|---|---|
| Suicide | Alcoholism |
| Prisons | Prozac |
| Criminals | Gambling |
| Neuroses | |

---

## Reading 14.1

# CALIFORNIA PRISONS AS GERIATRIC WARDS

*Miles Corwin*

---

*In the 1990s a belief spread across the country that a good way to solve the crime problem would be to build more prisons and lock up more people. Politicians increasingly ran on "get tough" platforms, and several states passed so-called three strikes and you're out laws. These laws aspire to put a person convicted of three major felonies in prison for life. Does this approach make sense? What would it cost? What would be the consequences to our penal system? What kinds of people are in our prisons now? Are we locking up (and turning loose) the people we should? Criminologists, judges, lawyers, and other people concerned with our criminal justice system are debating these questions.* Los Angeles Times *reporter Miles Corwin gives us insight to some of the issues in this look inside the California penal system.*

Every time S. M. Cohen takes a pill, visits a doctor, has a blood test, receives a shot of insulin or undergoes surgery, Californians pay for it.

Cohen, 67, an inmate at the California Institution for Men, has cancer and diabetes and takes 11 pills a day. He has blood tests twice a month and sees three specialists. His medical bills are more than $125,000 a year. His recent heart bypass cost $76,000.

Cohen, who is doing nine years for grand theft, is housed in a unit inmates call "Old Man's Dorm," a cavernous hall filled with grizzled prisoners who mill about using canes, walkers, wheelchairs, and crutches. The highlight of their week is the Wednesday night bingo game.

Soon the state's population of geriatric prisoners will increase dramatically because of the new "three strikes" law, which is aimed at putting three-time felons behind bars for life. And there will be many more facilities such as Old Man's Dorm—the only unit in the state designed for low-security, geriatric prisoners.

Californians will have to decide how long elderly offenders should be kept behind bars. This decision will highlight a philosophical divide between those who believe society's need for justice is paramount, and others who argue that many elderly prisoners should be given early release because they no longer are dangerous and it is expensive to imprison them.

The predicted wave of elderly "lifers" will "create an economic disaster for the state," said Dr. Armand Start, a national expert on prison medical care. Funding eventually will have to be siphoned off from other state programs such as education, he said.

California will not be able to build prisons fast enough to keep pace with the influx of elderly prisoners, said Start, a professor at the University of Wisconsin Medical School. Dangerous first-time offenders eventually will have to be paroled early, he said, to make room for older, less violent prisoners serving mandatory life terms.

Only about 2 percent of the state's prisoners are over 55. And Old Man's Dorm—officially known as Elm Hall—has long been viewed as a curiosity. But now Elm Hall, which is essentially a convalescent hospital behind bars, offers a glimpse into the future of the state prison system. There are some younger, disabled inmates, but most of the 150 prisoners are elderly men with bifocals who never miss the weekly meetings of their senior citizens' group.

There are no cellblocks because the prisoners need wheelchair and ambulance access. And despite chronic prison overcrowding, there are no bunk beds because most of the old-timers do not have the strength to climb to the top bed. Many are recovering from strokes, heart attacks, or crippling bouts of arthritis.

Elm Hall is a large cinderblock dormitory with rows of cots and lockers. A nurse is on duty until 10 P.M., to make sure no one slips in the shower or cannot get out of bed. Dorm living is better than the hard time of a cellblock, but most prisoners share one great fear.

"Everyone is terrified of dying here," said Cohen, who has a pencil-thin mustache and neatly trimmed gray hair. "That's what bothers us the most. We've seen people die in here, and it's the loneliest death in the world."

Elm Hall was designed in the early 1960s—with wheelchair ramps, bathtubs and a heating system—to house the state's small population of elderly prisoners. At that time the old-timers mingled easily with the mainline population. But during the 1980s, when young gangbangers began pouring into the prison system, the clash of generations created friction.

Many old-timers say they have only contempt for the young prisoners. Old-timers call themselves "convicts," and refer contemptuously to the young prisoners as "inmates." The inmates, the old-timers complain, "don't know how to do time." They snitch on each other; they steal out of lockers; they cut in front of people in chow lines. One elderly prisoner said that if, in his day, you cut in front of someone in line you would get a shiv in the belly.

William Ross, 81, a bald diminutive man with a faint resemblance to George Burns, is the dean of Old Man's Dorm. Ross, who is serving a two-year sentence for check kiting, looks out of place amid the young, tatooed prisoners pumping up at the weight pile.

One night last month, Ross was walking from the canteen, after buying cream, coffee and sugar. Two young inmates ran up behind him and tried to rip off his bag of supplies. But a few elderly prisoners, who were tailing him, began shouting for a guard. Ross held tight to the bag and the thieves were scared off.

"We old-timers stick together," said Ross, squinting through thick glasses. "When we head out to the yard we try to go in groups. If somebody has to go out alone, the others try to keep an eye out for him."

Years ago, in the days when it was not so smoggy and they could see Mt. Baldy, prisoners founded the Mountain View Senior Citizens Group. Members still collect aluminum cans and sometimes sell candy and doughnuts to raise money. If an old-timer is about to be released, but does not have enough money to get home or needs a new suit of clothes, group members will chip in with the cash. The family of one man who died in prison wanted to give him a proper burial, but they did not have the money. The group raised $1,000.

One Elm Hall inmate had such a long criminal history, officials say, he started out robbing stagecoaches. When he was released, he was 91 and still chain-smoking unfiltered Lucky Strikes.

Another renowned resident was Ronnie Fairbanks, whom officials listed as 90 when he last entered prison. But they could not be sure. Fairbanks had given authorities 21 birth dates since he was first arrested for larceny in 1928. He also had used 88 aliases during his criminal career, which included more than 40 arrests. Fairbanks often boasted that he never committed a violent crime. He was last arrested for stealing a suitcase at Los Angeles International Airport and was released from Chino in 1988. He has not been back. Yet.

"He may have been 90, but he was in great shape," said Diana Smith, program administrator for Elm Hall. "He jogged every day and kept himself in top physical condition. He was a very charming guy, very genteel, always polite."

Not every prisoner is eligible for Elm Hall, Smith said. When Charles Manson becomes a senior citizen, she said, he will not be welcome. Prison officials will not admit notorious criminals, child molesters or anyone with a violent history.

Building more facilities such as Elm Hall is not the best way to deal with aging inmates, many prison experts say. When criminals hit middle age they begin "burning out," said Start, who heads the National Center for Correctional Health Care Studies. Criminals in their 40s and 50s—like the rest of the population—are affected by aging. They become less impetuous, less hostile, less prone to violent behavior, he said.

A federal Bureau of Justice Statistics report found age to be the single most reliable indicator in predicting recidivism. About 22 percent of prisoners ages 18 to 24 return to prison within a year of their release; for prisoners over 55 the rate is less than 2 percent.

The 20 new prisons that are planned in California will not keep pace with the deluge of new inmates generated by the "three strikes" law, said Jonathan Turley, a law professor at George Washington University. In other states with "three strikes" measures, Turley said, officials had to release violent first-time offenders early because of overcrowding. They needed to make room for the growing population of elderly prisoners with life sentences. Turley predicted that California prison administrators eventually will have to do the same.

In Texas, which has a "three strikes" measure and a number of other mandatory sentencing provisions, inmates last year served an average of only 11 percent of their sentences. (California inmates serve about 50 percent of their sentences.) The Texas law recently has been changed, requiring inmates to serve a higher proportion of their terms. But young violent offenders will still have to be paroled early to make room for older prisoners serving mandatory life sentences, Texas prison experts say.

"This system makes no sense," said Turley, who heads a Washington, D.C., project that evaluates the parole status of elderly prisoners. "We'll be keeping in prison the inmates who are at a point in their life when they are least likely to commit another crime. And we'll be releasing those young offenders who are most dangerous. This system works against everything we've learned about recidivism."

Geriatric prisoners cost institutions about $60,000 a year—three times more than younger inmates, according to recent studies. On average, inmates over 55 suffer three chronic illnesses while incarcerated. Some do most of their time in prison infirmaries. Others run up huge bills at private hospitals because most prisons contract with outside facilities for major medical work.

But David Beatty, a spokesman for the National Victim Center, said victims' right advocates are more concerned with the "human cost" of releasing repeat offenders than "the financial cost" of keeping them in prison. Even if the odds are slim that a geriatric prisoner, if released, would commit another crime, Beatty said, "most victims aren't willing to take that chance. They feel more comfortable with these kinds of people locked up forever."

No one in the Department of Corrections can predict how many elderly prisoners to expect in coming years. Prison officials say they have been preoccupied with other problems such as AIDS, tuberculosis, and gangs. They plan to research the issue this year.

California's "three strikes" measure calls for a minimum of 25 years to life for any felon who has committed two prior violent or serious felonies—crimes that range from murder to home burglary. In today's political climate 25 years to life often is the equivalent of life without possibility of parole, said Heather MacKay, a staff attorney for the Prison Law Office, a Bay Area prison reform group. The parole board, she said, "because of the prevailing political climate, often is reluctant to grant any paroles at all."

While California gears up for a deluge of geriatric prisoners, many states are moving in the opposite direction. They are involved in "compassionate release" programs for elderly prisoners who do not appear to be a threat to society.

Law professor Turley heads The Project for Older Prisoners, which lobbies parole boards in dozens of states on the issue of early release for elderly prisoners. Of all the prisoners evaluated, the project has only recommended 10 percent—about 60—for early release. Not a single one the project has helped free has returned to prison, he said.

He founded the project five years ago when he represented Quenton Brown, a Louisiana man with an IQ of 51. Brown had robbed a bread store of $117 and a cherry pie. When police arrested him he was crouched beneath a house across the street, eating a piece of pie.

Turley heard about Brown's case after he had served 18 years of a 30-year sentence. Brown, then 67, was suffering from emphysema and bleeding ulcers and was released after Turley argued his case before the parole board. It did not make sense to continue to incarcerate inmates like Brown, Turley argued, when young, violent prisoners were being released under a court order to ease the overcrowding in Louisiana prisons.

Working with law student volunteers, Turley evaluates the criminal histories of prisoners and rejects those who were convicted of sex crimes or violent offenses. Many elderly prisoners, he said, belong behind bars for the rest of their lives. But how much of a threat is an inmate who is recovering from a stroke, has only a few months to live and wants to spend his final days with his family?

About 40 percent of Medicare dollars are spent in the last two weeks of a patient's life. Because some prisoners' families have insurance, Turley said, it makes more sense to release some of them than to stick taxpayers with exorbitant medical bills.

An elderly prisoner who costs the state $60,000 a year could be housed in a private nursing home for much less. Yet California will be building 20 prisons, at a cost of about $100,000 per cell, to accommodate a new wave of prisoners, many of them senior citizens.

Turley said he does not recommend "wholesale release of older prisoners." What he does advocate is legislation that will allow states, through special parole laws, to give consideration to low-risk, high-cost elderly inmates. But few California politicians are willing to support any measures calling for early release of prisoners—regardless of their age.

"Politicians are now running the prison system by sound bite," Turley said. "The truth of the matter is, by the time we interview inmates who are in their 60s, 70s and 80s, most of them are statistically less dangerous than the law students I drive to the prison with."

---

# QUESTIONS 14.1

**1.** In what ways do the demographics of the prison system mirror the demographics of the population in general?

**2.** What is the single most reliable indicator in predicting recidivism? Explain why this is true.

**3.** Critique the "three strikes" laws—what's good about them, and what's bad about them?

**4.** If you were redesigning our criminal justice system, especially the incarceration aspect of it, what changes would you make?

## Reading 14.2

# HARSH BRAND OF JUSTICE IN BELFAST

*Shawn Pogatchnik*

*Formal social control means laws, police, courts, and prisons and includes concepts like testimony, evidence, rights, and due process. But what if the formal system of social control doesn't work? As this article from the* Los Angeles Times *shows, other methods of gaining conformity may emerge.*

Like so many wild-eyed young men of Catholic west Belfast, Seamus Clark liked to drink hard and speed recklessly about town in stolen cars. Worse, his friends and family say, was the fact that "Shamy" had a proud mouth around the wrong people.

That combination made Clark a prime candidate for punishment—not by Northern Ireland's police and legal system, but at the Byzantine hands of local Irish Republican Army gun squads.

In December 1987, a month after neighborhood vigilantes pulled him from a stolen car and beat him unconscious, five masked IRA men came knocking on his door. They ordered his parents and siblings upstairs, turned up the volume on the television, pinned him to the floor and fired five bullets into his knees and ankles.

Last April the faceless men returned, clubbed Clark severely with bats and gave him 48 hours to get out of Belfast. Today, 22-year-old Clark is hiding out in a small town south of the city and faces amputation of his failing right leg.

### Trial by Gun

Clark's story, which is not unusual, illustrates the quip popular in west Belfast circles that "there's no law, but there's order."

In a bizarre outgrowth of Northern Ireland's two decades of civil unrest, paramilitary gun squads throughout the province have shot more than 1,500 people and in other ways assaulted at least another 350 as punishment for "antisocial behavior"— the local umbrella term for car stealing, shoplifting, drug trafficking, rape, prostitution and a range of other crimes.

Such trial-by-gun-squad is not confined to IRA-dominated Roman Catholic neighborhoods. Police spokesmen say that local "hoods," as well as political opponents, are regularly dealt with on the Protestant, working-class side of town by members of the Ulster Defense Association, Ulster Volunteer Force, Red Hand Commandos, and other loyalist gangs.

But the IRA's prominent role in punishment shootings was singled out for criticism last month by Peter Brooke, Britain's secretary of state for Northern Ireland. He said that British security forces wanted to get the IRA "off the backs of the people."

"There is no trial, no legal process," he said. "Instead there is the knock on the door at night, the cold fear of the victim and the weeks and months in the hospital. Or the final and enforced departure from home and family ties into exile . . . or the hood, the bullet in the back of the head and the sad funeral."

The province's official police force, the Royal Ulster Constabulary, has recorded more than 100 punishment shootings so far this year, plus another 65 punishment beatings. This form of punishment can involve the use of "hurley bats," used in Gaelic hockey, or "breeze blocking," in which a cinder-block is dropped onto a person's legs until they are heard to break.

## "Six Pack" Punishment

Punishment shootings vary in intensity—first-time offenders often being shot once or twice through the fleshy thigh, while others get a "six-pack" of bullets through the elbows, knees and ankles.

Compared to the carnage of a typical American city, the numbers of such shootings may seem relatively minor. But in Northern Ireland, with a total population of 1.58 million, the numbers are significant.

Nearly everyone in the insular world of west Belfast knows someone—often several people—who have been "done by the Rah" (slang for IRA).

Sinn Fein, the IRA's legal political wing that commands majority support within the high-unemployment Catholic ghettos, maintains that such "kneecapping" and other punishments serve as a useful and locally accepted deterrent.

"Punishment shootings and beatings are the end of a very long process, which begins essentially with complaints being made by concerned citizens," said Richard MacAuley, a Sinn Fein spokesman, seated inside one of the party's six public advice centers. "There's an expectation, in the absence of confidence in the police force, that republicans will fill the gap."

He cited one highly publicized example recently, when a woman was gang-raped by seven men in the Divis Flats, a Catholic area of high-rises just west of downtown Belfast.

"About a hundred local women organized a picket to demand that something be done," MacAuley said. "They didn't go to the RUC (constabulary) barracks in the area. They came to this building to demand that the republican movement do something."

The perpetrators fled the country before they could be shot, MacAuley said, noting that "if they come back to Belfast, one would expect very stern and harsh action to be taken against them."

## De Facto Police

How IRA gunmen gradually gained their role as de facto police in many nationalist areas of Northern Ireland, and particularly in west Belfast, is a matter of debate. But it followed the Protestant-Catholic riots and introduction of British troops onto the streets in 1969.

Since then, the Royal Ulster Constabulary has remained more than 90 percent Protestant and has been viewed by many Catholics here as the enemy.

Police bases in west Belfast resemble modern-day Fort Apaches, bolstered by corrugated iron walls behind which bristle surveillance cameras and listening equipment.

When constabulary members patrol the area, they do so in heavily armored Land Rovers, rarely leaving the relative protection of the vehicles at night. If on foot, they wear flak jackets and always are escorted by several heavily armed British army soldiers who guard against possible ambush.

Stopping "ordinary" crime in such hostile territory, police spokesmen concede, has been low on the constabulary's list of priorities in comparison with rooting out the terrorist threat. To some extent, IRA leaders appear to have opportunistically filled a perceived vacuum in those communities.

A sampling of local opinion suggests that a majority of residents within republican areas condone, if not openly support, the IRA's brand of frontier justice. One mother of three—a nurse at the local hospital—was blunt: "Kneecapping's too good for those wee hoods," she said, rocking her daughter in her arms.

However, a growing movement composed of victims of punishment shootings and their families is making a stand against the gunmen.

After her own 24-year-old son Paddy was kneecapped by local IRA men last July, Nancy Gracey founded Families Against Intimidation and Terror. The group is trying to protect would-be victims by getting them out of Belfast, or Northern Ireland if necessary, and bringing together kneecap victims and their families.

"We've a 21-year-old boy in our group who's had nine bullets in him since he was 17 years of age. The IRA took a hacksaw to the arm of another wee fella," Gracey said.

Community activists note that west Belfast teenagers and young adults have been weaned on a violent culture in which authority figures encourage their participation in politically motivated crime, such as throwing bricks at army patrols and hijacking vehicles to make road barricades and car bombs.

"If kneecapping somebody worked, solved the problem, maybe I could understand," said Carmel McCavana, a director of the west Belfast Parent Youth Support Group, which coordinates programs for "at risk" youth, many of whom already bear scars from IRA punishment attacks.

"OK, in the short term kneecapping put somebody in the hospital for a few weeks," she said. "But in the long term, we know several young people who've gotten right behind the wheel of a lifted (stolen) car again, before their wounds have even healed.

"One young lad I know of actually was driving with his crutch on the accelerator," she said. "The young people here have very little fear . . . and very little hope. They need help."

Help for a repeat criminal offender is hard to come by in this community, whose sense of siege traditionally is relieved through violent means.

Even Seamus Clark's family has mixed feelings about the role of IRA strongmen in their neighborhood.

"The IRA shooting its own, to me, is wrong. But someone's got to keep order," said his mother, 62-year-old Alice Clark. "The kids run wild on the streets here.

"My son was doing something he shouldn't have done. Everyone normally gets for their first offense one shot, no matter what they do. I'd have accepted that, so I would," she said.

One of the bullets used in the first attack still sits on her mantle, flattened from ricocheting off the floor and into a sacred-heart ornament on the wall.

"But the Rah shot him five times. That," she emphasized, "that wasn't fair."

# QUESTIONS 14.2

**1.** Which type of social control—formal or informal—is likely to work better in most situations? Why?

**2.** We used the terms *deterrence* and *brutalization* in this chapter as arguments that are used in connection with capital punishment. Do those terms have any relevance to the type of informal law enforcement used in Northern Ireland?

**3.** Do you think the law enforcement as described in this article is working? What are its advantages and disadvantages?

**4.** Construct for a place like Belfast a system of social control that you would be comfortable with—that would be acceptable to you. Describe it.

# NOTES

1. This discussion of the labeling aspect of deviation is drawn from Howard S. Becker, *Outsiders* (New York: Free Press, 1963), chapter 1. Also see his chapter 8 for a discussion of rule creators and rule enforcers. The case of the long-haired basketball player was described in *Sports Illustrated,* 16 February 1976.

2. For the complete report, see "On Being Sane in Insane Places," by D. L. Rosenhan, in *Science,* 19 January 1973, pp. 250–258. *Saturday Review* also has a good summary of the research in its 24 February 1973 issue, pp. 55–56.

3. The idea of primary and secondary deviance was first developed by Edwin Lemert. See his *Human Deviance, Social Problems and Social Control* (Englewood Cliffs, N.J.: Prentice-Hall, 1967), especially pp. 17–18 and 40–64. Also see the discussion in Marshall Clinard and Robert Meier, *Sociology of Deviant Behavior,* 5th ed. (New York: Holt, Rinehart & Winston, 1979), pp. 75–76.

4. Kai Erikson deals with these ideas in detail in his *Wayward Puritans* (New York: John Wiley, 1966), chapter 1, and in "Notes on the Sociology of Deviance," in *The Other Side,* edited by Howard S. Becker (New York: Free Press, 1964), especially pp. 9–15.

5. This is summarized from *Psychology 73/74 Encyclopedia* (Guilford, Conn.: Dushkin, 1973), pp. 166–169.

6. A more detailed discussion of mental illness, as well as of alcoholism, narcotics, and suicide, appears in Clinard and Meier, *Sociology of Deviant Behavior* (see note 3).

7. For figures in these paragraphs, see *Uniform Crime Reports, Statistical Abstracts,* and other Census Bureau documents.

8. These figures on suicides come from *Demographic Yearbook* (a United Nations publication) and *Vital Statistics of the United States* (from the Department of Health, Education, and Welfare).

9. Émile Durkheim, *Suicide,* translated by John A. Spaulding and George Simpson (New York: Free Press, 1951).

10. See "The British Gas Suicide Story and Its Criminological Implications," by Ronald Clarke and Pat Mayhew, in Michael Tonry and Norval Morris, *Crime and Justice: An Annual Review of Research* (Chicago:

University of Chicago Press, 1988), pp. 79–114.

11. The conflict perspective starts, of course, with Marx. Today, the major spokesman is Richard Quinney. He has written a number of books, including *Critique of Legal Order: Crime Control in Capitalist Society* (Boston: Little, Brown, 1974) and *Criminology,* 2d ed. (Boston: Little, Brown, 1979).

12. In his book, *Crimes Without Victims* (Englewood Cliffs, N.J.: Prentice-Hall, 1965), Edwin Schur discusses abortion, narcotics use, and homosexuality as victimless crimes.

13. See "Murder, Capital Punishment, and Television: Execution Publicity and Homicide Rates," by William Bailey, *American Sociological Review* 55 (October 1990), pp. 628–633.

14. See "Institutional and Post Release Behavior of Furman-Commuted Inmates in Texas," by James Marquart and Jonathan Sorensen, *Criminology* 26, no. 4 (1988), pp. 677–693.

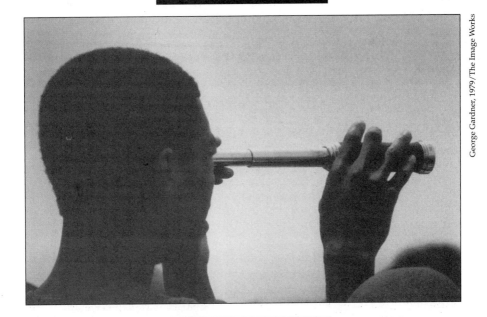

## CHAPTER FIFTEEN

George Gardner, 1979/The Image Works

# Sociology: Another Perspective

**A Sociological Perspective**

**Sociological Perspectives of Sociology**

**Sociologists—What Are They and What Do They Do?**

**Sociology and You**

THIS BOOK HAS attempted to provide you with an understanding of sociology. We looked at theories, at methods, and especially at the concepts sociologists use to gain knowledge and understanding of the human condition. At this point, we should have a reasonably good idea of what sociologists *do*.

But there is another perspective to examine. To me, an important aspect of a profession or discipline is what the occupants *think* about what they do. This is a more informal (an insider's) view and examines such issues as: How do these people see themselves? What do they feel they should be accomplishing? What are their internal conflicts and arguments? What is their level of job satisfaction and morale? Think of doctors and police officers again. They interest me partly because of what they do, but I am fascinated about what they *think* about what they do: job satisfaction, attitude toward patients or clients, morale, amount of stress and anxiety related to the job, the type of internal issues and conflicts that they argue over, and so on. The cumulative effect of these matters must affect one's personality and give one a particular perspective, a way of looking at the world.

This last chapter will examine the insider's view: What do sociologists think about what they do? How do sociologists see themselves, and what are some of the issues on which they disagree? How does this affect the sociologists' perspectives?

## A SOCIOLOGICAL PERSPECTIVE

Members of a discipline develop a particular way of looking at the world. They share these views as individuals, although the similarities of their perspectives are most readily seen when a group of people from the same discipline come together. Becoming a sociologist involves learning the meaning and use of a series of concepts and techniques and acquiring a body of knowledge. Going through this learning—a socialization process—provides one with a particular viewpoint. Members of any discipline can go through this process, and it affects the way they think, the way they talk, the way they behave.

The sociological perspective has a number of aspects. One is the tendency to see human behavior in the context of a group. To put it another way, sociologists understand people's behavior to be a product of their group affiliations. Consequently, sociologists focus their studies on these group affiliations. For example, let's look at Joe Blow as some of his friends and colleagues might look at him. What stands out to his physician is Joe's height, weight, pulse rate, blood pressure, and medical history. What his banker sees is his investment portfolio, cash reserves, and loan collateral. His minister might see him as a soul to be saved, a moral state to be improved. His professor might see him as a mind to be molded, a vessel to be filled with knowledge, or an intellect to be challenged. A sociologist looks at him and sees a white, middle-class, 21-year-old man who is a member of the Democratic Party.

## PETER BERGER (1929–   )

Peter Berger earned his Ph.D. in Sociology at the New School for Social Research. Berger's interests have ranged from religion to economic culture. In his studies of religion (*The Social Reality of Religion*), he has suggested, controversially, that sociologists should adopt a "methodologically atheist" stance and not seek to discuss whether religion is anything more than a social creation. In both *Invitation to Sociology* and *Sociology Reinterpreted: An Essay on Method and Vocation*, Berger has written on what sociology is and what sociologists do.

"Sociology, from its beginnings, has been a very peculiar discipline, in that it discovered its object along with the methods of studying it: Sociology was originally not just an approach to the study of society, but it was part and parcel of the discovering of the phenomenon 'society' as such." Berger has also researched the impact that globalization has had on indigenous cultures. Peter Berger is a professor of sociology and director of the Institute for the Study of Economic Culture at Boston University.

Sociologists are typically more interested in characteristics of categories and groups than in specific individuals, and in this sociologists differ from psychologists, who are more interested in the *individual* and his or her personality, motivations, emotions, and behavior. For the sociologist, the focus in Joe's case would probably be on the characteristics of middle-class whites, of 21-year-old men, or of Democrats in general rather than on Joe as an individual. The sociologist usually studies numbers of people (groups, categories, and societies) and their patterns of behavior (norms, roles, and institutions).

Other aspects of the sociological perspective are discussed by Peter Berger in his book *Invitation to Sociology*. Berger feels that there is a "sociological consciousness" that has several facets.[1] A *debunking* tendency of sociologists is reflected in their study of things taken for granted. Berger suggests that in analyzing the structure and institutions of societies, it is often necessary to "unmask the pretensions and the propaganda with which men cloak their actions with each other." There is a constant attempt, then, to find the underlying explanations for phenomena instead of just accepting the handy or traditional explanations that people readily give. For example, sociologists would view it as important to put commonsense explanations aside and determine what the real relationship is between narcotics legislation and drug use or between use of the death penalty and the homicide rate.

Although most of sociology deals with the *respectable,* a part of sociology is devoted to the study of the *unrespectable.* Early sociologists were reformers and muckrakers. They believed that sociologists should intervene and attempt to ameliorate problem conditions to ensure social progress. This preoccupation with social problems and with the unrespectable has abated somewhat today, but its influence in sociology is still strong.

Berger suggests that sociologists tend to be *relativist* and *cosmopolitan* in their outlook. Sociologists are involved in studying patterns of human interaction in a variety of cultures under many different conditions. They learn that behavior, ideas, and institutions are relative, relevant to specific cultures and specific locations. To make judgments about or evaluate certain behaviors, sociologists must, then, apply a broader, more cosmopolitan perspective; they must have a worldview, and this becomes part of their sociological consciousness.

Any similarities in outlook that are shared by sociologists emerge, to a great extent, as a result of their training. There are also many differences among them; some are evident when we examine how sociologists look at sociology.

## SOCIOLOGICAL PERSPECTIVES OF SOCIOLOGY

The discipline of sociology encompasses specific themes or traditions. In combination, these traditions constitute a heritage from which all sociologists draw. A particular sociologist, however, will be influenced more strongly by one tradition or perspective than by others. Just as a physician specializes in particular aspects of medicine and a lawyer concentrates in particular aspects of law, so the sociologist specializes in specific areas of sociology (race, social class, or deviant behavior, for example) and follows a particular theme or tradition that he or she personally feels to be most relevant.

There has always been in sociology, as in other disciplines, a debate over who we are and what we should be doing. This is a continual self-analysis, which attracts the interest and attention of many sociologists. It seems appropriate to describe this as "sociological perspectives on sociology." It is an interesting debate and probably reflects the growing pains of an area of study that is still trying to define itself and its place in the world of knowledge.

As we mentioned in Chapter One, science is often divided into **pure science** and **applied science.** Into which of these categories does sociology properly fall? Sociology is seen by some as a *pure* science in that its practitioners seek knowledge for knowledge's sake alone, without regard for the possible uses the knowledge might have. In other words, sociology is a science, not a vehicle for social action. In recent years, however, a growing number of sociologists have encouraged the development of an *applied* sociology. In a variety of ways, sociologists are putting their concepts and techniques to practical use. In general, these efforts involve attempts to deal with the problems of society or with the individual segments of society: industry, education, labor, government, and community. Applied sociology is often designed to encourage desired social change or to forestall undesired social change. Applied sociology has been referred to as **action sociology,** and those involved are sometimes called *social engineers.* Philip Hauser points out that with the proliferation of economic and social planning and the increase in the welfare functions of government, the need for sociologists will continue to increase for the performance of both scientific and social engineering tasks.[2]

Sociology may also be divided into **quantitative** and **qualitative** approaches. The quantitative approach, as we described in Chapter One, focuses on hypothesis testing, uses survey data, involves statistics and frequency distributions of a variable, and seeks prediction. The qualitative approach focuses on hypothesis generating, is more interpretative, and seeks understanding. A colleague of mine, Charles Varano, recently completed a qualitative study of a West Virginia town faced with a crisis—the closing of the town's major economic asset, its steel mill. In an amazing development, the steel company sold the mill to the workers! Professor Varano is from Southern California but he moved to the town, Weirton, West Virginia, for a year and visited on subsequent summers. He interviewed hundreds of people—mill workers, managers, local politicians, city workers, local merchants, religious leaders. He became very close to people working in and outside the mill, their families, children, and friends. "Whether visiting or sharing meals, mowing lawns, or helping shovel snow, I was able to talk with many people, of all ages informally and at length during my stay." What emerged from this qualitative study was an ethnography, a participant-observation study, that produced significant insights about the community and the chances that its experiment might be successful.[3]

As you might imagine, there is an ongoing debate about the relative usefulness of quantitative versus qualitative methods. Consistent with the quantitative approach, some sociologists take a purely **scientific perspective.** For them, sociology is a social science, and they emphasize the procedures and techniques of science. Such sociologists see themselves as white-coated individuals who gain knowledge by applying scientific methods to social phenomena. They are objective, detached, free of any values or biases that might affect their work. They are, perhaps, bundles of sensory neurons moving in space, observing and recording the facts of social interaction. Sociologists today are developing sophisticated mathematical models and statistical techniques and using ever larger computers to aid them in the scientific analysis of human behavior.[4]

These individuals are often called **positivists.** They adhere pretty much to the classical view of science as outlined in Chapter One. Their role as scientist is based on the model that has worked so well for those in the physical and biological sciences, and its adoption by sociologists probably reflects the hope that our success will be as great as theirs. At the same time, the positivist model in sociology is criticized by some for several reasons. First, there is the criticism of science in general. There is concern that blind faith in and uncritical acceptance of science has led people to ignore the faults of science. Questions are raised about a series of ethical considerations: Shouldn't scientists be responsible for the possible harmful uses of their discoveries? Should scientists manipulate people without their knowledge, or even *with* their knowledge? Whom do the scientists represent—themselves, science in general, society, or those people (government agencies, large organizations, private concerns) who are paying their research bills? Failure to receive satisfactory answers to these questions has led many to be disillusioned with science.[5]

A second form of criticism of the positivist model is more specific to sociology. The complaint here is that the information sociologists collect is not valid, not accurate. This is a serious criticism. Derek Phillips states in his book *Knowledge From What?* that 90 percent of the data collected by sociologists is produced by means of interviews and questionnaires. Sociologists have failed to take into account, he maintains, that the questionnaire or interview situation is a *social* situation, that statements of the people studied are affected by the questionnaire, by their response to the interviewer and the interviewer's characteristics (sex, age, race, social class, demeanor, and so on), by the nature of the interaction between the interviewer and the interviewee, and by numerous other factors that have been left uncontrolled.

Phillips's view encourages us to look more closely at the qualitative approach. The basic problem with quantitative research is that it is based on people's *reports* of their own behavior rather than on *actual observation* of their behavior. This is the issue of reactive versus nonreactive research, which we discussed in Chapter One. Phillips suggests that sociologists should instead concentrate on participant-observation and hidden-observation techniques. **Participant observation,** as we mentioned earlier, involves the investigator joining or living with the subjects being studied; the researcher becomes part of the group. Studies of primitive tribes, delinquent gangs, large organizations, and class and ethnic structure of communities, as well as many other topics, have been made using this technique. Participant observation allows the study of some events that could not otherwise be studied, and it can provide many insights that escape the more formal interview and questionnaire methods. **Hidden observation** refers to the use of records or traces left behind by people, such as checking the nose prints on the glass or the wear on the floor tile to determine the popularity of a museum exhibit, or to the actual observation of people without their knowing.[6] The key element in both participant and hidden observation is that we *watch* people's behavior rather than *ask them* about it. Concerns about the *accuracy* of our information, then, have led to disillusionment with some of the scientific aspects of sociology.

These problems, as well as other beliefs regarding the most appropriate path to knowledge, have led some sociologists to see the discipline not as science but as *social philosophy.* For them, sociologists are basically theorists who create grand and often highly complex and abstract generalizations about the nature of the universe. They frequently have remarkable insights about human behavior. However, from the scientist's viewpoint, the social philosopher's statements are often too general and abstract to be put to empirical test. Others see sociologists-as-social-philosophers as true scholars rather than as mere scientific technicians; they are widely read masters of many disciplines, and students may gather adoringly at their feet.

Some sociologists take a **humanistic perspective.** For them, the sociologist is interested in and concerned about human welfare, values, and conduct; he or she wants to improve the lot of people in general. An ultimate goal for the humanist is the self-realization and full development of the cultivated person. There might occasionally be a tendency for the humanistic and scientific perspectives to be at odds with each other. The humanist is interested in bettering the condition of people and in developing the individual to the fullest. The objective of science is the gaining of empirical knowledge about the world, without regard for the possible uses of such knowledge. Don Martindale suggests that humanism is a system of values describing "what ought to be" and modes of conduct designed to secure them; science is the value-free pursuit of knowledge, of "what is," renouncing all concern with what ought to be. It has also been suggested that the scientist is more interested in the *means*—gaining knowledge; the humanist, in the *ends*—improving the lot of people.[7]

Is sociology basically concerned with gathering empirical data about people's behavior in a scientific sense, or with the study of those conditions that will enable individuals to realize their fullest potential in a humanistic sense? It is impossible to put all of sociology into one camp or the other. Most sociologists probably use a combination of scientific and humanistic viewpoints. Peter Berger expresses the view of many, however, when he argues that sociology must be used for humanity's sake. Social science, like other sciences, can be and sometimes is dehumanizing and even inhuman. It should not be. When sociologists pursue their tasks with insight, sensitivity, empathy, humility, and a desire to *understand* the human condition, rather than with a cold and humorless scientism, then indeed, the "sociological perspective helps illuminate man's social existence."[8]

## SOCIOLOGISTS—WHAT ARE THEY AND WHAT DO THEY DO?

What makes one a sociologist? From what we have said in this chapter, a simple answer would be, One who has a sociological perspective. Also, a sociologist is a person who has earned an advanced degree in sociology. Most professional sociologists have master's degrees or doctorates, and many students who receive bachelor's degrees in sociology go into other fields, so the exact number of sociologists practicing sociology for a living is difficult to determine.

Most sociologists spend most of their time teaching. Opportunities for teaching sociology are found predominantly at the college and university level, but increasing numbers of high schools are adding sociology courses to their curricula. Some sociology courses, especially at the high-school level, have more of a practical, how-to-do-it focus. These courses might deal with dating, marriage and family, or personal adjustment.

Many sociologists are involved in research, either full-time or in combination with their teaching. Full-time researchers might run their own research agencies, but more often they work for state or federal agencies.

They might, for example, be responsible for collecting and analyzing population and census data. They might collect police arrest statistics and analyze and combine these into crime reports. Sociologists, especially at the larger universities, are about equally involved in research and teaching.

Sociologists work in a number of capacities for a variety of private, state, and federal agencies. Criminologists are hired by institutions, such as prisons and reformatories, and by departments of correction. Population experts are valuable in a policy-making capacity for a number of agencies. Sociologists act as consultants to cities for recreation programs, city planning, urban renewal, mass transit, and so on. The services of specialists in marriage and the family are used by schools, churches, and social agencies, and some sociologists are involved in private marriage counseling. Additional areas in which sociologists are involved in applied research include evaluation research (analyzing whether a project is working the way it is supposed to), legislative consulting, housing and environmental planning, military research (problems of military installations as communities, patterns of leadership), health services, consumer and advertising research, and so on. Applied sociology is likely to continue to grow as we become more and more an informational society.

## SOCIOLOGY AND YOU

What good has this time spent in sociology done for you? Admittedly it is difficult to measure the knowledge you have acquired. If the subject were typing and you had progressed from 2 words a minute to 80, there would be tangible evidence of success. Because the subject is sociology, if you were to go out and solve eight social problems during summer vacation, that might prove that the course had taught you something. The problem, however, is that we are not teaching a skill or a technique but a viewpoint. And it is sometimes difficult to measure the immediate usefulness of a new viewpoint or a new way of looking at things.

After a semester of introductory sociology, you should know more about the world you live in. Principally, you should have discovered from the course a new way of looking at the world and at yourself. As C. Wright Mills suggested in *The Sociological Imagination,* the study of sociology should show you a way of evaluating yourself as more than an isolated individual in a sea of humanity. The study of sociology should help you place yourself and history in perspective so you can more accurately identify and evaluate the factors that affect your behavior and the behavior of others. You should be more critical, now, and more able to evaluate aspects of the world that heretofore you may have taken for granted.

"Who am I?" and "How did I get this way?" are questions that are difficult to answer to anyone's satisfaction. Perhaps a sociological viewpoint will help deal with these questions; perhaps it won't. If the sociological viewpoint only helps to make you more aware of your social environment and of some of the forces at work there, the time spent has been worthwhile, for this will provide you with a greater understanding of and compassion for others.

# TERMS FOR STUDY

action sociology (456)

applied science (456)

hidden observation (458)

humanistic perspective (459)

participant observation (458)

positivist (457)

pure science (456)

qualitative research (457)

quantitative research (457)

scientific perspective (457)

 **INFOTRAC COLLEGE EDITION**
**Search Word Summary**

To learn more about the topics from this chapter, you can use the following words to conduct an electronic search on InfoTrac College Edition, an online library of journals. Here you will find a multitude of articles from various sources and perspectives:

**www.infotrac-college.com/wadsworth/access.html**

Atheism

Sociologists

Social groups

Socialization

criminologists

# NOTES

1. The discussion in this and the following paragraphs is drawn from Peter L. Berger's *Invitation to Sociology* (Garden City, N.Y.: Doubleday, 1963), chapter 2 ("Sociology as a Form of Consciousness").

2. Philip M. Hauser, "On Actionism in the Craft of Sociology," *Sociology Inquiry* 39 (Spring 1969), pp. 139–147. For more on applied sociology, see Arthur Shostak, ed., *Sociology in Action* (Homewood, Ill.: Dorsey, 1966), and Alvin Gouldner and S. M. Miller, eds., *Applied Sociology* (New York: Free Press, 1965).

3. Charles Varano, *Forced Choices: Class, Community, and Worker Ownership* (Albany, N.Y.: State University of New York Press, 1999).

4. A thorough and interesting discussion of several sociological perspectives, including

"abstracted empiricism," is included in C. Wright Mills's excellent book, *The Sociological Imagination* (New York: Oxford University Press, 1959).

5. A number of works deal with these issues. See, for example, Gideon Sjoberg, ed., *Ethics, Politics, and Social Research* (Cambridge, Mass.: Schenkman, 1967); Ralph Beals, *Politics of Social Research* (Chicago: Aldine, 1968); and Irving L. Horowitz, ed., *The Rise and Fall of Project Camelot* (Cambridge, Mass.: M.I.T. Press, 1967). Most methodology texts contain a discussion of scientific ethics.

6. Derek Phillips, *Knowledge From What?* (Chicago: Rand McNally, 1971). Also see John Lofland, *Analyzing Social Settings* (Belmont, Calif.: Wadsworth, 1971) and Eugene Webb, Donald Campbell, Richard

Schwartz, and Lee Sechrest, *Unobtrusive Measures: Nonreactive Research in the Social Sciences* (Chicago: Rand McNally, 1966).

7. This discussion of humanistic and scientific thought follows from Don Martindale's analysis in his *Social Life and Cultural Change* (Princeton, N.J.: Van Nostrand, 1962), pp. 443–462.

8. See Peter Berger's chapter, "Sociology as a Humanistic Discipline," in *Invitation to Sociology* (see note 1). Also see R. P. Cuzzort's discussion of Berger in Cuzzort's book, *Humanity and Modern Sociological Thought* (New York: Holt, Rinehart & Winston, 1969), chapter 10.

# GLOSSARY

## A

**Acculturation**   The process of assimilating, blending in, and taking on the characteristics of another culture.

**Achieved statuses**   Positions in society that are earned or achieved in some way.

**Action sociology**   The practical use of sociological concepts and techniques to deal with the problems of society; also called *applied sociology.*

**Aggregate**   A number of people clustered together in one place.

**Alienation**   A feeling of isolation, meaninglessness, and powerlessness; may occur when workers see themselves as only a small and insignificant part of a large, impersonal bureaucracy.

**Altruistic suicide**   The result of strong group attachments—the individual commits suicide to benefit the group and to follow group norms.

**Amalgamation**   Refers to biological (rather than cultural) mixing.

**Annihilation**   Elimination of one group by another.

**Anomic suicide**   Occurs in situations in which norms are confused or are breaking down, as in economic depressions or periods of rapid social change.

**Anomie**   A state of normlessness; a condition in which the norms and rules governing people's aspirations and moral conduct have disintegrated.

**Anticipatory socialization**   Occurs when a person adopts the values, behavior, or viewpoints of a group he or she would like to, but does not yet, belong to.

**Apollonian ceremony**   Sober, restrained, dignified, self-controlled religious ceremony; contrast with a Dionysian ceremony.

**Applied science**   Concerned with the practical and utilitarian uses of knowledge, with making knowledge useful to people.

**Ascribed statuses**   Positions automatically conferred on individuals through no choice of their own; statuses people are born with.

**Assimilation**   Mixing and merging of unlike cultures so that the two groups develop a common culture.

**Audience**   A collection of people at (watching, listening to) a public event; a passive crowd.

**Authoritarian personality**   A type of personality especially prone to prejudice; tends to be ethnocentric, to be rigidly conformist, and to worship authority and strength.

**Authority**   Legitimate, socially approved power.

## B

**Birth rate**   Number of live births per 1,000 people in the population.

**Bureaucracy**   (formal organization) An organization in which the activities of some people are systematically planned by other people to achieve some purpose; usually involves such characteristics as division of labor, hierarchy of authority, system of rules, impersonality, and technical efficiency.

## C

**Capitalism**   An economic system that emphasizes the ideas of private property, the profit motive, and free enterprise and free competition.

**Caste system**   A system of stratification in which lines or boundaries between levels (castes) are firmly drawn; mobility between castes is difficult if not impossible.

**Category**   A number of people who have a particular characteristic in common.

**Cause-effect relationship**   Describes a situation in which an event *B* (effect) always follows a previous event *A* (cause), and in the absence of *A*, *B* does not appear.

**Centralization**   The tendency of people to gather around some central or pivotal point in a city.

**Charisma**   A quality in an individual that sets him or her apart from others—the person is viewed as superhuman and capable of exceptional acts.

**Charismatic authority**   Power stemming from the unique personality of a particular person.

**Chronological age**   Age based on time from date of birth, for example, 8 years, 3 months or 15 years, 6 months, and so on. (Compare with *developmental age*.)

**Church**   A large and highly organized religious organization that represents and supports the *status quo*, is respectable, and in which membership is automatic; one is born into a church.

**Class consciousness**   An awareness of one's position or class in society and of the circumstances, interests, and concerns that that class has in common; from Karl Marx.

**Class model**   A theory describing the distribution of power that holds that economic activities lead to the development of two classes: the bourgeoisie, or capitalists, who own the means of production, and the proletariat, who own only their own labor; from Karl Marx.

**Class system**   A form of stratification emerging in more complex and industrialized societies. These societies require a more educated and skilled work force and an extensive division of labor. This system is more open—movement from one level to another is more feasible than in caste or estate systems.

**Closed group**   A group into which it is difficult if not impossible to gain membership.

**Collective behavior**   Spontaneous, unstructured, and unstable group behavior.

**Complementary needs theory**   A mate selection theory that holds that people select partners who make up for, balance, or supply characteristics they themselves don't have.

**Compound family**   The family resulting from polygamy; marriages involving multiple spouses—several wives or husbands or both—at the same time.

**Conflict theory**   A sociological theory that holds that the most appropriate way to understand society is to focus on the consequences of the conflict, competition, and discord that are common in social interaction; also used to explain social change; from Karl Marx.

**Control**   (1) A product of knowledge—if information about an event is available, then regulation (control) of the consequences of the event are possible. (2) The attempt by scientists to manipulate (control) factors that can affect what they are studying.

**Counterculture**   A subculture that has norms and values that are in opposition to or in conflict with norms and values of the dominant culture.

**Crazes**   The relatively short-term obsessions that members of a society or members of groups have toward specific behaviors or objects.

**Criminal behavior**   Behavior prohibited by law and subject to formal punishment.

**Crowd**   A temporary collection of people in close physical contact reacting together to a common stimulus.

**Cult**   A religious organization that is a small, short-lived, often local group that is frequently built around a dominant leader. The cult is smaller, less organized, and more transitory than the sect.

**Cultural lag** Phenomenon that occurs when related segments of society change at different rates.

**Cultural pluralism** A pattern of interaction in which unlike cultures maintain their own identity and yet interact with each other relatively peacefully.

**Cultural relativism** Opposite of ethnocentrism; suggests that each culture be judged from its own viewpoint without imposing outside standards of judgment.

**Culture** The complex set of learned and shared beliefs, customs, skills, habits, traditions, and knowledge common to the members of a society; the social heritage of a society.

**Cyclic theories** Theories of social change that hold that societies repeatedly pass back and forth through a series of stages (for example, order and disorder or prosperity and decline).

# D

**Decentralization** The tendency of people to move outward and away from the center of the city.

**Definition of the situation** The suggestion, first offered by W. I. Thomas, that reality is socially structured, with the result that people respond more to the meaning a situation has for them than to the objective features of that situation.

**Democratic model** Reflects the functionalist view that the institution of education provides equal opportunity for all students and rewards ability and hard work regardless of the student's origin.

**Democratic system** A system of government in which the power of the state is limited by the wishes of the people, who are seen as having direct input into the decision-making process.

**Demographic transition** The movement of a population from high birth and death rates to high birth and low death rates (and rapid population growth) to low birth and death rates.

**Demography** The study of human population—its distribution, composition, and change.

**Denomination** A religious organization that has less universality than the church and appeals to a smaller category of people; a racial, ethnic, or social-class grouping that is conventional and respectable.

**Dependent variable** The "effect"; that quantity or aspect of a study whose change the researcher wants to understand.

**Developmental age** Age based on a student's level of maturation or behavior, which would suggest a "readiness" to be placed at a particular grade level. (Compare with *chronological age*.)

**Deviant behavior** Behavior contrary to generally accepted norms; its limits are often established by custom or public opinion, sometimes by law.

**Differential association** Sutherland's theory of crime, which says that behavior, normal and deviant alike, is learned during interaction with others.

**Diffusion** The spread of objects or ideas from one society to another or from one group to another within the same society, resulting in changes in society.

**Dionysian ceremony** Ecstatic, emotional, nonrational religious ceremony characterized by loss of self-control; contrast with an Apollonian ceremony.

**Disaster** A situation in which there is a basic disruption of the social context within which individuals and groups function; a radical departure from the pattern of normal expectations.

**Discrimination** Actual behavior resulting in unfavorable and unequal treatment of individuals or groups.

**Dysfunction** An act that leads to change or destruction of some unit (organism, social group or institution, and so on).

# E

**Education** One of the five main institutional areas, the major focus of which is the process of socialization—learning the knowledge, skills, and appropriate ways of behaving to exist in a larger social world.

**Ego** In Freud's view, the acting self; the mediator between the id and superego.

**Egoistic suicide** Occurs when personal relationships are secondary, distant, and not group-oriented; when problems appear, the absence of emotional support from others makes suicide more likely.

**Elite model** A theory describing the distribution of power that holds that power is controlled by an elite who rule because of their superior organization, their personal attributes, or their control of valued resources such as wealth, government authority, and communication facilities.

**Elite-pluralism** A theory describing the distribution of power that holds that although power is spread across society, the impossibility of mass political participation means that small elite groups develop and make major decisions.

**Empiricism** Knowledge gained through experience and sense observation.

**Endogamy** Marriage within a certain group.

**Estate system** A form of stratification found in advanced agricultural societies in which land is the most important economic resource. Two (landowners, peasants) or three (landowners, clergy, peasants) estates exist, with the landowners, who might represent less than 5 percent of the population, owning much of the land and receiving most of the income. There is no movement or mobility between levels.

**Ethnicity** Refers to people bound together by cultural ties.

**Ethnocentrism** A type of prejudice that maintains that one's own culture's ways are right and other cultures' ways, if different, are wrong.

**Evolutionary theories** Theories of social change that see change as linear—as going in one direction, and that direction is usually toward progress and greater complexity.

**Exchange theory** A theory of mate selection that proposes the idea of a marketplace in which the seeker wants to maximize his chances for a happy marriage and looks for a good deal in which his or her own assets and liabilities are compared with those of the potential partner; the relationship is continued on the assumption that one will get more out of it than it will cost.

**Exogamy** Marriage outside a certain group.

**Experiment** Type of research in which the researcher introduces a stimulus to a subject or group and then observes the response; the researcher manipulates the independent variable (exercise, for example) and watches for change in the dependent variable (heart disease, for example).

**Expulsion** Removal of a group from the territory in which it resides.

**Extended family** More than two generations of the same family living together in close association or under the same roof.

# F

**Fads** The relatively short-term obsessions that members of a society or members of specific groups have toward specific behaviors or objects.

**Family of orientation** The family into which one is born.

**Family of procreation** The family of which one is a parent.

**Fashions** Temporary attachments to specific behaviors, styles, or objects; similar to fads but more widespread and of longer duration.

**Fecundity** The physical or biological ability to reproduce.

**Felony** Serious crime that is punished by a year or more in a state prison or by death.

**Fertility** Refers to the number of children born.

**Folk society** Robert Redfield's term used to describe isolated villages in nonindustrialized countries that are small, homogeneous, relatively self-contained, and largely based on subsistence activities.

**Folkways** Norms that are less obligatory than mores, the "shoulds" of society; sanctions for violation are mild.

**Function** An act that contributes to the existence of a unit, such as breathing and eating for the life of the organism.

**Functional analysis** A sociological theory that focuses on the structures that emerge in society and on the functions that these structures perform in the operation of society as a whole.

**Functional disorders** Illnesses caused by environmental factors—stress, rejection, unhappy family, and so on.

**Functional theories** Theories of social change that hold that there are tremendous forces resisting change but that these forces might be overcome slowly and adaptively because of changes outside the system, growth by differentiation, and internal innovations.

# G

**Gemeinschaft** Primary, closely knit society in which relationships are personal and informal and there is a commitment to or identification with the community.

**General fertility rate** The number of births per 1,000 women age 15 through 44.

**Generalized other** The sum of the viewpoints and expectations of the social group or community to which an individual belongs.

**Genocide** The intentional extermination of a whole ethnic or racial group.

**Gesellschaft** Secondary society based on contractual arrangements, bargaining, a well-developed division of labor, and rational thought rather than emotion.

**Group** A number of people who have shared or patterned interaction and who feel bound together by a "consciousness of kind" or a "we" feeling.

**Group marriage** The marriage of two or more men to two or more women at the same time.

# H

**Hawthorne effect** The fact that people respond to the process of being observed.

**Hidden curriculum** Refers to the subtle, unintended activities that occur in the process of schooling whereby students are taught the values, attitudes, manners, and personality characteristics appropriate to their future place in society.

**Hidden observation** The use of records or traces left behind by people, or the actual observation of people without their knowing.

**Horizontal groups** Groups whose members come predominantly from one social-class level, such as associations of doctors or carpenters.

**Horizontal social mobility** Movement from one occupation to another within the same social class; also used to refer to spatial or geographical mobility.

**Human ecology** The adaptation of people to their physical environment, their location in space.

**Humanistic perspective** A perspective in which the individual (sociologist) is interested in and concerned about human welfare, values, and conduct; there is a desire to improve the lot of people in general, with the ultimate goal being the self-realization and full development of the cultivated person.

**Hypothesis**   A testable statement, or proposition used to guide an investigation.

# I

**I**   Mead's concept of the subjective, acting self that initiates spontaneous and original behavior; is contrasted with the "me."

**Id**   In Freud's view, the primitive part of the personality made up of inborn, instinctual, antisocial drives.

**Idealistic culture**   Sorokin's term describing a culture midway between the sensate and ideational; the culture emphasizes logic and rationality.

**Ideal norms**   How a person should behave in a particular situation. (Compare with *real norms*.)

**Ideational culture**   Sorokin's term describing a culture that emphasizes feelings and emotions and is subjective, expressive, and religious.

**Incest taboo**   A prohibition that forbids or limits sexual contact between certain family members.

**Independent variable**   The "cause"; that quantity or aspect of the study that seems to produce change in another variable.

**Influence**   Subtle, informal, indirect power based on persuasion rather than coercion and force.

**Institutionalization**   The process in which values, beliefs, norms, and roles regarding a society's basic problems or functions become organized, formalized, and generally understood.

**Interest group**   A number of people who organize to influence decisions on issues important to them.

**Invasion**   The penetration of one group or function into a territory dominated by another group or function.

**Invention**   Change introduced through the creation of a new object or idea.

**Involuntary groups**   Groups in which membership is automatic and the participant has no choice regarding joining, such as one's family or the army platoon to which one is assigned.

# J

**Juvenile delinquency**   Refers to young people (under 21 in some states, under 18 in others) who commit criminal acts or who are wayward, truant, uncontrollable, or runaways.

# K

**Keynoting**   The process of focusing the behavior of an ambivalent crowd on a particular activity or issue.

# L

**Labeling**   The public stamping, typing, or categorizing of others.

**Labeling theory**   A theory that sees deviant behavior as a product of group definitions; becoming deviant involves a labeling process, and one is deviant because a particular group labels one as such.

**Latent function**   The unintended and less obvious, often unrecognized function.

**Legal-rational authority**   Power attached to positions rather than to people; positions are defined by a formal set of rules and procedures; typical of governments and businesses in modern societies.

**Life expectancy**   A person's expectation at birth of length of life, based on risks of death for people born in that year.

**Lobbyists**   People in the business of knowing, influencing, and persuading the right people in government.

**Looking-glass self**   A person's attitudes toward self derived from the person's interpretations of how others evaluate that person.

# M

**Manifest function**   The intended and recognized function of something.

**Marginal person**   A person who is caught between two antagonistic cultures; a product of both but a true member of neither.

**Mass hysteria**   A type of collective behavior in which a particular behavior, fear, or belief sweeps through a large number of people such as a crowd, a city, or a nation.

**Material culture**   The real, tangible things that a society creates and uses: screwdriver, house, classroom, supersonic jet, computer.

**Matriarchal society**   Society in which decisions are made by females.

**Me**   Mead's concept that describes the objective part of self, represents others' attitudes and viewpoints, and judges and evaluates; is contrasted with the "I."

**Mechanical solidarity**   Durkheim's description of the type of social cohesion likely to exist in small, homogeneous, preindustrial societies: uniformity among people and a lack of differentiation or specialization.

**Migration**   A permanent change of residence with the consequent relocation of one's interests and activities.

**Milling**   The excited, restless, physical movement of the people in a crowd through which communication occurs, leading to a definition of the situation and possible collective action.

**Minority status**   Social position in which people are treated as lower in social ranking and are subject to domination by other segments of the population.

**Misdemeanors**   Crimes that are less serious than felonies, punishable by up to a year in a county jail.

**Mob**   A focused, acting crowd that is emotionally aroused and intent on taking aggressive action.

**Monogamy**   Marriage with one person at a time.

**Mores**   Obligatory norms, the "musts" of society; sanctions are harsh if mores are violated.

**Mortality**   The number of deaths occurring in a population.

**Multiculturalism**   Formally (laws, institutions, and governmental policy) treating all groups equally in the educational and political process.

**Multinational**   A large corporation whose influence goes beyond national boundaries, often with connections throughout the world.

**Multiple causation**   The situation in which an event is caused by more than one factor.

**Mysticism**   Information or knowledge gained by mystical means: intuition, revelation, inspiration, magic, visions, spells; tends to be private knowledge experienced only by the mystic.

# N

**Nativism**   The attempt to improve a people's existence by eliminating all foreign persons, objects, and customs.

**Neoevolutionary theories**   Theories of social change that see societies as moving through a series of evolutionary stages but in a multilinear fashion; societies are influenced by varied factors, and not all cultures change at the same direction or at the same speed.

**Neuroses**   The mildest and most common of mental illnesses, in which the person is anxious, nervous, compulsive but generally able to function adequately in society.

**Nonmaterial culture** The abstract creations of a society, such as customs, laws, ideas, values, beliefs.

**Nonreactive research** Research in which records, physical traces, and signs left by people are studied; observation of subjects without direct intrusion, subject is not aware of being studied. (Compare with *reactive research.*)

**Norms** The accepted or required behavior for a person in a particular situation.

**Nuclear family** A married couple and their children.

# O

**Oligarchy** Rule by a few; Robert Michels's "Iron Law of Oligarchy" describes the tendency in large organizations for an increasingly conservative elite to emerge and take over, with abuse of power a likely consequence.

**Open group** A group to which membership is easy to gain.

**Operational definitions** Precise definitions of terms or conditions, also known as "working definitions," that allow others to know exactly the dimensions of what was studied.

**Organic disorders** Illnesses caused by brain injury, hereditary factors, or physiological deterioration.

**Organic solidarity** Durkheim's description of the type of social cohesion likely to exist in modern industrialized societies: a high degree of specialization and division of labor, and increasing individualism.

# P

**Panic** Nonadaptive or nonrational flight resulting from extreme fear and loss of self-control.

**Participant observation** Research in which the researcher is or appears to be a participant in the activity or group that is being studied.

**Patriarchal society** Society in which decisions are made by males.

**Peer group** Consists of people of relatively the same age, interests, and social position with whom one has reasonably close association and contact.

**Personality** The sum total of the physical, mental, emotional, social, and behavioral characteristics of an individual.

**Pluralistic model** A theory describing the distribution of power that holds that power is diffuse, or spread across many diverse interest groups.

**Political party** A number of people who organize to gain legitimate governmental power.

**Polyandry** A form of polygamy in which one woman has several husbands at a time.

**Polygamy** Plural marriage; the practice of having more than one husband or wife at a time.

**Polygyny** A form of polygamy in which one man has several wives at a time.

**Population pyramid** A pictorial or graphed profile of the age and gender characteristics of a population.

**Positivist** A term describing a perspective in which sociologists gain knowledge by closely following the scientific perspective; apply scientific methods to social phenomena; are objective and detached and try to be free of any values or biases that might affect their work.

**Power** The ability of one party (either an individual or group) to affect the behavior of another party.

**Prejudice** Favorable or unfavorable attitudes toward a person or group that are not based on actual experience.

**Prestige** One's distinction in the eyes of others; one's reputation.

**Primary deviance** Beginning, often minor, acts of deviant behavior that do not affect the individual's self-concept—

the individual does not see himself or herself as a deviant. (Compare with *secondary deviance*.)

**Primary group** A group in which contacts between members are intimate, personal, and face to face.

**Primary socialization** The first socialization an individual undergoes in childhood; one must undergo this to become a member of society; it ends with the establishment of the generalized other.

**Process theory** A theory of mate selection that proposes that through a series of social and psychological processes, the field of potential mates is progressively narrowed down by physical attraction, religious and racial differences, role and value similarities, psychological influences, and so on.

**Propaganda** Attempts to influence and change the public's viewpoint on an issue.

**Pseudocharisma** Refers to public figures who appear to have charisma, but who actually have a packaged and carefully created image manufactured by public-relations techniques.

**Psychoses** More severe of the mental illnesses, in which the person loses touch with reality, might have hallucinations and delusions, or might withdraw into a dream world.

**Public** A number of people who have an interest in, and difference of opinion about, a common issue.

**Public opinion** The ideas and beliefs held by a public on a given issue.

**Pure science** Attempts to discover facts and principles about the universe without regard for possible uses the knowledge might have.

# Q

**Qualitative research** An approach to gathering information that uses participant observation, hidden observation, historical records, and nonreactive techniques to generate

hypotheses; seeks interpretation and understanding; sometimes called *fieldwork* or *field research*; emphasis is more on subjective than objective interpretations.

**Quantitative research** An approach to gathering information that uses survey data, statistics, and frequency distributions to test hypotheses; emphasis is more on objective than subjective interpretations.

# R

**Race** People related by common descent or heredity; usually identified by hereditary physical features.

**Racism** The belief that people are divided into distinct hereditary groups that are innately different in their social behavior and mental capacities and that because of these differences can be ranked as superior and inferior; this ranking leads to differential treatment in access to society's resources such as jobs, living conditions, wealth, and power.

**Random sample** A sample of subjects in which every element in the total population being studied has an equal chance of being selected; the small sample of subjects is then theoretically representative of the larger population from which it was selected.

**Rationalism** A method of gaining knowledge through the use of reason and logic.

**Reactive research** Research in which subjects react or respond to a questionnaire or an interviewer. (Compare with *nonreactive research*.)

**Real norms** How a person actually behaves in a particular situation. (Compare with *ideal norms*.)

**Reference groups** Groups that serve as models for our behavior; groups whose perspectives we take as our own and use to mold our behavior.

**Reform movement** A movement that seeks modifications in certain aspects of society.

**Religion**   A unified system of beliefs, feelings, and behaviors related to things defined as sacred.

**Replacement level**   The average number of births necessary per woman over her lifetime for the population eventually to reach zero growth.

**Replication**   The repetition of studies (research efforts) in similar and unique circumstances to see if the results are consistent with earlier results.

**Resocialization**   A process in which major modifications or reconstructions are made of a person's patterns of behavior, personality, and self-concept.

**Resource control**   A type of power generated by the possession or control of valued items, such as an energy source, major communication facilities, and so on.

**Revolutionary movement**   A movement that seeks a complete change in the social order.

**Riot**   A situation in which mob behavior becomes increasingly widespread and destructive.

**Role**   The behavior of one who occupies a particular status or position in society.

**Role and value theories**   Theories of mate selection that hold that people who share common values and common definitions about roles are more likely to select each other.

**Role conflict**   Conflict that occurs when a person occupies several statuses or positions that have contradictory role requirements.

**Role requirements**   Norms specifying the expected behavior of people holding a particular position in society.

**Role strain**   Strain that occurs when there are differing and conflicting expectations regarding one's status or position.

**Rumor**   Unconfirmed, although not necessarily false, person-to-person communications.

# S

**Sanction**   The punishment one receives for violation of a norm or the reward granted for compliance with a norm.

**Schooling**   Refers to that part of education involving formal instruction in a classroom setting.

**Science**   The gaining of knowledge through sense observation; understanding through description by means of measurement, making possible prediction and thus adjustment to or control of the environment.

**Scientific perspective**   A perspective in which the individual (sociologist) emphasizes the procedures and techniques of science; the view that knowledge is properly gained by applying scientific methods to social phenomena.

**Secession**   The formal withdrawal of a group of people from a political, religious, or national union.

**Secondary deviance**   Deviant acts that represent a defense, attack, or adjustment to the problems created by reactions to primary deviation; the self-concept changes so that the individual begins to see and define self as deviant. (Compare with *primary deviance.*)

**Secondary group**   Group in which contacts between members are more impersonal than in a primary group; interaction is more superficial and is probably based on utilitarian goals.

**Secondary socialization**   Takes over where primary socialization leaves off; involves internalizing knowledge of new facets and sectors of life and of special skills and techniques.

**Sect**   Small religious organization that is less organized than the church and in which membership is voluntary; Members of sects usually show greater depth and fervor in their religious commitment than do members of churches.

**Secularization** Becoming more worldly or unspiritual, as opposed to becoming more religious or spiritual.

**Segregation** (1) Group conflict context: the setting apart of one group. (2) Ecological context: the clustering together of similar people.

**Self** One's awareness of and ideas and attitudes about one's own personal and social identity.

**Self-fulfilling prophecy** Occurs when a false definition of a situation evokes a new behavior that makes the originally false conception come true.

**Sensate culture** Sorokin's term describing a culture that emphasizes the senses and is objective, scientific, materialistic, profane, and instrumental.

**Sex ratio** Number of males per 100 females.

**Significant others** People (mother, father, older brother, and so on) whose viewpoints and expectations the individual considers particularly important.

**Slavery** A form of stratification usually based on economics in which some human beings are owned by others.

**Social change** Significant alterations in the social relationships and cultural ideas of a society.

**Social control** The processes, planned or unplanned, by which people are made to conform to collective norms.

**Social differentiation** The process of defining, describing, and distinguishing among different categories of people.

**Social disorganization** The condition resulting when norms and roles break down and customary ways of behaving no longer operate.

**Social inequality** A consequence of social differentiation in which categories of people are ranked or valued at different levels, based on such factors as wealth, race, ethnicity, age, or gender.

**Social institutions** Organized systems of social relationships that embody some common values and procedures and meet some basic needs of society.

**Social interaction** The process of being aware of others when we act and of modifying our behavior in accordance with others' responses.

**Socialism** An economic system that emphasizes the ideas that property should be collectively owned; that goods and services should be produced according to peoples' needs, regardless of profitability; that centralized planning should determine what is produced; and that wages and benefits should be equally distributed.

**Socialization** The social process whereby one learns the expectations, habits, skills, values, beliefs, and other requirements necessary for effective participation in social groups.

**Social mobility** Refers to movement (up, down, or sideways) within the social-class structure.

**Social movement** A group of people acting with some continuity to promote or to resist a change in their society or group.

**Social organization** The social fabric of society; the integrated set of norms, roles, cultural values, and beliefs through which people interact with each other, individually and in groups.

**Social reproduction model** Reflects the conflict view that the institution of education works to reproduce the social class of the student; that is, schooling functions in such a way that the social class of the student replicates the social class of the family from which the student comes.

**Social stratification** The division of people in society into layers or ranks. The source of the ranking can be one or a combination of factors: wealth, power, prestige, race, sex, age, religion, and so on.

**Social structure** The network of norms, roles, statuses, groups, and institutions through which people relate to each other in society.

**Society**   A number of people living in a specific area who are relatively organized, self-sufficient, and independent, and who share a common culture.

**Sociobiology**   The systematic study of the biological basis of social behavior in all kinds of organisms, including humans.

**Sociology**   The science or study of the origin, development, organization, and functioning of human society, including social behavior, groups, institutions, and social organization.

**State**   The political body organized to govern within society; the center of legitimate power.

**Status**   A position in society or in a group.

**Status inconsistency**   Situation that occurs when the factors that determine an individual's rank in society are not consistent with each other—for example, a college-educated carpenter or a black court justice.

**Status symbols**   Objects (often material possessions) or behaviors that one uses to show (or exaggerate) one's place on the social-class ladder.

**Stereotyping**   Using a simplified and standardized image or label to describe a group or category of people.

**Strain for consistency**   William Graham Sumner's view that there is a fundamental resistance to change in all societies in that people are basically conservative and seek stability.

**Subcultures**   Groups or segments of society that share many of the characteristics of the dominant culture, but that have some of their own specific customs or ways that tend to separate them from the rest of society.

**Succession**   The complete displacement or removal of an established group; the end product of invasion.

**Superego**   In Freud's view, that part of the personality made up of an internalized set of rules and regulations forming the conscience.

**Survey research**   Research that involves the collection of information from subjects through use of questionnaires or interviews; can be descriptive or causal-explanatory.

**Symbolic interaction**   A sociological theory that focuses on the *process* of person-to-person interaction, on how people develop viewpoints about themselves and others, and on forms of interpersonal communication.

## T

**Total fertility rate**   An estimate of the number of children born per 1,000 women.

**Totalitarian system**   A system of government in which the state has absolute and total control, power is highly centralized in one party or group, and dissent is not allowed.

**Tracking**   A procedure whereby students are placed in certain curricular or ability patterns based on test scores and on input from teachers, parents, and school administrators.

**Traditional authority**   Power that is granted according to custom; leaders are selected by inheritance.

## U

**Urban society**   Robert Redfield's term to describe modern Western industrialized nations; opposite of folk society.

## V

**Values**   General opinions and beliefs that people have about which ways of behaving are proper and acceptable, and about which ways are improper and unacceptable.

**Variable**   A condition or trait that changes or has different values (such as I.Q., temperature, or weight).

**Vertical group**   A group whose members come from a variety of social classes.

**Vertical social mobility**   Movement up or down the social-class ladder.

**Voluntary groups**   Groups that have open membership; people may join or not as they wish.

# W

**Wealthfare system**   According to Turner and Starnes, a tax system that has been constructed through the use of power and influence that allows the upper classes to maintain their privileged position by permitting them to avoid taxation and by giving them and the businesses they own direct cash payments and subsidies.

**Welfare system**   Ostensibly, according to Turner and Starnes, a humanitarian effort to keep people from starving by giving monetary support; it actually perpetuates poverty by keeping payments low, by forcing people to work in any available job, and by diffusing any collective power the poor may have.

# REFERENCES: AUTHOR BIOGRAPHIES

Abercrombie, Nicholas, Stephen Hill, and Bryan S. Turner, The Penguin Dictionary of Sociology (New York: Penguin Books, 1994).

Berger, Peter, and Hansfried Kellner, *Sociology Reinterpreted: An Essay on Method and Vocation* (New York: Anchor Press/Doubleday, 1981).

Coser, Lewis, *Masters of Sociological Thought: Ideas in Historical and Social Context* (New York: Harcourt Brace Jovanovich, 1971).

*Encyclopedia Britannica Online*, 1999–2000.

Erikson, Kai, *Encounters* (New Haven: Yale University Press, 1989).

Flanagan, William G., *Urban Sociology: Images and Structure* (Boston: Allyn and Bacon, 1999).

Jary, David, and Julia Jary, *The Harper Collins Dictionary of Sociology* (New York: Harper Collins, 1993).

Lilly, Robert J., Francis T. Cullen, and Richard A. Ball, *Criminological Theories: Context and Consequences* (Thousand Oaks, Calif.: Sage, 1995).

Marshall, Gordon, *The Concise Oxford Dictionary of Sociology* (New York: Oxford University Press, 1996).

Meek, Ronald L., *Marx and Engels on Malthus: Selections from the Writings of Marx and Engels Dealing with the Theories of Thomas Malthus* (New York: International Publishers, 1954).

Mills, C. Wright, *The Sociological Imagination* (New York: Oxford University Press, 1959).

Ritzer, George, *Modern Sociological Theory* (New York: McGraw-Hill, 1996).

Straus, Murray A., Richard J. Gelles, and Suzanne K. Steinmetz, *Behind Closed Doors: Violence in the American Family* (New York: Anchor Books, 1980).

*The Encyclopedia Dictionary of Sociology*, 3d ed. (Guilford, Conn.: Dushkin, 1986).

# INDEX

## A

Abortion, 172–176, 260, 267, 272, 409, 435
Aburdene, Patricia, 379–380, 382
Acculturation, 205
Adorno, T. W., 228
Age
  chronological, 324
  developmental, 325
  at marriage, 242–243
  of U.S. population, 202–203, 341
Aggregate, 109
Aging
  physical, 203
  and the prison system, 443–447
  social, 203
Agrarian societies, 142–143
AIDS, 348, 370
Alcohol, 107, 110
Alcoholism, 192, 430–431
Alger, Horatio, 156
Ali, Lorraine, 215
Alienation, 119, 192, 294
Allen, Paul, 146
Allport, Gordon, 205, 229
Amalgamation, 204
Ambrose, Kenneth, 281–282
American Indians, 191–192, 204, 219–223
Amish, 79–81, 204
Anderson, Elijah, 205, 257
Andreas, Dwayne, 303
Annihilation, 204
Anomie, 381
Anson, Ofra, 290
Anti-Semitism, 195
Anticipatory socialization, 39–40, 157
Apollonian, 269–270
Applied science, 8, 456
Arapesh, 82, 198, 233
Aristotle, 5
Asahara, Shoko, 274
Asch, Solomon, 114, 125, 138–139
Asian Americans, 187, 195–197
Assimilation, 204
Audience, 398
Authoritarian personality, 185

Authority, 296
  charismatic, 296–297
  legal-rational, 296
  traditional, 296
Autokinetic effect, 114

## B

Bagger, Hope, 202
Bailey, William, 452
Ballantine, Jeanne, 67
Barber, Benjamin, 332
Barnes, Carole, 26–30
Baron, James, 139
Barrett, Michael J., 322–323, 332
Barrington, B. L., 321
Bayes, Thomas, 212
Beaker people, 3–4
Beals, Ralph, 461
Beatty, David, 446
Beck, E. M., 393
Becker, Howard, 393, 426, 451
Beeghley, Leonard, 138, 177–178
Bell Curve, 183
Bellah, Robert, 284
Bensman, Joseph, 312
Berger, Peter, 38–39, 43, 51, 67, 455, 459, 461–462
Berry, Brewton, 181, 204, 228–229
Bettelheim, Bruno, 67
Bierstedt, Robert, 87, 103, 371, 393
Birth control, 268, 272, 352, 362–366
Birth order, 37
Birth rate, 72, 159, 189, 240, 335–337, 351, 354
  world, 336–337, 351
Blacks, 186–191, 204, 208–215, 244, 257–261, 320, 371–372, 434
Blau, Peter, 139
Bonaparte, Napoleon, 371
Bonneh, Dan, 290
Boskoff, Alvin, 393
Botstein, Leon, 323
Bourgeoisie, 149, 299, 377
Bowles, Samuel, 317, 332
Boyer, Peter J., 416
Brainwashing, 40